introduction to MICROWAVES

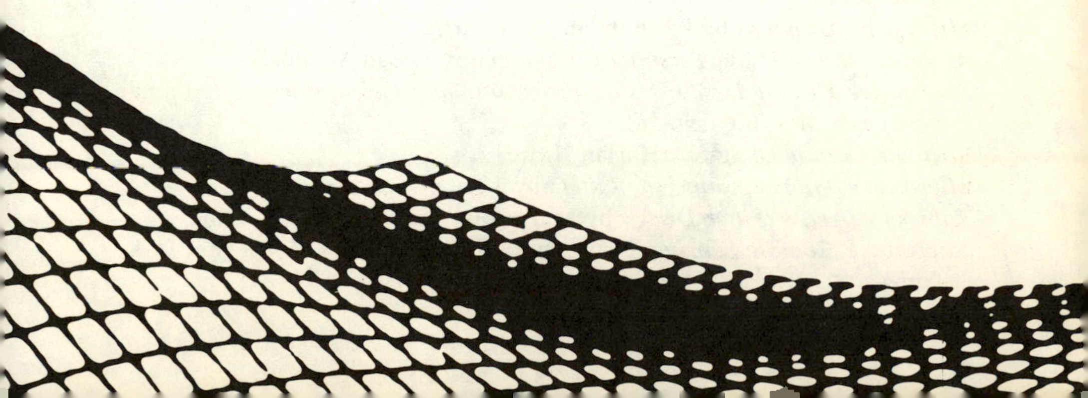

The Artech House Microwave Library

Introduction to Microwaves by Fred E. Gardiol
Microwaves Made Simple: Principles and Applications by W. Stephen Cheung and Frederic H. Levien
Microwave Tubes by A. S. Gilmour, Jr.
Electric Filters by Martin Hasler and Jacques Neirynck
Nonlinear Circuits by Martin Hasler and Jacques Neirynck
Microwave Technology by Erich Pehl
Receiving Systems Design by Stephen J. Erst
Microwave Mixers by Stephen A. Maas
Feedback Maximization by B.J. Lurie
Applications of GaAs MESFETs by Robert Soares, et al.
GaAs Processing Techniques by Ralph E. Williams
GaAs FET Principles and Technology by James V. DiLorenzo and Deen D. Khandelwal
Dielectric Resonators, Darko Kajfez and Pierre Guillon, eds.
Modern Spectrum Analyzer Theory and Applications by Morris Engelson
Design Tables for Discrete Time Normalized Lowpass Filters by Arild Lacroix and Karl-Heinz Witte
Microwave Materials and Fabrication Techniques by Thomas S. Laverghetta
Handbook of Microwave Testing by Thomas S. Laverghetta
Microwave Measurements and Techniques by Thomas S. Laverghetta
Principles of Electromagnetic Compatibility by Bernhard E. Keiser
Linear Active Circuits: Design and Analysis by William Rynone, Jr.
The Design of Impedance-Matching Networks for Radio-Frequency and Microwave Amplifiers by Pieter L.D. Abrie
Microwave Filters, Impedance Matching Networks, and Coupling Structures by G.L. Matthaei, Leo Young, and E.M.T. Jones
Analysis, Design, and Applications of Fin Lines by Bharathi Bhat and Shiban Koul
Microwave Engineer's Handbook, 2 vol., Theodore Saad, ed.
Handbook of Microwave Integrated Circuits by R.K. Hoffmann
Microwave Integrated Circuits, Jeffrey Frey and Kul Bhasin, eds.
Computer-Aided Design of Microwave Circuits by K.C. Gupta, Ramesh Garg, and Rakesh Chadha
Microstrip Lines and Slotlines by K.C. Gupta, R. Garg, and I.J. Bahl
Advanced Mathematics for Practicing Engineers by Kurt Arbenz and Alfred Wohlhauser
Microstrip Antennas by I.J. Bahl and P. Bhartia
Antenna Design Using Personal Computers by David M. Pozar
Microwave Circuit Design Using Programmable Calculators by J. Lamar Allen and Max Medley, Jr.
Stripline Circuit Design by Harlan Howe, Jr.
Microwave Transmission Line Filters by J.A.G. Malherbe
Tables for Active Filter Design by Mario Biey and Amedeo Premoli
Microwave Remote Sensing, 3 vol., by F.T. Ulaby, R.K. Moore, and A.K. Fung

introduction to MICROWAVES

fred e. GARDIOL

artech house

ARTECH HOUSE, INC.
610 Washington Street
Dedham, MA 02026

International Standard Book Number: 0-89006-134-3
Library of Congress Catalog Card Number: 83-72774

A translation from the French of *Hyperfrequences, Traite d'Electricite,* Vol. XIII, © 1981 Editions Giorgi, Presses Polytechniques Romandes, Lausanne, Suisse.

CONTENTS

FOREWORD

General Description of the Field of Microwaves

This textbook deals with the upper boundary, on the frequency scale, of the field of electrical engineering. It considers radioelectric signals having extremely fast variations, with periods between picoseconds and nanoseconds. Microwaves are mostly encountered in three kinds of applications:

1. radar, used for detection and measurements;
2. radiocommunications, for point-to-point links, most particularly for satellite and space communications;
3. heating, drying, cooking of many different types of materials.

The first two types of applications are in the field of information acquisition and transfer. In radar, microwaves permit acquisition of certain information, which is contained within the echo signal from a target. The location and characteristics of the latter are determined by comparing the signal received after reflection from the emitted signal. In communications, microwaves provide the support to the information which is to be transmitted over very great distances. As for the last kind of application, it utilizes a most remarkable property of microwaves, that of in-depth heating. This is an energy transfer application, which has found a number of uses in everyday life, from the kitchen to the medical treatment of various ailments. In the future, microwaves may also play a role in the long distance transfer of electrical power.

Microwaves are thus located across domains of information and power. The more conventional applications belong to the first domain.

It was during World War II, at the beginning of the 1940s, that microwaves first appeared as a new field in electrical engineering. In the same way as for other areas, the evolution of microwaves was rich in new developments and spectacular breakthroughs. While the principal domains of applications have by now become well established, deep changes still occur occasionally, mostly at the technological level. Waveguides progressively gave way to planar circuits, less bulky and easier to realize. Generating tubes, while remaining quite necessary for high power levels, have lost their preeminence elsewhere to semiconductors; first to diode devices, progressively replaced in turn by transistors.

It may sound somewhat reckless to write a book dealing with a topic which is still evolving. Will the data presented still be valid when the book is published? If such a risk does exist, it obviously also exists for all the areas of interest to engineers. After all, the engineer's task is to pursue the evolution of new techniques and methods in order to provide improved solutions. Still, short term obsolescence should not be feared too seriously; any topic may be treated at various levels of complexity, some of which grow old faster than others.

For instance, let us consider a radar. Its principle is based on electromagnetic echoes. Waves propagate and are reflected according to unchanging physical laws governing the universe. Of course, our knowledge of these laws is incomplete. Nevertheless, the models which have been developed to describe the real world are sufficiently accurate, so that differences are most often undetectable – when the adequate model is selected. The velocity of light and the ratio of received power to emitted power are specified by physics and geometry, not by the technical characteristics of the system. Even the internal layout remains practically unchanged; a radar must contain an emitter, an antenna, a receiver, together with a connecting network and data processing circuits. It is only on the component level, where the technological aspect dominates, that obsolescence becomes real. Still, by enclosing each component within a "black box," characterized by a scattering matrix, it remains possible to reason in a correct manner, even when the actual contents of the box are unknown. Of course, one must distinguish between those who make the boxes and those who use them. A self-consistent and reasonably up-to-date book may be written for the users.

Since the contents of the black boxes evolve with time, no book, no matter how well prepared, can long remain informative as to the latest technological developments. The engineer must therefore have other sources of information which allow him to keep abreast in the specialized domain he has selected. A number of technical journals are available for this very purpose, while conferences, symposia, and workshops are organized, covering most areas of interest. However, in order to actually understand and digest the information read or heard, a basic minimal level of knowledge is a prerequisite. The technical teaching, to which this book is associated, aims at providing this basic knowledge which is needed to ensure a continuous personal updating.

General Organization of the Book

The first chapter provides a short description of the microwave field: its frequency bands, periods, wavelengths. The specific properties of microwaves are reviewed and a concise historical note traces the most significant steps. Maxwell's equations in phasor notation are introduced and boundary conditions described.

The main applications of microwaves were traditionally associated with metallic pipes, the waveguides. More recently, the inventory of transmission lines was enriched by the addition of planar lines (microstrip) and dielectric waveguides (optical fibers). The study of guided wave propagation along uniform structures is covered in Chapter 2.

The production and measurement of signals having stable frequencies requires resonant circuits. At microwaves, these circuits are made of cavities and resonators, most often metallic boxes loosely coupled to one or several transmission lines. Chapter 3 first considers a closed cavity, develops its equivalent circuit, and then studies the frequency response of a coupled cavity.

Microwave signals are generated and amplified through interactions of electromagnetic fields with electrically charged particles. These rather complex phenomena can only be described briefly here, providing a basic understanding of the operating principles. The main tubes and semiconductors used to produce and to amplify microwave signals are presented in Chapter 4.

A microwave signal is characterized by its power and frequency, or by a set of both quantities (spectrum). The usual measurement techniques are introduced in Chapter 5, which describes at the same time the most significant sources of error.

Any system consists in the assembly of a number of components: its global behavior results from the properties of all its components. Each one of them is characterized in terms of its scattering matrix. The main constraints imposed by linearity, reciprocity, symmetry, losslessness are derived in Chapter 6. The chapter goes on with a general catalog of the components most often encountered in microwaves.

The techniques used to measure components, in terms of the signals reflected by them or transmitted through them are developed in Chapter 7.

The book ends with a description of the main applications of microwaves at the present time: radar; communications (microwave links, satellites); industrial and consumer applications for heating, measurement, and control; medical applications and biological effects. The possibility of using microwaves in the future for the transfer of power is briefly mentioned. In each case, the particular aspects of the application and the main problems encountered are presented.

In the English version, Chapter 9 is added to cover basic line theory (in the original French text, this topic is covered in a companion volume, *Electromagnétisme*).

A series of appendices, mostly of mathematical and of physical nature, completes this volume in Chapter 10. In particular, the analogies between microwaves and acoustics are listed.

Comments Related to the Use of Different Models of the Real World

The microwave field covers quite a wide range of applications of very diverse nature. Many different problems are found in the field. On the fundamental level, basic theories quickly lead to rather complex mathematics. In rather sharp contrast, many practical applications are still based on the traditional "cut and try" experimental approach. These are the two extremes, between which both the theory and the experiment take a part in the practical realization of devices and systems. This book tries to survey all of this variety. It reviews the fundamental theory, describes measurement techniques, and outlines in a nutshell the most significant applications.

The same variety is encountered in the use of models at different levels of complexity. The study of waveguides and cavities is done by considering the electrical and magnetic fields, described in terms of Maxwell's equations. The same macroscopic model is also adequate to consider the interactions between electromagnetic fields and an electron beam, as they occur in most generating and amplifying vacuum tubes. For other active components, such as the solid state sources, the use of quantum physics is required. The characterization and measurement of components is done in terms of circuit theory, using the scattering matrix formalism and Kirchhoff's laws. At microwaves, the terms of the scattering matrix are most often transcendental functions.

Rules of Presentation

This book is divided into chapters (Ch.), marked by an arabic numeral (Ch. 2). Every chapter is divided into sections (Sec.) marked by two arabic numerals separated by a dot (Sec. 4.2). Every section is in turn divided into sub-sections (§) marked by three arabic numerals separated by two dots (§ 2.3.11). Internal references specify the chapter, the section, or the sub-section.

The bibliographic references are numbered continuously and marked by a single arabic numeral between square brackets: e.g., [56] .

A term appears in *medium italic type* the first time it is defined within the text. An important passage is emphasized by composition in ***bold italic type***.

A critical or complex sub-section is indicated by the sign ■ in front of its numerical index. A sub-section which is not a basic requirement for the understanding of what follows is marked by the sign □ in front of its numerical index.

The equations are numbered continuously within each chapter and marked by two arabic numerals between parentheses and separated by a dot (2.153). An equation is emphasized by its number in bold type (**1.3**). The figures and tables are numbered continuously within each chapter and marked by two arabic numbers, preceded by Fig. (Fig. 2.54) or Table (Table 1.2).

Acknowledgments

The whole team of the Microwave Group, Laboratory for Electromagnetism and Acoustics of the Ecole Polytechnique Fédérale, Lausanne, Switzerland, contributed to the elaboration and production of this volume. Particularly worth mentioning are the contributions of Messrs. Jean-Claude Besson (coordination); Jean-Francois Zurcher, Jean-Dominique Decotignie, and Kurt Hofer (figures); Viron Teodoridis (equations); and Thomas Sphicopoulos (problems). The manuscript of the French text was typed by Mrs. Alix Wend. At various stages of the process, the text was checked and commented on by colleague professors, assistants, and my students in Lausanne and Oran. I wish to thank all those who actively participated in this task.

The original French version of this book, *Hyperfréquences*, is (lucky) number 13 in a series of 22 volumes devoted to the field of electrical engineering: *Traité d'Electricité de l'Ecole Polytechnique Fédérale de Lausanne.* My colleagues and I are particularly grateful to Professor Jacques Neirynck, who provided the original spark which started the whole project, and who has been busy ever since supervising the elaboration of the volumes. He is most ably seconded in this task by Mrs. Claire-Lise Delacrausaz (Presses Polytechniques Romandes) and Mr. Allen Kilner (assembly).

Most of my translation into English has been checked and commented on by Dr. K.N. Tripathi of the University of Delhi.

Lausanne, June 1983 *F. Gardiol*

CHAPTER ONE

GENERAL BACKGROUND

1.1 MAIN DEFINITIONS

1.1.1 Definition: Microwaves

The frequency range extending from 300 MHz up to 300 GHz is generally known as *microwaves,* which thus characterize signals having between 300 million and 300 billion periods per second (from $3 \cdot 10^8$ to $3 \cdot 10^{11}$ Hz, fig. 1.1). These limits are to some extent arbitrary: they permit to position the microwave domain between the waves used for radio and TV diffusion (lower ranges) and infrared rays (higher ranges).

1.1.2 Denominations: UHF, SHF, and EHF Decades

A finer designation of the frequency bands divides them into decades (Fig. 1.2). The microwave range more or less covers the three decades called *ultra-high frequencies* (UHF), *supra-high frequencies* (SFH), and *extra-high frequencies* (EHF). The use of ***superlatives*** can however be confusing: Is extra really larger than supra? Are both of them larger than ultra? And then, what comes beyond extra? As no suitable next higher term was found, the upper ranges are known as infrared, visible rays, and ultraviolets (Fig. 1.1).

Microwaves are divided into still narrower bands, corresponding to specific waveguide sizes (Tables 2.14 and 2.19).

1.1.3 Energy of a Microwave Photon

In quantum physics, electromagnetic radiation is considered to be a flow of photons of energy hf, where f is the frequency and h is Planck's constant ($4.14 \cdot 10^{-15}$ eV or $6.63 \cdot 10^{-34}$ Js). A microwave photon thus has an energy located in the range between roughly $1.2 \cdot 10^{-6}$ and $1.2 \cdot 10^{-3}$ eV.

1.1.4 Order of Magnitude of the Periods

The period T of a microwave signal, defined as the inverse of the frequency f is located between 3 ns (nanoseconds) and 3 ps (picoseconds).

1.1.5 Wavelength

The *wavelength* of a microwave signal for an electromagnetic wave propagating across space at the velocity of light $c_0 = 2.997925 \ldots 10^8$ m/s is defined by $\lambda = c_0/f = c_0 T$, somewhere between 1 m and 1 mm.

1.1.6 Remark

The term *microwaves,* used to define a range of frequencies, denotes the ***smallness*** of the wavelength encountered, as compared to those utilized for radio and television. The use of the comparative prefix micro may lead to confusions here too: one would expect ***microwaves*** to denote wavelengths in the ***micrometer*** range and not, as is actually the case, in the ***millimeter to meter*** ranges.

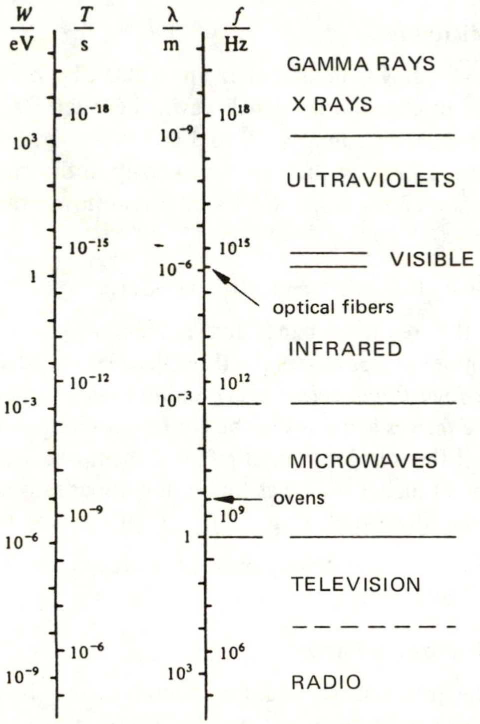

Fig. 1.1 Subdivisions of the electromagnetic spectrum.

Table 1.2 Division, nomenclature, and allocation of microwaves.

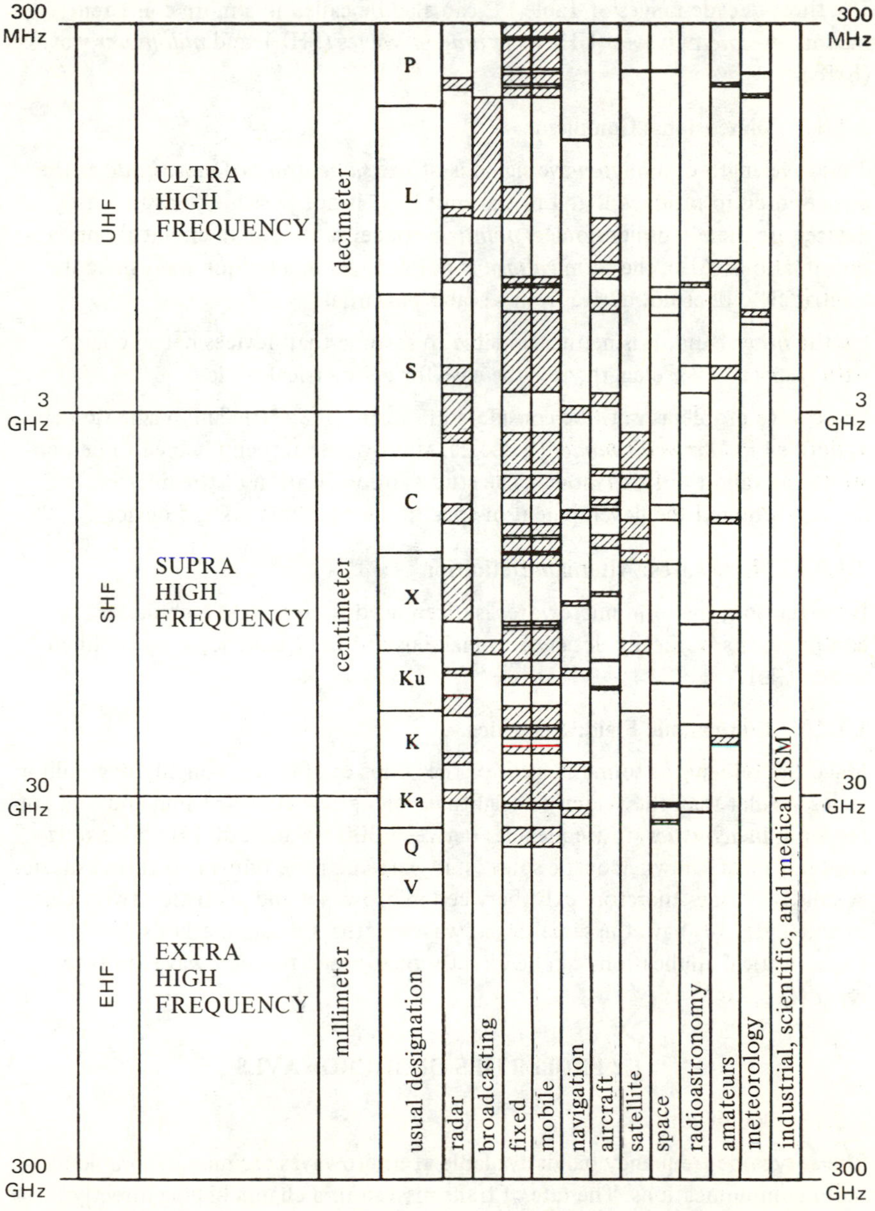

1.1.7 Alternate Names: Decimeter, Centimeter, and Millimeter Waves

The three decade ranges of Table 1.2 can also be called in a more consistent manner *decimeter* waves (UHF), *centimeter* waves (SHF), and *millimeter* waves (EHF).

1.1.8 Dimensional Comment

The wavelength of a microwave signal is of the ***same order of magnitude*** as the devices used to produce it and to transmit it. It is not possible to assume that devices are merely dimensionless ***points*** in space, as is done in circuit theory approximations. Also, the term ***voltage*** is not defined in a unique way, since the electric field does not derive from a scalar potential.

On the other hand, it is neither possible to assume that devices become large with respect to wavelength, as is the case for geometrical optics.

Microwave problems must be considered in terms of electric and magnetic fields, as defined in ***Maxwell's model*** (§ 1.4.2). Microwaves actually helped to demonstrate the validity of this model. The study and utilization of the microwave domain required the development of specific new techniques and devices.

1.1.9 Microwaves: Alternate Definition

By extension, the term microwaves is often used to designate techniques and components specifically designed to make use of the frequency range defined under 1.1.1.

1.1.10 Comparable Field: Acoustics

Since the velocity of sound in air is of the order of 300 m/s, roughly one million times smaller than the velocity of light, wavelengths between 1 mm and 1 m correspond, in ***acoustics***, to frequencies between 300 Hz and 300 kHz. This range contains most known acoustic sources, in particular the human voice and music. A similarity does therefore exist between microwaves and acoustics, two fields in which devices have the same size as wavelengths. Certain methods developed for acoustical applications can be directly ***transposed*** to microwaves and *vice versa* (Sec. 10.5).

1.2 PROPERTIES OF MICROWAVES

1.2.1 Bandwidth

The very wide frequency bands available at microwaves are most favorable for radio communications. The rate of transmission of a channel being directly proportional to its bandwidth, a simple calculation shows that over the 300 MHz to 300 GHz frequency range, 999 times more information can be transmitted over a specified time period than in all the lower frequency bands taken together. As a result, the use of microwaves permits meeting the increasing need for communications channels.

This particular property is directly related to the signal frequency. Following the same line of reasoning, even much larger amounts of information could, in principle, be transmitted over the infrared and visible spectra, using lasers and optical fiber systems (Sec. 2.10 and § 8.2.15).

1.2.2 Transparency of the Ionosphere

The ionosphere forms a group of ***ionized*** layers of ***electron plasma***, which surround the earth at altitudes ranging from 50 to 10,000 km (§ 8.2.6). These ionized layers are produced by the impact of solar radiation, so that their parameters (density, height) fluctuate widely between day and night, exhibiting seasonal variations and changes related to the solar activity.

The electromagnetic propagation within the ionosphere is similar to that in a waveguide (Ch. 2). Signals at frequencies lower than 10-40 MHz (cut-off frequency) are partially or totally reflected. This property is used to realize multiple reflection links at short waves. Higher frequency signals travel across the ionosphere, but experience distortion, which decreases with frequency. Microwave signals, well above the ionospheric cut-off, are hardly affected at all, at sufficiently small power levels. For these reasons, microwaves are utilized for satellite communications and space transmissions. Radioastronomy deals with microwave radiation emanating from distant stars, galaxies, and quasars (§ 8.6.4).

1.2.3 Partial Transparency of the Atmosphere

The various atmospheric components (oxygen, nitrogen, water vapor, carbon dioxide) and the many elements in suspension (water droplets, ice crystals, dust, smoke) do not significantly affect signals having frequencies smaller than about 10 GHz. Higher frequency signals, on the other hand, experience several unwanted effects: absorption, depolarization, and scintillation (Sec. 8.2).

1.2.4 Electromagnetic Noise

The noise picked up by an antenna directed skyward, in the absence of signal, goes through a relatively flat minimum over the 1 to 10 GHz frequency range. The received noise power is the product of the equivalent noise temperature (in Kelvin) by Boltzmann's constant ($k_B = 1.3804 \cdot 10^{-23}$ J/K) and by the receiver's bandwidth (Sec. 7.6). Over the 1-10 GHz band, the noise temperature decreases below 10 Kelvin. In practical terms, this means that it is within this frequency range that signals of the lowest amplitude can be detected, and thus the most sensitive receivers can be designed. For instance, signals of extremely low levels received after transmission across planetary space are often in the neighborhood of 3 GHz. A further requirement is for the receiver itself not to produce too much noise, which would irremediably degrade the input signal-to-noise ratio.

1.2.5 Directivity of Antennas

The width of the beam radiated by an antenna is directly proportional to the ratio of the wavelength to the antenna's largest dimension. When transmitting a signal from one point to another (microwave link), or when determining the origin of a reflection (radar), a narrow beamwidth is required. It is then either necessary to have a large size antenna, which is often not convenient for mechanical reasons, or to utilize a signal of high frequency. Microwaves are well suited for such applications.

Even narrower beamwidths are obtained using signals in the visible range (laser). As the beam is then extremely narrow, it must be aimed quite accurately towards the detector, and pointing problems may become significant.

1.2.6 Reflection on Targets

The effective reflection area of an object depends in a very sensitive manner on the ratio of the object's size to the wavelength (Sec. 8.1). When the reflecting element is much smaller than the wavelength, the reflection becomes vanishingly small. On the other hand, when the wavelength becomes much smaller than the object, the effective reflection area for a metallic element is approximately its cross section transverse to the beam. Centimeter waves thus detect objects of meter size, but are not affected by raindrops (§ 1.2.3). The latter can, on the other hand, perturb the detection at millimeter waves.

1.2.7 Interaction with Matter

When an electromagnetic wave impinges on a material sample, it is preferentially absorbed at particular frequencies, which are the resonant frequencies of the material. The resonances observed at microwaves depend on the molecular composition of matter. This effect is put to good use in chemical and physical analyses (§ 8.5.15).

In particular, water strongly absorbs all microwaves, a specific property which leads to two types of applications:

1. microwave heating, utilized for the cooking of food, the drying and thermal processing of numerous materials, and the medical treatment of a number of ailments by hyperthermia (Sec. 8.3);
2. the detection and measurement of moisture contained within materials.

1.2.8 Non-Ionizing Radiation

The energy of a molecular bond is larger, by several orders of magnitude, than the energy belonging to a microwave photon (§ 1.1.3). This means that a single photon, at microwave frequencies, does not possess sufficient energy ***to break a chemical link***, for instance by photoelectric effect. Microwaves are thus a *non-ionizing* form of *radiation.* At frequencies far above microwaves, a single photon can have enough energy to extract an electron and produce ionization: this happens in the visible spectrum, ultraviolet (suntan), X and gamma rays (Fig. 1.1).

1.2.9 Stable Oscillation Frequencies

The most stable known atomic oscillators, hydrogen, cesium, and rubidium exhibit extremely stable oscillations within the microwave range. As a result, all atomic clocks and frequency standards make use of microwaves (§ 5.3.11).

1.2.10 Remark

The set of properties listed in the previous paragraphs makes microwaves a privileged field for a large number of applications, such as satellite communications and radar, but also heating and measurement. Some of these applications simply could not exist without microwaves. For others, microwaves provide the best compromise among various requirements to be satisfied simultaneously. A more detailed description of the most significant applications is given in Chapter 8.

1.3 HISTORICAL LANDMARKS

1.3.1 First Experiments in Radio Communications

The spark generators used by Hertz and Marconi for their first radio transmissions in 1888 and 1890 were non-coherent emitters, generating a very broad spectrum of noise extending well into the microwave range. The latter did thus play some part – not recognized as such then – during the very first radio transmissions.

1.3.2 First Waveguide

In 1894, Sir Oliver Lodge surrounded a spark generator with a metal tube and noted that the resulting radiated signal exhibited directional properties. This effect was at the time considered a laboratory curiosity, and it did not lead to practical applications. With the advent of vacuum tubes, the further development of radio was hence oriented towards lower frequencies.

1.3.3 Communications Experiments

The first practical experiments carried out with the purpose of using microwaves for the transmission of information are credited to George Southworth, at the ***Bell Telephone Laboratories***, in the US during the 1920s and 1930s. Wave propagation along copper water pipes was investigated. Since then, users of microwaves have often been nicknamed ***plumbers***.

1.3.4 Development of Radar

The advent of microwave techniques is quite closely linked to the development of radar during World War II, mostly in Great Britain and the US. The basic principle, already outlined by Sir Robert Wattson-Watt around 1930, makes use of the electromagnetic echo produced by a target. A short signal pulse is launched, the time between its departure and the return of a reflected signal is measured. To ensure an adequate detection of targets, signals of increasingly higher frequencies were found to be needed. The magnetron was developed for this purpose, being the first tube having a high power generating capability at microwaves.

The design and the industrial fabrication of radar systems started around 1940, mostly in the US, and in particular in the Boston area. Radar played a most significant role during World War II, among others in the well known Battle of Britain. Some of the research and development work was de-classified at the end of the war. Particularly worth mentioning is the 25-volume set published by the ***Massachusetts Institute of Technology*** (Radiation Laboratory Series), covering all aspects of radar design, including the foundations of microwaves. Books of this series are still often consulted [1].

1.3.5 Development of Microwave Tubes

Microwave generators were at first vacuum tubes, specifically designed for use within radar systems. The magnetron, mentioned in the previous section, is based on the interactions of an electron flow within crossed electric and magnetic fields, an effect studied since the 1920s in several countries. It became operative at the beginning of World War II. The klystron was invented in 1935 by two brothers, Russell and Sigurd Varian. Many different kinds of microwave generators were developed later on, the most significant ones being the backward-wave oscillator (BWO) and the traveling wave tube (TWT). A detailed description of generators and amplifiers is available in Chapter 4.

1.3.6 Advent of Ferrite Devices

The first non-reciprocal linear passive device appeared in 1956, a gyrator invented by C. Lester Hogan. The many isolators and circulators developed since then are mostly devices for protection, decoupling, and control. They are nowadays found in most microwave systems (Sec. 6.7).

1.3.7 Satellite Communications

In 1962 ***Telstar*** was launched, the first communications satellite placed in a low earth orbit. Three years later, in 1965, the satellite ***Early Bird*** was placed in a geostationary orbit (remaining over a fixed location on the equator). Since then, successive generations of satellites are playing an increasing role in communications, mostly for international links, but also within domestic networks (§ 8.2.13).

1.3.8 Solid State Two-Ports

Active semiconductor devices appeared on the market in the 1960s, replacing vacuum tubes as microwave sources for low and medium power levels. The first device of this kind developed was the Gunn diode, based on a physical phenomenon discovered by J.B Gunn in 1962. It was followed by junction diodes making use of avalanche and transit time (Ch. 4).

1.3.9 Microwave Transistors

Diode generators and amplifiers discussed in the previous section are increasingly being replaced, since the end of the 1970s, by transistors, either bipolar or MESFETs (Sec. 4.8).

1.3.10 Microwave Printed Circuits

Over the same time span, metallic waveguides are often superseded by more compact circuits in microstrip, slot line, or coplanar line, realized by means of printed circuit techniques (Sec. 2.11).

1.4 BASIC ASSUMPTIONS

■ 1.4.1 Sine Wave Regime, Phasor-Vectors

The theoretical developments presented in this book consider exclusively time-dependent ***sine wave*** signals, represented in complex notation by means of ***phasor-vectors***. More complex signal patterns can in principle always be developed over a finite or infinite sum of sine waves (Fourier series, Fourier transform) [215].

The selection of sine waves as the basis for decomposition results from the fact that, in all systems of linear equations the eigenfunctions are sine waves, which move along within the system ***without deformation***, even though their amplitude and phase do vary with position. In every dispersive system, nonsinusoidal signals, on the contrary, change shape while propagating, either through amplitude distortion or through phase distortion.

The actual physical field $X(\boldsymbol{r},t)$, function of space and time, is related to the corresponding phasor-vector $\underline{X}(\boldsymbol{r})$ by the relationship

$$X(\boldsymbol{r},t) = \mathrm{Re}[\sqrt{2}\ \underline{X}(\boldsymbol{r}) \exp(\mathrm{j}\omega t)] \tag{1.1}$$

where vector $\boldsymbol{r}$ represents the position in space, t the time and $\omega = 2\pi f$ the angular frequency corresponding to frequency f. Underlining means that a quantity is complex (phasor or phasor-vector). Bold face italics denote vectors. The modulus of the phasor-vector $\underline{X}(\boldsymbol{r})$ is the effective value of the corresponding physical field.

■ 1.4.2 Maxwell's Equations in Complex Notation

Maxwell's equations, in term of phasor-vectors in a linear, homogeneous, possibly lossy, isotropic (§ 1.4.6) medium are given by [214]

$$\nabla \times \underline{E} = -j\omega \underline{B} = -j\omega \underline{\mu}\, \underline{H} \qquad \text{V/m}^2 \qquad (1.2)$$

$$\nabla \times \underline{H} = \underline{J} + j\omega \underline{D} = (\sigma + j\omega \underline{\epsilon})\underline{E} \qquad \text{A/m}^2 \qquad (1.3)$$

$$\nabla \cdot \underline{E} = \underline{\rho}/\underline{\epsilon} \qquad \text{V/m}^2 \qquad (1.4)$$

$$\nabla \cdot \underline{H} = 0 \qquad \text{A/m}^2 \qquad (1.5)$$

with

$\underline{E}$ electric phasor-vector

$\underline{H}$ magnetic phasor-vector

$\underline{B}$ induction phasor-vector

$\underline{D}$ displacement phasor-vector

$\underline{J}$ current density phasor-vector

$\underline{\rho}$ charge-density phasor

σ conductivity

$\underline{\epsilon} = \epsilon_0 \underline{\epsilon}_r = \epsilon' - j\epsilon'' = \epsilon_0(\epsilon_r' - j\epsilon_r'') = \epsilon'(1 - j \tan \delta)$

complex permittivity

$\underline{\mu} = \mu_0 \underline{\mu}_r = \mu' - j\mu'' = \mu_0(\mu_r' - j\mu_r'')$

complex permeability

The terms $\underline{\epsilon}_r$ and $\underline{\mu}_r$ are respectively the *relative* permittivity and permeability of the medium. The *loss tangent* $\tan \delta = \epsilon''/\epsilon'$ characterizes the attenuation of the electric field in a lossy material.

In many metals and semiconductors, the current density $\boldsymbol{J}$ is many orders of magnitude larger than the displacement current $\partial \boldsymbol{D}/\partial t$. It is then often assumed that, in first approximation, $\sigma \cong \infty$, the conductor being then a *perfect electric conductor* (pec). However, for physical reasons the current density cannot become infinite: the above condition means that the electric field $\boldsymbol{E}$ must vanish within a pec.

Similarly, in ferromagnetic materials the permeability μ becomes quite large, and is often assumed to be infinite, as a first-order approximation. One then has a *perfect magnetic conductor* (pmc). In this instance, the induction field $\boldsymbol{B}$ cannot become infinite, so that the magnetic field $\boldsymbol{H}$ must vanish within a pmc.

■ 1.4.3 Boundary Conditions

On the interface separating two different materials, ***none of which is a perfect***

electric conductor, the tangential components of the electric and of the magnetic field are continuous, as expressed by

$$\boldsymbol{n} \times (\underline{\boldsymbol{E}}_1 - \underline{\boldsymbol{E}}_2) = 0 \qquad \text{V/m} \qquad \textbf{(1.6)}$$

$$\boldsymbol{n} \times (\underline{\boldsymbol{H}}_1 - \underline{\boldsymbol{H}}_2) = 0 \qquad \text{A/m} \qquad \textbf{(1.7)}$$

where $\boldsymbol{n}$ is the unit vector normal to the interface, directed from medium 2 towards medium 1, and where indices 1 and 2 specify, respectively, the medium within which the field is defined. When $\omega \neq 0$, the boundary conditions for the normal components of the phasor-vectors are automatically satisfied when (1.6) and (1.7) are met.

On the surface of a ***perfect electric conductor*** (*pec*, $\sigma = \infty$), the electric phasor-vector must meet the condition

$$\boldsymbol{n} \times \underline{\boldsymbol{E}} = 0 \qquad \text{V/m} \qquad \textbf{(1.8)}$$

This condition is approximately satisfied on a metallic surface (short-circuit).

On the surface of a ***perfect magnetic conductor*** (*pmc*, $\mu = \infty$) without surface current, the magnetic phasor-vector must meet the condition

$$\boldsymbol{n} \times \underline{\boldsymbol{H}} = 0 \qquad \text{A/m} \qquad \textbf{(1.9)}$$

Magnetic materials no longer have, at microwaves, a sufficiently large permeability to satisfy the above condition. It can nevertheless be used to take into account geometrical symmetry (open-circuit plane), or to approximately represent the interface between a very high permittivity dielectric and one having a much smaller permittivity.

Boundary conditions on the normal components of the electric field yield:

$$\boldsymbol{n} \cdot (\underline{\epsilon}_{r1} \underline{\boldsymbol{E}}_1 - \underline{\epsilon}_{r2} \underline{\boldsymbol{E}}_2) = \underline{\rho}_s / \epsilon_0 \qquad \text{V/m} \qquad (1.10)$$

Surface charges $\underline{\rho}_s$ can only be found on the interface between two materials when at least one of the two has a non-zero conductivity σ.

Similarly, the magnetic field must satisfy

$$\boldsymbol{n} \cdot (\underline{\mu}_{r1} \underline{\boldsymbol{H}}_1 - \underline{\mu}_{r2} \underline{\boldsymbol{H}}_2) = 0 \qquad \text{A/m} \qquad (1.11)$$

1.4.4 Remark: Magnetic Field in a Perfect Electric Conductor

Since the electric field $\boldsymbol{E}$ must vanish inside of a perfect electric conductor (§ 1.4.2), it follows from (1.2) that the magnetic field must also vanish there when $\omega \neq 0$.

At the edge of a perfect electric conductor flows a surface current $\boldsymbol{A}$, defined by the boundary condition on the tangential magnetic field:

$$n \times \underline{H} = -\underline{A} \qquad \text{A/m} \qquad (1.12)$$

where the normal vector $\boldsymbol{n}$ is pointed towards the conductor, *i.e.*, towards the outside of a waveguide or of a cavity.

1.4.5 Constants for Vacuum

In vacuum, and approximately in air, the electric displacement $\underline{D}$ is related to the electric field $\underline{E}$ by the *electric constant* ϵ_0

$$\epsilon_0 = \frac{|\underline{D}|}{|\underline{E}|} = 8.854 \ldots 10^{-12} \qquad \text{As/Vm} \qquad (1.13)$$

Similarly, the *magnetic constant* μ_0 is the ratio of the magnetic induction $\underline{B}$ by the magnetic field $\underline{H}$ in all nonmagnetic materials

$$\mu_0 = \frac{|\underline{B}|}{|\underline{H}|} = 4\pi \cdot 10^{-7} \qquad \text{Vs/Am} \qquad (1.14)$$

The *velocity of light in vacuum* c_0 is directly related to the product of these two constants by

$$c_0 = \frac{1}{\sqrt{\epsilon_0 \mu_0}} = 2.997\,925 \ldots 10^8 \qquad \text{m/s} \qquad (1.15)$$

$$\cong 300.000 \text{ km/s} = 3 \cdot 10^8 \text{ m/s} = 300 \text{ m/}\mu\text{s} = 300 \text{ mm/ns}$$

The *characteristic impedance of vacuum* Z_0 is defined as:

$$Z_0 = \sqrt{\mu_0/\epsilon_0} \cong 120\pi = 376.6 \ldots \qquad \Omega \qquad (1.16)$$

1.4.6 Definitions: Isotropy and Anisotropy

Inside the great majority of materials utilized in practical applications, phasor vectors $\underline{D}$ and $\underline{E}$ are parallel to one another, or ***colinear***. The same occurs for $\underline{B}$ and $\underline{H}$. Such a material is then called *isotropic.* The permittivity $\underline{\epsilon}$ and the permeability $\underline{\mu}$ are both ***scalar quantities*** (§ 1.4.2).

For a family of magnetic dielectrics called ferrites (Sec. 6.7), the phasor-vectors $\underline{B}$ and $\underline{H}$ of a microwave signal ***are not parallel***. These are *anisotropic* materials, for which the permeability becomes a tensor $\underline{\overline{\overline{\mu}}}$, represented by a 3 × 3 matrix. Within certain anisotropic materials, the reciprocity theorem [214] is no longer applicable (Sec. 6.7).

1.4.7 Polarization of a Field

The polarization of any vector field is defined by the geometrical locus of the tip of the vector, as it varies during one period T. Equation (1.1) may be written in the following form:

$$X(r,t) = \mathrm{Re}[\sqrt{2}\,\underline{X}] \cos \omega t - \mathrm{Im}[\sqrt{2}\,\underline{X}] \sin \omega t$$

$$= X(r,0) \cos \omega t + X(r,T/4) \sin \omega t \tag{1.17}$$

This is actually the equation of an ellipse having two conjugate half-axes $X(r,0)$ and $X(r,T/4)$, which defines the plane of the ellipse as well as its dimension. In the most general case, therefore, a field is *elliptically polarized.*

There are two limiting cases:

1. when the two vectors are directed parallel to each other, or when one of the two vanishes, the ellipse shrinks to a straight line. The field is then *linearly polarized,* and the condition

$$\underline{X} \times \underline{X}^* = 0 \tag{1.18}$$

   is satisfied. The asterisk * denotes the complex conjugate.

2. when the two vectors have the same length and are perpendicular to each other, the ellipse becomes a circle, and the field is *circularly polarized.* This occurs when

$$\underline{X} \cdot \underline{X} = 0 \tag{1.19}$$

   and the field may be described by

$$\underline{X} = |\underline{X}|\,(e_o - \mathrm{j} e_{T/4}) \tag{1.20}$$

   where e_o and $e_{T/4}$ are unit vectors along the two directions $X(r,0)$, and $X(r,T/4)$, respectively.

Any field may be expressed by the sum of two linearly polarized fields: this is actually done in (1.17). Any field may also be decomposed into two circularly polarized fields, rotating in opposite directions. The long half-axis of the ellipse is then the sum of the two radii, and the short axis their difference (which vanishes for a linear polarization).

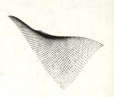

CHAPTER TWO

TRANSMISSION LINES AND WAVEGUIDES

2.1 DEFINITIONS AND CLASSIFICATION

2.1.1 Basic Assumption: Uniformity Along the Direction of Propagation

The propagation characteristics of a transmission line or of a waveguide are determined from the study of the electromagnetic fields within the general structure depicted in Figure 2.1.

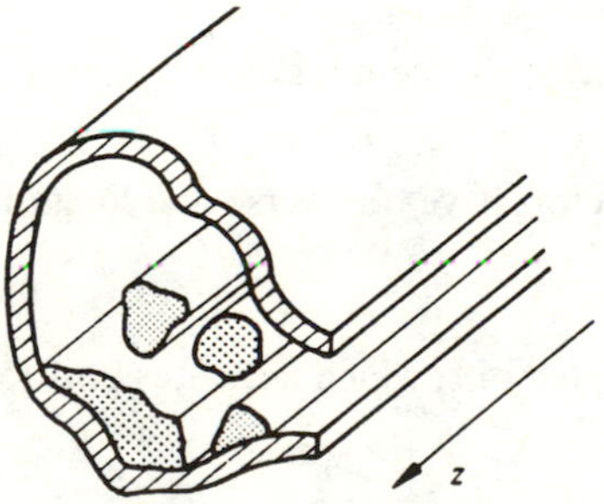

Fig. 2.1 Waveguide of arbitrary cross section, uniform along the direction of propagation.

The cross section of this structure is arbitrary. It can contain several different propagation media (air, dielectrics, ferrites), in which case it is called *inhomogeneous.* It can also contain, or be surrounded by, metallic conductors. The direction of propagation is rectilinear: the longitudinal coordinate axis z is located along this direction. A displacement along this axis does not modify either the geometry, nor the material properties: these quantities are thus independent of the longitudinal coordinate z (translation-invariance). The transmission line or waveguide is then called *uniform.*

2.1.2 Selection of a Coordinate System

The system of coordinates used to study the propagation must be well adapted to the transverse geometry of the line. Boundaries and interfaces should be described by simple geometrical expressions, so that boundary conditions for the fields can be satisfied. In all uniform structures, the coordinate system contains the coordinate z, defined in the previous section, and two coordinates located in the transverse plane. For problems exhibiting a symmetry of revolution, the circular cylindrical coordinate system is the obvious choice. A rectangular waveguide is best studied in a rectangular reference base. Other geometries are treated in elliptical, parabolic cylindrical coordinates, or still in reference systems obtained by the application of conformal mapping in the transverse plane.

Transverse planes are always perpendicular to the direction of propagation. Transverse and longitudinal dependencies of the fields are thus independent of each other, and the differential operator ***del*** or ***nabla*** (§10.1.2) can be expressed by the sum of a transverse part (index t) and a longitudinal one:

$$\nabla = \nabla_t + e_z \frac{\partial}{\partial z} \qquad \text{m}^{-1} \qquad (2.1)$$

In the rectangular coordinate system, the transverse part of the operator can be further separated, in terms of the coordinates x and y:

$$\nabla_t = e_x \frac{\partial}{\partial x} + e_y \frac{\partial}{\partial y} \qquad \text{m}^{-1} \qquad (2.2)$$

In all other coordinate systems, the transverse operator cannot be separated in this manner (§ 10.1.6).

Similarly, the phasor-vectors have transverse and longitudinal components:

$$\underline{X} = \underline{X}_t + e_z \underline{X}_z \qquad (2.3)$$

The same is true for the vector r, which indicates position:

$$r = r_t + e_z z \qquad \text{m} \qquad (2.4)$$

2.1.3 Consequence: Separation of Variables

Within every system of cylindrical coordinates (which correspond to uniform lines), ***Maxwell's equations*** are partially separable. The general solution for the electromagnetic fields is then the sum of several terms, each term being the product of a function of the transverse coordinate r_t by a function of the longitudinal variable z. For any component $\underline{X}_i$ of one of the fields, one has:

$$\underline{X}_i(r_t, z) = \sum_k \underline{f}_{ki}(r_t)\, \underline{g}_{ki}(z) \qquad (2.5)$$

The transverse and longitudinal dependencies of the fields can then be studied independently of each other.

2.1.4 Longitudinal Dependence of the Fields

Applying the method of separation of variables, and then grouping the resulting terms, a wave equation involving the direction of propagation is obtained for every field component in an isotropic structure:

$$\frac{\partial^2 \underline{X}_i(r_t, z)}{\partial z^2} - \underline{\gamma}^2 \underline{X}_i(r_t, z) = 0 \qquad (2.6)$$

The complex constant $\underline{\gamma}^2$ results from the separation process. Its value depends on frequency, material properties and the particular solution considered for the fields (type of mode, § 2.1.8).

The detailed treatment for a particular situation is given later on in paragraph 2.2.4 (TE mode within a closed metallic waveguide). Equation (2.6) is satisfied by the general form of solution (§ 9.3.4):

$$\underline{X}_i(r_t, z) = \underline{X}_{i+}(r_t) \exp(-\underline{\gamma} z) + \underline{X}_{i-}(r_t) \exp(+\underline{\gamma} z) \qquad (2.7)$$

The two terms correspond, respectively, to the forward and to the backward waves, functions $\underline{X}_{i+}(r_t)$ and $\underline{X}_{i-}(r_t)$ defining their transverse dependencies in the plane $z = 0$.

2.1.5 Propagation Factor and Related Quantities

The meaning of the *propagation factor* $\underline{\gamma} = \alpha + j\beta$ is developed here.

The real part α, expressed in Neper per meter (Np/m), is called *attenuation per unit length.* It represents the damping of the wave traveling along the transmission line. Its inverse $\delta = 1/\alpha$ is the *skin depth,* the distance over which the signal amplitude decreases to 1/e of its initial value.

The imaginary part β, measured in radians per meter (rad/m), is the *phaseshift per unit length*, which indicates the phase variation of the wave along the direction of propagation. The *wavelength* along the line or waveguide, denoted by λ_g, is inversely proportional to β:

$$\lambda_g = 2\pi/\beta \qquad \text{m} \qquad (2.8)$$

The two *propagation velocities,* the *phase* and *group velocities,* are respectively defined by the relations

$$v_\varphi = \omega/\beta \qquad \text{m/s} \qquad (2.9)$$

$$v_g = (\partial\beta/\partial\omega)^{-1} \qquad \text{m/s} \qquad (2.10)$$

An observer following a zero of the field moves at the phase velocity, the envelope of a modulated signal at the group velocity.

2.1.6 Definition: Transmission Line

Transmission line is the preferred term to describe transmission systems with two or more metallic conductors electrically insulated from one another (for instance the two-wire line and the coaxial line of Table 2.4).

2.1.7 Definition: Waveguide

The propagation in a *waveguide* is generally ensured by successive reflections on the guide boundaries. These are conducting walls in the case of metallic waveguides. Dielectric waveguides and optical fibers utilize the total reflection on the interface between two dielectric materials (Table 2.4).

2.1.8 Definition: Modes of Propagation

The resolution of Maxwell's equations (1.2 to 1.5), subject to the transverse boundary conditions (1.6 to 1.9) in the structure of Figure 2.1 is an ***eigenvalue*** problem. A set of solutions is obtained, which are called the *modes of propagation.* They are the eigenvectors of the problem, each one associated with an eigenvalue (propagation factor, transverse wavenumber).

The modes of propagation form an infinite set, which is discrete for a closed structure (surrounded by a metal boundary), to which must be added a continuous spectrum of radiating modes when the structure is open (fields extending to infinity).

Each mode possesses specific propagation properties: attenuation and phase-shift per unit length, propagation velocities, cutoff frequency. When propagation of a signal takes place at the same time over several modes, the difference in the propagation velocities may produce distortions of the signal. To avoid this unwanted effect in radar and in communications, the shape and the dimensions of the line are adjusted so that only one mode can propagate at the signal frequency. The possible existence of several propagating modes limits the available frequency band in all transmission lines. However, a significant exception must be noted: in fiber optics, multimode operation is actually used. The length of the line or the bit rate of the signal are reduced to maintain distortion within acceptable limits.

The presence (or absence) of a particular mode of propagation on a transmission line depends on the excitation, *i.e.,* on the boundary conditions at both ends of the line (generator and load).

2.1.9 Classification of the Modes of Propagation

The presence or the absence of longitudinal field components affects the propagation behavior of the modes. Four mode categories can exist, as shown in Table 2.2.

Table 2.2 The four types of propagation mode.

E_z	H_z	*Name*	*Acronym*	*Other Denomination*
$=0$	$=0$	Transverse electromagnetic	TEM	
$=0$	$\neq 0$	Transverse electric	TE	H
$\neq 0$	$=0$	Transverse magnetic	TM	E
$\neq 0$	$\neq 0$	Hybrid		EH or HE

2.1.10 Classification of Lines and Waveguides

A large number of different structures can be used to transmit electromagnetic signals. They can be

1. either open (radiation can take place) or closed (fields entirely enclosed within a conducting envelope);
2. either homogeneous (one single propagation medium without transverse dependence of the material properties), or inhomogeneous (several different propagating media, or a single medium having a continuous variation of the material properties in the transverse plane);
3. either conductorless, or possessing one or more conductors.

These three criteria delineate 10 categories, outlined in Table 2.3. The most significant waveguides and transmission lines within each category are indicated in the corresponding boxes. The cross sections of these lines are shown in Figure 2.4.

2.1.11 Definitions: Loaded Waveguide and Loaded Line

When a waveguide or a transmission line (for instance a coaxial line) is ***partially filled*** with dielectric (Table 2.4), they are respectively called *loaded waveguide* or *loaded line.*

Table 2.3 Classification of lines and waveguides.

		Conductorless	*One conductor*	*Two conductors*
Open	Homogeneous	Waves in space	Uniconductor line	Two wire line Stripline
	Inhomogeneous	Dielectric guides Fiber optics Waves in inhomogeneous medium	Goubau line	Microstrip Slot line Coplanar line Insulated two-wire line
Closed	Homogeneous	Impossible	Metallic Waveguide	Coaxial line
	Inhomogeneous	Impossible	Loaded metallic Waveguide	Loaded coaxial line

2.2 PROPAGATION WITHIN A CLOSED HOMOGENEOUS METALLIC WAVEGUIDE

2.2.1 Assumptions

The following idealized situation is considered (Fig. 2.5):

1. the structure is uniform along the direction of propagation z, as specified under 2.1.1,
2. the cross section has an arbitrary shape. It can be simply connected (hollow waveguide) or multiply connected (several conductors),
3. the material which completely fills the waveguide is isotropic, linear, and homogeneous. Its properties $\underline{\epsilon}$, $\underline{\mu}$, and σ do not depend on the position within the waveguide, nor upon the amplitude of the signal.
4. the envelope, which entirely surrounds the waveguide, is formed of a perfect conductor, either electric (pec, approximation of a metal), or magnetic (pmc, corresponding to a plane of symmetry),
5. the waveguide does not contain any electrical space charge ($\underline{\rho} = 0$).

■ 2.2.2 Separation of Maxwell's Equations into Longitudinal and Transverse Components

Since the structure is uniform lengthwise, (2.1) and (2.3) are applied to Maxwell's equations. The divergence equations (1.4) and (1.5) then take the form:

$$\nabla_t \cdot \underline{E}_t + \frac{\partial \underline{E}_z}{\partial z} = 0 \qquad \text{V/m}^2 \qquad (2.11)$$

$$\nabla_t \cdot \underline{H}_t + \frac{\partial \underline{H}_z}{\partial z} = 0 \qquad \text{A/m}^2 \qquad (2.12)$$

Table 2.4 Cross sections of the main transmission lines and waveguides.

Homogeneous		Inhomogeneous	
two-conductor line		insulated two-conductor line	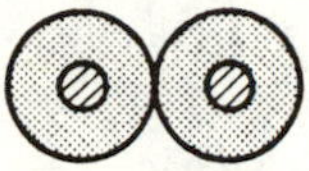
one-conductor line		Goubau line	
		microstrip	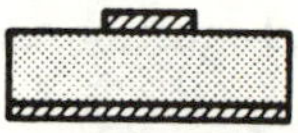
stripline		slot line	
		coplanar line	
coaxial line	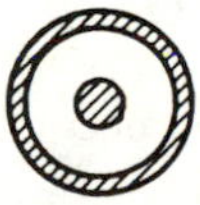	loaded coaxial line	
metallic waveguides	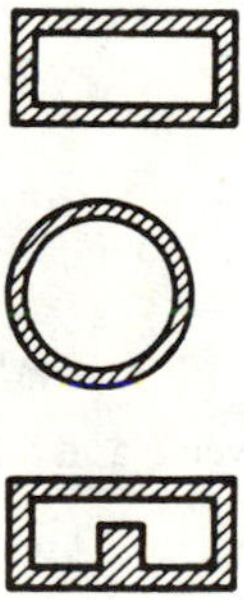	loaded metallic waveguides	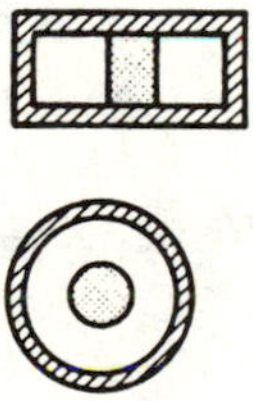
		dielectric waveguides	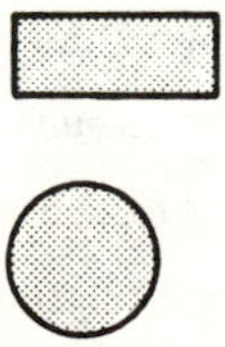
		optical fibers	

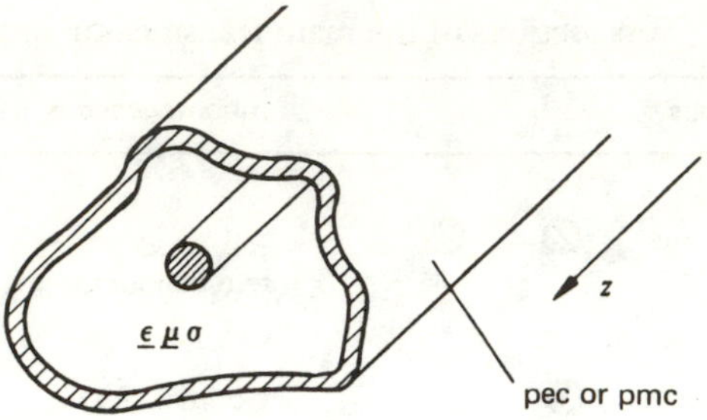

Fig. 2.5 Model of metallic waveguide.

The curl equations are developed in the same manner, yielding for (1.2):

$$\left(\nabla_t + \boldsymbol{e}_z \frac{\partial}{\partial z}\right) \times (\underline{\boldsymbol{E}}_t + \boldsymbol{e}_z\underline{E}_z) = -j\omega\underline{\mu}\underline{\boldsymbol{H}}_t - j\omega\underline{\mu}\boldsymbol{e}_z\underline{H}_z \qquad \text{V/m}^2 \qquad (2.13)$$

Carrying out the multiplication within the left-hand term, four components are obtained. The first one is purely longitudinal, the next two are transverse, while the last one is identically zero (cross product of two parallel vectors).

$$\nabla_t \times \underline{\boldsymbol{E}}_t + \boldsymbol{e}_z \frac{\partial}{\partial z} \times \underline{\boldsymbol{E}}_t + \nabla_t \times \boldsymbol{e}_z\underline{E}_z + \boldsymbol{e}_z \frac{\partial}{\partial z} \times \boldsymbol{e}_z\underline{E}_z \qquad \text{V/m}^2 \qquad (2.14)$$

Arranging the transverse components, this equation becomes:

$$\nabla_t \times \underline{\boldsymbol{E}}_t + \boldsymbol{e}_z \times \left(\frac{\partial \underline{\boldsymbol{E}}_t}{\partial z} - \nabla_t\underline{E}_z\right) = -j\omega\underline{\mu}\underline{\boldsymbol{H}}_t - j\omega\underline{\mu}\boldsymbol{e}_z\underline{H}_z \qquad \text{V/m}^2 \qquad (2.15)$$

The same procedure is applied to (1.3):

$$\nabla_t \times \underline{\boldsymbol{H}}_t + \boldsymbol{e}_z \times \left(\frac{\partial \underline{\boldsymbol{H}}_t}{\partial z} - \nabla_t\underline{H}_z\right) = (j\omega\underline{\epsilon} + \sigma)\underline{\boldsymbol{E}}_t + (j\omega\underline{\epsilon} + \sigma)\boldsymbol{e}_z\underline{E}_z \qquad \text{A/m}^2 \qquad (2.16)$$

Relations (2.15) and (2.16) each possess a transverse and a longitudinal part. The identification of the latter ones yield respectively:

$$\nabla_t \times \underline{\boldsymbol{E}}_t = -j\omega\underline{\mu}\boldsymbol{e}_z\underline{H}_z \qquad \text{V/m}^2 \qquad (2.17)$$

$$\nabla_t \times \underline{\boldsymbol{H}}_t = (j\omega\underline{\epsilon} + \sigma)\boldsymbol{e}_z\underline{E}_z \qquad \text{A/m}^2 \qquad (2.18)$$

The transverse relations are obtained by cross-multiplication of (2.15) and (2.16) by $\boldsymbol{e}_z$, which provides, making use of simple vector relations:

$$\nabla_t\underline{E}_z - \frac{\partial \underline{\boldsymbol{E}}_t}{\partial z} = -j\omega\underline{\mu}\boldsymbol{e}_z \times \underline{\boldsymbol{H}}_t \qquad \text{V/m}^2 \qquad (2.19)$$

$$\nabla_t\underline{H}_z - \frac{\partial \underline{\boldsymbol{H}}_t}{\partial z} = (j\omega\underline{\epsilon} + \sigma)\boldsymbol{e}_z \times \underline{\boldsymbol{E}}_t \qquad \text{A/m}^2 \qquad (2.20)$$

2.2.3 Requirements for the Existence of a TEM Mode ($\underline{E}z = 0, \underline{H}z = 0$)

In a Transverse Electromagnetic (TEM) mode, the fields are, by definition, purely transverse. As a result, $\underline{E}_z$ and $\underline{H}_z$ both vanish everywhere (Table 2.2). Introducing this condition into (2.11), (2.12), (2.17) and (2.18) yields:

$$\nabla_t \times \underline{E}_t = 0 \qquad \text{V/m}^2 \qquad (2.21)$$

$$\nabla_t \cdot \underline{E}_t = 0 \qquad \text{V/m}^2 \qquad (2.22)$$

$$\nabla_t \times \underline{H}_t = 0 \qquad \text{A/m}^2 \qquad (2.23)$$

$$\nabla_t \cdot \underline{H}_t = 0 \qquad \text{A/m}^2 \qquad (2.24)$$

The first two equations define a two-dimensional problem of electrostatics without space charges, the next two a two-dimensional problem of magnetostatics without currents [214].

Both problems admit as solutions those of the two-dimensional Laplace's equation.

For a simply connected waveguide (hollow metal tube, without internal conductor), Laplace's equation only admits the trivial solution $\underline{E}_t = 0$. The boundary condition imposes a constant potential on the metal tube. Solutions of Laplace's equation require that the potential be constant everywhere inside the tube. Thus, an empty hollow waveguide cannot propagate a TEM mode.

Quite on the contrary, on a two-conductor structure, a TEM mode can propagate, since different potentials may exist on the two conductors. The wave equation is obtained by differentiating (2.19) with respect to z, first letting $\underline{E}_z = 0$, and then making use of (2.20):

$$\frac{\partial^2 \underline{E}_t}{\partial z^2} = j\omega\underline{\mu}\boldsymbol{e}_z \times \frac{\partial \underline{H}_t}{\partial z} = -j\omega\underline{\mu}(j\omega\underline{\epsilon} + \sigma)\boldsymbol{e}_z \times \boldsymbol{e}_z \times \underline{E}_t$$

$$= j\omega\underline{\mu}(j\omega\underline{\epsilon} + \sigma)\underline{E}_t \qquad \text{V/m}^3 \qquad (2.25)$$

This is the wave equation of a plane wave propagating along the z-axis in an infinite, homogeneous medium characterized by $\underline{\epsilon}$, $\underline{\mu}$, and σ [214]. The propagation of a TEM mode on a homogeneous multiconductor line only depends upon the propagation medium, it is independent of the geometry and the line dimensions. The propagation factor $\underline{\gamma}$ is given by:

$$\underline{\gamma} = \alpha + j\beta = \sqrt{j\omega\underline{\mu}(j\omega\underline{\epsilon} + \sigma)} \qquad \text{m}^{-1} \qquad \textbf{(2.26)}$$

As the phaseshift per unit length β is defined for all frequencies starting from $\omega = 0$, the TEM mode can propagate at any frequency on a multiconductor line. More generally, a line having N conductors can propagate $N - 1$ linearly independent TEM modes.

2.2.4 Relations for a TE Mode ($\underline{E}_z = 0, \underline{H}_z \neq 0$), Wave Numbers

For a TE (Transverse Electric) mode to propagate in a waveguide, a differential relation should be obtained in terms of $\underline{H}_z$ only, since $\underline{E}_z = 0$. The solutions must satisfy the boundary conditions on the sides of the waveguide. The transverse divergence of (2.20) is taken, and then developed making use of (2.17):

$$\nabla_t^2 \underline{H}_z - \frac{\partial}{\partial z} \nabla_t \cdot \underline{\boldsymbol{H}}_t = (j\omega\underline{\epsilon} + \sigma) \nabla_t \cdot (\boldsymbol{e}_z \times \underline{\boldsymbol{E}}_t)$$

$$= -(j\omega\underline{\epsilon} + \sigma)\boldsymbol{e}_z \cdot \nabla_t \times \underline{\boldsymbol{E}}_t$$

$$= j\omega\underline{\mu}(j\omega\underline{\epsilon} + \sigma)\underline{H}_z \qquad \text{A/m}^3 \qquad (2.27)$$

The *wave number* $\underline{k}$ is then defined by:

$$\underline{k}^2 = -j\omega\underline{\mu}(j\omega\underline{\epsilon} + \sigma) \qquad \text{m}^{-2} \qquad \mathbf{(2.28)}$$

The second term of the left-hand side of (2.27) is transformed, with the aid of (2.12), so that a relation containing only the $\underline{H}_z$ component is obtained:

$$\nabla_t^2 \underline{H}_z + \frac{\partial^2 \underline{H}_z}{\partial z^2} + \underline{k}^2 \underline{H}_z = 0 \qquad \text{A/m}^3 \qquad (2.29)$$

The first term of the equation involves the differentiation of $\underline{H}_z$ with respect to transverse coordinates, the second term with respect to the longitudinal coordinate. For a non-zero solution to exist, the variations in the two orthogonal directions must be independent from one another, and separation of variables can be applied:

$$\nabla_t^2 \underline{H}_z + \underline{p}^2 \underline{H}_z = 0 \qquad \text{A/m}^3 \qquad \mathbf{(2.30)}$$

$$\frac{\partial^2 \underline{H}_z}{\partial z^2} - \underline{\gamma}^2 \underline{H}_z = 0 \qquad \text{A/m}^3 \qquad \mathbf{(2.31)}$$

$$\underline{p}^2 - \underline{\gamma}^2 = \underline{k}^2 \qquad \text{m}^{-2} \qquad \mathbf{(2.32)}$$

The first equation only involves the transverse dependence of the mode. It defines the *transverse wave number* $\underline{p}$. In presence of boundary conditions, solutions of equation (2.30) are only found for particular values of $\underline{p}$, which are the eigenvalues of the TE mode problem (§ 2.1.8). The transverse wave number $\underline{p}$ is specified by the guide cross section (shape and size) and by the transverse distribution of the fields for the mode considered; it is independent of the medium filling the guide. In a homogeneous guide (Sections 2.2 to 2.7), the transverse wave number is always real and positive (§ 2.2.23).

Equation (2.31) is the longitudinal wave equation (2.6) for the $\underline{H}_z$ component. The remaining equation (2.32) provides the link between the propagation factor $\underline{\gamma}$, the transverse wave number $\underline{p}$ and the wave number $\underline{k}$.

2.2.5 Transverse Fields of the TE Mode

The transverse components of the electromagnetic fields can be expressed in terms of the longitudinal magnetic field component $\underline{H}_z$. Taking the transverse gradient of (2.12) and replacing ∇_t^2 by $-p^2$ (making use of 2.30, which applies not only to $\underline{H}_z$, but also to other components which derive from it), one obtains:

$$\nabla_t{}^2 \underline{H}_t = -\nabla_t \frac{\partial \underline{H}_z}{\partial z} = -p^2 \underline{H}_t \qquad \text{A/m}^3 \qquad (2.33)$$

and therefore:

$$\underline{H}_t = \frac{1}{p^2} \nabla_t \frac{\partial \underline{H}_z}{\partial z} \qquad \text{A/m} \qquad (2.34)$$

The transverse electric field component is obtained by introducing the value of $\underline{H}_t$ just defined in (2.34) into (2.19), deriving with respect to z and making use of (2.31):

$$\underline{E}_t = \frac{j\omega\underline{\mu}}{p^2} \boldsymbol{e}_z \times \nabla_t \underline{H}_z \qquad \text{V/m} \qquad (2.35)$$

2.2.6 Relations between Transverse Components for the TE Mode

The transverse components of the fields are also related to one another. The first such relation is given by (2.19):

$$\frac{\partial \underline{E}_t}{\partial z} = j\omega\underline{\mu}\boldsymbol{e}_z \times \underline{H}_t \qquad \text{V/m}^2 \qquad (2.36)$$

A second equation linking the two fields is obtained from two equations in Section 2.2.5. Equation (2.34) is derived with respect to z, the double derivative is replaced by the squared propagation factor (2.31). Equation (2.35) is cross-multiplied by $\boldsymbol{e}_z$. Identifying the terms thus obtained yields the desired relation:

$$\frac{\partial \underline{H}_t}{\partial z} = -\frac{\underline{\gamma}^2}{j\omega\underline{\mu}} \boldsymbol{e}_z \times \underline{E}_t \qquad \text{A/m}^2 \qquad (2.37)$$

2.2.7 Definitions: Voltage and Current on an Equivalent Transmission Line

The two equations (2.36) and (2.37) are formally analogous to line equations (§ 9.3.3). Line theory can therefore be employed to pursue the study of waveguides. To do it, the transverse field components are expressed as products of a function of the longitudinal coordinate, multiplied by a function having a transverse dependence (§ 2.1.3):

$$\underline{E}_t(r_t,z) = \underline{U}_e(z)\underline{E}_T(r_t) \qquad \text{V/m} \qquad (2.38)$$

$$\underline{H}_t(r_t,z) = \underline{I}_e(z)\underline{H}_T(r_t) \qquad \text{A/m} \qquad (2.39)$$

The functions of z resulting from the separation are the *voltage* $\underline{U}_e(z)$ and the *current* $\underline{I}_e(z)$ on an equivalent transmission line.

2.2.8 Commentary

In order to avoid possible confusion later on, the subscript t is used here to represent the transverse component of a field, and the subscript capital T to specify ***its part which depends only upon the transverse coordinate.*** The very dimensions of the two terms are different: while the transverse components have dimensions of fields, their transversely-dependent parts have as dimension the inverse of a length.

2.2.9 Equivalent Transmission Line for TE Modes

Relation (2.36) is developed, making use of (2.38) and (2.39):

$$\frac{\partial}{\partial z}\underline{E}_t = \underline{E}_T \frac{d\underline{U}_e}{dz} = j\omega\underline{\mu}\,\boldsymbol{e}_z \times \underline{H}_t = j\omega\underline{\mu}\,\underline{I}_e\boldsymbol{e}_z \times \underline{H}_T \qquad \text{V/m}^2 \qquad (2.40)$$

Two equations are obtained by identification of the transversely-dependent and longitudinally-dependent parts of (2.40). In this way, one obtains the first transmission line equation (9.10), which links the current to a variation of the voltage

$$\frac{d\underline{U}_e}{dz} = -j\omega\underline{\mu}\,\underline{I}_e = -\underline{Z}'\underline{I}_e \qquad \text{V/m} \qquad (2.41)$$

$$\underline{E}_T = -\boldsymbol{e}_z \times \underline{H}_T \qquad \text{m}^{-1} \qquad (2.42)$$

In a similar way, (2.37) is developed:

$$\frac{\partial}{\partial z}\underline{H}_t = \underline{H}_T \frac{d\underline{I}_e}{dz} = -\frac{\underline{\gamma}^2}{j\omega\underline{\mu}}\boldsymbol{e}_z \times \underline{E}_t = -\frac{\underline{\gamma}^2}{j\omega\underline{\mu}}\underline{U}_e\boldsymbol{e}_z \times \underline{E}_T \qquad \text{A/m} \qquad (2.43)$$

After separation, the second line equation (9.11) is obtained.

$$\frac{d\underline{I}_e}{dz} = -\frac{\underline{\gamma}^2}{j\omega\underline{\mu}}\underline{U}_e = -\underline{Y}'\underline{U}_e \qquad \text{A/m} \qquad (2.44)$$

$$\underline{H}_T = \boldsymbol{e}_z \times \underline{E}_T \qquad \text{m}^{-1} \qquad (2.45)$$

2.2.10 Note

The separation into functions of the transverse and longitudinal coordinates is arbitrary: $\underline{U}_e$ could quite well be multiplied by any arbitrary constant $\underline{K}$, if at the same time $\underline{E}_T$ was divided by that constant $\underline{K}$. This does not affect (2.38) at all, but it does modify the following relations (2.41) to (2.45). The simplest, and hence most logical, choice was made here, letting $\underline{K} = 1$. In this manner, the amplitudes of $\underline{E}_T$ and $\underline{H}_T$ are chosen equal to each other, and the ratio of $\underline{U}_e$ to $\underline{I}_e$ becomes the quotient of the transverse electric field to the transverse magnetic field.

2.2.11 Equivalent TE Mode Circuit of an Infinitesimal Length of Waveguide

The equivalent circuit of a section of transmission line of length dz is shown in Figure 2.6 (§ 9.3.2). For a TE mode in waveguide, the series impedance $\underline{Z}'$ is given by (2.41). The shunt admittance $\underline{Y}'$, defined in (2.44), is expressed as follows, developing $\underline{k}$ in terms of (2.28):

$$\underline{Y}' = \frac{\underline{\gamma}^2}{\mathrm{j}\omega\underline{\mu}} = \frac{p^2 - \underline{k}^2}{\mathrm{j}\omega\underline{\mu}} = \frac{p^2}{\mathrm{j}\omega\underline{\mu}} + \frac{\mathrm{j}\omega\underline{\mu}(\mathrm{j}\omega\underline{\epsilon} + \sigma)}{\mathrm{j}\omega\underline{\mu}} \qquad \mathrm{S/m} \qquad (2.46)$$

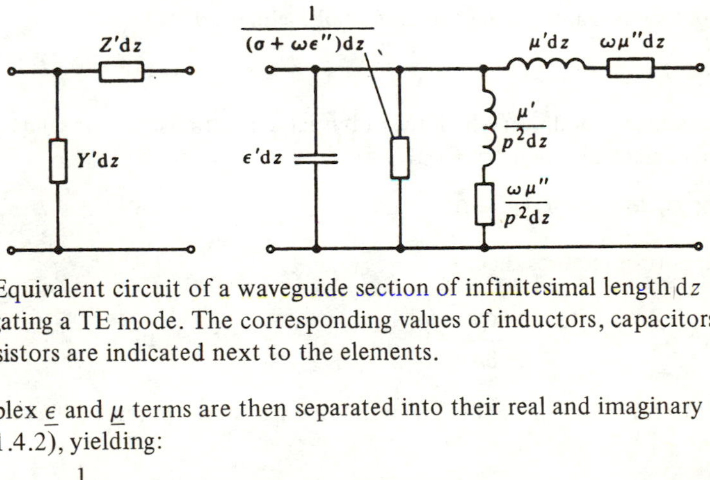

Fig. 2.6 Equivalent circuit of a waveguide section of infinitesimal length dz propagating a TE mode. The corresponding values of inductors, capacitors, and resistors are indicated next to the elements.

The complex $\underline{\epsilon}$ and $\underline{\mu}$ terms are then separated into their real and imaginary parts (§ 1.4.2), yielding:

$$\underline{Y}' = \frac{1}{\mathrm{j}\omega\mu'/p^2 + \omega\mu''/p^2} + \mathrm{j}\omega\epsilon' + \omega\epsilon'' + \sigma \qquad \mathrm{S/m} \qquad (2.47)$$

2.2.12 Link with Line Theory for a TE Mode, Equivalent Impedance

The study of propagation in a waveguide can now be carried out using standard transmission line theory (Chap. 9). The currents and voltages on the equivalent line take the form:

$$\underline{U}_e = \underline{U}_{e+} \exp(-\underline{\gamma} z) + \underline{U}_{e-} \exp(+\underline{\gamma} z) \qquad \text{V} \qquad (2.48)$$

$$\underline{I}_e = (1/\underline{Z}_e)\,[\underline{U}_{e+} \exp(-\underline{\gamma} z) - \underline{U}_{e-} \exp(+\underline{\gamma} z)] \qquad \text{A} \qquad (2.49)$$

with the propagation factor $\underline{\gamma}$ and the *equivalent characteristic impedance* $\underline{Z}_e$ given by:

$$\underline{\gamma} = \sqrt{\underline{Z}'\underline{Y}'} = \sqrt{j\omega\underline{\mu}(\underline{\gamma}^2/j\omega\underline{\mu})} \qquad \text{m}^{-1} \qquad (2.50)$$

$$\underline{Z}_e = \sqrt{\underline{Z}'/\underline{Y}'} = j\omega\underline{\mu}/\underline{\gamma} \qquad \Omega \qquad (2.51)$$

where $\underline{U}_{e+}$ and $\underline{U}_{e-}$ are the voltages of the forward and backward waves at $z = 0$, respectively, which are imposed by the boundary conditions at both ends of the guide (source and load).

2.2.13 Definition: Transverse Potential $\underline{\Psi}$ of a TE Mode

Equation (2.18) shows that, when $\underline{E}_z$ vanishes, the transverse magnetic field derives from a scalar potential, defined by:

$$\underline{\boldsymbol{H}}_T = -\nabla_t \underline{\psi} \qquad \text{m}^{-1} \qquad (2.52)$$

This potential $\underline{\psi}$ must satisfy the Helmholtz equation, obtained from (2.30):

$$\nabla_t^2 \underline{\psi} + p^2 \underline{\psi} = 0 \qquad \text{m}^{-2} \qquad (2.53)$$

The boundary condition which the potential $\underline{\psi}$ must meet at the guide walls on a perfect electric conductor (pec) is

$$\boldsymbol{n} \times \underline{\boldsymbol{E}}_T = -(\boldsymbol{n} \cdot \underline{\boldsymbol{H}}_T)\boldsymbol{e}_z = 0 \qquad \text{m}^{-1} \qquad (2.54)$$

which in turn requires that:

$$-\boldsymbol{n} \cdot \underline{\boldsymbol{H}}_T = \boldsymbol{n} \cdot \nabla_t \underline{\psi} = \frac{\partial \underline{\psi}}{\partial n} = 0 \qquad \text{m}^{-1} \qquad (2.55)$$

On a perfect magnetic conductor (pmc), the transverse potential itself must vanish:

$$\underline{\psi} = 0 \qquad 1 \qquad (2.56)$$

2.2.14 Relationship between the Transverse Potential $\underline{\Psi}$ and the Longitudinal Magnetic Component for a TE Mode

Expressing $\underline{\boldsymbol{H}}_T$ (2.34) in terms of (2.39), making use of (2.52) and integrating, one obtains, after setting the arbitrary integration constant equal to zero:

$$-\underline{I}_e\underline{\psi} = \frac{1}{p^2}\frac{\partial \underline{H}_z}{\partial z} \qquad \text{A} \qquad (2.57)$$

Taking the derivative with respect to z and making use of (2.31) and of (2.44) yields:

$$\underline{H}_z = \frac{p^2}{j\omega\underline{\mu}}\underline{U}_e\underline{\psi} \qquad \text{A/m} \qquad (2.58)$$

The longitudinal magnetic component is then proportional to the equivalent line voltage; its transverse dependence is the one of the transverse potential $\underline{\psi}$.

2.2.15 TM Mode Relationships ($\underline{H}_z = 0, \underline{E}_z \neq 0$)

The equations for a Transverse Magnetic mode are the dual of those for the TE mode. Developments are quite analogous, so that only the main results are presented here. Transverse and longitudinal wave equations become

$$\nabla_t^2\underline{E}_z + p^2\underline{E}_z = 0 \qquad \text{V/m}^3 \qquad (2.59)$$

$$\frac{\partial^2\underline{E}_z}{\partial z^2} - \underline{\gamma}^2\underline{E}_z = 0 \qquad \text{V/m}^3 \qquad (2.60)$$

$$p^2 - \underline{\gamma}^2 = \underline{k}^2 \qquad \text{m}^{-2} \qquad (2.32)$$

2.2.16 Transverse Fields of the TM Mode

The transverse field components are functions of $\underline{E}_z$:

$$\underline{E}_t = \frac{1}{p^2}\nabla_t\frac{\partial \underline{E}_z}{\partial z} \qquad \text{V/m} \qquad (2.61)$$

$$\underline{H}_t = -\frac{j\omega\underline{\epsilon} + \sigma}{p^2}\,e_z \times \nabla_t\underline{E}_z \qquad \text{A/m} \qquad (2.62)$$

2.2.17 Relations between Transverse Components of the TM Mode

The approach used in Section 2.2.6 yields here:

$$\frac{\partial \underline{H}_t}{\partial z} = -(j\omega\underline{\epsilon} + \sigma)e_z \times \underline{E}_t \qquad \text{A/m}^2 \qquad (2.63)$$

$$\frac{\partial \underline{E}_t}{\partial z} = \frac{\underline{\gamma}^2}{j\omega\underline{\epsilon} + \sigma}\,e_z \times \underline{H}_t \qquad \text{V/m}^2 \qquad (2.64)$$

2.2.18 TM Mode Equivalent Line

The transverse and longitudinal dependencies are separated by means of equations (2.38) and (2.39). The same equations (2.42) and (2.45) are obtained, as for TE modes, while the longitudinal relations here become:

$$\frac{d\underline{I}_e}{dz} = -(j\omega\underline{\epsilon} + \sigma)\underline{U}_e = -\underline{Y}'\underline{U}_e \qquad \text{A/m} \qquad (2.65)$$

$$\frac{d\underline{U}_e}{dz} = -\frac{\underline{\gamma}^2}{j\omega\underline{\epsilon} + \sigma}\underline{I}_e = -\underline{Z}'\underline{I}_e \qquad \text{V/m} \qquad (2.66)$$

2.2.19 TM Mode Equivalent Circuit of a Line Segment of Infinitesimal Length

Here, the impedance per unit length $\underline{Z}'$ must be developed to determine the series elements of the equivalent circuit

$$\underline{Z}' = \frac{\underline{\gamma}^2}{j\omega\underline{\epsilon} + \sigma} = \frac{p^2 - \underline{k}^2}{j\omega\underline{\epsilon} + \sigma}$$

$$= \frac{1}{j\omega\epsilon'/p^2 + (\omega\epsilon'' + \sigma)/p^2} + j\omega\mu' + \omega\mu'' \qquad \Omega/\text{m} \qquad (2.67)$$

The shunt admittance $\underline{Y}'$ is obtained from (2.65). The equivalent circuit for a length of line dz is shown in Figure 2.7. It is the dual of the one for TE modes.

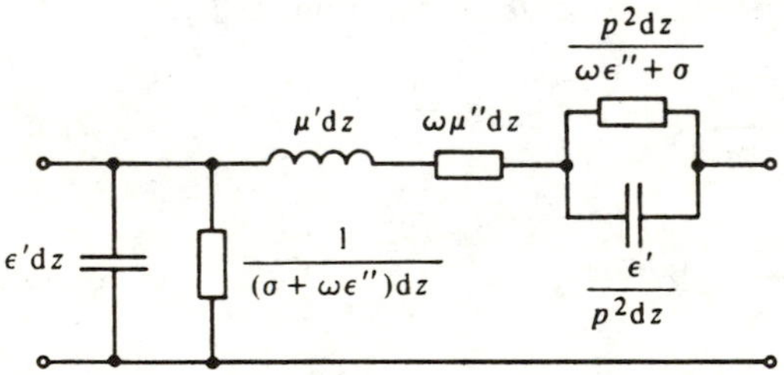

Fig. 2.7 Equivalent circuit of a waveguide section of infinitesimal length dz propagating a TM mode. The corresponding values of inductors, capacitors, and resistors are indicated next to the elements.

2.2.20 Link with Line Theory for a TM Mode

The only different term among those defined in Section 2.2. 12 is the equivalent characteristic impedance $\underline{Z}_e$, which for a TM mode becomes:

$$\underline{Z}_e = \sqrt{\underline{Z}'/\underline{Y}'} = \underline{\gamma}/(j\omega\underline{\epsilon} + \sigma) \qquad \Omega \qquad (2.68)$$

2.2.21 Definition of the Transverse Potential $\underline{\Phi}$ of a TM Mode

The transverse electric field of a TM mode derives from a scalar potential $\underline{\phi}$, because its curl vanishes:

$$\underline{E}_{\mathrm{T}} = -\nabla_{\mathrm{t}}\underline{\phi} \qquad \mathrm{m}^{-1} \qquad (2.69)$$

This potential is a solution of Helmholtz's equation:

$$\nabla_{\mathrm{t}}^2\underline{\phi} + \underline{p}^2\underline{\phi} = 0 \qquad \mathrm{m}^{-2} \qquad (2.70)$$

in the presence of boundary conditions at the guide walls (§ 1.4.3):

1. on a perfect electric conductor (pec)

$$\underline{\phi} = 0 \qquad 1 \qquad (2.71)$$

2. on a perfect magnetic conductor (pmc)

$$\boldsymbol{n} \cdot \nabla_{\mathrm{t}}\underline{\phi} = \frac{\partial \underline{\phi}}{\partial n} = 0 \qquad \mathrm{m}^{-1} \qquad (2.72)$$

2.2.22 Relationship between the Transverse Potential $\underline{\Phi}$ and the Longitudinal Electric Field of a TM Mode

These two quantities are related by an equation obtained using (2.61), (2.38), (2.69) and (2.66):

$$\underline{E}_z = \frac{\underline{p}^2}{\mathrm{j}\omega\underline{\epsilon} + \sigma}\, \underline{I}_{\mathrm{e}}\underline{\phi} \qquad \mathrm{V/m} \qquad (2.73)$$

The longitudinal electric field component of a TM mode is proportional to the current on the equivalent line and to the transverse potential $\underline{\phi}$.

■ 2.2.23 Properties of the Transverse Wave Number $\underline{p}$

The transverse wave number $\underline{p}$ has specific properties, which can be determined in the general case, in a waveguide of arbitrary cross section. This is done using the two-dimensional divergence theorem (§ 10.2.2), applied to the guide cross section:

$$\int_S \nabla \cdot \underline{A}\, \mathrm{d}A = \oint_C \underline{A} \cdot \boldsymbol{n}\, \mathrm{d}l \qquad (2.74)$$

A particular transverse vector $\underline{A}$ is selected for the demonstration: it is defined as follows:

$$\underline{A} = \underline{\theta}^* \nabla_{\mathrm{t}}\underline{\theta} \qquad \mathrm{m}^{-1} \qquad (2.75)$$

where $\underline{\theta}$ is a transverse potential, either $\underline{\psi}$ for a TE mode or $\underline{\phi}$ for a TM mode, $\underline{\theta}^*$ being its complex conjugate. With the aid of (2.53) or (2.70), the transverse divergence of $\underline{A}$ becomes:

$$\nabla_t \cdot \underline{A} = \nabla_t \underline{\theta}^* \cdot \nabla_t \underline{\theta} + \underline{\theta}^* \nabla_t^2 \underline{\theta} = |\nabla_t \underline{\theta}|^2 - p^2 |\underline{\theta}|^2 \qquad \text{m}^{-2} \qquad (2.76)$$

The value obtained is introduced into (2.74):

$$\int_S |\nabla_t \underline{\theta}|^2 \,\mathrm{d}A - p^2 \int_S |\underline{\theta}|^2 \,\mathrm{d}A = \oint_C \underline{\theta}^* \nabla_t \underline{\theta} \cdot \boldsymbol{n} \,\mathrm{d}l \qquad 1 \qquad (2.77)$$

The right-hand side term vanishes when boundary conditions are applied, either for a TE or for a TM mode, on either a pec or a pmc boundary. The transverse wave number p is then defined, in terms of the transverse potentials, by:

$$p^2 = \frac{\int_S |\nabla_t \underline{\theta}|^2 \,\mathrm{d}A}{\int_S |\underline{\theta}|^2 \,\mathrm{d}A} \qquad \text{m}^{-2} \qquad (2.78)$$

This is the quotient of two real positive quantities: the transverse wave number in a closed homogeneous waveguide is therefore ***always a real quantity***. It is not a function of the medium filling the waveguide (homogeneous and isotropic, § 2.2.1). The positive real value for p will always be taken later on.

2.2.24 Normalization of Transverse Functions $\underline{E}_T$ and $\underline{H}_T$

For the sake of consistency, the total power transmitted on the equivalent transmission line must be the same as that propagating within the waveguide, so that:

$$\underline{S}_z = \int_S (\underline{E} \times \underline{H}^*) \cdot \boldsymbol{e}_z \mathrm{d}A = \underline{U}_e \underline{I}_e^* \int_S \underline{E}_T \times (\boldsymbol{e}_z \times \underline{E}_T^*) \cdot \boldsymbol{e}_z \,\mathrm{d}A$$

$$= \underline{U}_e \underline{I}_e^* \int_S |\underline{E}_T|^2 \,\mathrm{d}A = \underline{U}_e \underline{I}_e^* \qquad \text{VA} \qquad (2.79)$$

This equation is satisfied when

$$\int_S |\underline{E}_T|^2 \,\mathrm{d}A = \int_S |\underline{H}_T|^2 \,\mathrm{d}A = \int_S |\nabla_t \underline{\theta}|^2 \,\mathrm{d}A = 1 \qquad 1 \qquad (2.80)$$

Furthermore, application of (2.78) yields:

$$\int_S |\underline{\theta}|^2 \,\mathrm{d}A = \frac{1}{p^2} \qquad \text{m}^2 \qquad (2.81)$$

with $\underline{\theta} = \underline{\phi}$ for a TM mode and $\underline{\theta} = \underline{\psi}$ for a TE mode.

2.2.25 Definitions: Wave Impedance, Waveguide Impedances

The equivalent voltage and current $\underline{U}_e$ and $\underline{I}_e$ defined by (2.38) and (2.39) are quantities proportional to the transverse electric and magnetic fields in the waveguide. They do not actually represent the physical voltage and the current which would exist at some location within the waveguide. The ratio $\underline{Z}_e$ of these two quantities is called the *wave impedance.*

Some authors have defined, in addition, a guide voltage $\underline{U}_g$ and a guide current $\underline{I}_g$ by the following expressions

$$\underline{U}_g = \max \int_A^B \underline{E} \cdot \mathrm{d}\boldsymbol{l} \qquad \mathrm{V} \qquad (2.82)$$

$$\underline{I}_g = \int_{C'} \underline{A} \cdot \boldsymbol{e}_z \, \mathrm{d}l \qquad \mathrm{A} \qquad (2.83)$$

where A and B are two points on the transverse plane of the guide and C' is a portion of the waveguide boundary; for instance, the portion where the surface current flows towards z. In a hollow waveguide, one finds that $\underline{U}_g \underline{I}_g^* \neq \underline{S}_z$ (longitudinal component of the power transmitted through the guide). This means that three different *waveguide impedances* can be defined as follows [2]:

$$\underline{Z}_{UI} = \underline{U}_g / \underline{I}_g \qquad \Omega \qquad (2.84)$$

$$\underline{Z}_{PI} = \underline{S}_z / |\underline{I}_g|^2 \qquad \Omega \qquad (2.85)$$

$$\underline{Z}_{PU} = |\underline{U}_g|^2 / \underline{S}_z \qquad \Omega \qquad (2.86)$$

The values taken by the three waveguide impedances are different. The ratios of two of them depend on the waveguide type: an uncertainty results when one wishes to connect waveguides of different shapes. In addition, higher-order modes appear locally next to discontinuities (§ 6.1.4). The study of the fields near a junction between two waveguides leads to problems very difficult to treat by theoretical methods. Most often, experimental methods are used to match discontinuities.

2.2.26 Assumption: Lossless Waveguide

When the waveguide is filled with a ***lossless material***, $\sigma = 0, \epsilon'' = 0, \mu'' = 0$. The wave number $\underline{k}$ then becomes a real number:

$$k = \omega \sqrt{\epsilon \mu} = \omega / c \qquad \mathrm{m}^{-1} \qquad (2.87)$$

where c is the velocity of light within the medium filling the waveguide.

The study of wave propagation always considers a non-dissipative system (perfect conductors and lossless medium) first. This assumption is valid in the following paragraphs, and in Sections 2.4 to 2.6, 2.9, and 2.10. The effect of losses is taken into account later on as a first order perturbation (Sec. 2.7).

2.2.27 Definition: Diagram of Dispersion

For a lossless waveguide, the propagation factor is given by:

$$\underline{\gamma} = \alpha + j\beta = \sqrt{p^2 - k^2} = \sqrt{p^2 - (\omega/c)^2} \qquad \text{m}^{-1} \tag{2.88}$$

The plot of α and β versus frequency is called the *diagram of dispersion.* An example is shown in Figure 2.8 for a particular rectangular waveguide. It must be noted that an infinite number of such curves exist, each one corresponding to a particular value of p.

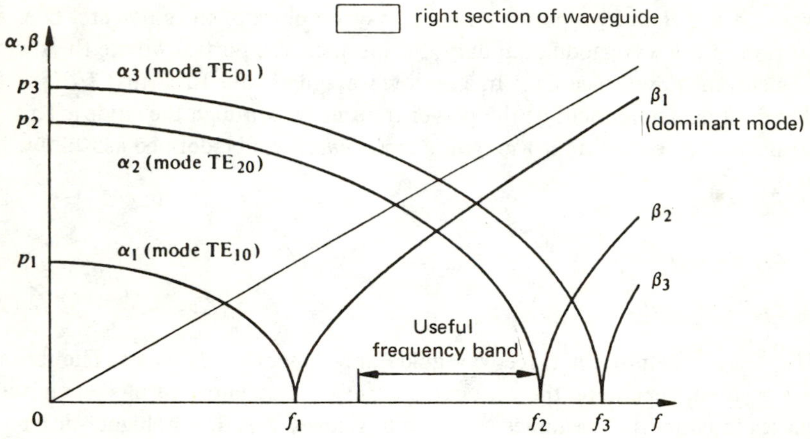

Fig. 2.8 Dispersion diagram for a rectangular waveguide having a width over height ratio of 2.25.

2.2.28 Definition: Cutoff Frequency and Wavelength

For every mode, there exists one particular frequency at which the propagation factor vanishes ($\underline{\gamma} = 0$): it is called the *cutoff frequency* f_c of the mode. Its value is obtained from (2.88):

$$f_c = \frac{\omega_c}{2\pi} = \frac{pc}{2\pi} = \frac{p}{2\pi\sqrt{\epsilon\mu}} = \frac{c}{\lambda_c} \qquad \text{Hz} \tag{2.89}$$

The *cutoff wavelength* λ_c is similarly defined as $2\pi/p$.

2.2.29 Waveguide Fields at Cutoff

When $\underline{\gamma}$ is equal to zero at the cutoff frequency, one of the two transverse fields vanishes. For a TE mode, this happens to the transverse magnetic field (2.34), while the transverse electric field disappears at the TM mode cutoff (2.61). In both instances, the power transmitted goes to zero at the cutoff frequency of a mode (2.79). It is not possible to transmit a signal along the waveguide making use of this particular mode at cutoff.

2.2.30 Mode below Cutoff: Evanescent Mode

When $f < f_c$ for a given mode, the term appearing under the root of (2.88) is real positive, and the propagation factor $\underline{\gamma}$ becomes purely real. This corresponds to an attenuation of the wave, without phaseshift:

$$\alpha = \sqrt{p^2 - (\omega/c)^2} \qquad \text{m}^{-1} \qquad \textbf{(2.90)}$$

The curve representing $\alpha(\omega)$ within the diagram of dispersion is an ellipse (Fig. 2.8). A mode below cutoff is called *evanescent*. Its wave impedance is purely imaginary. For a TE mode (2.51):

$$\underline{Z}_e = j\omega\mu/\alpha \qquad \Omega \qquad (2.91)$$

it is an inductive reactance: a TE mode below cutoff stores magnetic energy. For a TM mode, the wave impedance below cutoff is (2.68):

$$\underline{Z}_e = \alpha/j\omega\epsilon \qquad \Omega \qquad (2.92)$$

A TM mode below cutoff stores electric energy, it is represented by a capacitive reactance.

2.2.31 Mode above Cutoff

When $f > f_c$ for a given mode, the term under the root of (2.68) is real negative, so that the propagation factor is pure imaginary: the signal then propagates without attenuation:

$$\beta = \sqrt{k^2 - p^2} = \sqrt{(\omega/c)^2 - p^2} \qquad \text{m}^{-1} \qquad \textbf{(2.93)}$$

The curve representing the function $\beta(\omega)$ in the diagram of dispersion is a hyperbola (Fig. 2.8).

2.2.32 Definition: Guide Wavelength

A wave propagating along a waveguide displays a periodicity in the longitudinal direction, its period being the *guide wavelength* λ_g (2.8):

$$\lambda_g = \frac{2\pi}{\beta} = \frac{2\pi}{\sqrt{k^2 - p^2}} = \frac{2\pi}{k}\,\frac{1}{\sqrt{1 - (p/k)^2}}$$

$$= \frac{\lambda}{\sqrt{1 - (\lambda/\lambda_c)^2}} \qquad \text{m} \qquad \textbf{(2.94)}$$

with $\lambda = 2\pi/k$ the wavelength within medium ϵ, μ. The guide wavelength λ_g is always larger than the wavelength λ in the unbounded propagating medium.

2.2.33 Phase and Group Velocities, Phase Distortion

The two velocities of propagation, defined in Section 2.1.5, for a propagating mode become, respectively:

$$v_\varphi = \frac{\omega}{\beta} = \frac{\omega}{\sqrt{(\omega/c)^2 - p^2}} = \frac{c}{\sqrt{1 - (pc/\omega)^2}} \quad \text{m/s} \qquad (2.95)$$

$$v_g = \left(\frac{\partial\beta}{\partial\omega}\right)^{-1} = \frac{\sqrt{(\omega/c)^2 - p^2}}{\omega/c^2} = c\sqrt{1 - (pc/\omega)^2} \quad \text{m/s} \qquad (2.96)$$

It must be noted that $v_\varphi v_g = c^2$. Both velocities are frequency-dependent, the waveguide is then called *dispersive,* meaning that signals of different frequencies propagate at different velocities. A modulated signal then undergoes *phase distortion*, which becomes particularly severe near cutoff. In information transfer applications, use of the frequency band located between f_c and 1.25 f_c is generally avoided.

2.2.34 Wave Impedance

For a mode above cutoff (propagating), the wave impedance is a real quantity, corresponding to a transfer of active power. In the case of a TE mode, it is given by (2.51)

$$Z_e = \frac{j\omega\mu}{\gamma} = \frac{\omega\mu}{\beta} = \sqrt{\frac{\mu}{\epsilon}}\,\frac{\lambda_g}{\lambda} \quad \Omega \qquad (2.97)$$

For a TM mode, its value is (2.68)

$$Z_e = \frac{\gamma}{j\omega\epsilon} = \frac{\beta}{\omega\epsilon} = \sqrt{\frac{\mu}{\epsilon}}\,\frac{\lambda}{\lambda_g} \quad \Omega \qquad (2.98)$$

2.2.35 Frequency Ranges, Dominant Mode, Higher Order Modes

The diagram of dispersion of a waveguide can be divided in four ranges of frequency, based on their properties for operation (Fig. 2.8):

1. $f < f_1$ no mode can propagate, the fields decay with distance (§ 2.2.30). Such a guide can be utilized as an attenuator (§ 6.3.8), as part of filters or for isolation purposes.
2. $f_1 < f < 1.25\, f_1$ only one mode, called the *dominant mode,* can propagate. It is, however, highly dispersive over this frequency range, which is therefore not used to transmit information.

3. $1.25\, f_1 < f < f_2$ a single mode, the dominant mode, can propagate. The dispersion is generally acceptable within this range, if the waveguide is not too long.

4. $f > f_2$ several modes, the dominant mode and one or more *higher order modes,* can propagate over this range (multimode operation). The propagation velocities and attenuations are, however, different for different modes, so that the signal is distorted. Whenever feasible, it is preferable to use another frequency band, otherwise, precautions must be taken so the higher order modes are not excited.

2.2.36 Definition: Degenerate Modes

In certain waveguide geometries, several modes with different transverse potentials $\underline{\theta}$ (§ 2.2.23) have the same transverse wave number p. They are called *degenerate modes.* The transverse field distributions of these modes are different, but the corresponding curves in the dispersion diagram are merged together: same attenuation below cutoff, same cutoff frequency, same phaseshift above cutoff.

In addition, any linear combination of degenerate modes is also a degenerate mode, only a few of them are independent or orthogonal.

2.2.37 Comment: TEM Mode

The TEM mode in a two-conductor line can be considered as a special case of either a TE or a TM mode having $p = 0$.

2.2.38 Comment: Hybrid Mode

In a waveguide filled with a homogeneous material, all possible solutions can be separated into independent TE and TM modes, to which is added a TEM mode for a two-conductor line. There is no hybrid mode in this geometry.

2.3 RECTANGULAR WAVEGUIDE

2.3.1 Definition of the Geometry

A waveguide of rectangular cross section is considered, having a width a and a height b, with $a > b$ (Fig. 2.9). The rectangular coordinates corresponding to this geometry are shown in the figure. The guide walls are metallic, supposed to be perfectly conducting in the lossless approximation.

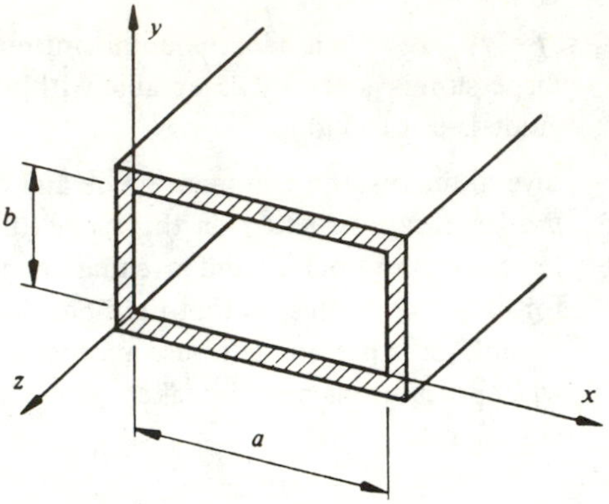

Fig. 2.9 Rectangular waveguide.

2.3.2 Solution of Helmholtz's Equation

The transverse wave equation to be solved is:

$$\nabla_t^2 \underline{\theta} + p^2 \underline{\theta} = 0 \qquad \mathrm{m}^{-2} \qquad (2.99)$$

where $\underline{\theta} = \underline{\phi}$ for a TM mode and $\underline{\theta} = \underline{\psi}$ for a TE mode. Boundary conditions to be satisfied on the conducting walls, for TM modes, are:

$$\underline{\phi} = 0 \qquad 1 \qquad (2.71)$$

and for TE modes

$$\boldsymbol{n} \cdot \nabla_t \underline{\psi} = \frac{\partial \underline{\psi}}{\partial n} = 0 \qquad 1 \qquad (2.55)$$

In rectangular coordinates, the transverse differential operator is given by (2.2), so that the transverse Laplace operator becomes:

$$\nabla_t^2 = \frac{\partial^2}{\partial x^2} + \frac{\partial^2}{\partial y^2} \qquad \mathrm{m}^{-2} \qquad (2.100)$$

The method of separation of variables in rectangular coordinates is used [214], letting

$$\underline{\theta}(x,y) = \underline{X}(x)\,\underline{Y}(y) \qquad 1 \qquad (2.101)$$

Introducing this product of functions into (2.99) and then dividing by $\underline{\theta}$ yields:

$$\frac{1}{X}\frac{\mathrm{d}^2 X}{\mathrm{d}x^2} + \frac{1}{Y}\frac{\mathrm{d}^2 Y}{\mathrm{d}y^2} + p^2 = 0 \qquad \mathrm{m}^{-2} \qquad (2.102)$$

The first term is a function of x only, the second one of y only. For a non-zero solution to exist, both terms must be taken as constants, so that:

$$\frac{d^2 X}{dx^2} + \underline{u}^2 X = 0 \qquad \text{m}^{-2} \qquad (2.103)$$

$$\frac{d^2 Y}{dy^2} + \underline{v}^2 Y = 0 \qquad \text{m}^{-2} \qquad (2.104)$$

$$\underline{u}^2 + \underline{v}^2 = p^2 \qquad \text{m}^{-2} \qquad (2.105)$$

The solutions are circular functions, which all correspond to positive real values of u^2 and of v^2. Only they can satisfy the boundary conditions (2.71) or (2.55), which require periodic functions. In this manner the general solution is obtained:

$$\underline{\theta} = (\underline{A}_1 \sin ux + \underline{A}_2 \cos ux) \cdot (\underline{B}_1 \sin vy + \underline{B}_2 \cos vy) \qquad 1 \qquad (2.106)$$

The constants are then determined from the boundary conditions and the normalization relations (2.80) or (2.81).

2.3.3 Transverse Potential of TM Modes

This condition

$$\underline{\theta} = \underline{\phi} = 0 \qquad 1 \qquad (2.71)$$

is satisfied for the guide walls when $\underline{A}_2 = 0, \underline{B}_2 = 0$ and when, in addition, $u = m\pi/a$, $v = n\pi/b$, yielding:

$$\phi_{mn} = C_{mn} \sin \frac{m\pi x}{a} \sin \frac{n\pi y}{b} \qquad 1 \qquad (2.107)$$

The terms m and n are positive integers, both different from zero, thus satisfying the condition:

$$mn \neq 0 \qquad 1 \qquad (2.108)$$

These solutions correspond to values of p (discrete set) as given by (2.105):

$$p_{mn} = \sqrt{\left(\frac{m\pi}{a}\right)^2 + \left(\frac{n\pi}{b}\right)^2} \qquad \text{m}^{-1} \qquad (2.109)$$

The normalization constant C_{mn} is evaluated from (2.81):

$$C_{mn} = 2/(p_{mn}\sqrt{ab}) \qquad 1 \qquad (2.110)$$

The potential of the first TM mode, which is the TM_{11}, is shown in Figure 2.10. The potentials of the other TM_{mn} modes are similar, containing, respectively, m half-periods across the width and n half-periods across the waveguide height.

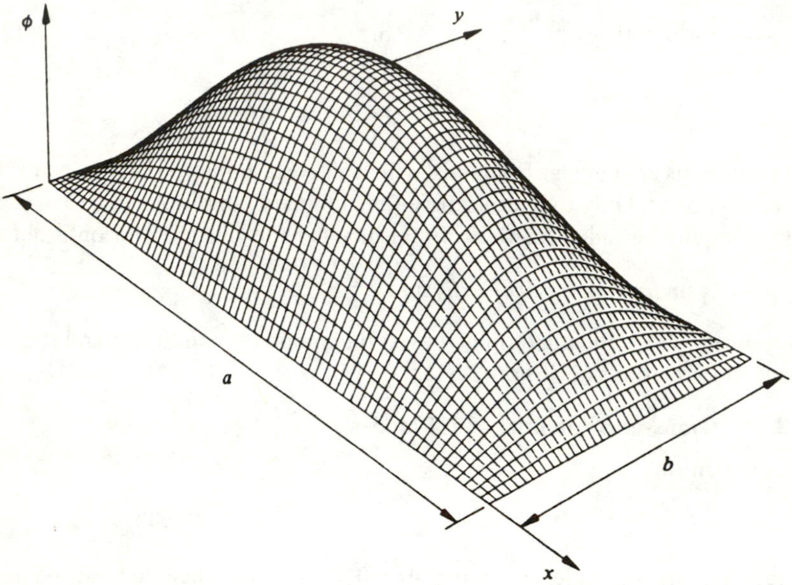

Fig. 2.10 Transverse potential of the TM_{11} mode.

2.3.4 Transverse Fields of TM Modes

The transverse electric field is derived from potential ϕ (2.69), the transverse magnetic field being perpendicular (2.45). They are respectively given by:

$$\boldsymbol{E}_T = -\nabla_t\phi = -C_{mn}\left(\boldsymbol{e}_x \frac{m\pi}{a}\cos\frac{m\pi x}{a}\sin\frac{n\pi y}{b} + \boldsymbol{e}_y \frac{n\pi}{b}\sin\frac{m\pi x}{a}\cos\frac{n\pi y}{b}\right) \quad \mathrm{m}^{-1} \tag{2.111}$$

$$\boldsymbol{H}_T = \boldsymbol{e}_z \times \boldsymbol{E}_T = C_{mn}\left(\boldsymbol{e}_x \frac{n\pi}{b}\sin\frac{m\pi x}{a}\cos\frac{n\pi y}{b} - \boldsymbol{e}_y \frac{m\pi}{a}\cos\frac{m\pi x}{a}\sin\frac{n\pi y}{b}\right) \quad \mathrm{m}^{-1} \tag{2.112}$$

The field lines of the TM_{11} mode are depicted in Figure 2.11.

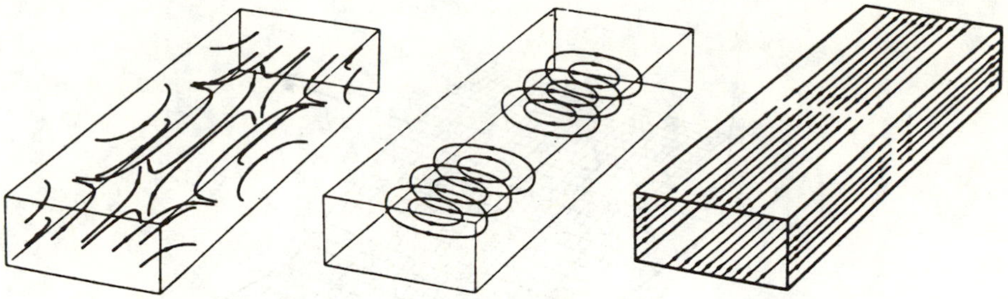

Fig. 2.11 Display of the fields for a TM_{11} mode.

2.3.5 General Remark for TM Modes

The transverse electric field lines are the steepest descent lines on the potential hill of the TM mode. The magnetic field lines are the level curves or equipotentials of ϕ. The longitudinal electric field is proportional to the potential ϕ itself (§ 2.2.22).

2.3.6 Transverse Potential of TE Modes

The boundary condition for this case which must be satisfied on the guide's walls requires that the normal derivatives of the potential vanish on the metal surfaces, as follows:

$$\frac{\partial \psi}{\partial x} = 0 \qquad \text{when } x = 0 \text{ and } x = a \qquad \mathrm{m}^{-1} \tag{2.113}$$

$$\frac{\partial \psi}{\partial y} = 0 \qquad \text{when } y = 0 \text{ and } y = b \qquad \mathrm{m}^{-1} \tag{2.114}$$

These conditions are met by (2.106), letting $\underline{A}_1 = 0, \underline{B}_1 = 0, u = m\pi/a$, and $v = n\pi/b$, yielding the potential:

$$\psi_{mn} = C_{mn} \cos \frac{m\pi x}{a} \cos \frac{n\pi y}{b} \qquad 1 \tag{2.115}$$

The terms m and n are positive integers, but one of the two can vanish (if both were to vanish simultaneously, the transverse potential would be constant, and then the transverse fields identically zero – this is a trivial solution). One must therefore have:

$$m + n \neq 0 \qquad 1 \tag{2.116}$$

The corresponding transverse wave numbers are also given by

$$p_{mn} = \sqrt{\left(\frac{m\pi}{a}\right)^2 + \left(\frac{n\pi}{b}\right)^2} \qquad \mathrm{m}^{-1} \tag{2.109}$$

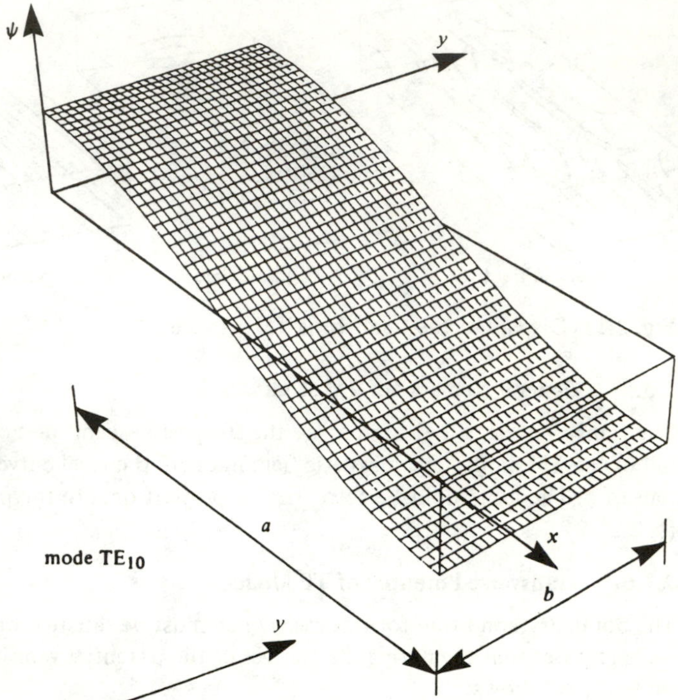

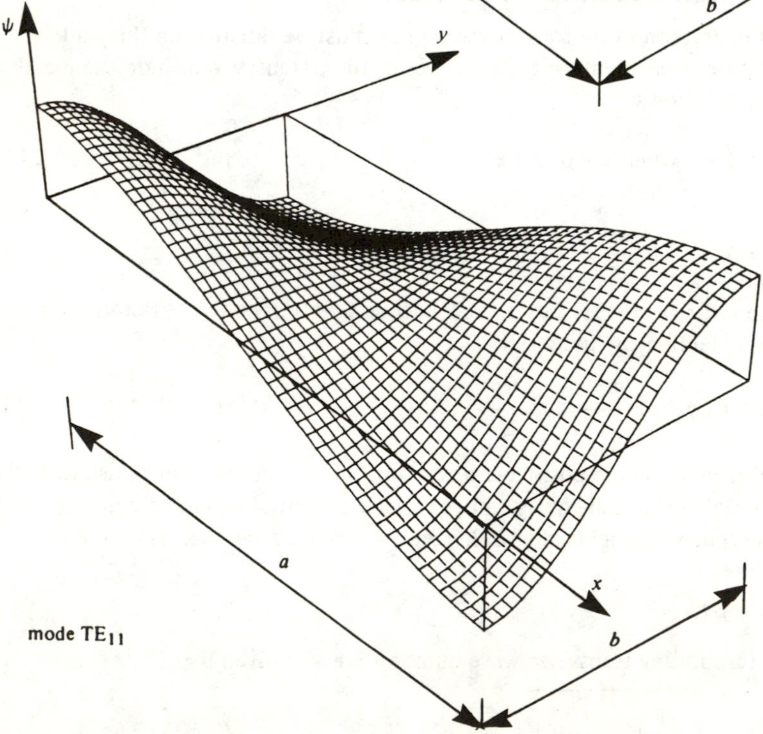

Fig. 2.12 Transverse potential of the TE_{10} and TE_{11} modes.

The normalization constant C_{mn} is also evaluated in terms of (2.81), yielding two possible situations:

1. when m or n vanishes

$$C_{mn} = \sqrt{\frac{2}{ab}}\ \frac{1}{p_{mn}} \qquad 1 \qquad (2.117)$$

2. when m and n are finite

$$C_{mn} = \frac{2}{\sqrt{ab}}\ \frac{1}{p_{mn}} \qquad 1 \qquad (2.118)$$

The transverse potentials of modes TE_{10} and TE_{11} are presented in Figure 2.12. Those of other TE_{mn} modes are comparable, containing m half-periods along x and n half-periods along y, respectively.

2.3.7 Transverse Fields for TE Modes

The magnetic field is derived from potential ψ (2.52), the transverse electric field being perpendicular (2.42):

$$\boldsymbol{H}_{\mathrm{T}} = -\nabla_{\mathrm{t}}\psi = C_{mn}\left(\boldsymbol{e}_x \frac{m\pi}{a} \sin\frac{m\pi x}{a} \cos\frac{n\pi y}{b} + \boldsymbol{e}_y \frac{n\pi}{b} \cos\frac{m\pi x}{a} \sin\frac{n\pi y}{b}\right) \qquad \mathrm{m}^{-1} \qquad (2.119)$$

$$\boldsymbol{E}_{\mathrm{T}} = -\boldsymbol{e}_z \times \boldsymbol{H}_{\mathrm{T}} = C_{mn}\left(\boldsymbol{e}_x \frac{n\pi}{b} \cos\frac{m\pi x}{a} \sin\frac{n\pi y}{b} - \boldsymbol{e}_y \frac{m\pi}{a} \sin\frac{m\pi x}{a} \cos\frac{n\pi y}{b}\right) \qquad \mathrm{m}^{-1} \qquad (2.120)$$

The field lines of both TE_{10} and TE_{11} modes are shown in Figure 2.13.

2.3.8 General Remarks for TE Modes

The lines of transverse magnetic field are the lines of steepest descent on the potential hill ψ of mode TE. The electric field lines then become the level lines, or equipotentials of ψ. The longitudinal magnetic field is proportional to the potential ψ (§ 2.2.14).

2.3.9 Cutoff Frequencies

The cutoff frequencies of TE_{mn} and TM_{mn} modes are obtained from (2.89):

$$f_{mn} = \frac{pc}{2\pi} = \frac{c}{2}\sqrt{\left(\frac{m}{a}\right)^2 + \left(\frac{n}{b}\right)^2} \qquad \mathrm{Hz} \qquad \mathbf{(2.121)}$$

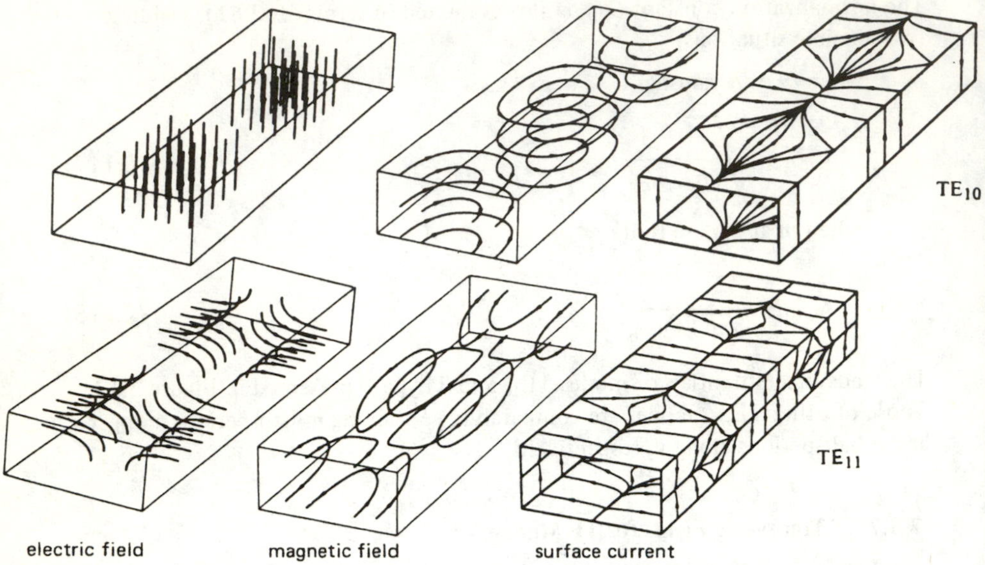

Fig. 2.13 Display of the fields for the TE_{10} and TE_{11} modes.

When both m and n are non-zero, the modes TE_{mn} and TM_{mn} are degenerate (§ 2.2.36)

The lowest cutoff frequency belongs to the TE_{10} mode, thus it is the dominant mode (§ 2.2.35). The first higher order mode can either be the TE_{01} mode (when $b > a/2$) or the TE_{20} (when $b < a/2$):

$$f_{10} = \frac{c}{2a} \qquad f_{01} = \frac{c}{2b} \qquad f_{20} = \frac{c}{a} \qquad \text{Hz} \tag{2.122}$$

In the latter case, the useful band of the waveguide extends from $1.25\, f_{10}$ up to $f_{20} = 2 f_{10}$. This is the widest band that can be obtained in a rectangular waveguide.

□ 2.3.10 Comment: Duality

The comparison of the transverse field lines of the TE_{11} mode in Figure 2.13 with those of the mode TM_{11} in Figure 2.11 shows the duality between the two modes: the electric lines become the magnetic ones and vice versa, and pec limits at the guide's edge must be interchanged with pmc limits on both symmetry axes at

$x = a/2$ and $y = b/2$. The same duality exists for all the degenerate TE_{mn} and TM_{mn} modes (§ 2.2.36, same values of m and n, with $mn \neq 0$).

2.3.11 Dominant TE_{10} Mode

To avoid multimode propagation (§ 2.2.35), the use of a waveguide is restricted to the part of the frequency range over which a single mode propagates and where dispersion is acceptable. The sizes of normalized waveguides are shown in Table 2.14 [3], with their main physical characteristics.

The guide wavelength is obtained with (2.94)

$$\lambda_{10} = \frac{2\pi}{\beta_{10}} = \frac{\lambda}{\sqrt{1 - (\lambda/2a)^2}} \qquad \text{m} \qquad \textbf{(2.123)}$$

The transverse potential here becomes (2.115)

$$\psi_{10} = C_{10} \cos \frac{\pi x}{a} = \frac{1}{\pi} \sqrt{2a/b} \cos \frac{\pi x}{a} \qquad 1 \qquad (2.124)$$

The field components are then given by

$$\underline{E}_t = -e_y \sqrt{2/ab} \sin \frac{\pi x}{a} [\underline{U}_{e+} \exp(-j\beta_{10} z) + \underline{U}_{e-} \exp(+j\beta_{10} z)] \qquad \text{V/m} \qquad (2.125)$$

$$\underline{H}_t = e_x \sqrt{2/ab} \frac{\beta_{10}}{\omega\mu} \sin \frac{\pi x}{a} [\underline{U}_{e+} \exp(-j\beta_{10} z) - \underline{U}_{e-} \exp(+j\beta_{10} z)] \qquad \text{A/m} \qquad (2.126)$$

$$\underline{H}_z = \frac{1}{j\omega\mu} \frac{\pi}{a} \sqrt{2/ab} \cos \frac{\pi x}{a} [\underline{U}_{e+} \exp(-j\beta_{10} z) + \underline{U}_{e-} \exp(+j\beta_{10} z)] \qquad \text{A/m} \qquad (2.127)$$

2.3.12 Note

At the center of the upper and lower guide walls, at $x = a/2$, the magnetic field is purely transverse (Fig. 2.13). This implies that the surface current is purely longitudinal at that location: it is permissible to machine a narrow longitudinal slot along the wall without disturbing the current, and therefore, without producing radiation. This property is used for measurement purposes (Sec. 7.2) and matching (§ 6.3.21).

In the same manner, vertical nonradiating slots can be cut in the side walls.

Any other slot orientation will cut across lines of current and thus give rise to radiated fields: this property is taken advantage of to design antennas.

Table 2.14 Characteristics of the normalized rectangular waveguides (the frequencies indicated for the dominant modes are the recommended minimum and maximum values).

Type 153 IEC–	Frequency band in GHz		Dimensions		Attenuation in dB/m		
	Dominant mode						
	from	to	a/mm	b/mm	@f/GHz	theory	max.
R 3	0.32	0.49	584,2	292.1	0.386	0,00078	0,0011
R 4	0,35	0.53	533,4	266.7	0.422	0,00090	0,0012
R 5	0,41	0.62	457,2	228.6	0,49	0,00113	0.0015
R 6	0,49	0,75	381.0	190,5	0,59	0,00149	0,002
R 8	0,64	0,98	292.1	146.05	0,77	0.00222	0.003
R 9	0,76	1,15	247.65	123,82	0,91	0,00284	0,004
R 12	0,96	1,46	195,58	97,79	1.15	0,00405	0,005
R 14	1,14	1,73	165,10	82,55	1.36	0,00522	0,007
R 18	1,45	2,20	129,54	64,77	1,74	0,00749	0,010
R 22	1,72	2,61	109,22	54,61	2,06	0,00970	0,013
R 26	2,17	3,30	86,36	43,18	2,61	0,0138	0,018
R 32	2,60	3,95	72,14	34,04	3,12	0,0189	0,025
R 40	3,22	4,90	58,17	29,083	3,87	0,0249	0,032
R 48	3,94	5,99	47,55	22,149	4,73	0,0355	0,046
R 58	4,64	7,05	40,39	20,193	5,57	0,0431	0,056
R 70	5,38	8,17	34,85	15,799	6,46	0,0576	0,075
R 84	6,57	9,99	28,499	12,624	7,89	0,0794	0,103
R 100	8,20	12,5	22,860	10,160	9,84	0,110	0,143
R 120	9,84	15,0	19,050	9,525	11,8	0,133	
R 140	11,9	18,0	15,799	7,898	14,2	0,176	
R 180	14,5	22,0	12,954	6,477	17,4	0,238	
R 220	17,6	26,7	10,668	4,318	21,1	0,370	
R 260	21,7	33,0	8,636	4,318	26,1	0,435	
R 320	26,4	40,0	7,112	3,556	31,6	0,583	
R 400	32,9	50,1	5,690	2,845	39,5	0,815	
R 500	39,2	59,6	4,775	2,388	47,1	1,060	
R 620	49,8	75,8	3,759	1,880	59,9	1,52	
R 740	60,5	91,9	3,099	1,549	72,6	2,03	
R 900	73,8	112	2,540	1,270	88,6	2,74	
R 1 200	92,2	140	2,032	1,016	111	3,82	
R 1 400	114	173	1,651	0,826	136,3	5,21	
R 1 800	145	220	1,295	0,648	174,0	7,50	
R 2 200	172	261	1,092	0,546	206,0	9,70	
R 2 600	217	330	0,864	0,432	260,5	13,76	

2.3.13 Note

Propagation of the dominant mode in a rectangular waveguide may also be considered as the multiple reflections of a plane wave on two parallel metal boundaries. This simple treatment permits the study of TE_{mo} and TE_{on} modes.

□ 2.3.14 Guide Voltage for the Dominant TE_{10} Mode

The definition (2.82) yields the expression for the guide voltage:

$$\underline{U}_g(z) = \underline{U}_e(z)\max \int_A^B \underline{E}_T \cdot \mathrm{d}l = \underline{U}_e(z)\sqrt{\frac{2}{ab}}\, b = \underline{U}_e(z)\sqrt{\frac{2b}{a}} \qquad \text{V} \qquad (2.128)$$

□ 2.3.15 Guide Current for the Dominant TE_{10} Mode

The guide current is determined from integration of the transverse magnetic field on the upper guide wall (2.83):

$$\underline{I}_g(z) = \underline{I}_e(z) \int_0^a \underline{H}_x \mathrm{d}x = \underline{I}_e(z)\sqrt{2/ab} \int_0^a \sin\frac{\pi x}{a}\,\mathrm{d}x$$

$$= \underline{I}_e(z)\,\frac{2}{\pi}\,\sqrt{2a/b} \qquad \text{A} \qquad (2.129)$$

□ 2.3.16 Guide Impedances for the Dominant TE_{10} Mode

The product $\underline{U}_g\underline{I}_g^*$ does not give the power transmitted through the waveguide:

$$\underline{U}_g\underline{I}_g^* = \underline{U}_e\underline{I}_e^*\sqrt{2b/a}\;\; 2/\pi\sqrt{2a/b} = \underline{U}_e\underline{I}_e^*\, 4/\pi \neq \underline{U}_e\underline{I}_e^* \qquad \text{W} \qquad (2.130)$$

As a result, the three guide impedances defined by (2.84), (2.85), and (2.86) take different values. For a forward wave they are:

$$\underline{Z}_{UI} = \underline{U}_g/\underline{I}_g = \frac{\underline{U}_{e+}}{\underline{Y}_e\underline{U}_{e+}}\sqrt{2b/a}\;\;\pi/2\;\sqrt{b/2a} = \underline{Z}_e\,\pi b/2a \qquad \Omega \qquad (2.131)$$

$$\underline{Z}_{PU} = \frac{|\underline{U}_g|^2}{\underline{U}_e\underline{I}_e^*} = \frac{|\underline{U}_{e+}|^2}{\underline{Y}_e|\underline{U}_{e+}|^2}\, 2b/a = \underline{Z}_e\, 2b/a \qquad \Omega \qquad (2.132)$$

$$\underline{Z}_{PI} = \frac{\underline{U}_e\underline{I}_e^*}{|\underline{I}_g|^2} = \frac{\underline{Y}_e|\underline{U}_{e+}|^2}{\underline{Y}_e^2|\underline{U}_{e+}|^2}\,\frac{\pi^2}{4}\,\frac{b}{2a} = \underline{Z}_e\,\frac{\pi^2 b}{8a} \qquad \Omega \qquad (2.133)$$

These three values are related to each other by a multiplicative constant. When matching a transition between two rectangular waveguides of different cross sections, any one of the three definitions can be used. But obviously, the same definition must be used for the two guides.

2.3.17 Particular Case: Square Waveguide, Space Degeneracy

When $a = b$, the TE_{10} and the TE_{01} modes have the same cutoff frequency and the same dispersion diagram; *i.e.,* they are degenerate. The transverse field distribution of one mode is obtained simply by rotating the distribution of the other mode by 90°. This is called a *space degeneracy* (sd). Any linear combination of the two modes is still a mode of propagation in the waveguide (§ 2.2.36).

The first higher order modes become the TE_{11} and the TM_{11} modes, which are also degenerate modes, albeit not space degenerate, for which (2.110) yields

$$f_{11} = \frac{c}{2} \sqrt{1/a^2 + 1/a^2} = \frac{c}{2a} \sqrt{2} = \sqrt{2}\, f_{10} = \sqrt{2}\, f_{01} \qquad \text{Hz} \qquad (2.134)$$

The useful frequency band of the square waveguide is much narrower than the one of a rectangular waveguide having $a \geqslant 2b$.

2.4 CIRCULAR WAVEGUIDE

2.4.1 Definition of the Geometry

A waveguide having a circular cross section of radius a, as shown in Figure 2.15, is considered. Due to the symmetry, the circular cylindrical coordinate system (ρ, ϕ, z) is clearly the one best suited to study it.

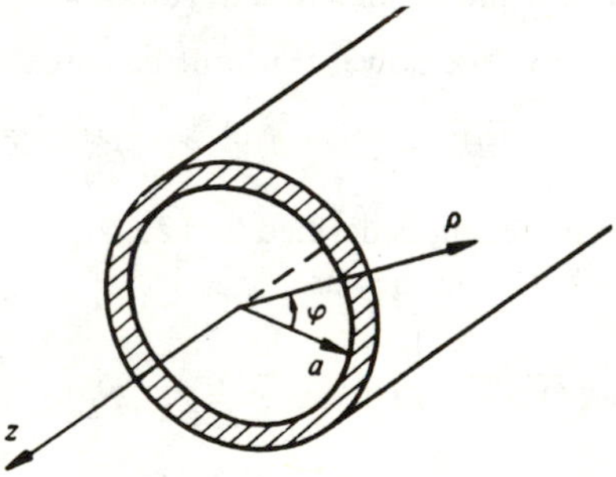

Fig. 2.15 Circular waveguide.

■ 2.4.2 Resolution of the Helmholtz Equation

The equation to be solved, (2.53) or (2.70), becomes, in circular cylindrical coordinates,

$$\frac{1}{\rho} \frac{\partial}{\partial \rho} \left(\rho \frac{\partial \theta}{\partial \rho} \right) + \frac{1}{\rho^2} \frac{\partial^2 \theta}{\partial \varphi^2} + p^2 \theta = 0 \qquad \text{m}^{-2} \qquad (2.135)$$

The method of separation of variables also applies in this system, letting

$$\underline{\theta}(\rho,\varphi) = \Phi(\varphi)\, R(\rho) \tag{2.136}$$

Introducing this product into (2.135) and dividing by θ/ρ^2 yields

$$\frac{\rho}{R}\left(\frac{\mathrm{d}R}{\mathrm{d}\rho} + \rho\,\frac{\mathrm{d}^2 R}{\mathrm{d}\rho^2}\right) + p^2\rho^2 = -\frac{1}{\Phi}\,\frac{\mathrm{d}^2\Phi}{\mathrm{d}\varphi^2} = m^2 \tag{2.137}$$

The left-hand term is a function of ρ only, the middle one of φ only; both must be equal to a constant m^2 for a non-zero solution to be obtained. The differential equation for φ then admits as solutions:

$$\Phi = B_1 \sin m\varphi + B_2 \cos m\varphi \tag{2.138}$$

This solution must be periodic in φ $[\Phi(\varphi + 2\pi) = \Phi(\varphi)]$, so that m must be either a positive integer or zero.

The function $\underline{R}(\rho)$ must satisfy equation:

$$\rho^2\,\frac{\mathrm{d}^2 R}{\mathrm{d}\rho^2} + \rho\,\frac{\mathrm{d}R}{\mathrm{d}\rho} + (p^2\rho^2 - m^2)\,R = 0 \tag{2.139}$$

This is the Bessel differential equation, which admits as solutions Bessel functions of the first order J_m and of the second order N_m (Sec. 10.3). Since second order Bessel functions become singular at the origin, which is on the waveguide axis here, they cannot be solutions to the present problem. Indeed, there are no space charges within the waveguide. The solution then takes the form:

$$\underline{\theta}(\rho,\varphi) = \mathrm{J}_m(p\rho)\,[\underline{B}_1 \sin m\varphi + \underline{B}_2 \cos m\varphi] \tag{2.140}$$

2.4.3 Transverse Potential of TM Modes

For the condition

$$\underline{\theta} = \underline{\phi} = 0 \tag{2.71}$$

to be satisfied on the metal wall of the guide, one must have:

$$\mathrm{J}_m(p_{mn}^{\mathrm{TM}}\, a) = 0 \tag{2.141}$$

Solutions thus correspond to the zeros of the first order Bessel functions (Table 10.3). The first zero appears for $p_{01}^{\mathrm{TM}} = 2.405/a$, the corresponding transverse potential becoming:

$$\phi_{01} = C_{01}^{\mathrm{TM}}\, \mathrm{J}_0(2.405\,\rho/a) \tag{2.142}$$

It is sketched on Figure 2.16.

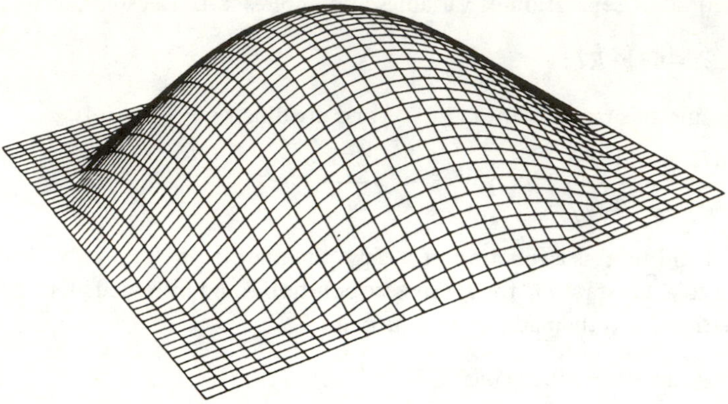

Fig. 2.16 Transverse potential ϕ_{01} of the mode TM_{01}.

The general expression for the potential of mode TM_{mn} is

$$\phi_{mn} = C_{mn}^{TM} \, J_m (p_{mn}^{TM} \rho) \begin{Bmatrix} \cos m\varphi \\ \sin m\varphi \end{Bmatrix} \qquad 1 \qquad (2.143)$$

The brackets mean that two solutions may exist: a solution in $\sin m\phi$ and one in $\cos m\phi$. When $m \neq 0$, these two solutions are space degenerate modes (§ 2.3.17). The normalization constant C_{mn}^{TM} is evaluated using (2.81), yielding after calculation of the integral (§ 10.3.8):

$$C_{mn}^{TM} = \frac{\sqrt{2 - \delta(m)}}{a \, p_{mn}^{TM} \sqrt{\pi} \; J'_m \, (p_{mn}^{TM} a)} \qquad 1 \qquad (2.144)$$

with $\delta(0) = 1$ and $\delta(m) = 0$ if $m \neq 0$. The prime $'$ denotes derivation of the function with respect to its argument.

2.4.4 Transverse Fields of TM Modes

They are obtained from derivation of (2.143), making use of (2.69) and of (2.45):

$$\boldsymbol{E}_{\mathrm{T}} = C_{mn}^{TM} \left(-\boldsymbol{e}_\rho p_{mn}^{TM} J'_m (p_{mn}^{TM} \rho) \begin{Bmatrix} \cos m\varphi \\ \sin m\varphi \end{Bmatrix} - \boldsymbol{e}_\varphi \frac{m}{\rho} J_m (p_{mn}^{TM} \rho) \begin{Bmatrix} -\sin m\varphi \\ \cos m\varphi \end{Bmatrix} \right) \qquad \mathrm{m}^{-1} \qquad (2.145)$$

$$H_T = C_{mn}^{TM} \left(e_\rho \frac{m}{\rho} J_m(p_{mn}^{TM} \rho) \begin{Bmatrix} -\sin m\varphi \\ \cos m\varphi \end{Bmatrix} - e_\varphi p_{mn}^{TM} J'_m(p_{mn}^{TM} \rho) \begin{Bmatrix} \cos m\varphi \\ \sin m\varphi \end{Bmatrix} \right) \quad \text{m}^{-1} \qquad (2.146)$$

The fields of modes TM_{01}, TM_{11} and TM_{21} are depicted in Figure 2.17.

2.4.5 Transverse Potential of TE Modes

The boundary condition to be satisfied at the guide walls ($\rho = a$) is given by:

$$\left.\frac{\partial \theta}{\partial \rho}\right|_{\rho=a} = \left.\frac{\partial \psi}{\partial \rho}\right|_{\rho=a} = 0 \quad \text{m}^{-1} \qquad (2.147)$$

It is met when

$$J'_m(p_{mn}^{TE} a) = 0 \quad 1 \qquad (2.148)$$

The solutions correspond to the zeros of the derivatives of first order Bessel functions, *i.e.*, to the extrema of the latter (Table 10.8). The potential of mode TE_{mn} is given by:

$$\psi_{mn} = C_{mn}^{TE} J_m(p_{mn}^{TE} \rho) \begin{Bmatrix} \cos m\varphi \\ \sin m\varphi \end{Bmatrix} \quad 1 \qquad (2.149)$$

The normalization constant C_{mn}^{TE}, evaluated with (2.81) yields, taking into account the expressions of Section 10.3.8

$$C_{mn}^{TE} = \frac{\sqrt{2 - \delta(m)}}{a\, p_{mn}^{TE} \sqrt{\pi} \left[1 - \left(\dfrac{m}{p_{mn}^{TE} a} \right)^2 \right]^{1/2} J_m(p_{mn}^{TE} a)} \quad 1 \qquad (2.150)$$

The normalization constants for TE and TM modes having the same set of subscripts (m, n) are different. The same holds true for the transverse wave numbers.

2.4.6 Transverse Fields of TE Modes

To obtain them, (2.52) and (2.42) are applied to (2.149):

$$H_T = C_{mn}^{TE} \left(-e_\rho p_{mn}^{TE} J'_m(p_{mn}^{TE} \rho) \begin{Bmatrix} \cos m\varphi \\ \sin m\varphi \end{Bmatrix} - e_\varphi \frac{m}{\rho} J_m(p_{mn}^{TE} \rho) \begin{Bmatrix} -\sin m\varphi \\ \cos m\varphi \end{Bmatrix} \right) \quad \text{m}^{-1} \qquad (2.151)$$

$$E_T = C_{mn}^{TE}\left(-e_\rho \frac{m}{\rho} J_m(p_{mn}^{TE}\rho)\begin{Bmatrix}-\sin m\varphi\\ \cos m\varphi\end{Bmatrix}\right.$$

$$\left.+e_\varphi p_{mn}^{TE} J'_m(p_{mn}^{TE}\rho)\begin{Bmatrix}\cos m\varphi\\ \sin m\varphi\end{Bmatrix}\right) \quad m^{-1} \tag{2.152}$$

The fields of modes TE_{01}, TE_{11} and TE_{21} are presented in Figure 2.17.

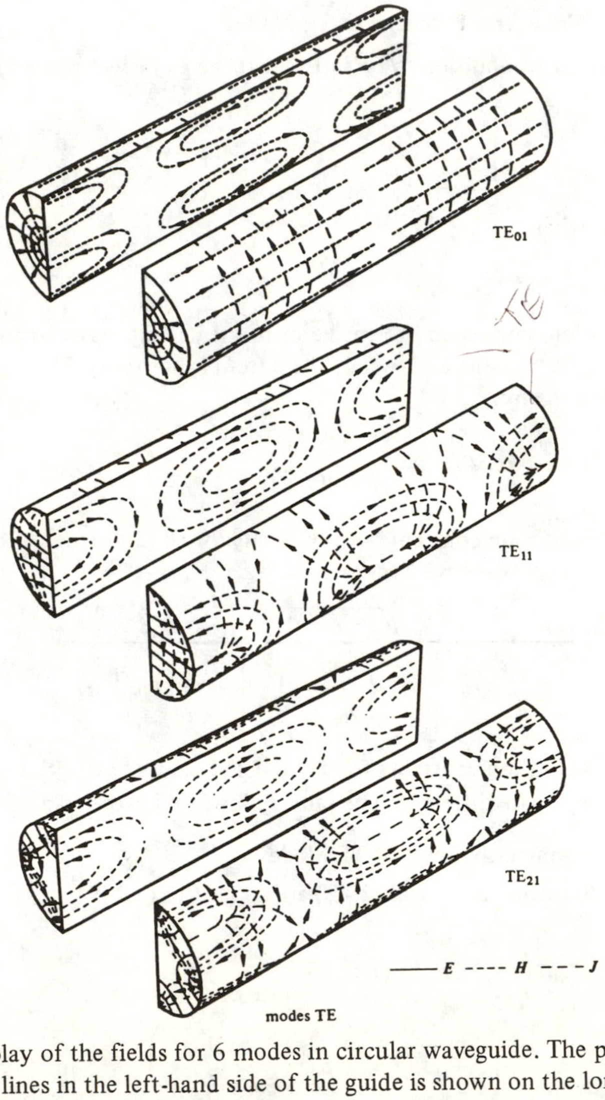

Fig. 2.17 Display of the fields for 6 modes in circular waveguide. The projection of the field lines in the left-hand side of the guide is shown on the longitudinal cross section.

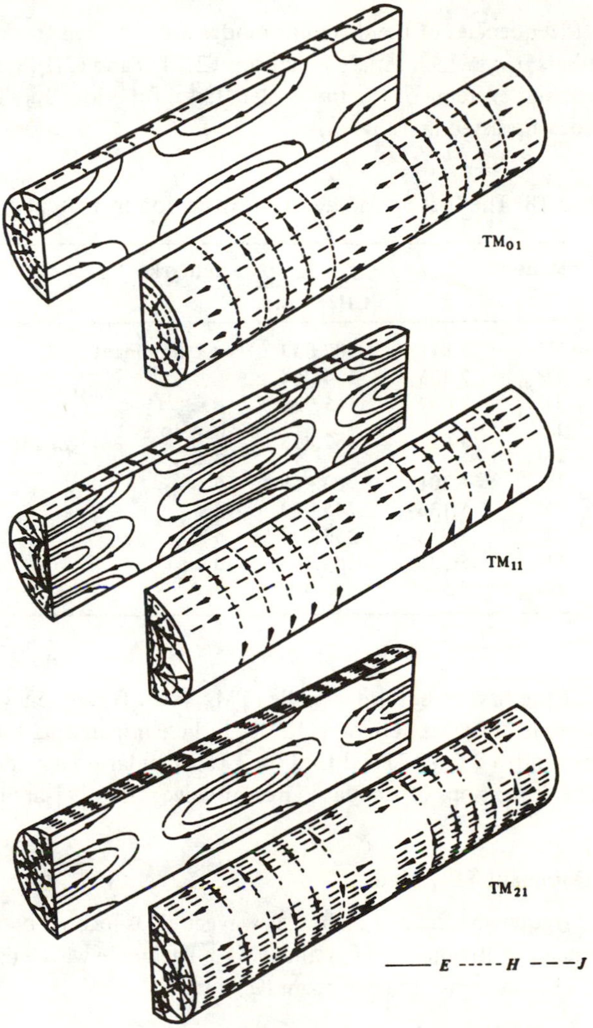

modes TM

2.4.7 Note

The eigenvalues of TE and TM modes having the same subscripts (m, n) are different, it is therefore necessary to always specify whether a transverse wave number p_{mn} corresponds to a TE or to a TM mode. The same is true also for the normalization constants, which are given by the different expressions (2.144) and (2.150).

2.4.8 Cutoff Frequencies

The cutoff frequencies of the different modes are obtained from the transverse wave numbers p_{mn}, which satisfy equations (2.141) and (2.148); all are related to the zeros and extrema of the first order Bessel functions J_m. The first propagating modes are listed in Table 2.18.

Table 2.18 The first ten modes of propagation in a circular waveguide.

Mode	$p_{mn}a$	$\frac{f_c}{\text{GHz}} \cdot \frac{a}{\text{mm}}$	remarks	
TE_{11}	1.841	87.843	d.s. dominant	
TM_{01}	2.405	114.754		
TE_{21}	3.054	145.720	d.s.	
TE_{01}	3.832	182.828	low-loss	degenerate
TM_{11}	3.832	182.828	d.s.	degenerate
TE_{31}	4.201	200.459	d.s.	
TM_{21}	5.136	245.043	d.s.	
TE_{41}	5.318	253.723	d.s.	
TE_{12}	5.331	254.386	d.s.	
TM_{02}	5.520	263.390		

The ratio of the first higher order mode (TM_{01}) cutoff frequency to that of the dominant mode (TE_{11}) is here only 1.306, to be compared to 1.414 for the square waveguide (§ 2.3.17) and to 2 for a rectangular waveguide having $a \geqslant 2b$. The dimensions of standard circular waveguides [4] are listed in Table 2.19.

2.4.9 Dominant TE_{11} Mode

The useful frequency range of a circular waveguide is located between the cutoff frequencies of the modes TE_{11} and TM_{01}. The guide wavelength of the dominant TE_{11} mode is obtained from (2.94):

$$\lambda_g = \frac{\lambda}{\sqrt{1 - (0.293\,\lambda/a)^2}} \qquad \text{m} \qquad (2.153)$$

Its transverse potential is given by (2.149). It is shown in Figure 2.20.

$$\psi_{11} = 0.8871\, J_1(1.841\,\rho/a) \begin{Bmatrix} \cos\varphi \\ \sin\varphi \end{Bmatrix} \qquad 1 \qquad (2.154)$$

Table 2.19 Characteristics of circular waveguides.

Type	radius	cutoff frequencies in GHz			Attenuation in dB/m, TE_{11} mode		
153 IEC–	a/mm	TE_{11} mode	TM_{01} mode	TE_{01} mode	@ f/GHz	theory	max.
C 3.3	323.9	0.27	0.35	0.56	0.325	0.00067	0.0009
C 4	276.7	0.32	0.41	0.66	0.380	0.00085	0.0011
C 4.5	236.4	0.37	0.48	0.77	0.446	0.00108	0.0014
C 5.3	201.9	0.43	0.57	0.90	0.522	0.00137	0.0018
C 6.2	172.5	0.51	0.66	1.06	0.611	0.00174	0.0023
C 7	147.39	0.60	0.78	1.24	0.715	0.00219	0.0029
C 8	125.92	0.70	0.91	1.45	0.838	0.00278	0.0036
C 10	107.57	0.82	1.07	1.70	0.980	0.00352	0.0046
C 12	91.88	0.96	1.25	1.99	1.147	0.00447	0.0058
C 14	78.50	1.20	1.46	2.33	1.343	0.00564	0.0073
C 16	67.05	1.31	1.71	2.73	1.572	0.00715	0.0093
C 18	57.29	1.53	2.00	3.19	1.841	0.00906	0.012
C 22	48.93	1.79	2.34	3.74	2.154	0.0115	0.015
C 25	41.81	2.10	2.74	4.37	2.521	0.0140	0.018
C 30	35.71	2.46	3.21	5.12	2.952	0.0184	0.024
C 35	30.52	2.88	3.76	5.99	3.455	0.0233	0.030
C 40	25.99	3.38	4.41	7.03	4.056	0.0297	0.039
C 48	22.22	3.95	5.16	8.23	4.744	0.0375	0.049
C 56	19.05	4.61	6.02	9.60	5.534	0.0473	0.062
C 65	16.27	5.40	7.05	11.2	6.480	0.0599	0.078
C 76	13.894	6.32	8.26	13.2	7.588	0.0759	0.099
C 89	11.912	7.37	9.63	15.3	8.850	0.0956	0.124
C 104	10.122	8.68	11.3	18.1	10.42	0.1220	0.150
C 120	8.737	10.00	13.1	20.9	12.07	0.1524	
C 140	7.544	11.6	15.2	24.2	13.98	0.1893	
C 165	6.350	13.8	18.1	28.8	16.61	0.2459	
C 190	5.563	15.8	20.6	32.9	18.95	0.3003	
C 220	4.762	18.4	24.1	38.4	22.14	0.3787	
C 255	4.165	21.1	27.5	43.9	25.31	0.4620	
C 290	3.563	24.6	32.2	51.2	29.54	0.5834	
C 330	3.175	27.7	36.1	57.6	33.20	0.6938	
C 380	2.781	31.6	41.3	65.7	37.91	0.8486	
C 430	2.387	36.8	48.1	76.6	44.16	1.0650	
C 495	2.184	40.2	52.5	83.7	48.26	1.2190	
C 580	1.790	49.1	64.1	102	58.88	1.643	
C 660	1.583	55.3	72.3	115	66.41	1.967	
C 765	1.384	63.5	82.9	132	76.15	2.413	
C 890	1.194	73.6	96.1	153	88.30	3.011	

$$\underline{E}_{\mathrm{T}} = 0.8871 \left[-\boldsymbol{e}_\rho \frac{1}{\rho} \mathrm{J}_1(1.841\,\rho/a) \begin{Bmatrix} -\sin\varphi \\ \cos\varphi \end{Bmatrix} + \boldsymbol{e}_\varphi \frac{1.841}{a} \mathrm{J}_1'(1.841\,\rho/a) \begin{Bmatrix} \cos\varphi \\ \sin\varphi \end{Bmatrix} \right] \ \mathrm{m}^{-1} \tag{2.155}$$

$$\underline{H}_{\mathrm{T}} = 0.8871 \left[-\boldsymbol{e}_\rho \frac{1.841}{a} \mathrm{J}_1'(1.841\,\rho/a) \begin{Bmatrix} \cos\varphi \\ \sin\varphi \end{Bmatrix} - \boldsymbol{e}_\varphi \frac{1}{\rho} \mathrm{J}_1(1.841\,\rho/a) \begin{Bmatrix} -\sin\varphi \\ \cos\varphi \end{Bmatrix} \right] \ \mathrm{m}^{-1} \tag{2.156}$$

$$\underline{H}_z = \frac{3}{\mathrm{j}\omega\underline{\mu}a^2}\,\underline{U}_{\mathrm{e}}\,\mathrm{J}_1(1.841\,\rho/a) \begin{Bmatrix} \cos\varphi \\ \sin\varphi \end{Bmatrix} \ \mathrm{A/m} \tag{2.157}$$

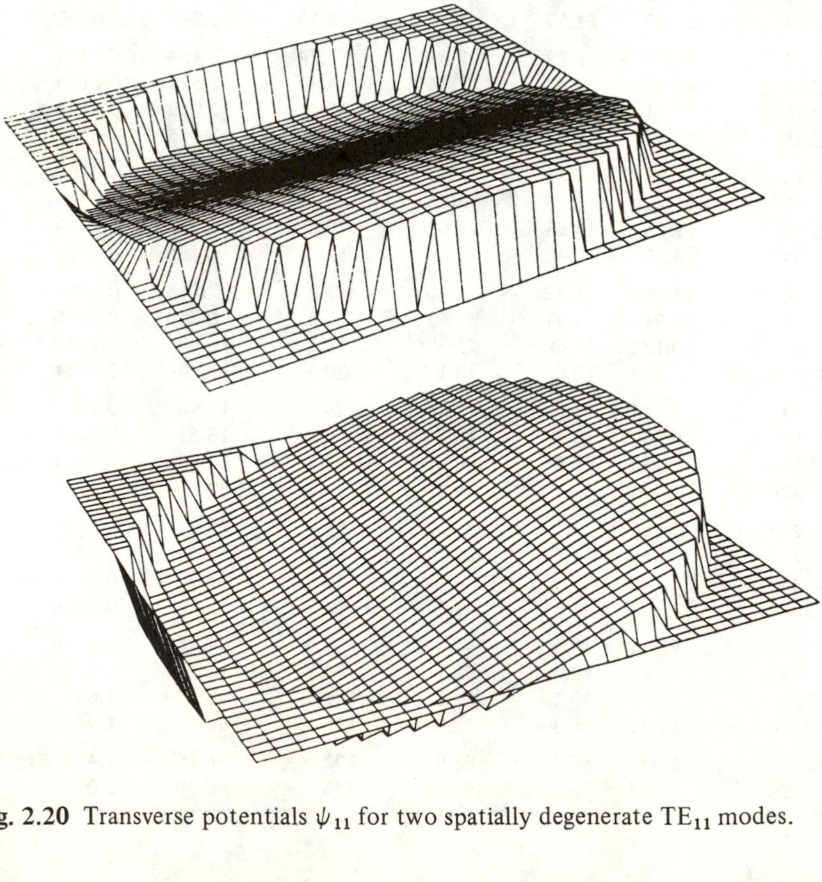

Fig. 2.20 Transverse potentials ψ_{11} for two spatially degenerate TE_{11} modes.

2.4.10 Particular Case: TE_{01} Mode

Another mode of the circular cylindrical waveguide possesses particular characteristics: it is the TE_{01} mode (third higher order mode, Table 2.18). The transverse potential of this mode is obtained from (2.149):

$$\psi_{01} = 0.366\, J_0(3.832\, \rho/a) \qquad 1 \qquad (2.158)$$

The field components are determined from relations (2.151) and (2.152):

$$\boldsymbol{E}_T = \boldsymbol{e}_\varphi\, 1.4\, J_0'(3.832\, \rho/a) = -\boldsymbol{e}_\varphi\, 1.4\, J_1(3.832\, \rho/a) \qquad m^{-1} \qquad (2.159)$$

$$\boldsymbol{H}_T = -\boldsymbol{e}_\rho\, 1.4\, J_0'(3.832\, \rho/a) = \boldsymbol{e}_\rho\, 1.4\, J_1(3.832\, \rho/a) \qquad m^{-1} \qquad (2.160)$$

$$\underline{H}_z = \frac{5.367}{j\omega\mu a^2}\, \underline{U}_e\, J_0(3.832\, \rho/a) \qquad A/m \qquad (2.161)$$

On the guide wall, for $\rho = a$, the magnetic field is purely longitudinal. The current flowing on the guide wall therefore has only an azimuthal component (1.12). Annular slots can be cut in the waveguide walls, without disturbing the fields within the guide (non-radiating slots). Furthermore, the amplitude of the longitudinal magnetic field, and thus the current, decays with increasing frequency at constant power (2.161). This means that the attenuation in an actual waveguide (with lossy walls) decreases as frequency increases (§ 2.7.6).

2.5 OTHER HOLLOW WAVEGUIDES

2.5.1 Introduction

In addition to rectangular and circular waveguides, which are most often found in microwave systems, several other cross sections are occasionally encountered in specific applications for instance, to increase the frequency bandwidth or the power handling capability, or to handle components inserted into the guide [5]. Also, certain intermediate cross sections appear along transitions between waveguides of different shapes. Elliptical and ridge waveguides are the two particular types encountered fairly often in practice.

2.5.2 Elliptical Waveguide

Waveguides of elliptical cross section possess a remarkable mechanical property: they more or less retain their original shape when bent or twisted (when changes in guide direction or modification of the plane of polarization are made). Elliptical waveguides are often used to connect antennas for this reason. The study of the field distribution is carried out in an elliptical-hyperbolic coordinate system, with solutions obtained in terms of Mathieu functions [6].

2.5.3 Ridge Waveguide

The introduction of one or two metallic ridges in a rectangular waveguide (Fig. 2.21) permits the increase of the frequency bandwidth. The ridges reduce the cutoff frequency of the dominant mode, without significantly affecting the higher order modes. Cutoff frequency ratios up to 3 or 4 are obtained in this manner.

However, both the electric and the magnetic fields are concentrated next to the ridge. As a result, the power handling capability of the waveguide is significantly reduced: the breakdown field is attained for a much lower power level than in a rectangular waveguide. The attenuation due to the metal losses (§ 2.7.3) is also larger than in a rectangular waveguide due to the concentration of the magnetic field.

The shape of ridge waveguides does not allow for an exact analytical resolution of the Helmholtz equation. Several approximate methods were therefore developed for this study [7,8].

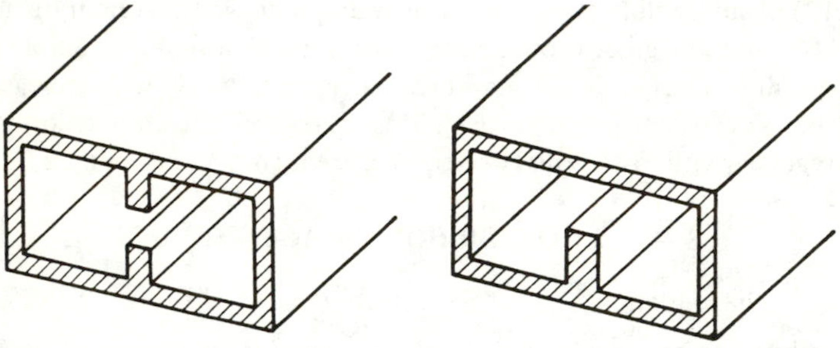

Fig. 2.21 Ridge waveguides.

2.6 TWO CONDUCTOR LINES: TEM MODES

2.6.1 General Considerations

As mentioned in Section 2.2.3, the TEM mode on a two-conductor line propagates at all frequencies from dc upward. In contrast to hollow waveguides, it does not have any cutoff which would restrict operation to the sole very high frequency ranges. The propagation factor is a function of the properties of the homogeneous material filling the line and of the frequency only (2.26). In a lossless structure (§ 2.2.26) it becomes:

$$\underline{\gamma} = \sqrt{j\omega\mu j\omega\epsilon} = j\omega\sqrt{\mu\epsilon} = j\frac{\omega}{c_0}\sqrt{\epsilon_r\mu_r} \qquad \text{m}^{-1} \tag{2.162}$$

Similarly, the wave impedance Z_e, defined as the ratio of the two transverse fields (§ 2.2.25) is independent of the transmission line shape:

$$Z_e = \sqrt{j\omega\mu / j\omega\epsilon} = \sqrt{\mu/\epsilon} \qquad \Omega \qquad (2.163)$$

This wave impedance must, however, not be confused with the waveguide or line impedance, which depends on the geometry. The latter impedance must be employed when matching.

The transverse fields of the TEM mode are solutions of the two-dimensional Laplace's equation (§ 2.2.3), in the presence of the boundary conditions on the two metallic conductors.

Since it has no cutoff frequency, the TEM mode on a two-conductor line is ***always*** the dominant mode.

2.6.2 Coaxial Line

The electrostatic and magnetostatic studies of this cross section (Fig. 2.22) are carried out in [214].

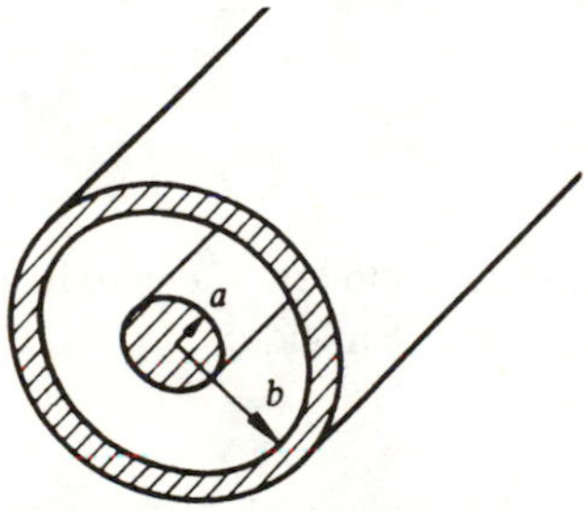

Fig. 2.22 Coaxial line: a circular cylindrical conductor is coaxially located within a conducting tube.

The transverse potential $\underline{\phi}$ is obtained, considering the TEM mode as a particular case of TM mode:

$$\phi^{\mathrm{TEM}} = \frac{\ln(\rho/a)}{\sqrt{2\pi \ln(b/a)}} \qquad 1 \qquad (2.164)$$

As there is no cutoff frequency, $p = 0$ (§ 2.2.37). The normalization condition (2.81) cannot be used, however condition (2.80) for $\underline{\boldsymbol{E}}_{\mathrm{T}}$ remains valid. The transverse fields are given by:

$$\underline{\boldsymbol{E}}_{\mathrm{T}} = -\boldsymbol{e}_{\rho} \frac{1}{\rho\sqrt{2\pi \ln(b/a)}} \qquad \mathrm{m}^{-1} \qquad (2.165)$$

$$\underline{H}_T = -e_\varphi \frac{1}{\rho\sqrt{2\pi}\ln(b/a)} \qquad \text{m}^{-1} \qquad (2.166)$$

The capacity per unit length is

$$C' = \frac{2\pi\epsilon}{\ln(b/a)} \qquad \text{F/m} \qquad (2.167)$$

The inductance per unit length is

$$L' = \frac{\mu}{2\pi}\ln(b/a) \qquad \text{H/m} \qquad (2.168)$$

For high frequencies, the current is concentrated on the surfaces of the conductors (skin effect), so that only the field between the conductors needs to be considered. The line impedance is then obtained from equation (9.15).

$$Z_c = \sqrt{Z'/Y'} = \frac{1}{2\pi}\sqrt{\frac{j\omega\mu}{j\omega\epsilon}}\ln(b/a)$$

$$= Z_e \frac{1}{2\pi}\ln(b/a) \qquad \Omega \qquad (2.169)$$

In contrast to what happens in hollow waveguides (§ 2.3.16), the three definitions of waveguide impedance coincide for a TEM mode.

2.6.3 Stripline

In this structure, depicted in Figure 2.23, the transverse fields remain in the vicinity of the center conductor, between two ground planes connected to each other. The electrostatic study of the problem was carried out by conformal mapping for the limiting case $b = 0$ (Schwartz-Christoffel transform) yielding [9]:

$$\underline{Z}_c = \underline{Z}_e \frac{1}{4}\frac{K(1-u)}{K(u)} \qquad \Omega \qquad (2.170)$$

with

$$u = [\cosh(\pi w/4h)]^{-1} \qquad 1 \qquad (2.171)$$

and where $K(u)$ is the first order elliptical integral [10]. Approximate relations take into account the center conductor thickness.

The propagation study assumed infinitely wide ground planes. In practice, however, the field is found to decrease quite rapidly away from the center conductor, so that the ground plane width can be reduced to approximately 5 w without significantly affecting TEM mode propagation.

As this structure is ***open***, a continuous spectrum of radiating modes must be added to the discrete spectrum of guided modes.

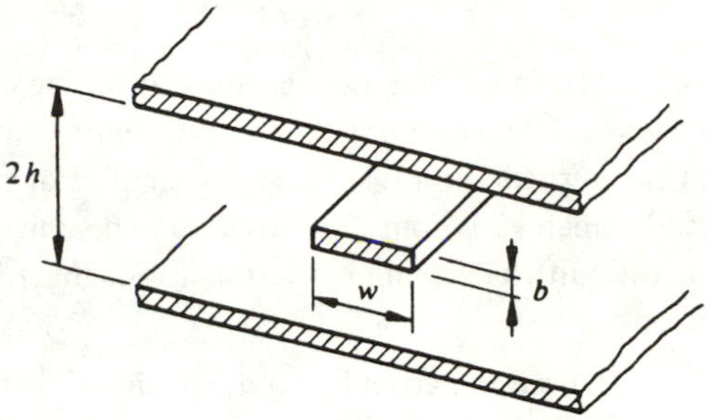

Fig. 2.23 Stripline.

2.6.4 Advantage: Frequency Bandwidth

Since a two-conductor line has no lower frequency cutoff, it can be utilized over a very broad frequency range, from $f = 0$ up to the cutoff of the first TE mode.

2.6.5 Disadvantage: Connections

It is more difficult to connect electrically two conductors than a single one; hollow waveguides are more easily joined to one another than coaxial lines, which require high precision, and hence expensive, connectors in which the effects of discontinuities are compensated for by elaborate matching elements. Furthermore, the center conductor has to be mechanically supported to ensure the uniformity of the cross section.

2.6.6 Disadvantage: Field Concentration

In a two-conductor line, the fields tend to concentrate next to the conductors, mainly near the one having the smaller cross section. This limits the power-handling capability of the line. The breakdown field is reached for a lower power level than in hollow waveguides of same cross section. The heating up of the center conductor also limits the power handling. In hollow waveguides, on the other hand, the fields spread more evenly, resulting in larger power-handling capabilities than in two conductor lines, for similar sizes.

2.7 PERTURBATION METHODS

2.7.1 Definition

When a system is only slightly different from another one, for which a solution is available, the study can be carried out considering the effect of the difference, called *perturbation.* The basic equations of the two systems are compared, yielding an expression for the difference between the significant parameters of the two systems.

The relationship obtained in this manner can most often be evaluated only by making approximations. The method is then ***approximate***, providing accurate results only when the perturbation is sufficiently small. For this reason, it is mostly used to study small variations close to a known solution. Since the range of validity is very difficult to determine, such methods must be used with a great deal of caution.

In waveguides, they are used in particular to determine the attenuation due to the ohmic losses within the metal walls, the effect produced by the introduction of a dielectric (insulating material) and by small variations in the guide cross section (for instance, the effect of rounded edges).

■ 2.7.2 Perturbation of the Propagation Factor

The known (unperturbed) system is taken to be a lossless waveguide that is either a vacuum, or is filled with air. Maxwell's equations in this system, (1.2) and (1.3), separating the longitudinal dependence for a ***forward wave*** (2.7), yield:

$$\nabla_t \times \underline{E}_0 + j\omega\mu_0\underline{H}_0 = \underline{\gamma}_0 \boldsymbol{e}_z \times \underline{E}_0 \qquad \text{V/m}^2 \qquad (2.172)$$

$$\nabla_t \times \underline{H}_0 - j\omega\epsilon_0\underline{E}_0 = \underline{\gamma}_0 \boldsymbol{e}_z \times \underline{H}_0 \qquad \text{A/m}^2 \qquad (2.173)$$

The subscript 0 refers to the unperturbed system. The corresponding equations are written for the ***perturbed system*** also, in which the perturbation may be due to source currents $\underline{J}_m$ and $\underline{J}_e$, respectively magnetic and electric:

$$\nabla_t \times \underline{E} + j\omega\mu_0\underline{H} + \underline{J}_m = \underline{\gamma} \boldsymbol{e}_z \times \underline{E} \qquad \text{V/m}^2 \qquad (2.174)$$

$$\nabla_t \times \underline{H} - j\omega\epsilon_0\underline{E} - \underline{J}_e = \underline{\gamma} \boldsymbol{e}_z \times \underline{H} \qquad \text{A/m}^2 \qquad (2.175)$$

A relationship for $\underline{\gamma}$ is obtained from these equations for the fields in the two systems. One takes the complex conjugates of (2.172) and (2.173), which are then dot-multiplied by $\underline{H}$ and $-\underline{E}$ respectively. Equations (2.174) and (2.175) are dot-multiplied by $\underline{H}_0^*$ and $-\underline{E}_0^*$. The addition of the four expressions obtained in this manner, when similar terms have been grouped, yields:

$$\nabla_t \cdot (\underline{E}_0^* \times \underline{H} + \underline{E} \times \underline{H}_0^*) + \underline{H}_0^* \cdot \underline{J}_m + \underline{E}_0^* \cdot \underline{J}_e$$

$$= (\underline{\gamma} + \underline{\gamma}_0^*)\, \boldsymbol{e}_z \cdot (\underline{E}_0^* \times \underline{H} + \underline{E} \times \underline{H}_0^*) \qquad \text{VA/m}^3 \qquad (2.176)$$

This expression is then integrated over the waveguide cross section S. The divergence theorem (10.52), and the boundary condition on the guide walls, which is a pec in the unperturbed system, are made use of:

$$\underline{\boldsymbol{n}} \times \underline{E}_0^* = 0 \qquad \text{V/m} \qquad (2.177)$$

In this manner, the perturbation formula is obtained:

$$\underline{\gamma} + \underline{\gamma}_0^* = \frac{\oint_C (\underline{E} \times \underline{H}_0^*) \cdot \boldsymbol{n}\, \mathrm{d}l + \int_S (\underline{J}_e \cdot \underline{E}_0^* + \underline{J}_m \cdot \underline{H}_0^*)\, \mathrm{d}A}{\int_S (\underline{E} \times \underline{H}_0^* + \underline{E}_0^* \times \underline{H}) \cdot \boldsymbol{e}_z \mathrm{d}A} \qquad \text{m}^{-1} \qquad \mathbf{(2.178)}$$

This expression is exact; no approximation whatsoever was made while deriving it. However, the perturbed fields $\underline{E}$ and $\underline{H}$ are unknown: certain assumptions about them must be made to be able to determine $\underline{\gamma}$ with this expression.

■ 2.7.3 Application: Effect of Wall Losses in a Waveguide

In this case, a boundary condition is modified: the guide walls are made of a real metal, which has a large, but not infinite, conductivity σ. There are no perturbing source currents within the waveguide:

$$\underline{J}_m = 0 \qquad \text{V/m}^2 \qquad \underline{J}_e = 0 \qquad \text{A/m}^2 \qquad (2.179)$$

The magnetic field is not perturbed to first order, so that:

$$\underline{H} \cong \underline{H}_0 \qquad \text{A/m} \qquad (2.180)$$

As for the electric field, its average value within the guide is hardly affected. On the other hand, the tangential component no longer vanishes on the guide walls, but is specified there by:

$$\underline{E}_{\tan} = \underline{Z}_m \underline{A} = -\underline{Z}_m (\boldsymbol{n} \times \underline{H}) \cong -\underline{Z}_m (\boldsymbol{n} \times \underline{H}_0) \qquad \text{V/m} \qquad (2.181)$$

where $\underline{A} = -\boldsymbol{n} \times \underline{H}$ is the surface current density on the guide wall (1.12) and $\underline{Z}_m$ the characteristic impedance of the metal, determined by

$$\underline{Z}_m \cong \sqrt{\mathrm{j}\omega\mu'_m/\sigma} = (1+\mathrm{j})\sqrt{\omega\mu'_m/2\sigma} = (1+\mathrm{j})R_m \qquad \Omega \qquad (2.182)$$

The subscript m indicates metal related quantities.

Introducing these field values into (2.178), one obtains for a propagating wave ($\alpha_0 = 0$):

$$\underline{\gamma} + \underline{\gamma}_0^* = \alpha + \mathrm{j}(\beta - \beta_0) = \frac{(1+\mathrm{j})R_\mathrm{m} \oint_C |\underline{H}_0|^2 \mathrm{d}l}{2 \int_S \boldsymbol{P}_0 \cdot \boldsymbol{e}_z \, \mathrm{d}A} \qquad \mathrm{m}^{-1} \qquad (2.183)$$

with $\boldsymbol{P}_0$, the real part of Poynting vector, representing the power density of an electromagnetic wave. It is given by

$$\boldsymbol{P}_0 = \mathrm{Re}[\underline{\boldsymbol{E}}_0 \times \underline{\boldsymbol{H}}_0^*] \qquad \mathrm{W/m^2} \qquad (2.184)$$

The integral in the denominator of (2.183) is the total active power transmitted along the waveguide. Making use of the normalization of paragraph 2.2.24:

$$\int \boldsymbol{P}_0 \cdot \boldsymbol{e}_z \, \mathrm{d}A = \underline{U}_\mathrm{e} \underline{I}_\mathrm{e}^* \qquad \mathrm{W} \qquad (2.185)$$

For a TM mode, with the aid of (2.39), (2.45) and (2.69) one finds:

$$\underline{\boldsymbol{H}}_0 = -\underline{I}_\mathrm{e} \boldsymbol{e}_z \times \nabla_\mathrm{t} \underline{\phi} \qquad \mathrm{A/m} \qquad (2.186)$$

and thus

$$\alpha + \mathrm{j}(\beta - \beta_0) = \frac{(1+\mathrm{j})R_\mathrm{m}}{2Z_\mathrm{e}} \oint_C |\nabla_\mathrm{t} \underline{\phi}|^2 \mathrm{d}l \qquad \mathrm{m}^{-1} \qquad (2.187)$$

For a TE mode, making use of (2.39), (2.52) and (2 58) one finds:

$$\underline{\boldsymbol{H}}_0 = -\underline{I}_\mathrm{e} \nabla_\mathrm{t} \underline{\psi} + \boldsymbol{e}_z \frac{p^2}{\mathrm{j}\omega\mu_0} \underline{U}_\mathrm{e} \underline{\psi} \qquad \mathrm{A/m} \qquad (2.188)$$

and thus

$$\alpha + \mathrm{j}(\beta - \beta_0) = \frac{(1+\mathrm{j})R_\mathrm{m}}{2} \left(\frac{1}{Z_\mathrm{e}} \oint_C |\nabla_\mathrm{t} \underline{\psi}|^2 \mathrm{d}l + \frac{p^4 Z_\mathrm{e}}{\omega^2 \mu_0^2} \oint_C |\underline{\psi}|^2 \mathrm{d}l \right) \qquad \mathrm{m}^{-1} \qquad (2.189)$$

or, still, replacing $\underline{Z}_\mathrm{e}$ by its value drawn from (2.97):

$$\alpha = \beta - \beta_0 = \frac{R_\mathrm{m}}{2\omega\mu_0\beta_0} \left(\beta_0^2 \oint_C |\nabla_\mathrm{t} \underline{\psi}|^2 \mathrm{d}l + p^4 \oint_C |\underline{\psi}|^2 \mathrm{d}l \right) \qquad \mathrm{m}^{-1} \qquad (2.190)$$

2.7.4 Note

At the cutoff frequency of a mode, the waveguide does not carry any power over this mode (§ 2.2.29). Equation (2.183) then becomes singular, as its denominator vanishes. The perturbation method for the propagation factor is not valid at cutoff or close to it (§ 2.7.1). Another formulation, specifically derived for the cutoff frequency is therefore needed (§ 2.7.9).

2.7.5 Application: Dominant Mode in a Rectangular Waveguide

The expression obtained in (2.124) for $\underline{\psi}$ of the TE_{10} mode is introduced into (2.190). The two integrals are evaluated, yielding:

$$\alpha = \beta - \beta_0 \cong \frac{R_m}{\omega\mu_0\beta_0 a}\left[2p^2 + (a/b)k^2\right] \qquad \text{m}^{-1} \tag{2.191}$$

To determine how the attenuation varies with frequency, all frequency-dependent quantities in (2.191) are replaced by their expressions: R_m (2.182), k (2.28) and β (2.93), yielding eventually:

$$\alpha \cong \sqrt{\frac{\pi\mu_m}{2\sigma\mu_0}\sqrt{\epsilon/\mu_0}}\;\frac{1}{a^{3/2}}\left(\frac{2(f/f_{10})^{-1/2} + (a/b)(f/f_{10})^{3/2}}{\sqrt{(f/f_{10})^2 - 1}}\right) \text{Np/m} \tag{2.192}$$

The attenuation within a copper waveguide is shown in Figure 2.24 as a function of the normalized frequency $f \cdot a$, with a/b as parameter. As previously mentioned, there is a singularity at the cutoff frequency: the attenuation goes to infinity there, which is meaningless, outside of the validity domain for perturbation methods. The attenuation decreases as frequency increases, goes through a minimum (usually outside of the useful frequency range of the guide), then increases at higher frequencies.

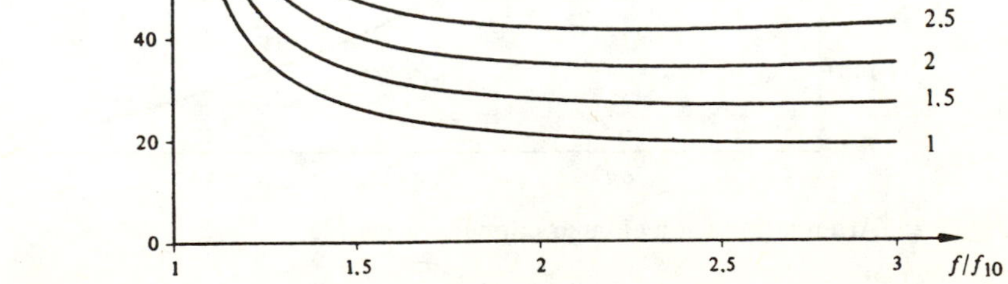

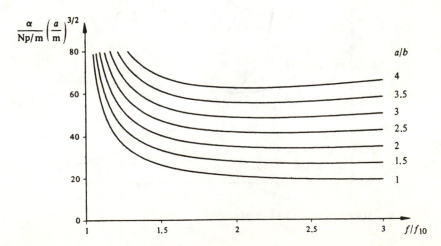

Fig. 2.24 Attenuation of a rectangular waveguide as a function of frequency and of the a/b ratio for the dominant TE_{10} mode.

The attenuation also increases as the waveguide height is reduced; therefore, a/b ratio must be selected to be as small as possible, remembering the constraint imposed by the appearance of higher order modes (§ 2.3.9). The value $a/b = 2$ is, therefore, most often taken.

2.7.6 Application: Modes in Circular Waveguide

Introducing the values of the transverse potentials $\underline{\psi}$, respectively obtained for the dominant TE_{11} (§ 2.4.9) and the higher order TE_{01} (§ 2.4.10) modes, one obtains:

$$\alpha_{11} \cong \frac{R_m}{aZ_0}\,(1 - f_{11}^2/f^2)^{-1/2}\left(\frac{f_{11}^2}{f^2} + \frac{1}{(1.841)^2 - 1}\right) \quad \text{Np/m} \tag{2.193}$$

$$\alpha_{01} \cong \frac{R_m}{aZ_0}\,\frac{f_{01}^2}{f\sqrt{f^2 - f_{01}^2}} \quad \text{Np/m} \tag{2.194}$$

The attenuation dependencies of the two modes are compared in Figure 2.25 in terms of frequency. While the attenuation of the dominant TE_{11} mode decreases and goes through a minimum, the one of the TE_{01} does not possess any minimum, but keeps decreasing steadily as frequency increases, roughly as $\alpha \sim f^{-1/2}$. This peculiar behavior is due to the decrease of the surface current, mentioned in Section 2.4.10.

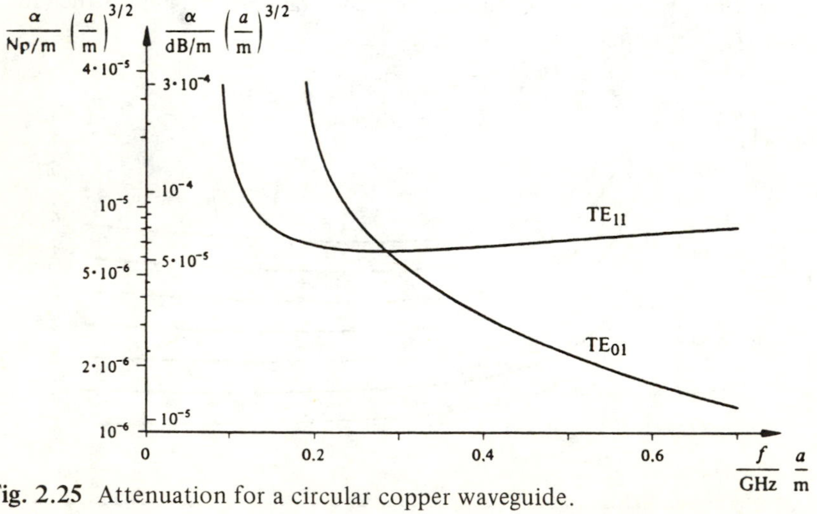

Fig. 2.25 Attenuation for a circular copper waveguide.

It is thus possible, at least in principle, to obtain attenuations as small as desired by operating at frequencies far above the waveguide cutoff. However, since a higher order mode is involved, it is difficult to be certain that only this particular mode will be excited: geometrical irregularities and bends along the waveguide induce coupling to other propagating modes, which exhibit significant losses. Circular waveguides are in operation on the TE_{01} mode for communications [11]. Their use was considered for high power transfer [12].

2.7.7 Application: TEM Mode in Coaxial Line

The attenuation on a coaxial line is determined in quite a similar manner: introducing the transverse potential $\underline{\phi}$ (2.164) into the equation for TM modes, and yielding after some calculations [13]:

$$\alpha^{TEM} = \frac{R_m}{2\sqrt{\mu_0/\epsilon}} \, \frac{1/a + 1/b}{\ln(b/a)} \qquad \text{Np/m} \qquad \textbf{(2.195)}$$

The attenuation increases as the square root of the frequency (R_m term). Its dependence upon the dimensions is shown, as a function of the b/a ratio, in Figure 2 26. It goes through a minimum for $b/a = 3.5911$ . . .

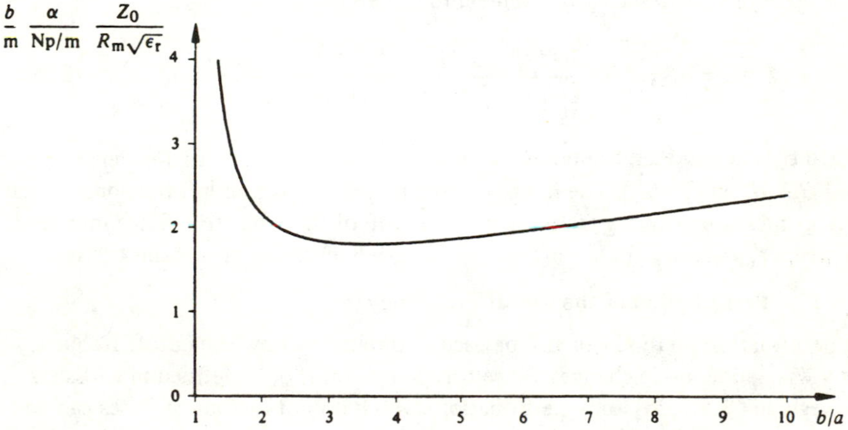

Fig. 2.26 Attenuation of a coaxial line as a function of the b/a ratio.

2.7.8 Introduction of Dielectric in a Waveguide

A dielectric insert of cross section ΔS and of relative permittivity $\underline{\epsilon}_r$ is introduced into a waveguide. The uniformity along the direction of propagation must be maintained (Fig. 2.27).

Within the dielectric cross section ΔS, Maxwell's equation (2.173) becomes:

$$\nabla_t \times \underline{H} - j\omega\epsilon_0\underline{\epsilon}_r\underline{E} = \underline{\gamma}\boldsymbol{e}_z \times \underline{H} \qquad \text{A/m}^2 \qquad (2.196)$$

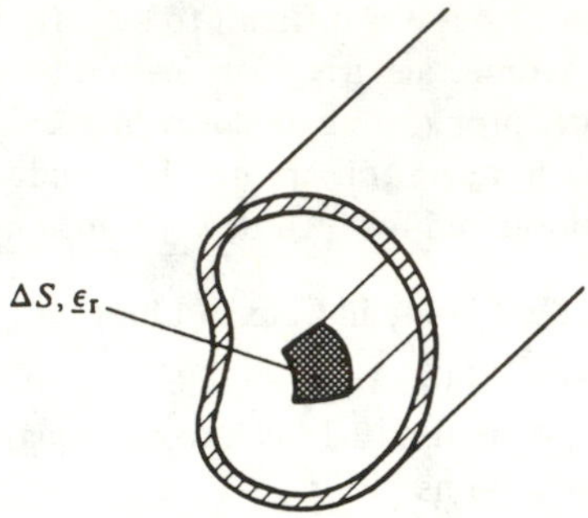

Fig. 2.27 Hollow waveguide loaded with a longitudinal bar of dielectric.

Comparing it with (2.175), one obtains the value of the source current representing the dielectric insert:

$$= \mathrm{j}\omega\epsilon_0(\underline{\epsilon}_\mathrm{r} - 1)\underline{E} \qquad \mathrm{A/m^2} \qquad (2.197)$$

and therefore, introducing this value into (2.178):

$$\underline{\gamma} + \underline{\gamma}_0^* = \alpha + \mathrm{j}(\beta - \beta_0) = \frac{\mathrm{j}\omega\epsilon_0 \int_{\Delta S}(\underline{\epsilon}_\mathrm{r} - 1)\underline{E} \cdot \underline{E}_0^* \, \mathrm{d}A}{\int_S (\underline{E} \times \underline{H}_0^* + \underline{E}_0^* \times \underline{H}) \cdot \boldsymbol{e}_z \, \mathrm{d}A} \qquad \mathrm{m^{-1}} \qquad (2.198)$$

When $\underline{\epsilon}_\mathrm{r}$ is not too large and $\Delta \underline{S} \ll \underline{S}$, one can assume that, to first order, $\underline{E} \cong \underline{E}_0$ and $\underline{H} \cong \underline{H}_0$ in (2.198). The integral in the numerator is then proportional to the energy increase resulting from the introduction of the dielectric. Since in material media $\epsilon_\mathrm{r}' \geqslant 1$, the phase shift per unit length β increases or remains constant.

2.7.9 Perturbation of the Cutoff Frequency

A perturbation method can also be used to determine how the cutoff frequency of a waveguide mode changes. Since this parameter is only defined in a lossless waveguide (§ 2.2.28), only perturbations which do not introduce losses can be considered. This happens for instance when a lossless dielectric is introduced into the guide (Fig. 2.27), or when the waveguide cross section is slightly modified (cep wall also in the modified structure). As the propagation factor vanishes at cutoff, Maxwell's equations (1.2) and (1.3) become for the unperturbed system at the cutoff angular frequency $\omega_{\mathrm{c}0}$:

$$\nabla_\mathrm{t} \times \underline{E}_0 + \mathrm{j}\omega_{\mathrm{c}0}\mu_0\underline{H}_0 = 0 \qquad \mathrm{V/m^2} \qquad (2.199)$$

$$\nabla_\mathrm{t} \times \underline{H}_0 - \mathrm{j}\omega_{\mathrm{c}0}\epsilon_0\underline{E}_0 = 0 \qquad \mathrm{A/m^2} \qquad (2.200)$$

In the system perturbed by source currents $\underline{J}_\mathrm{e}$ and $\underline{J}_\mathrm{m}$ one obtains similarly, at the cutoff angular frequency ω_c:

$$\nabla_t \times \underline{E} + j\omega_c\mu_0\underline{H} + \underline{J}_m = 0 \qquad \text{V/m}^2 \qquad (2.201)$$

$$\nabla_t \times \underline{H} - j\omega_c\epsilon_0\underline{E} - \underline{J}_e = 0 \qquad \text{A/m}^2 \qquad (2.202)$$

The ensuing procedure is exactly the same as for the propagation factor (§ 2.7.2). The cutoff frequencies are different in the two systems, so that:

$$\nabla_t \cdot [\underline{E}_0^* \times \underline{H} + \underline{E} \times \underline{H}_0^*] + j(\omega_c - \omega_{c0})[\epsilon_0\underline{E}_0^* \cdot \underline{E} + \mu_0\underline{H}_0^* \cdot \underline{H}] + \underline{E}_0^* \cdot \underline{J}_e + \underline{H}_0^* \cdot \underline{J}_m = 0 \qquad \text{VA/m}^3 \qquad (2.203)$$

After integration over the cross section of the unperturbed waveguide, and making use of the boundary condition on $\underline{E}_0$ (2.177) one obtains the perturbation equation:

$$\omega_c - \omega_{c0} = \frac{j\oint_C(\underline{E} \times \underline{H}_0^*) \cdot \boldsymbol{n}\, dl + j\int_{\Delta S}(\underline{J}_e \cdot \underline{E}_0^* + \underline{J}_m \cdot \underline{H}_0^*)\, dA}{\int_S (\epsilon_0\underline{E}_0^* \cdot \underline{E} + \mu_0\underline{H}_0^* \cdot \underline{H})\, dA} \qquad \text{rad/s} \qquad (2.204)$$

2.7.10 Application: Introduction of Dielectric into the Waveguide

The procedure follows as in Section 2.7.8, considering however a strictly lossless dielectric, for the reason outlined in the previous paragraph. The electric perturbation current $\underline{J}_e$ is defined by (2.97) and introduced into (2.204), letting also $\underline{E} \cong \underline{E}_0$ and $\underline{H} \cong \underline{H}_0$:

$$\omega_c - \omega_{c0} = \frac{-\omega_c\int_{\Delta S}(\epsilon_r - 1)\,|\underline{E}_0|^2\, dA}{2\int_S |\underline{E}_0|^2\, dA} \qquad \text{rad/s} \qquad (2.205)$$

While the integral in the numerator has the same form as in the relation giving the propagation factor (2.198), here the denominator integral is proportional to twice the energy stored in a unit length of waveguide. Since the relative permittivity ϵ_r is larger than unity for all material media, the cutoff frequency decreases when a dielectric is introduced.

2.7.11 Application: Modification of the Guide Boundaries

The perturbed waveguide considered has a cross section slightly smaller than the original one (for instance, a rectangular waveguide having slightly rounded corners). The contour C follows the original section, the contour C' the modified one (Fig. 2.28).

It is here the line integral within (2.204) that must be considered. This term vanishes on the contour C', due to the boundary condition on the cep wall. On the other hand it is not equal to zero on the part of the contour C which does not coincide with C'. Integrating around the perturbation ΔS, one obtains, applying the two-dimensional Poynting relation [214]:

$$\oint_C (\underline{E} \times \underline{H}_0^*) \cdot \boldsymbol{n}\, \mathrm{d}l = \oint_C (\underline{E} \times \underline{H}_0^*) \cdot \boldsymbol{n}\, \mathrm{d}l - \oint_{C'} (\underline{E} \times \underline{H}_0^*) \cdot \boldsymbol{n}\, \mathrm{d}l$$

$$\cong \mathrm{j}\omega_\mathrm{c} \int_{\Delta S} (\epsilon_0 |\underline{E}_0|^2 - \mu_0 |\underline{H}_0|^2) \mathrm{d}A \quad \text{VA/m} \qquad (2.206)$$

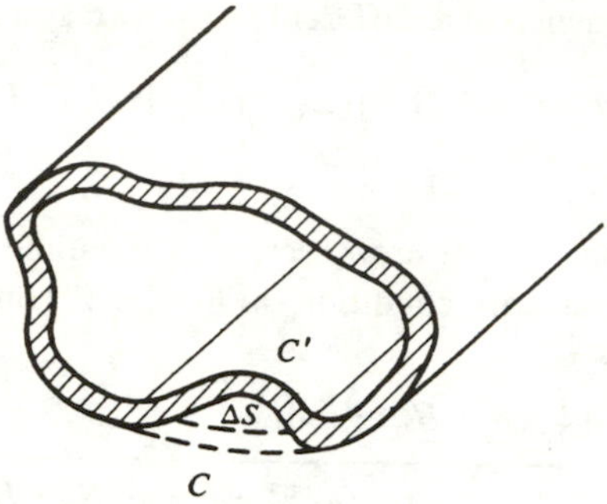

Fig. 2.28 Alteration of the waveguide cross section.

Introducing this relation into (2.204) one obtains, in the absence of other perturbations:

$$\omega_\mathrm{c} - \omega_\mathrm{co} \cong -\omega_\mathrm{c} \frac{\int_{\Delta S} (\epsilon_0 |\underline{E}_0|^2 - \mu_0 |\underline{H}_0|^2)\, \mathrm{d}A}{\int_S (\epsilon_0 |\underline{E}_0|^2 + \mu_0 |\underline{H}_0|^2)\, \mathrm{d}A} \quad \text{rad/s} \qquad \textbf{(2.207)}$$

The cutoff frequency decreases if the cross section is reduced in a region where the electric energy is predominant. It increases when the section is reduced at a location of larger magnetic energy. Variations in the opposite direction occur when the waveguide cross section is enlarged.

2.8 INHOMOGENEOUS LINES AND WAVEGUIDES

2.8.1 Definitions

The above name designates a waveguide or a transmission line, uniform along the direction of propagation (Section 2.1), but involving a transverse variation of the material properties of the propagating medium or media. The electromagnetic fields of a propagating wave extend over a region possessing properties which vary across the transverse plane.

Most often, the structure is composed of at least two homogeneous materials (air, dielectric). In this case, the fields within each one of the two media are first determined, and then the continuity conditions for the fields across the respective borders are applied (§ 1.4.3). In particular, the longitudinal dependence of the fields, and thus the propagation factor $\underline{\gamma}$, must be the same everywhere. While this resolution scheme is straightforward, its practical application often presents serious difficulties. Most often, only numerical computer techniques

are able to yield accurate results. For many practical applications, however, simple approximate relations prove to be sufficient.
Other structures possess a continuous transverse variation of the permittivity $\underline{\epsilon}_r(\boldsymbol{r}_t)$ or of the permeability $\underline{\mu}_r(\boldsymbol{r}_t)$. It becomes necessary then to solve Maxwell's equations with space-varying coefficients. When this problem cannot be solved directly, however, it can always be reduced to the previous one, replacing the continuous variation by a staircase approximation.

2.8.2 Absence of TEM Modes on Inhomogeneous Lines

The propagation factor $\underline{\gamma}$ of a TEM wave depends only upon frequency and material properties of the medium in which it propagates (2.26). As a result, in an inhomogeneous line, this term is a function of position: this is inconsistent with the continuity of the fields, so that no TEM mode can propagate in an inhomogeneous line. This fact is quite general, it is true for any inhomogeneous line.

2.8.3 Definition: Quasi-TEM Mode

At first sight, it may come as a surprise to realize that an insulated telephone cable, which is an inhomogeneous line, cannot transmit a TEM mode. As a matter of fact, the dominant mode of many inhomogeneous two-conductor lines still behaves like a TEM mode at low frequencies. This is true in particular for the insulated two-wire pair, the microstrip and the loaded coaxial line. Their dominant mode is therefore called *quasi-TEM.* The longitudinal components of the fields, while non-zero, have amplitudes considerably smaller than the transverse ones, and thus produce little effect. The propagation of this mode can therefore be determined approximately from electrostatics. The inhomogeneous structure is replaced by an equivalent homogeneous one, which is studied as in Section 2.6. The effective permittivity of the homogeneous equivalent medium is determined by the electrostatic study of the structure [30]. This simplified model is only valid at frequencies below an upper boundary, which depends upon the geometry and dimensions of the line.

2.8.4 Possibility of Existence of a TM Mode

An inhomogeneous line, uniform along the direction of propagation, is considered (Sec. 2.1). At an arbitrary point on the interface between the two media, a local coordinate system is defined: the y-axis normal to the surface, the x-axis tangent to it and perpendicular to the direction of propagation z (Fig. 2.29). When the tangential component of the transverse electric field exists, it must be continuous across the separating surface (1.6), so that:

$$\underline{E}_{x1} = \underline{E}_{x2} \qquad \text{V/m} \qquad (2.208)$$

The subscripts 1 and 2 refer to the media involved.

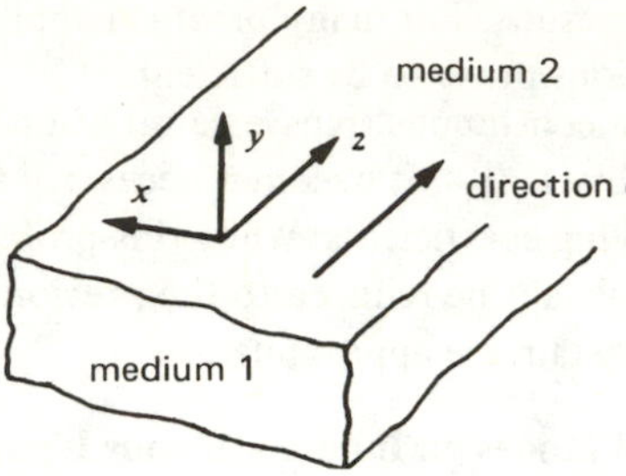

Fig. 2.29 Interface and definition of a local rectangular coordinate system in an inhomogeneous line.

Maxwell's equation (1.3) is used here for a forward wave (2.5), the fields of which have a longitudinal dependence in $\exp(-\underline{\gamma} z)$:

$$\underline{E}_{xi} = \frac{1}{(j\omega\underline{\epsilon}_i + \sigma_i)} \left(\frac{\partial \underline{H}_{zi}}{\partial y} - \frac{\partial \underline{H}_{yi}}{\partial z} \right) = \frac{1}{(j\omega\underline{\epsilon}_i + \sigma_i)} \left(\frac{\partial \underline{H}_{zi}}{\partial y} + \underline{\gamma}\underline{H}_{yi} \right) \quad \text{V/m} \qquad (2.209)$$

Introducing this expression into both sides of (2.208), and then taking into account the continuity of the normal magnetic field component [$\underline{H}_{y1} = \underline{H}_{y2} = \underline{H}_y$, (1.11)] one obtains:

$$\frac{1}{j\omega\underline{\epsilon}_1 + \sigma_1} \frac{\partial \underline{H}_{z1}}{\partial y} - \frac{1}{j\omega\underline{\epsilon}_2 + \sigma_2} \frac{\partial \underline{H}_{z2}}{\partial y} = -\left(\frac{1}{j\omega\underline{\epsilon}_1 + \sigma_1} - \frac{1}{j\omega\underline{\epsilon}_2 + \sigma_2} \right) \underline{\gamma}\underline{H}_y \quad \text{V/m} \qquad (2.210)$$

When $\underline{H}_y$ exists, the longitudinal magnetic component $\underline{H}_z$ must also be nonzero: the mode is not TM.

Consequently, an inhomogeneous line or waveguide can support a TM mode ($\underline{H}_z = 0$) only if:

1. the transverse electric field is normal to the interface,
2. the transverse magnetic field is tangential at all points to this surface.

2.8.5 Possibility of Existence of a TE Mode

When the tangential component of the transverse magnetic field is non-zero, it must be continuous across the interface:

$$\underline{H}_{x1} = \underline{H}_{x2} \quad \text{A/m} \qquad (2.211)$$

Proceeding in the same way as in the previous demonstration, equation (1.2) yields here:

$$\underline{H}_{xi} = \frac{-1}{j\omega\mu_0}\left(\frac{\partial \underline{E}_{zi}}{\partial y} + \underline{\gamma}\,\underline{E}_{yi}\right) \qquad \text{A/m} \qquad (2.212)$$

Introducing the continuity relation for $\underline{\epsilon E}_y$, one finds, assuming that there are no surface charges on the interface (1.10):

$$\frac{\partial \underline{E}_{z1}}{\partial y} - \frac{\partial \underline{E}_{z2}}{\partial y} = -\underline{\gamma}\,\underline{E}_{y1}\left(\frac{\underline{\epsilon}_{r1}}{\underline{\epsilon}_{r2}} - 1\right) \qquad \text{V/m}^2 \qquad (2.213)$$

When $\underline{E}_{y1}$ exists, the same is true for $\underline{E}_z$, so that the mode is not TE. Consequently, an inhomogeneous waveguide or transmission line can support a TE mode ($\underline{E}_z = 0$) only when

1. the transverse magnetic field is normal everywhere to the interface,
2. the transverse electric field is tangential everywhere to this surface.

2.8.6 Hybrid Modes

When the requirements of the two previous paragraphs are not met, the modes of propagation are hybrid ($\underline{H}_z \neq 0, \underline{E}_z \neq 0$).

2.9 SIMPLE CASE: DIELECTRIC PLATE

2.9.1 Description of the Geometry

A lossless dielectric plate, having a width $2d$ and a relative permittivity ϵ_r, is located parallel to the $y0z$ plane (Fig. 2 30). On both sides of the plate, *i.e.,* for $|x| > d$ the propagating medium is air. Since interfaces are parallel planes, the problem can be solved analytically.

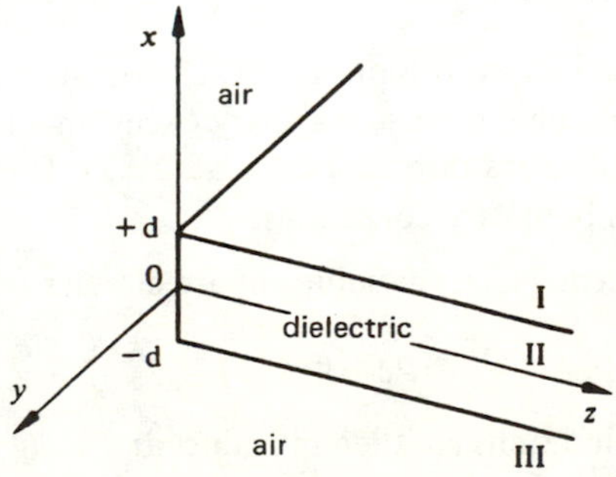

Fig. 2.30 Dielectric slab.

For the fields to be continuous across the boundaries, they must at least have the same phaseshift per unit length β in the three regions. From (2.32), the transverse wave numbers must then be different. Within the dielectric:

$$\underline{p}_{\mathrm{d}} = \sqrt{\omega^2 \epsilon_0 \mu_0 \epsilon_{\mathrm{r}} - \beta^2} \qquad \mathrm{m}^{-1} \qquad (2.214)$$

In both air regions

$$\underline{p}_{\mathrm{a}} = \sqrt{\omega^2 \epsilon_0 \mu_0 - \beta^2} \qquad \mathrm{m}^{-1} \qquad (2.215)$$

Subscripts a and d denote, respectively, air and dielectric.

The fields of a wave ***guided*** by the plate must decrease as one moves away from it: the only adequate solutions for this problem are those whose fields vanish when $|x|$ goes to infinity.

■ 2.9.2 TE Modes

Let us consider first a TE wave propagating towards positive z (forward wave), assuming that the field components are independent of the variable y. The longitudinal magnetic component satisfies the two-dimensional Helmholtz's equation (2.30), which becomes here:

$$\frac{\mathrm{d}^2 \underline{H}_z}{\mathrm{d}x^2} + \underline{p}^2_{\mathrm{a,d}} \underline{H}_z = 0 \qquad \mathrm{A/m}^3 \qquad (2.216)$$

When $\underline{p}^2$ is real positive, the solutions to this equation are circular functions (as in rectangular waveguide). Exponentials are obtained when $\underline{p}^2$ is real negative. Within air (regions I and III), the fields must decay to zero at infinity, a condition which is only satisfied by decreasing exponentials, and therefore:

$$\underline{H}_{z\mathrm{I}} = \underline{A} \exp(-|\underline{p}_{\mathrm{a}}|x - \underline{\gamma} z) \qquad \mathrm{A/m} \qquad (2.217)$$

$$\underline{H}_{z\mathrm{III}} = \underline{D} \exp(|\underline{p}_{\mathrm{a}}|x - \underline{\gamma} z) \qquad \mathrm{A/m} \qquad (2.218)$$

Introducing these two expressions into (2.216), one sees that $\underline{p}_{\mathrm{a}} = -\mathrm{j}|\underline{p}_{\mathrm{a}}|$: the transverse wave number is pure imaginary (while in a homogeneous waveguide it is always real, as demonstrated in Section 2.2.23). The $\underline{A}$ and $\underline{D}$ constants are determined from boundary conditions.

Within the dielectric plate, the solutions are circular functions:

$$\underline{H}_{z\mathrm{II}} = [\underline{B} \sin p_{\mathrm{d}} x + \underline{C} \cos p_{\mathrm{d}} x] \exp(-\underline{\gamma} z) \qquad \mathrm{A/m} \qquad (2.219)$$

The boundary conditions are then introduced:

1. continuity of the tangential magnetic field

when $x = d$ $\quad \underline{A} \exp(-|\underline{p}_a| d) = \underline{B} \sin(p_d d) + \underline{C} \cos(p_d d)$ A/m (2.220)

when $x = -d$ $\quad \underline{D} \exp(-|\underline{p}_a| d) = -\underline{B} \sin(p_d d) + \underline{C} \cos(p_d d)$ A/m (2.221)

2. continuity of the tangential electric field, obtained from (2.35):

$$\underline{E}_t = \frac{j\omega\mu}{\underline{p}^2} \boldsymbol{e}_z \times \boldsymbol{e}_x \frac{\partial \underline{H}_z}{\partial x} = \boldsymbol{e}_y \frac{j\omega\mu}{\underline{p}^2} \frac{\partial \underline{H}_z}{\partial x} \qquad \text{V/m} \tag{2.222}$$

The continuity of $\frac{1}{\underline{p}^2} \frac{\partial \underline{H}_z}{\partial x}$ must be ensured, yielding respectively at $x = \pm d$:

$$\frac{1}{|\underline{p}_a|} \underline{A} \exp(-|\underline{p}_a| d) = \frac{1}{p_d} (\underline{B} \cos p_d d - \underline{C} \sin p_d d) \qquad \text{A} \tag{2.223}$$

$$\frac{-1}{|\underline{p}_a|} \underline{D} \exp(-|\underline{p}_a| d) = \frac{1}{p_d} (\underline{B} \cos p_d d + \underline{C} \sin p_d d) \qquad \text{A} \tag{2.224}$$

The transverse component of the magnetic field $\underline{H}_x$ is then also continuous. The four constants $\underline{A}$, $\underline{B}$, $\underline{C}$ and $\underline{D}$ are eliminated by means of the following procedure: sums and differences of the two pairs of above equations are taken. Dividing then the two expressions involving $\underline{C}$, and then the two involving $\underline{B}$, one obtains in the two cases:

$$|\underline{p}_a| = p_d \tan(p_d d) \qquad \text{m}^{-1} \tag{2.225}$$

$$|\underline{p}_a| = -p_d \operatorname{cotan}(p_d d) \qquad \text{m}^{-1} \tag{2.226}$$

Equation (2.225) corresponds to modes symmetrical in $\underline{E}_y$, for which:

$$\underline{B} = 0 \quad \text{and} \quad \underline{A} = \underline{D} = \underline{C} \cos(p_d d) \exp(|\underline{p}_a| d) \qquad \text{A/m} \tag{2.227}$$

Equation (2.226) is the one for antisymmetrical modes in $\underline{E}_y$, having:

$$\underline{C} = 0 \quad \text{and} \quad \underline{A} = -\underline{D} = \underline{B} \sin(p_d d) \exp(|\underline{p}_a| d) \qquad \text{A/m} \tag{2.228}$$

The frequency-dependence of the propagation factor is determined by expressing $|\underline{p}_a|$ in terms of p_d with the aid of (2.214) and (2.215), and then introducing the value obtained into (2.225) and (2.226), yielding in both cases:

$$p_d \tan(p_d d) = \sqrt{\Delta p^2 - p_d^2} \qquad \text{m}^{-1} \qquad (2.229)$$

$$-p_d \operatorname{cotan}(p_d d) = \sqrt{\Delta p^2 - p_d^2} \qquad \text{m}^{-1} \qquad (2.230)$$

with

$$\Delta p^2 = p_d^2 - \underline{p}_a^2 = p_d^2 + |\underline{p}_a|^2 = \omega^2 \epsilon_0 \mu_0 (\epsilon_r - 1) \qquad \text{m}^{-1} \qquad (2.231)$$

It is not possible to write the dispersion relation in explicit form. In contrast with homogeneous waveguides, one does not obtain here a simple function of the form $\beta(\omega)$. For each mode, a transcendental equation must be solved. The resolution procedure can be graphical: in Figure 2.31, the two sides of equations (2.229) and (2.230) are drawn. The solutions are the intersections of the rising curves, representing the left-hand side of the equations, with the circle, which corresponds to the right-hand side, the radius increasing with frequency. As can be seen, there is always at least one intersection, corresponding to the first symmetrical mode (TE_{10}); this mode has no cutoff. However, the graphical method is tedious, and numerical computation techniques are most often utilized nowadays. The dispersion diagram for the first TE modes on a plate of thickness $2d$ having a relative permittivity $\epsilon_r = 9$ is presented in Figure 2.32. As frequency increases, the curves asymptotically tend towards those of the TE_{m0} modes in a rectangular waveguide of same width, $2d$, filled with the same dielectric.

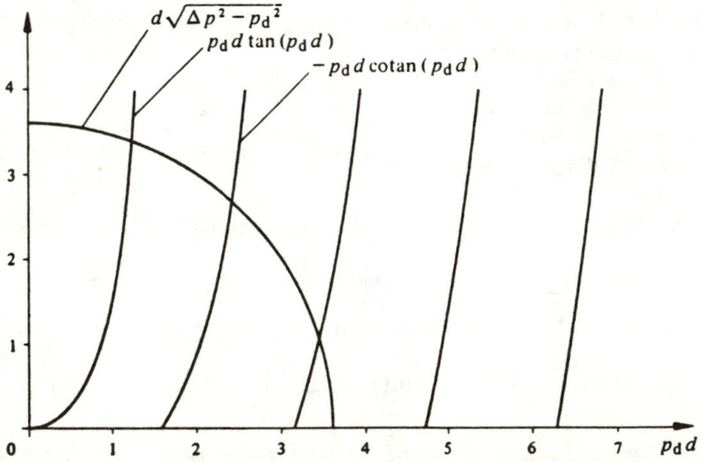

Fig. 2.31 Graphical display of the relations (2.229) and (2.230).

2.9.3 Field Distribution for the First TE Mode

The amplitude of the $\underline{E}_y$ field component of the first TE mode on a dielectric plate is sketched in Figure 2.33 for various frequencies, corresponding to a power transfer of 1 W per unit width. At low frequencies, the field decays very slowly away from the plate. As the frequency increases, the field tends to concentrate within the dielectric. This is a general fact, which is verified in all inhomogeneous guiding structures.

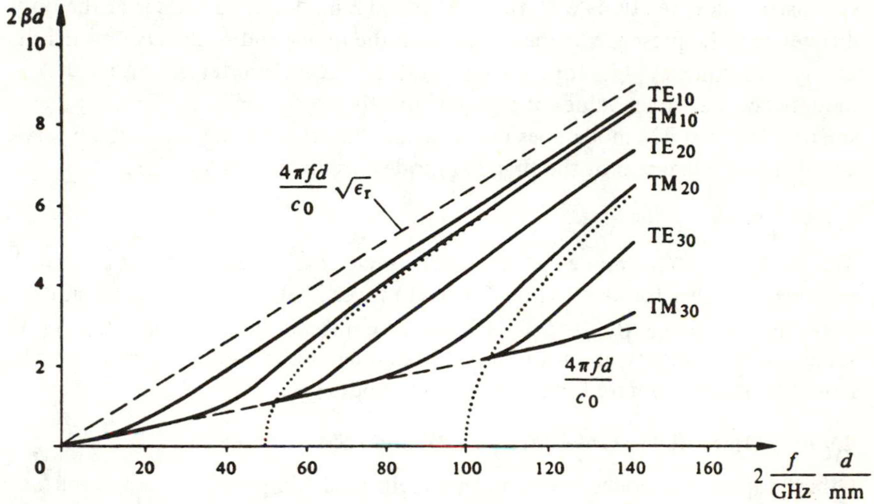

Fig. 2.32 Dispersion diagram for a dielectric slab of relative permittivity $\epsilon_r = 9$. The dotted lines correspond to a metallic waveguide.

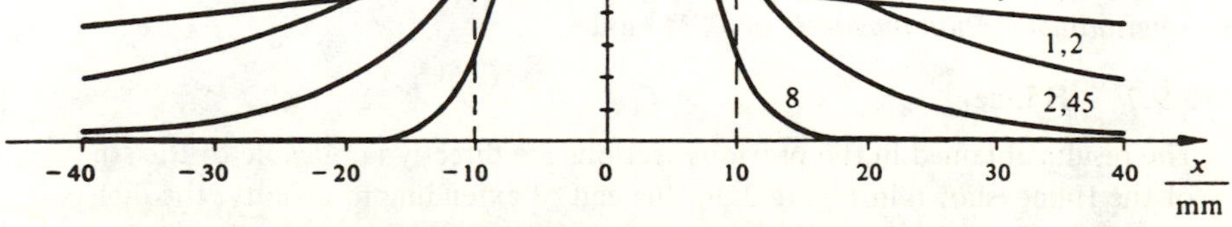

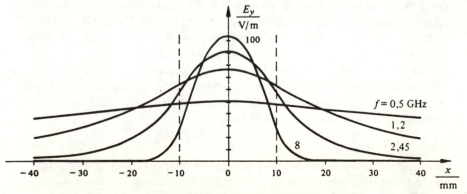

Fig. 2.33 Distribution of the electric field in a dielectric slab 20 mm thick of relative permittivity $\epsilon_r = 9$.

2.9.4 TM Modes

The study of y-independent TM modes is entirely similar to that of TE modes. In this case, there is a longitudinal component of the electric field $\underline{E}_z$, and transverse components $\underline{E}_x$ and $\underline{H}_y$. The dispersion equations then become:

$$p_d \tan(p_d d) = \epsilon_r \sqrt{\Delta p^2 - p_d^2} \qquad \text{m}^{-1} \qquad (2.232)$$

$$-p_d \operatorname{cotan}(p_d d) = \epsilon_r \sqrt{\Delta p^2 - p_d^2} \qquad \text{m}^{-1} \qquad (2.233)$$

Comparing these relations with those for the TE modes, it appears that the only difference is the presence of the ϵ_r factor in the right-hand term. This means that, at a given frequency, *i.e.*, for a constant value of Δp, the intersection for TM modes occurs at larger values of p_d, and thus the corresponding values for β are smaller. The first TM mode does not have any cutoff, just like the first TE mode. The dispersion diagram of the first TM modes is shown in Figure 2.32.

2.9.5 Note

The TE_{mo} and TM_{mo} modes both have the same cutoff frequency for a given m, which vanishes for $m = 1$ (Fig. 2.32). However, their respective dispersion curves do not merge ($\beta_{mo}^{\text{TE}} \neq \beta_{mo}^{\text{TM}}$). In an inhomogeneous line, modes having the same cutoff frequency are not necessarily degenerate: the situation is different from the one encountered in homogeneous guides (§ 2.2.36).

2.9.6 Other Higher Order Modes: LSE and LSM

Other propagating modes having y-dependent field components can be arbitrarily composed by superposition of two modes of the same family (TE or TM) propagating along directions within the $y0z$ plane at angles $+\theta$ and $-\theta$ to the z-axis. The study is similar to the one mentioned in Section 2.3.13, considering the dominant mode in a rectangular waveguide as the combination of two plane waves. In this manner, an infinity of modes can be formed. Containing at the same time $\underline{E}_z$ and $\underline{H}_z$ components, these modes are neither TE nor TM, but hybrid. However, a mode formed by the superposition of two TE modes has no $\underline{E}_x$ component: as the electric field is within the longitudinal section $y0z$, this mode is called the *longitudinal section electric* or *LSE* mode. Similarly, the mode formed with two TM modes has no $\underline{H}_x$ component, and is therefore called the *longitudinal section magnetic* or *LSM* mode.

□ 2.9.7 H Line

The results obtained in the previous sections are directly applicable to the study of the H-line, shown in Figure 2.34. Instead of extending to infinity, the dielectric plate is bounded by two metallic conductors (pec) at $y = 0$ and $y = b$ [15]. These two planes introduce additional conditions to the fields. The TE modes, considered in Section 2.9.2, have only an $\underline{E}_y$ component of the electric field, which is perpendicular to the conducting planes, also satisfying the additional boundary conditions. All these modes can propagate without any modification, with respect to Section 2.9.2.

The TM modes of Section 2.9.4 have, on the other hand, electric field components $\underline{E}_x$ and $\underline{E}_z$, which are both tangential to the conductors. They do not meet the boundary conditions on the metal planes, so that TM modes cannot propagate in the structure of Figure 2.34.

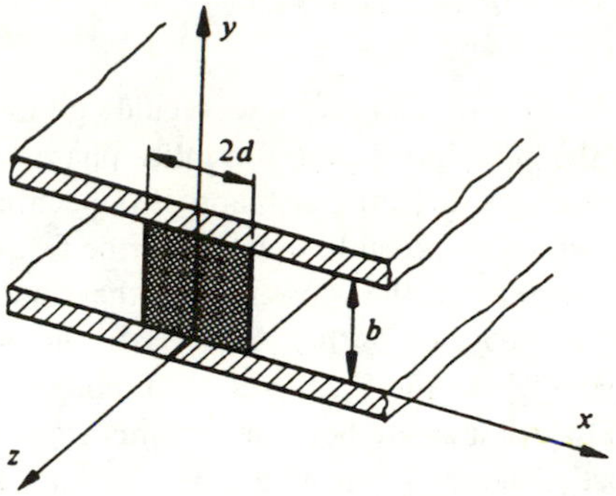

Fig. 2.34 The H-line.

The LSE and LSM modes defined in Section 2.9.6 can propagate whenever their y-dependencies are compatible with the spacing b between the two plates. Fields must therefore have a sine wave dependence, as in rectangular waveguides:

$$\left\{\begin{matrix}\sin\\ \cos\end{matrix}\right\}(n\pi y/b) \qquad 1 \qquad (2.234)$$

where n is a positive integer. For the LSE and LSM modes, one obtains the dispersion relations by solving, respectively, (2.229) and (2.230) or (2.232) and (2.233), and making use of the parameter p_{d} which is given by:

$$p_{\mathrm{d}} = \sqrt{\omega^2 \epsilon_0 \mu_0 \epsilon_{\mathrm{r}} - \beta^2 - (n\pi/b)^2} \qquad \mathrm{m}^{-1} \qquad (2.235)$$

The dominant mode in the structure is the first TE mode.

□ 2.9.8 Loaded Rectangular Waveguide

The study of propagation in a rectangular waveguide containing a dielectric slab (Fig. 2.35) follows the same lines as the previous problem (§ 2.9.7). The lateral boundaries, previously located at infinity, are here brought closer, to $x = 0$ and $x = a$, where two metallic planes (cep) are located. The tangential electric field components ($\underline{E}_z$ and $\underline{E}_y$) must vanish in both planes. This requirement cannot be satisfied by the decreasing exponentials of the previous problems. These are replaced by hyperbolic or circular functions, depending on frequency. The dominant mode (quasi-TE_{10}) must satisfy the relations [16]:

$$\tan p_{\mathrm{d}} h = \frac{p_{\mathrm{a}} p_{\mathrm{d}} (\tan p_{\mathrm{a}} d_1 + \tan p_{\mathrm{a}} d_2)}{p_{\mathrm{d}}^2 (\tan p_{\mathrm{a}} d_1 \ \tan p_{\mathrm{a}} d_2) - p_{\mathrm{a}}^2} \qquad \text{if } p_{\mathrm{a}}^2 > 0 \qquad 1 \tag{2.236}$$

$$\tan p_{\mathrm{d}} h = \frac{|\underline{p}_{\mathrm{a}}| p_{\mathrm{d}} (\tanh |\underline{p}_{\mathrm{a}}| d_1 + \tanh |\underline{p}_{\mathrm{a}}| d_2)}{p_{\mathrm{d}}^2 (\tanh |\underline{p}_{\mathrm{a}}| d_1 \ \tanh |\underline{p}_{\mathrm{a}}| d_2) - |\underline{p}_{\mathrm{a}}^2|} \qquad \text{if } p_{\mathrm{a}}^2 < 0 \qquad 1 \tag{2.237}$$

The insertion of a dielectric slab into a waveguide permits the modification of the electrical length; this principle is used to realize phaseshifters (§ 6.3.11). It was claimed that by placing a dielectric slab across the center of the waveguide (Fig. 2.35, $d_1 = d_2$), it would be possible to increase the frequency bandwidth and the power handling capability of the waveguide at the same time [17]. This claim was shown to be erroneous on both counts, based on an incomplete study of the problem: when higher order LSE and LSM modes are taken into account, the resulting bandwidth increase becomes unimportant [18]. As for the power increase, it is mostly theoretical: it would be necessary to ensure a perfect mechanical fit between the dielectric material and the metal at both ends of the slab. Any gap between the two actually reduces the power handling capability well below the one of the empty waveguide [19].

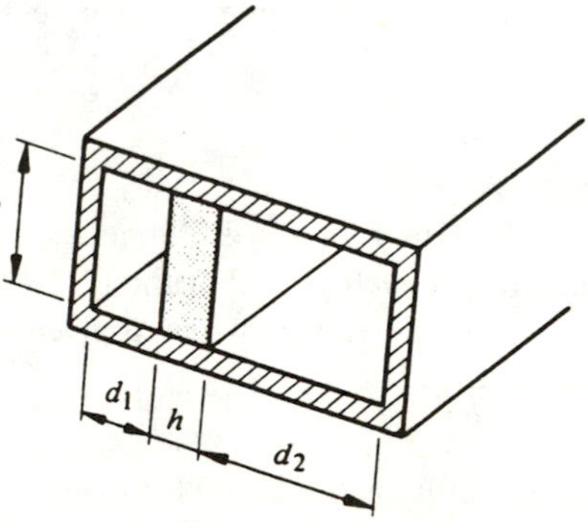

Fig. 2.35 Rectangular waveguide loaded with a dielectric slab.

2.10 CIRCULAR DIELECTRIC WAVEGUIDE, OPTICAL FIBER

2.10.1 Definition of the Geometry, Core, Cladding

An inner cylinder of lossless dielectric, named the *core*, is used to guide the waves. It has a circular section of radius a, and its permittivity $\epsilon_1(\rho)$ is larger than that of the surrounding medium. It is embedded in a homogeneous and isotropic material, the *cladding*, of permittivity ϵ_2. While the cladding itself is also bounded further along in the transverse direction, the fields of a guided mode decay sufficiently fast for the effect of the outer boundary to be negligible. The theoretical study therefore considers a cladding extending radially to infinity. Figure 2.36 represents the general structure of the optical fiber.

The main interest of the structure is that it guides signals without the need for metal conductors. Optical fibers are not affected by radioelectrical perturbations, and can be used to establish links in the vicinity of high voltages (§ 8.2.15).

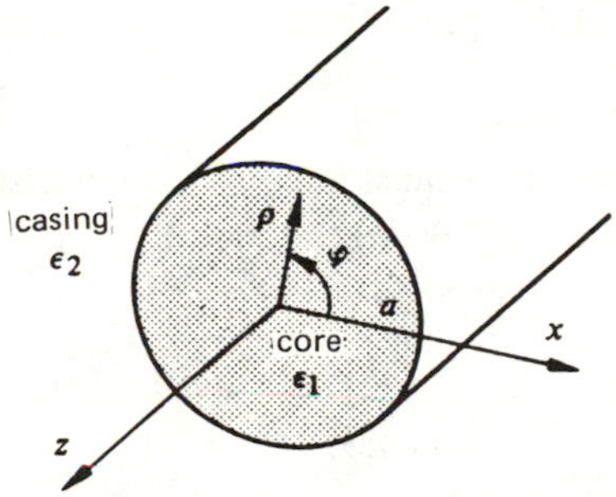

Fig. 2.36 Circular dielectric waveguide, optical fiber with step index.

2.10.2 Step Index Optical Fiber

The first kind of fiber to be manufactured has a homogeneous core (ϵ_1 is a constant). Such fibers are called *step index fibers.*

2.10.3 Equations to be Solved

Since the structure is circularly symmetrical, the differential equation is the same as in circular metallic waveguide (2.135). Two equations are actually obtained, one for medium 1 (core), the other for medium 2 (cladding) on the outside and extending to infinity (§ 2.10.1). The transverse wave numbers p_i are defined in the same way as for the dielectric plate (§ 2.9.1). Normalized parameters u, v and w are more commonly utilized in the literature. They are defined as:

$$u = ap_1 = a\sqrt{\omega^2 \epsilon_1 \mu_0 - \beta^2} \qquad 1 \qquad (2.238)$$

$$w = -\mathrm{j}a\underline{p}_2 = a\sqrt{\beta^2 - \omega^2 \epsilon_2 \mu_0} \qquad 1 \qquad (2.239)$$

They are related through equation:

$$v^2 = u^2 + w^2 = a^2(p_1^2 - \underline{p}_2^2) = a^2\omega^2\epsilon_1\mu_0 2\Delta \qquad 1 \qquad (2.240)$$

The relative difference Δ between the two permittivities is a significant parameter for the guiding process:

$$\Delta = (\epsilon_1 - \epsilon_2)/2\epsilon_1 \qquad 1 \qquad (2.241)$$

Within the core, the field components must remain bounded on the axis, a condition which specifies the same dependence as in a hollow metallic waveguide:

$$\underline{\theta}_1(\rho,\varphi) = \underline{B}_1 \mathrm{J}_m(u\rho/a) \begin{Bmatrix} \cos m\varphi \\ \sin m\varphi \end{Bmatrix} \qquad 1 \qquad (2.242)$$

where m is either zero or a positive integer.

Within the cladding, the fields must fall down and vanish at infinity, a condition which is met by the modified Bessel function $\mathrm{K}_m(w\rho/a)$ (§ 10.4.1). In this way one obtains:

$$\underline{\theta}_2(\rho,\varphi) = \underline{\beta}_2 \mathrm{K}_m(w\rho/a) \begin{Bmatrix} \cos m\varphi \\ \sin m\varphi \end{Bmatrix} \qquad 1 \qquad (2.243)$$

The field components tangential to the boundary between the two media at $\rho = a$ are continuous. Considering the solutions for the hollow metallic waveguide (Sec. 2.4), it is apparent that both the electric and the magnetic fields have radial and azimuthal components, except when $m = 0$. From Sections 2.8.4 and 2.8.5, it follows that only when m vanishes can TE and TM modes exist. In all other situations, modes of propagation are ***hybrid*** ($\underline{E}_z \neq 0, \underline{H}_z \neq 0$, § 2.1.9).

2.10.4 Definition: Numerical Aperture *NA*

By analogy with other optical systems, the *numerical aperture* of an optical fiber is defined by the sine of the largest incident angle which an impinging ray can have and be guided. The ray reaches the extremity of the fiber, assumed to be cut in a transverse plane. The numerical aperture *NA* is then given by:

$$NA = \sqrt{(\epsilon_1 - \epsilon_2)/\epsilon_0} = \sqrt{2\Delta\epsilon_1/\epsilon_0} \qquad 1 \qquad (2.244)$$

2.10.5 TE_{0n} Modes

Letting $m = 0$ in (2.242) and (2.243), the transverse potential $\underline{\psi}$ is expressed within both media (core and cladding):

$$\underline{\psi}_1(\rho,\varphi) = \underline{B}_1 \mathrm{J}_0(u\rho/a) \qquad \rho < a \qquad 1 \tag{2.245}$$

$$\underline{\psi}_2(\rho,\varphi) = \underline{B}_2 \mathrm{K}_0(w\rho/a) \qquad \rho > a \qquad 1 \tag{2.246}$$

The field components $\underline{E}_\phi$ and $\underline{H}_z$ exist and must be continuous at $\rho = a$. From (2.42), (2.52), and (2.58), it results that $\partial\underline{\psi}/\partial\rho$ and $p^2\,\underline{\psi}$ are both continuous, requiring that

$$u^2 \underline{B}_1 \mathrm{J}_0(u) = -w^2 \underline{B}_2 \mathrm{K}_0(w) \qquad 1 \tag{2.247}$$

$$u\underline{B}_1 \mathrm{J}_0'(u) = w\underline{B}_2 \mathrm{K}_0'(w) \qquad 1 \tag{2.248}$$

Combining the two relations and noting that $\mathrm{J}_0'(x) = -\mathrm{J}_1(x)$ and $\mathrm{K}_0'(x) = -\mathrm{K}_1(x)$ (§ 10.3.5 and 10.4.5), the dispersion relation for the TE_{0n} modes is obtained:

$$\frac{\mathrm{J}_1(u)}{u\,\mathrm{J}_0(u)} = -\frac{\mathrm{K}_1(w)}{w\,\mathrm{K}_0(w)} \qquad 1 \tag{2.249}$$

The dispersion diagrams are then determined by simultaneously solving (2.249) and (2.240), either graphically or numerically. Just as for the dielectric plate of Section 2.9, no explicit relation of the form $\beta(\omega)$ can be formulated here, either.

2.10.6 TM_{0n} Modes

The transverse potential $\underline{\phi}$ of the TM_{0n} mode has the same radial dependence as the $\underline{\psi}$ potential of the previous paragraphs (2.245) and (2.246). Here, the field components $\underline{H}_\phi$ and $\underline{E}_z$ are continuous at $\rho = a$, requiring the continuity of $\partial\underline{\phi}/\partial\rho$ and of $\underline{p}^2\underline{\phi}/\underline{\epsilon}$. Proceeding in the same manner as in Section 2.10.5, the dispersion relation for the TM_{0n} modes is obtained

$$\frac{\epsilon_1 \mathrm{J}_1(u)}{u\,\mathrm{J}_0(u)} = -\frac{\epsilon_2 \mathrm{K}_1(w)}{w\,\mathrm{K}_0(w)} \qquad 1 \tag{2.250}$$

■ 2.10.7 Hybrid Modes

When $m \neq 0$, the boundary conditions on the core's side impose the simultaneous existence of longitudinal components of the two fields. A hybrid mode is thus formed by the sum of one TE and one TM mode, which are ***coupled by the continuity requirements for the fields*** at $\rho = a$. The potentials have the general dependencies given in (2.242) and (2.243).

For the longitudinal components $\underline{E}_z$ and $\underline{H}_z$ to be continuous, it is necessary to have here, as in the two previous paragraphs:

$$u^2 \underline{B}_1^{\mathrm{TE}} \mathrm{J}_m(u) = -w^2 \underline{B}_2^{\mathrm{TE}} \mathrm{K}_m(w) \tag{2.251}$$

$$u^2/\epsilon_1 \underline{B}_1^{\mathrm{TM}} \mathrm{J}_m(u) = -w^2/\epsilon_2 \underline{B}_2^{\mathrm{TM}} \mathrm{K}_m(w) \tag{2.252}$$

The continuity of the azimuthal components $\underline{E}_\phi$ and $\underline{H}_\phi$ provides two requirements involving ***at the same time*** the two potentials $\underline{\psi}$ and $\underline{\phi}$:

$$\left.\left(\frac{\partial\phi_1}{\partial\varphi} - \rho\,\frac{\partial\psi_1}{\partial\rho}\right)\right|_a = \left.\left(\frac{\partial\phi_2}{\partial\varphi} - \rho\,\frac{\partial\psi_2}{\partial\rho}\right)\right|_a$$
$$= \pm\, m\underline{B}_1^{\mathrm{TM}} \mathrm{J}_m(u) + u\underline{B}_1^{\mathrm{TE}} \mathrm{J}'_m(u)$$
$$= \pm\, m\underline{B}_2^{\mathrm{TM}} \mathrm{K}_m(w) + w\underline{B}_2^{\mathrm{TE}} \mathrm{K}'_m(w) \tag{2.253}$$

$$\left.\left(\rho\,\frac{\partial\phi_1}{\partial\rho} + \frac{\partial\psi_1}{\partial\varphi}\right)\right|_a = \left.\left(\rho\,\frac{\partial\phi_2}{\partial\rho} + \frac{\partial\psi_2}{\partial\varphi}\right)\right|_a$$
$$= u\underline{B}_1^{\mathrm{TM}} \mathrm{J}'_m(u) \pm m\underline{B}_1^{\mathrm{TE}} \mathrm{J}_m(u)$$
$$= wB_2^{\mathrm{TM}} \mathrm{K}'_m(w) \pm mB_2^{\mathrm{TE}} \mathrm{K}_m(w) \tag{2.254}$$

with signs + and − corresponding to the two possible azimuthal dependencies, in sine and in cosine, of equations (2.242) and (2.243). The four continuity equations for the field components can be written in matrix form:

$$\begin{pmatrix} 0 & u^2 \mathrm{J}_m(u) & 0 & w^2 \mathrm{K}_m(w) \\ u^2 \mathrm{J}_m(u)/\epsilon_1 & 0 & w^2 \mathrm{K}_m(w)/\epsilon_2 & 0 \\ \pm\, m\mathrm{J}_m(u) & u\mathrm{J}'_m(u) & \mp\, m\mathrm{K}_m(w) & -w\mathrm{K}'_m(w) \\ u\mathrm{J}'_m(u) & \pm\, m\mathrm{J}_m(u) & -w\mathrm{K}'_m(w) & \mp\, m\mathrm{K}_m(w) \end{pmatrix} \begin{pmatrix} B_1^{\mathrm{TM}} \\ B_1^{\mathrm{TE}} \\ B_2^{\mathrm{TM}} \\ B_2^{\mathrm{TE}} \end{pmatrix} = 0 \tag{2.255}$$

For this set of equations to admit a non-zero solution, the determinant must vanish. This provides the dispersion relation for hybrid modes:

$$\left(\frac{\mathrm{J}'_m(u)}{u\mathrm{J}_m(u)} + \frac{\mathrm{K}'_m(w)}{w\mathrm{K}_m(w)}\right)\left(\frac{\epsilon_1 \mathrm{J}'_m(u)}{\epsilon_2 u\mathrm{J}_m(u)} + \frac{\mathrm{K}'_m(w)}{w\mathrm{K}_m(w)}\right)$$
$$= m^2\left(\frac{1}{u^2} + \frac{1}{w^2}\right)\left(\frac{\epsilon_1}{\epsilon_2}\,\frac{1}{u^2} + \frac{1}{w^2}\right) \tag{\textbf{2.256}}$$

This equation actually possesses two sets of solutions, corresponding respectively to HE and EH modes. When $m = 0$, one actually finds within (2.256) the dispersion relations of the TE and TM modes.

2.10.8 Weak Guidance Approximation, HE and EH Modes

For technological reasons resulting from the diffusion process generally utilized to manufacture optical fibers, the relative difference of permittivity Δ is very small:

$$\Delta \ll 1 \qquad 1 \qquad (2.257)$$

As a first order approximation, one can let $\epsilon_1/\epsilon_2 \cong 1$ in the dispersion relation (2.256), which then becomes:

$$\frac{J'_m(u)}{u J_m(u)} + \frac{K'_m(w)}{w K_m(w)} = \pm m \left(\frac{1}{u^2} + \frac{1}{w^2} \right) \qquad 1 \qquad (2.258)$$

For the upper sign in the equation, the recurrence formulas for the Bessel functions (§ 10.3.5 and 10.4.5) permit (2.258) to be written in the form:

$$- \frac{J_{m+1}(u)}{u J_m(u)} = \frac{K_{m+1}(w)}{w K_m(w)} \qquad 1 \qquad (2.259)$$

while for the lower sign:

$$\frac{J_{m-1}(u)}{u J_m(u)} = \frac{K_{m-1}(w)}{w K_m(w)} \qquad 1 \qquad (2.260)$$

Simultaneous solutions of (2.259) and (2.240) are called *EH-modes,* those of (2.260) and (2.240) *HE-modes.* The dispersion diagram for a step index fiber is shown in Figure 2.37.

2.10.9 Cutoff Frequencies

The modified Bessel function K_m decreases in an exponential fashion. When the parameter w is positive real, *i.e.,* when (2.239):

$$\beta > \omega \sqrt{\mu_0 \epsilon_2} \qquad \mathrm{m}^{-1} \qquad (2.261)$$

the fields actually decay within the cladding. The energy is then mostly contained within the core region. On the other hand, when w vanishes, which occurs for:

$$\beta = \omega \sqrt{\mu_0 \epsilon_2} \qquad \mathrm{m}^{-1} \qquad (2.262)$$

the transverse potential obtained with (2.135) has the dependence:

$$\underline{\theta}(\rho,\varphi) = (\underline{C}\rho + \underline{D}) \begin{Bmatrix} \sin m\varphi \\ \cos m\varphi \end{Bmatrix} \qquad 1 \tag{2.263}$$

and the fields in the cladding no longer decay away from the core. The energy is not contained within the core and there is no guiding effect anymore.

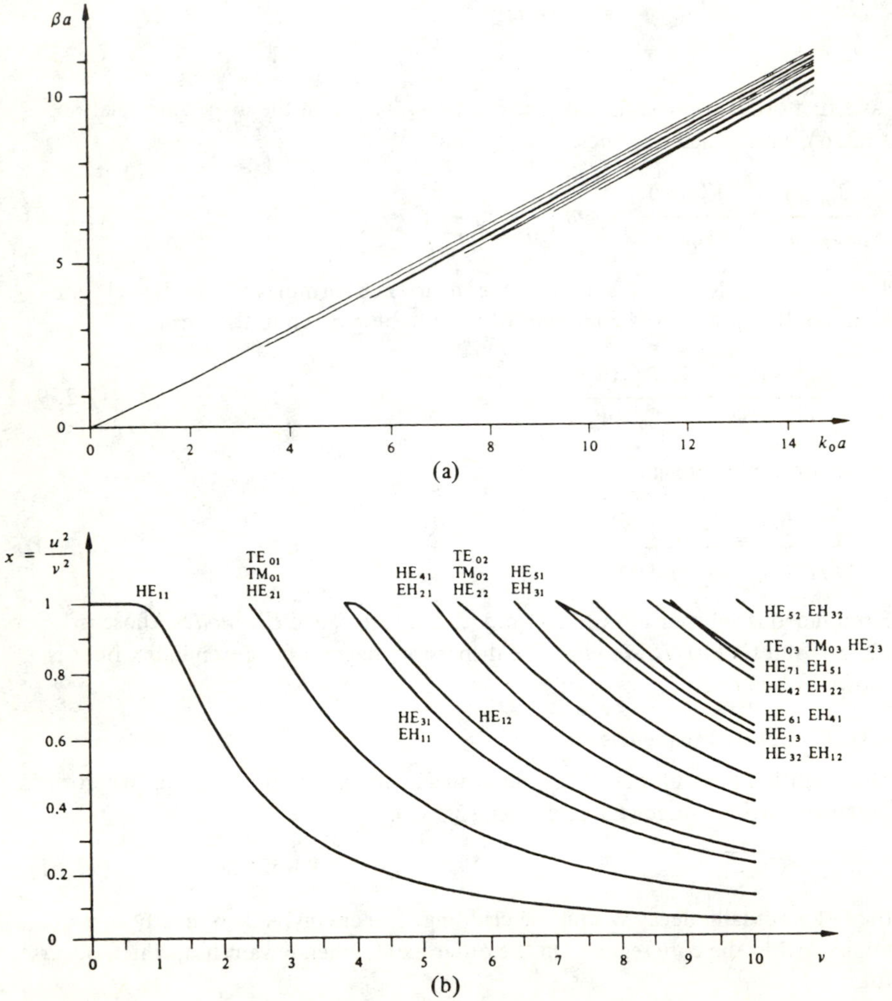

Fig. 2.37 Dispersion diagrams for an optical fiber with step index: (a) βa in terms of ka for $\Delta = 0.1$ and $\epsilon_1 = 1.53^2\,\epsilon_0$; (b) $x = u^2/v^2$ in terms of v for $\Delta = 0.1$.

The relation (2.262) provides the cutoff condition. Table 2.38 presents the normalized cutoff frequencies ν for the different modes.

2.10.10 Dominant Mode HE_{11}

Table 2.38 shows that only the HE_{11} mode does not possess a cutoff frequency: it is the *dominant mode* which can, in theory, propagate even down to zero frequency. In actual fact, when frequency decreases, the radial decay of the fields within the cladding becomes so slow that a gigantic cross section would be needed to contain the signal. Also, material losses would become quite significant. The property of field concentration within dielectrics, previously shown for the TE_{10} mode on the dielectric plate (Fig. 2.33), is again encountered for the optical fiber.

Table 2.38 Mode characteristics for a step index fiber.

Modes	characteristic equation	normalized cutoff frequency ν, $w = 0, u = \nu$: equation	solution	
TE_{0n}	$\frac{J_1(u)}{uJ_0(u)} + \frac{K_1(w)}{wK_0(w)} = 0$	$J_0(\nu) = 0$	j_{0n}	
TM_{0n}	$\frac{\epsilon_1 J_1(u)}{uJ_0(u)} + \frac{\epsilon_2 K_1(w)}{wK_0(w)} = 0$	$J_0(\nu) = 0$	j_{0n}	
HE_{mn}	$\frac{J_{m-1}(u)}{uJ_m(u)} - \frac{K_{m-1}(w)}{wK_m(w)} = 0$	$\nu J_1(\nu) = 0$	0 et j_{1n}	$m = 1$
		$\frac{J_{m-2}(\nu)}{J_{m-1}(\nu)} = 0$	$j_{m-2,n}$	$m \neq 1$
EH_{mn}	$\frac{J_{m+1}(u)}{uJ_m(u)} + \frac{K_{m+1}(w)}{wK_m(w)} = 0$	$\frac{\nu J_m(\nu)}{J_{m+1}(\nu)} = 0$	j_{mn}	

where j_{mn} is the n-th zero (different from zero) of the m-th order Bessel function J_m.

2.10.11 Unimodal and Multimodal Fibers

Higher order modes start propagating at frequencies above the normalized cutoff frequency $\nu = 2.405$ (Table 2.38). At lower frequencies, only the dominant mode HE_{11} can propagate: the fiber operates in an *unimodal* manner. When ν is larger than 2.405, several modes can propagate at the same time, the operating behavior is called *multimodal*.

The available sources for fiber optics (laser, LED) only provide signals at certain specific wavelengths. The most commonly used radiation is of the order of 850 nm (350 THz). For a signal of angular frequency ω, the radius a of a *unimodal fiber* core must be smaller than an upper limit, specified directly from (2.240):

$$a \leqslant 2.405/(\omega\sqrt{\epsilon_1\mu_0 2\Delta}) = 2.405/[\omega\sqrt{\mu_0(\epsilon_1 - \epsilon_2)}] = 2.405\, c_0/(\omega \cdot NA) \quad \text{m} \qquad (2.264)$$

The core of a unimodal fiber must have, at the most, a radius of a ***few micrometers***. As for *multimode fibers,* they typically have a core radius in the tens of micrometers range.

2.10.12 Note

The very small core dimension of a unimodal fiber makes its use most difficult in practice, in particular because of alignment problems at connections.

2.10.13 Dispersion in an Optical Fiber

In the same way as in a hollow metallic waveguide (§ 2.2.35), a signal propagating along an optical fiber undergoes a distortion produced by the dispersion of the phase velocity (guide dispersion), together with the differences of the propagation velocities of different modes (multimode dispersion). The latter effect is clearly absent in a unimodal fiber.

Table 2.39 Comparison of the dispersion for 3 kinds of sources.

parameters		light-emitting diode	multimode solid-state laser	monomode solid-state laser
λ	nm	900	850	850
$\delta\lambda$	nm	30	1	0,1
$\delta\lambda/\lambda$		0.033	$1.2\ 10^{-3}$	$1.2\ 10^{-4}$
$\Delta\tau_g$	ps/km	330	12	1,2
$\Delta\tau_m$	ps/km	1 800	80	8
$\Delta\tau_{pm}$	ps/km	25 000	25 000	25 000

step index fiber: $\Delta = 0.003$, $\epsilon_1 = 1.45^2\epsilon_0$, $\Delta\tau_g = 0.003\,\dfrac{\delta\lambda/\lambda}{c_0}$

To these two distortion sources, one must add another effect, due to the material itself (material dispersion). It occurs because the permittivity of the material varies with frequency $\epsilon(f)$. This effect does not exist in an empty hollow waveguide.

Table 2.39 shows the respective significance of these three sources of distortion, for signals generated by three types of oscillators [20]. The spread of the group delay time $\Delta\tau$ resulting from these different effects is presented for a step index fiber. The variations show how a pulse propagating along a fiber spreads with distance. The subscripts refer to the different effects: g for guide, m for material and pm for multimode propagation.

2.10.14 Graded Index Fibers

The distortions produced by the multimodal propagation can be significantly reduced by using fibers whose core permittivity continuously varies in the radial direction. These are the *graded index fibers.* The smallest multimode phase dispersion is obtained when the index variation is ***approximately quadratic*** (Fig. 2.40). When studying wave propagation in graded index fibers, it is no longer possible to separate the longitudinal components of $\underline{E}$ and $\underline{H}$. Two coupled differential equations in $\underline{E}_z$ and $\underline{H}_z$ are obtained [21]. Several resolution techniques, of an approximate nature, have been developed [22]. They can be categorized into two main families: ***field methods*** and ***ray geometrical*** methods. The latter ones are, however, only valid when the number of propagating modes is large. Most of the approximate formulations commonly used are based on ray approximations [23].

A large number of scientific books and articles [24 to 29] deal with dielectric waveguides and fiber optics, covering on one side the study of propagation, on the other their utilization for communications engineering and measurement purposes (§ 8.2.15).

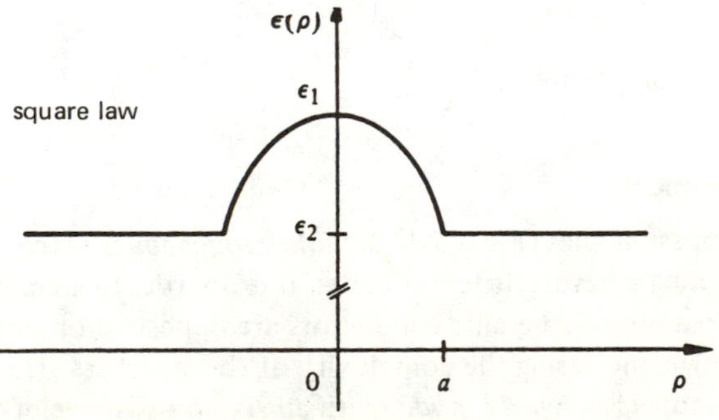

Fig. 2.40 Graded index fiber.

2.11 PLANAR LINES

2.11.1 Introduction

The printed circuit technique, which permits to manufacture in series connecting electrical networks by means of photolithographic methods (§ 2.11.22) has also extended into the microwave range. Known as ***microwave integrated circuits*** (***MIC***), this technique permits low power circuits and assemblies to be realized. These are far less bulky than their waveguide counterparts, which are also more complex to design.

2.11.2 Definition: Planar Circuits

The name of *planar circuits* designates conductor networks deposited over one or both faces of an insulating plate (dielectric substrate). At microwave frequencies, the dielectric substrate and the metallic conductors form one or several transmission lines, the properties of which depend upon size and material properties. The structures most commonly used are the microstrip (§ 2.11.4 to 17), the slot line (§ 2.11.18), and the coplanar line (§ 2.11.19), represented in Figure 2.41. For millimeter wave applications, these structures must be entirely shielded by metal boundaries (waveguide) to suppress radiation. In this manner the suspended substrate line (§ 2.11.20) and the finline (§ 2.11.21) are obtained.

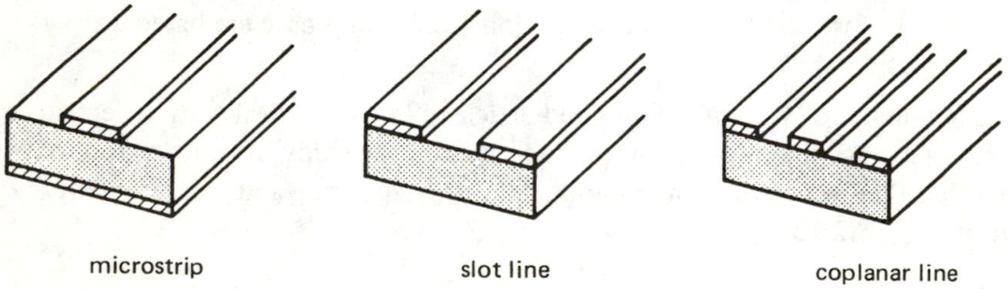

Fig. 2.41 Main planar lines.

2.11.3 Comments

Planar transmission lines (Fig. 2.41) are ***inhomogeneous***, as they consist of two propagation media having different material properties (generally air and one dielectric). In addition, metallic conductors are deposited on one or both faces of the dielectric, increasing the complexity of the structure. The dominant mode in these structures is a ***hybrid mode***, with all six components of the electromagnetic fields.

These lines are inherently more complex than the conductorless lines of the previous sections. The corresponding electromagnetic problem does not possess an exact analytical solution for the field distribution and the propagation characteristics. A large number of approximate computation methods were developed.

Since the commonly utilized lines are open (Fig. 2.41), they may radiate towards the surrounding space, an effect which is most often spurious, and which increases with frequency. It can be avoided to some extent by using high permittivity substrates, as fields tend to concentrate within the dielectric.

2.11.4 Microstrip Line

This is a structure having a thin narrow conductor ribbon deposited on one face of the dielectric substrate, the other face being entirely metallized (Fig. 2.42). The characteristic physical parameters of the line are:

1. the relative permittivity ϵ_r of the substrate. The use of high permittivity dielectrics tends to concentrate the fields within the substrate, lessening the radiation;
2. the substrate thickness h, generally a fraction of a millimeter;
3. the width w of the metal ribbon. This width is of the same order as the substrate thickness h ($0.1 \leqslant w/h \leqslant 10$). The characteristic line impedance Z_c is controlled by varying the width of the ribbon;
4. the thickness b of the ribbon, generally negligible ($b/h \ll 1$).

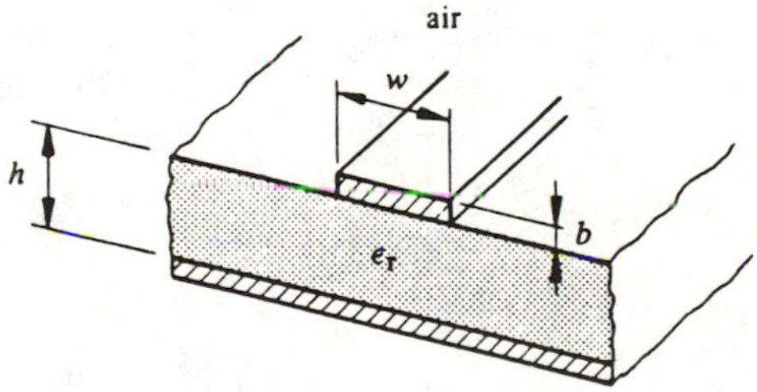

Fig. 2.42 Parameters of simple microstrip.

In the first approximation, a microstrip line can be considered as a half stripline (§ 2.6.3). The field distributions in the two structures do indeed possess similarities (Fig. 2.43).

The stripline is homogeneous, its mode of propagation is therefore TEM. On the inhomogeneous microstrip, in contrast, the dominant mode is hybrid. Nevertheless, the longitudinal field components $\underline{E}_z$ and $\underline{H}_z$ have far smaller amplitudes than the corresponding transverse components $\underline{E}_t$ and $\underline{H}_t$. The dominant mode in microstrip is, therefore, a ***quasi-TEM mode*** (§ 2.8.3).

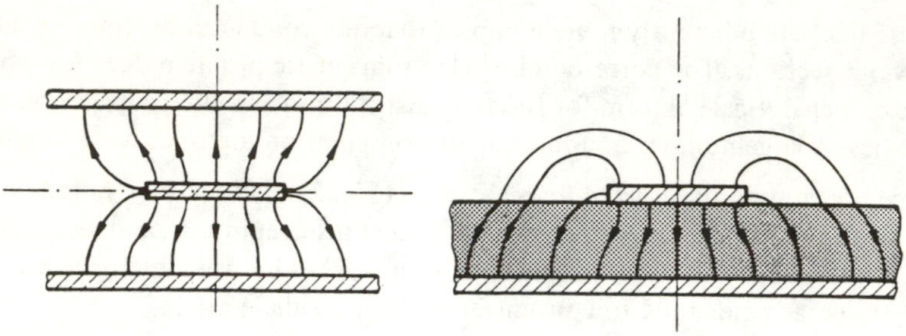

Fig. 2.43 Electric field in a stripline and in a microstrip.

2.11.5 Quasi-TEM Approximation

Since the longitudinal components $\underline{E}_z$ and $\underline{H}_z$ are quite small, they can be neglected when operating below a frequency limit (§ 2.11.16), even though the tangential field components are then discontinuous across the dielectric boundaries (§ 2.8.4 and § 2.8.5). The hybrid dominant mode is then replaced to first order by a TEM mode, the study of which can be carried out with electrostatics (§ 2.11.6 and 2.11.7).

2.11.6 Definition: Effective Permittivity ϵ_e for $b = 0$

The inhomogeneous structure is replaced by a line with conductors of the very same dimensions, but placed within a single homogeneous dielectric of *effective relative permittivity* ϵ_e (Fig. 2.44) The value of this permittivity is given by the capacitance of the inhomogeneous structure, yielding approximate relations [30]:

$$\epsilon_e \cong \frac{1}{2}(\epsilon_r + 1) + \frac{1}{2}(\epsilon_r - 1)\left[\left(1 + 12\frac{h}{w}\right)^{-1/2} + 0.04\left(1 - \frac{w}{h}\right)^2\right] \qquad \frac{w}{h} \leqslant 1 \qquad 1 \qquad (2.265)$$

$$\epsilon_e \cong \frac{1}{2}(\epsilon_r + 1) + \frac{1}{2}(\epsilon_r - 1)\left(1 + 12\frac{h}{w}\right)^{-1/2} \qquad \frac{w}{h} \geqslant 1 \qquad 1 \qquad (2.266)$$

The relative error for these approximations is less than 1% when $0.05 \leqslant w/h \leqslant 20$ and $\epsilon_r \leqslant 16$.

The phase velocity v_φ and the line wavelength λ_g are directly related to the effective permittivity

$$v_\varphi = c_0/\sqrt{\epsilon_e} \qquad \text{m/s} \qquad (2.267)$$

$$\lambda_g = \lambda_0/\sqrt{\epsilon_e} \qquad \text{m} \qquad (2.268)$$

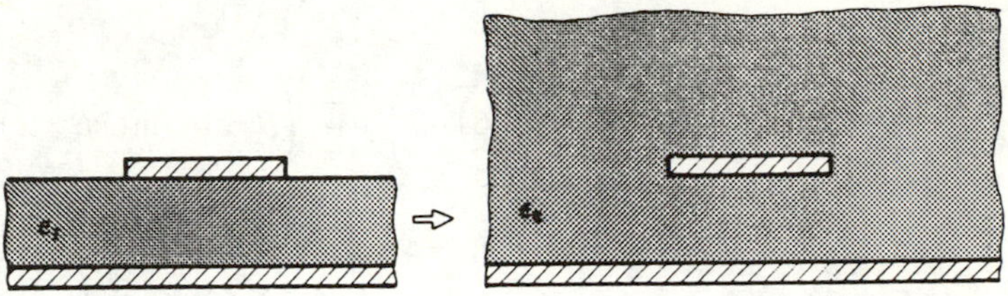

Fig. 2.44 Principle of the quasi-TEM approximation.

2.11.7 Characteristic Impedance of Microstrip for $b = 0$

When the metal ribbon has zero thickness, the application of conformal mapping provides an exact analytical solution [31]. However, the relationships obtained in this manner are of an implicit nature, requiring computer calculations to determine the characteristic impedance. Approximate expressions were derived by Schneider [31], and were later touched up by Hammerstad [30], so that the remaining error is less than 1% for $0.05 \leqslant w/h \leqslant 20$.

$$Z_c \cong \frac{Z_0}{2\pi\sqrt{\epsilon_e}} \ln(8h/w + w/4h) \qquad \frac{w}{h} \leqslant 1 \qquad \Omega \quad \textbf{(2.269)}$$

$$Z_c \cong \frac{Z_0}{\sqrt{\epsilon_e}} \left(\frac{w}{h} + 1.393 + 0.667 \ln\left(\frac{w}{h} + 1.444\right)\right)^{-1} \qquad \frac{w}{h} > 1 \qquad \Omega \quad \textbf{(2.270)}$$

where $Z_0 \cong 120\pi = 376.6\ldots\ \Omega$ is the characteristic impedance of the vacuum (§ 1.4.5).

These expressions permit the analysis of a microstrip line, *i.e.*, to determine ϵ_e and Z_c in terms of the line dimensions and the substrate permittivity.

2.11.8 Microstrip Line Synthesis for $b = 0$

The reverse operation of synthesis, *i.e.*, the determination of the w/h ratio providing a required characteristic impedance Z_c, is carried out using another approximate relation (also within 1%) derived by Wheeler [32]. For $w/h < 2$:

$$\frac{w}{h} \cong 4\left[\frac{1}{2}\exp(A) - \exp(-A)\right]^{-1} \qquad \frac{w}{h} \leqslant 2 \qquad 1 \qquad (2.271)$$

with

$$A = \pi\sqrt{2(\epsilon_r + 1)}\,(Z_c/Z_0) + \frac{\epsilon_r - 1}{\epsilon_r + 1}(0.23 + 0.11/\epsilon_r) \qquad 1 \qquad (2.272)$$

while for $w/h \geqslant 2$

$$\frac{w}{h} \cong \frac{\epsilon_r - 1}{\pi \epsilon_r}\left(\ln(B-1) + 0.39 - 0.61/\epsilon_r\right) + \frac{2}{\pi}\left(B - 1 - \ln(2B-1)\right) \quad 1 \qquad (2.273)$$

with

$$B = \frac{\pi}{2\sqrt{\epsilon_r}} \frac{Z_0}{Z_c} \quad 1 \qquad (2.274)$$

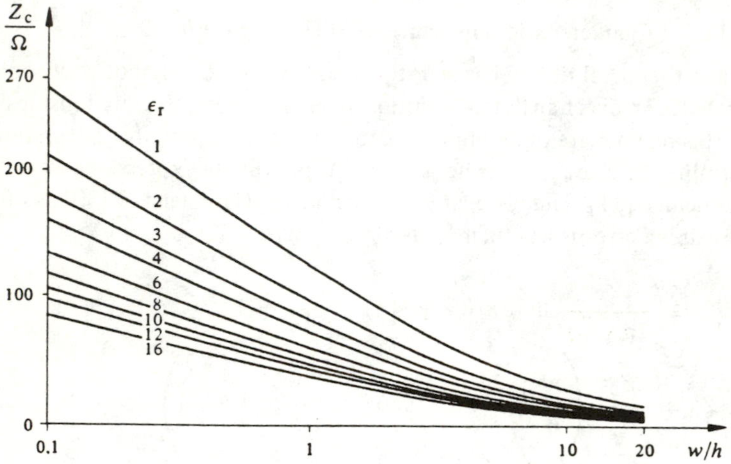

Fig. 2.45 Characteristic impedance of a microstrip.

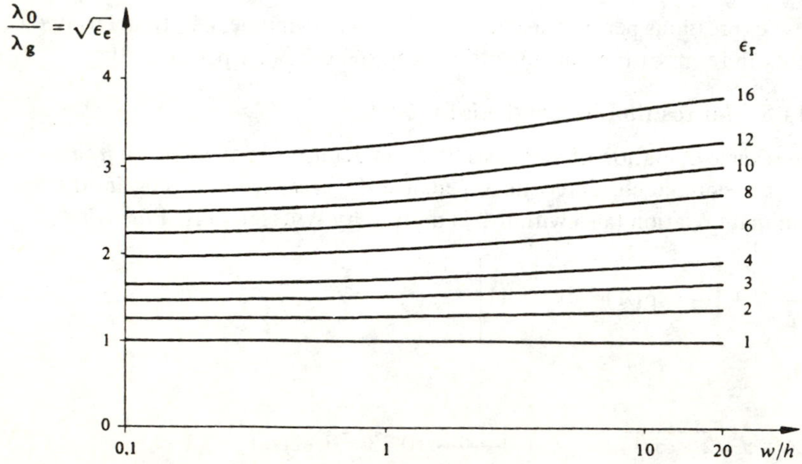

Fig. 2.46 Wavelength ratio on a microstrip.

Figures 2.45 and 2.46 represent, respectively, the characteristic impedance Z_c and the wavelength ratio λ_0/λ_g for microstrip, as a function of the w/h quotient and of the relative permittivity ϵ_r of the dielectric substrate.

2.11.9 Note

When selecting the size of a microstrip for a specified characteristic impedance, one determines the wavelength ***at the same time***. This effect must be taken into account when designing microstrip circuits.

2.11.10 Ribbon Having a Non-Zero Thickness b

To some extent, the actual thickness b of the upper conductor (ribbon) can be taken into account by replacing the real width w by an ***effective ribbon width*** w_e, which is slightly larger, in the calculations [9]:

$$w_e = w + \frac{b}{\pi}\left(1 + \ln\frac{2x}{b}\right) \qquad \text{m} \qquad (2.275)$$

with $x = h$ if $w > h/2\pi$ and $x = 2\pi w$ if $h/2\pi > w > 2b$.

2.11.11 Attenuation of Microstrip

Three distinct effects tend to dampen the signal propagating along a microstrip. First of all, as in waveguide, part of the signal heats up the conductor through ohmic losses. In addition, parts of the fields propagate through a dielectric which is not perfectly lossless, so that another part of the signal is transformed into heat in the substrate. Finally, as the line is not enclosed, some of the signal leaks away in the form of radiation, whenever discontinuities are present along the transmission line.

2.11.12 Attenuation due to Joule Losses

The losses within the conductors are evaluated with perturbation methods [33], yielding approximate expressions [34]:

$$\alpha_c \cong \frac{1}{h}\,\frac{10}{\pi \ln 10}\,\frac{R_m}{Z_c}\,\frac{32 - (w/h)^2}{32 + (w/h)^2}\left[1 + \frac{h}{w}\left(1 + \frac{\partial w_e}{\partial b}\right)\right] \qquad \frac{w}{h} \leqslant 1$$

$$\text{dB/m} \qquad (2.276)$$

$$\alpha_c \cong \frac{1}{h}\,\frac{20}{\ln 10}\,\frac{\epsilon_e Z_c R_m}{Z_0^2}\left(\frac{w}{h} + \frac{6h}{w}\left[\left(1 - \frac{h}{w}\right)^5 + 0.08\right]\right)\left(1 + \frac{h}{w}\left(1 + \frac{\partial w_e}{\partial b}\right)\right)$$

$$\frac{w}{h} \geqslant 1 \qquad \text{dB/m} \qquad (2.277)$$

where $\partial w_e/\partial b$ is determined from (2.275), yielding:

$$\frac{\partial w_e}{\partial b} = \frac{1}{\pi} \ln \frac{2h}{b} \quad \text{if } w/h \geqslant 1/2\pi \quad 1 \tag{2.278}$$

$$\frac{\partial w_e}{\partial b} = \frac{1}{\pi} \ln \frac{4\pi w}{b} \quad \text{if } w/h \leqslant 1/2\pi \quad 1 \tag{2.279}$$

and with $R_m \cong \sqrt{\omega\mu'/2\sigma}$ the characteristic impedance of metal (2.182).

These theoretical values apply to ideal plane metallic surfaces. For rough metal surfaces, significantly larger values are measured for the attenuation, up to twice the theoretical value. A rough approximation is generally quite sufficient for practical applications [35]:

$$\alpha_c \cong 8.686\, R_m/wZ_c \qquad \text{dB/m} \tag{2.280}$$

2.11.13 Attenuation due to Dielectric Losses

Dielectric losses are generally much smaller than conductor losses (except when low-grade substrates are used). The attenuation per unit length α_d resulting from substrate losses is given by [34]:

$$\alpha_d \cong 27.3 \frac{\epsilon_e - 1}{\epsilon_r - 1} \frac{\epsilon_r}{\epsilon_e} \frac{\tan\delta}{\lambda_g} \qquad \text{dB/m} \tag{2.281}$$

2.11.14 Radiation

Theoretically, a uniform microstrip does not radiate: radiation is related to the existence of unguided higher order modes, which may be excited at discontinuities. In practical applications, a line always presents some discontinuities, at least at its ends. Precautions must therefore be taken to avoid excessive radiation. Radiation losses measured at a characteristic impedance of 50 Ω were found to be approximately proportional to $(hf)^2/\sqrt{\epsilon_r}$ [34]. A 1 mm thick substrate can be used up to 3 GHz when $\epsilon_r = 2.5$, up to 4 GHz with $\epsilon_r = 10$. The upper frequency f_m, at which 1% of the power is radiated at the extremity of an open microstrip, is given by:

$$f_m(\text{GHz}) \cong 2.14\, \epsilon_r^{1/4}/h \ (\text{mm}) \tag{2.282}$$

This relation is presented in nomogram form in Figure 2.47.

For high frequency applications, a high permittivity and thin substrate must be selected. When this is not sufficient, for instance at millimeter waves, then the lines must be placed within an enclosure, to positively prevent any kind of radiation.

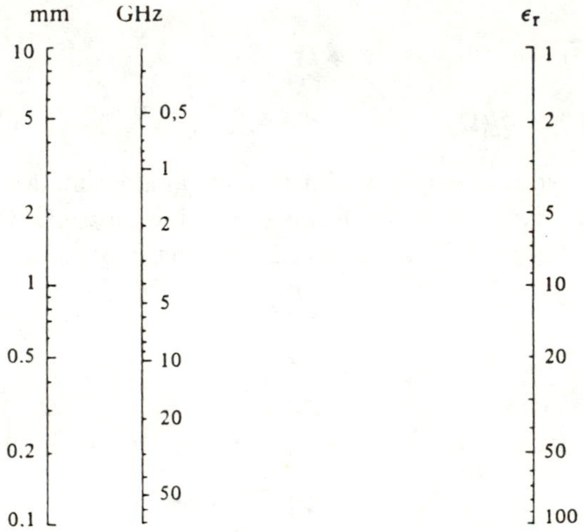

Fig. 2.47 Nomogram for the frequency limit f_{max} at which a microstrip starts radiating.

Radiation not only produces losses along the line, it can also bring about cross-talk between parallel channels, a spurious effect which may be even more obnoxious for communications.

2.11.15 Remark

As the current is concentrated in a thin and narrow conductor (ribbon), located on top of an imperfect dielectric material, the attenuation in a microstrip is significantly larger, for a given length, than in a waveguide. Consequently, a microstrip line is only used to make short interconnections, typically a few centimeters long, between other circuit elements.

2.11.16 Dispersion in Microstrip

Due to its quasistatic nature, the TEM approximation can only yield frequency independent values for the effective permittivity ϵ_e and the characteristic impedance Z_c. This approximation cannot take into account the field concentration within the substrate, which increases with frequency (§ 2.9.3). Approximate methods permit the evaluation of this effect and the calculation of a frequency-dependent effective permittivity as follows [36]:

$$\epsilon_e(f) \cong \epsilon_r - \frac{\epsilon_r - \epsilon_e}{1 + G(f/f_d)^2} \quad 1 \qquad (2.283)$$

where ϵ_e is the low-frequency effective permittivity defined in Section 2.11.6, f is the frequency, and the constants f_d and G are given by:

$$f_d = \frac{1}{2\mu_0} \frac{Z_c}{h} \qquad \text{Hz} \qquad (2.284)$$

$$G = 0.6 + 0.009\, Z_c(\Omega) \qquad 1 \qquad (2.285)$$

Figure 2.48 provides an easy way to determine f_d as a function of the characteristic impedance Z_c and of the thickness h of the substrate. One can quickly determine whether the correction of (2.283) is actually required; this is not the case when $f \ll f_d$.

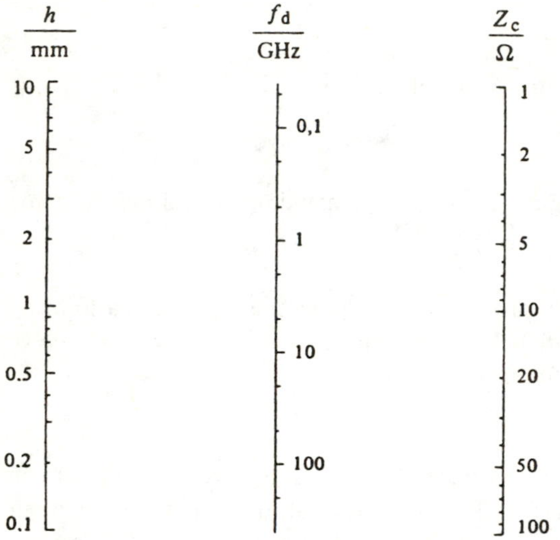

Fig. 2.48 Nomogram for the frequency f_d in the equation (2.284) for the dispersion on microstrip.

□ 2.11.17 Complete Study of Microstrip

When one leaves aside the simple quasistatic approximation and tries to determine the actual field distribution in the microstrip line, he is faced with an extremely complex mathematical problem. It is actually not possible to obtain an exact analytical solution for it. Many authors developed computer programs based on numerical computation methods. A comprehensive survey was carried out by Gupta [37], who classifies them into three categories:

1. coupled modes. The dominant (hybrid) mode is formed of a TEM component and an additional term, either TE or TM depending upon the authors. Such an approach permitted the evaluation of the effect of dispersion and the definition of the effective frequency-dependent permittivity (§ 2.11.16);

2. finite differences [214]. As this approach is based on the cutting of the cross section into small squares, it supposes a finite cross section. The microstrip must be enclosed in a metal box to use this method;
3. integral methods. The fields on a microstrip (open or boxed in) are represented by a development over an infinite series (Fourier series or transform) [215]. The coefficients of the terms are obtained by point matching, by application of a variational principle or some other approximate manner.

In practical terms, the complete study of microstrip only presents a limited interest. The TEM approximation is quite adequate for the design of most circuits, and their bandwidth is limited anyway by the onset of radiation (§ 2.11.14). The most interesting results of the complete study are related to higher order modes, beyond reach of the quasi-TEM approximation. The real propagation diagram of a microstrip is shown in Figure 2.49 [38]. The quasi-TEM approximation is seen to provide a linearization of the $\beta(\omega)$ characteristic for the dominant mode.

The field distribution of a wave propagating along a microstrip is depicted in Figure 2.50 [39].

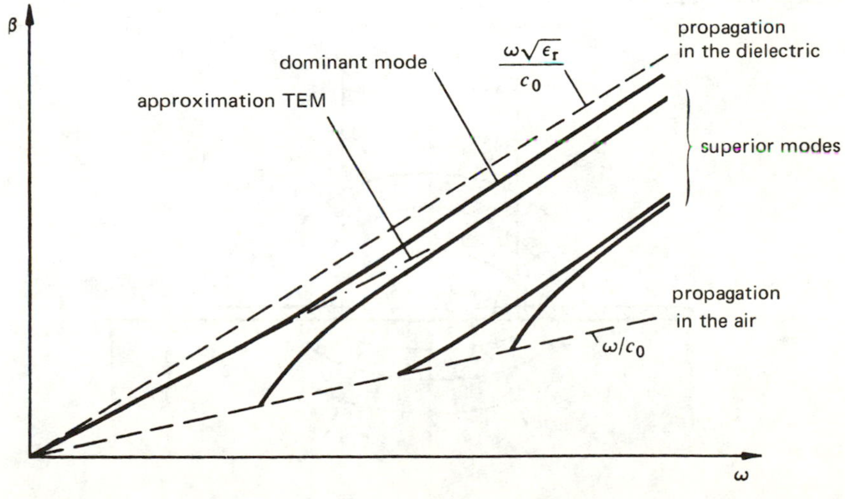

Fig. 2.49 Dispersion diagram of a microstrip line.

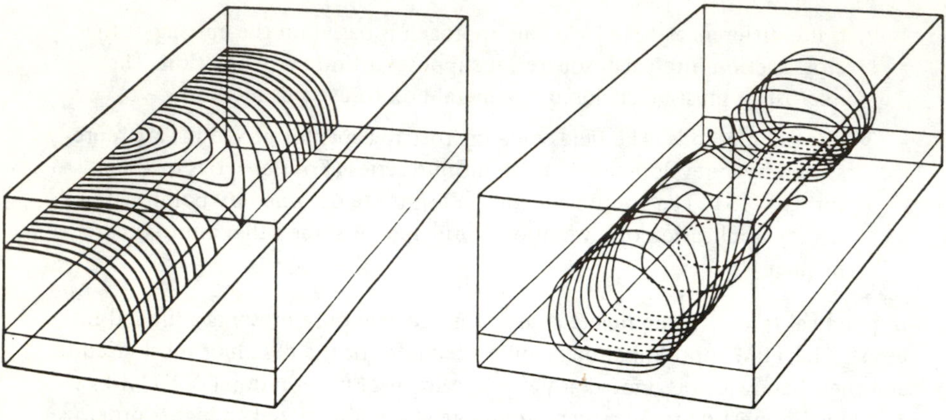

Fig. 2.50 Field lines of the dominant mode (quasi-TEM) on a microstrip line surrounded by a metallic enclosure.

2.11.18 Slot Line, Microslot

In the *slot* or *microslot line,* the two conductors of the transmission line are deposited over the same face of the dielectric substrate. The other face is not metallized (Fig. 2.51).

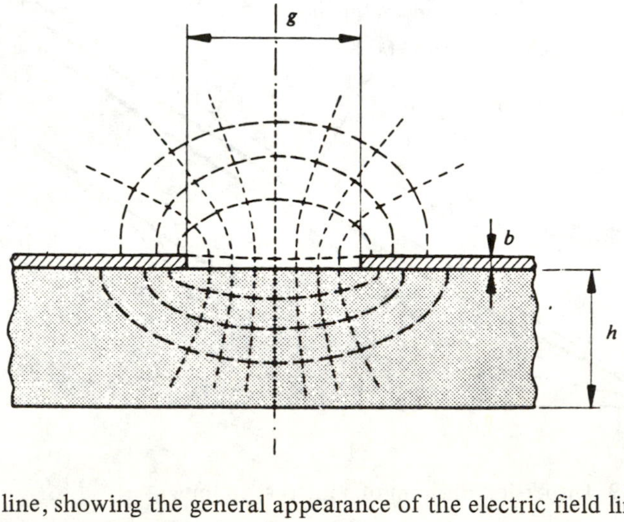

Fig. 2.51 Slot line, showing the general appearance of the electric field lines (— —) and magnetic field lines (- - -).

The electric field possesses one component within the interface, and the magnetic field possesses a normal component. Consequently, it results from Section 2.8.4 that the magnetic field must also have a longitudinal component. Due to the asymmetry produced by the presence of different propagation media, the electric field also has a longitudinal component (§ 2.8.5). As a result, the dominant mode of the slot is hybrid. It is not possible to neglect the longitudinal components, as was done in microstrip (§ 2.11.15).

A detailed analysis of the propagation in the microslot was carried out by Cohn [40] and approximate results are available in graphical form for several substrates [41]. An experimental mathematical formulation is based on these results. It provides a relative error smaller than 2% when $9.7 \leqslant \epsilon_r \leqslant 20$, letting also $b = 0$ [42]. For $0.02 \leqslant g/h \leqslant 0.2$, the approximate relations are:

$$\lambda_g/\lambda_0 = 0.923 - 0.448 \log \epsilon_r + 0.2\, g/h - (0.29\, g/h + 0.047) \log(h/\lambda_0 \times 10^2) \qquad 1 \qquad (2.286)$$

$$\begin{aligned} Z_c = {} & 72.62 - 35.19 \log \epsilon_r + 50\, h/g(g/h - 0.02)(g/h - 0.1) \\ & + \log(g/h \times 10^2)(44.28 - 19.58 \log \epsilon_r) \\ & - [0.32 \log \epsilon_r - 0.11 + g/h(1.07 \log \epsilon_r + 1.44)](11.4 - 6.07 \log \epsilon_r \\ & - h/\lambda_0 \times 10^2)^2 \qquad \Omega \end{aligned} \qquad (2.287)$$

while for $0.2 \leqslant g/h \leqslant 1$

$$\begin{aligned} \lambda_g/\lambda_0 = {} & 0.897 - 0.483 \log \epsilon_r + g/h(0.111 - 0.0022\, \epsilon_r) \\ & -(0.121 + 0.094\, g/h - 0.0032\, \epsilon_r) \log(h/\lambda_0 \times 10^2) \qquad 1 \end{aligned} \qquad (2.288)$$

$$\begin{aligned} Z_c = {} & 113.19 - 53.55 \log \epsilon_r + 1.25\, g/h(114.59 - 51.88 \log \epsilon_r) + 20(g/h - 0.2) \\ & (1 - g/h) - [0.15 + 0.23 \log \epsilon_r + g/h(-0.79 + 2.07 \log \epsilon_r)] \\ & \cdot [(10.25 - 5 \log \epsilon_r + g/h(2.1 - 1.42 \log \epsilon_r) - h/\lambda_0 \times 10^2)^2] \ \Omega \end{aligned} \qquad (2.289)$$

It must be mentioned that while the theoretical results of Cohn for the λ_g/λ_0 quotient are very close to the measured values, this is not true for the characteristic impedance Z_c, where discrepancies up to 30% have been reported [41].

2.11.19 Coplanar Line

Like the slot line, the *coplanar line* has all conductors on the same side of the substrate. Two slots of the same width are separated by a metallic ribbon (Fig. 2.52).

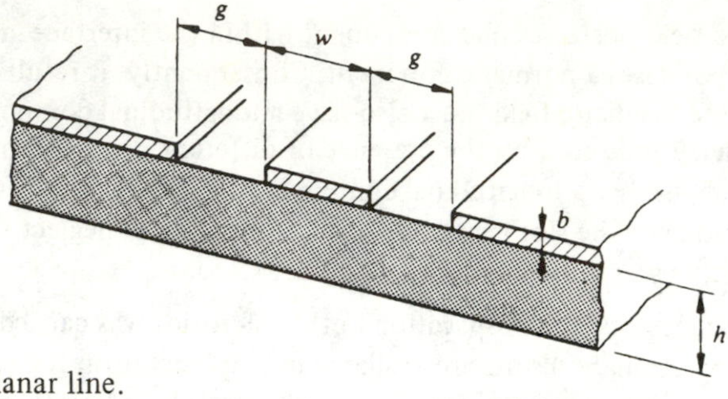

Fig. 2.52 Coplanar line.

This line was studied by Wen [43], who showed that a quasi-TEM approximation provides values corresponding fairly well with measurements for $\epsilon_r = 9.5$, 16, and 130. The conformal mapping used considers an infinitely thick substrate and conductors of zero thickness ($b = 0$). The characteristic impedance is given by:

$$Z_c \cong \frac{Z_0}{\sqrt{8(\epsilon_r + 1)}} \frac{K'(u)}{K(u)} \qquad \Omega \qquad (2.290)$$

with

$$u = (1 + 2g/w)^{-1} \qquad 1 \qquad (2.291)$$

where $K(u)$ is the complete first order elliptical integral [10].

2.11.20 Suspended Substrate Line

Radiation is avoided (§ 2.11.14) by enclosing a microstrip line within a waveguide below cutoff at the frequency of the signal, obtaining in this manner a *suspended substrate line* (Fig. 2.53).

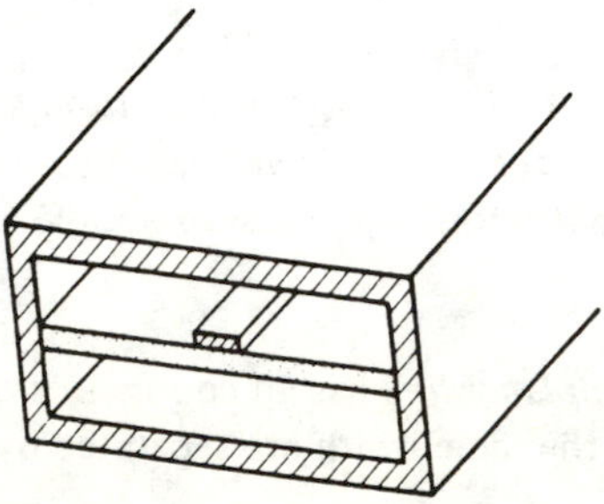

Fig. 2.53 Suspended substrate line.

The suspended substrate line was studied by Brenner [44], who made use of the finite-difference method [214] to obtain computation expressions. The guide must be narrow enough to prevent propagation of waveguide modes [45].

2.11.21 Finline

A ***slot line*** can similarly be placed within a waveguide below cutoff to suppress radiation. The resulting structure is called *finline* (Fig. 2.54). The study of this structure is carried out using Cohn's approach [39], leading to approximate relations [46, 47].

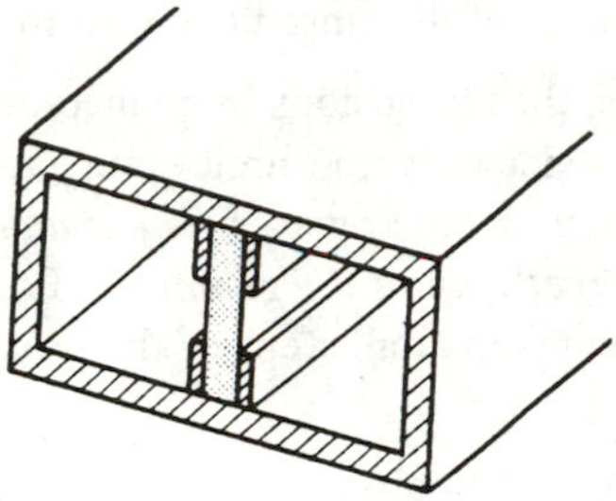

Fig. 2.54 Finline.

2.11.22 Manufacturing Procedures

A dielectric plate of constant thickness and electrical characteristics is metallized, on either one or both faces. ***Thin film*** techniques are generally used: evaporation, sputtering, or for plastic substrates, a thin metal sheet is glued or hot-pressed onto the substrate. ***Thick film*** processes (silk screening, conductive pastes) are generally not uniform enough for microwave applications. Most commercial manufacturers provide substrates already metallized.

The manufacturing procedure for planar circuits is basically the one used to realize standard printed circuits:

1. drafting of a ***scale mask*** of the network. The ***final mask*** is obtained by photographic reduction on a high-resolution film, using a precision camera;
2. mechanical and chemical cleaning of the metallized surfaces of the substrate, followed by oven drying;
3. deposition of a layer of photoresist. Several different processes may be used: dipping, spraying, centrifuge;
4. exposure. The light sensitive layer is exposed, through the reduced mask, to ultraviolet light. The latter must fall straight on the substrate, to accurately reproduce the layout;

5. development. For high precision, the temperature of the bath and the timing must be carefully controlled;
6. etching, by immersion into an agitated bath or by spraying the reagent onto the circuit. The etching agent used depends upon the metal to be removed;
7. chemical removal of the remaining photoresist;
8. electroplating: in some particular applications, a metal layer is subsequently plated (copper or gold). A low-current density electrolytic deposition process is used to obtain a uniform thickness;
9. the circuits must be carefully rinsed between the different operations.

In traditional electronics, the dimensions of connecting lines are relatively unimportant: they must provide only an ohmic connection with small enough losses. This is no longer true at microwaves. ***The characteristic impedance and the phaseshift depend directly upon the geometry***. For a device to operate correctly, great care must be taken at all steps of the procedure, in order to obtain the accuracy needed.

2.11.23 Materials for Planar Circuits

The dielectric substrate must satisfy a number of mechanical requirements (stability; machinability; availability; resistance to shock, vibration, chemical reagents; good adhesive properties to metal, etc.), and electrical requirements (low losses, homogeneity, isotropy, stability in time, etc.). Needless to say, these many requirements are often inconsistent with one another, so that most often only some compromise can be obtained.

Three different kinds of materials are used as dielectric substrates:

1. ceramics (alumina: Al_2O_3, beryllia: BeO): these materials are obtained by sintering. In their finished state, they are hard and brittle, difficult to machine. Their thermal conductivity can become large, allowing the heat dissipated to be evacuated from the circuit. Relative permittivities are within the 6-10 range, and the loss tan $\delta = \epsilon''/\epsilon'$ (§ 1.4.2) is very small, around $2\text{-}4 \cdot 10^{-4}$ at 10 GHz.

2. plastics: polytetrafluorethylene or PTFE (known under its trade name of Teflon by Dupont de Nemours), polystyrene, polyolefin, and polyester. The properties of this class of materials vary widely. Generally quite easy to machine, they each present some drawback. PTFE and polyolefin both have a poor mechanical stability. Polyolefin does not stand up well under high temperatures and there are soldering problems. Polystyrene has better mechanical behavior, but is not thermally compatible with copper. These materials have small relative permittivities (around 2 to 4) and a fairly low loss tangent, which is still greater than that for ceramics ($\cong 5 \cdot 10^{-4}$);

3. composite materials: finely divided ceramic powders are embedded within a plastic matrix (polystyrene, polyolefin, PTFE). Greater permittivities are obtained, while maintaining good machinability. The loss tangent is then larger ($\cong 10^{-3}$) and the material may become inhomogeneous.

There is no such thing as an ***ideal substrate***, which at the same time would fit all the mechanical and electrical requirements. In every practical application, the material having the properties which most closely resemble those desired will be chosen (a compromise). Properties of substrate materials are listed in Table 10.16 [48].

2.11.24 Comparison: Pros and Cons of Planar Lines

The design of microwave circuits (Ch. 6) provides some interesting problems of topology, most particularly when making use of waveguides. Three-dimensional assemblies must then be made with a number of waveguides, sticking out at odd angles, which need to be connected together. With the exception of some simple components, which are manufactured in series using molding or casting techniques, most elements require machining, brazing, or welding operations, which must be performed individually for each component. In addition, the waveguide sizes are determined by the operating frequency, precluding any attempt to reduce size and bulk.

Planar circuits, on the other hand, exist practically within a two-dimensional universe. Their manufacture is based on photographic processes, which can be repeated any number of times once a prototype has been designed and checked. The dimensions can be significantly reduced by using high permittivity substrates.

The insertion of circuit elements, among other solid-state devices, is easily implemented within planar circuits. Devices are generally connected in series on microstrip, in parallel on slot lines, while both kinds of connections can be made on coplanar lines. The insertion of devices is much more difficult in waveguides, where special parts must be made to insert and connect the devices, which furthermore must be encapsulated (Sec. 6.9).

The main restrictions on planar lines are their low power handling capability, which limits their applications to small signal levels, and the discontinuities produced by the connectors, which are difficult to avoid. To make the best possible use of planar circuits, one should ideally design the whole microwave assembly on a single substrate.

The rather large losses of planar lines are not significant: lines of this kind are never used to transmit a signal over any great distance, waveguide and coaxial lines are used for this purpose.

2.12 PROBLEMS

2.12.1 The cutoff frequency of a waveguide mode is 143 MHz. Determine the frequency, the phase and the group velocities, and the wave impedance, for both a TE and a TM mode, corresponding to $\beta = 4$ rad/m.

2.12.2 A 3 GHz signal undergoes a 228 degree phase shift when traveling along a ***dielectric-filled waveguide*** 2 cm long. It is also known that the cutoff frequency of the same waveguide, when ***empty***, is 9 GHz. What is the permittivity of the dielectric?

2.12.3 Determine the wave impedance and the three waveguide impedances in an air-filled rectangular waveguide of 2 X 4 cm cross section, for a 9 GHz signal propagating in the dominant mode. What is the phaseshift produced by a 1 m long section? How much time does a pulse of signal take to travel over 10 m?

H.ω X **2.12.4** Which modes can propagate at 15 GHz in a rectangular waveguide having a 3 X 1.5 cm cross section? Indicate all the degeneracies. Determine the phase and group velocities for all the modes listed.

2.12.5 Determine the size of a suitable air-filled waveguide for the 12.5 to 19 GHz frequency band. Determine also the largest attenuation one may expect when the walls are made of aluminum, for a 100 m length.

2.12.6 Determine the attenuation per unit length α in Np/m at 3 GHz for a 10 cm wide waveguide filled with water, whose complex permittivity at that frequency is $\underline{\epsilon}_r = 85\,(1 - j0.2)$.

2.12.7 Determine the cutoff frequencies for the dominant and the first higher order modes in the waveguide shown in Figure 2.55.

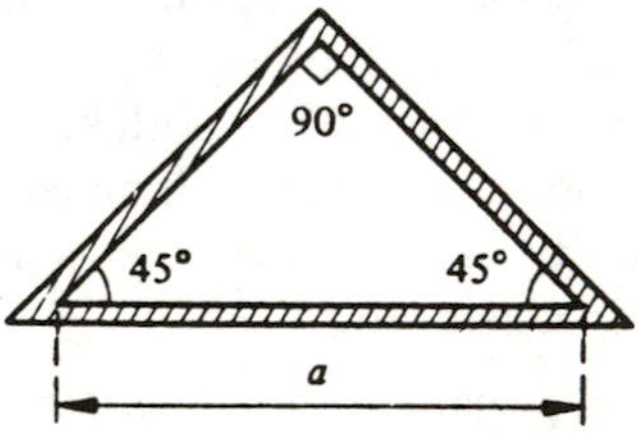

Fig. 2.55 Waveguide with a triangular cross section (rectangular isosceles triangle).

H.ω **2.12.8** What is the smallest radius that a circular waveguide must have to allow the propagation of the TE_{04} mode at 60 GHz?

2.12.9 To look at what happens inside a cavity resonating at 6 GHz, a hole is drilled into the wall and a circular metallic tube of 10 cm length soldered on top of it. What inner diameter must this metal tube have, to be certain that the signal radiated at its end is down by a power factor of 10^{-9} with respect to the signal inside of the cavity?

2.12.10 The cross section of a metallic waveguide is a circular sector with 60° opening angle and a radius a (Fig. 2.56). Determine the cutoff frequency for the dominant and the first higher order mode, and deduce the useful band for this waveguide.

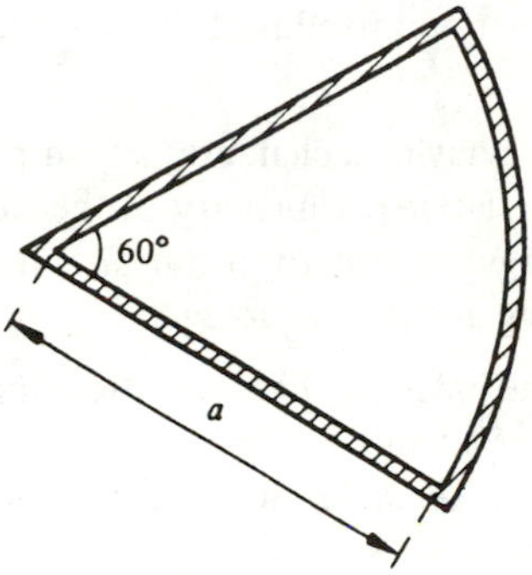

Fig. 2.56 Waveguide with a circular sector cross section.

2.12.11 The ridge waveguide of Figure 2.21 is considered. The ridge width is $a/10$, its height $b/3$ (a and b being, respectively, the width and the height of the rectangular waveguide). Using the perturbation method, determine approximately the cutoff frequency of the dominant mode, of the first higher order mode, and the useful frequency band for $a = 2b$. Compare them with the corresponding values for the empty waveguide.

2.12.12 A wave propagates on a dielectric plate 2 mm thick of relative permittivity $\epsilon_r = 36$, surrounded with air. Determine the cutoff frequencies of the higher order modes.

2.12.13 Determine approximately, using the perturbation method, the cutoff frequency of the loaded waveguide of Figure 2.35. Determine this frequency for $h/a = 0.2$, $b/a = 0\ 5$ and $\epsilon_r = 3$.

2.12.14 Making use of the dispersion relations for step index fibers, show that the straight line $\beta = (\omega/c_0)\sqrt{\epsilon_{r1}}$ is an asymptote for all the dispersion curves of the modes of propagation. Show also that this value is an upper bound.

2.12.15 A unimodal step index fiber must operate at $\lambda_0 = 850$ nm. What must the core diameter be when $\Delta = 0.01$ and $\epsilon_{r1} = 2.34$? What value does this diameter take if $\lambda_0 = 1.5\,\mu$m?

2.12.16 Which modes can propagate at $\lambda_0 = 850$ nm in a step-index fiber having $a = 5\,\mu$m, $\epsilon_{r1} = 2.34$ and $\Delta = 0.01$?

2.12.17 Determine the dimensions of a quarter-wave transformer (§ 9.6.3) in microstrip connecting a 12.5 Ω line to a 50 Ω line. The signal frequency is 3.8 GHz. A $h = 0.6$ mm thick substrate of relative permittivity $\epsilon_r = 9.5$ is available. The thickness b of the conductors is assumed to be negligible. Determine approximately the attenuation introduced by this transformer, considering copper conductors.

2.12.18 A microstrip line having a characteristic impedance Z_c of 50 Ω has a width $w = 1.8$ mm. The relative permittivity of the substrate is $\epsilon_r = 3$. What is the limiting frequency at which radiation starts? Is it necessary to correct for dispersion when operating at that frequency?

2.12.19 A microstrip line having a characteristic impedance Z_c of 50 Ω, on a substrate with $\epsilon_r = 4.5$ has a width of 2 mm. What would be the impedance of a line having a half-width? Calculate the effective permittivity for both lines.

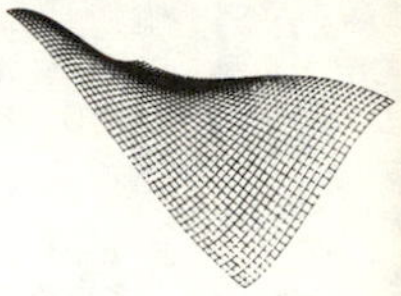

CHAPTER 3

RESONANT CAVITIES

3.1 INTRODUCTION AND GENERAL BACKGROUND

3.1.1 Definition

In a dielectric medium, either homogeneous or inhomogeneous, which is entirely surrounded by a conductor boundary (pec or pmc), Maxwell's equations only admit non-zero solutions at certain particular frequencies. The bounded enclosure is called *resonant cavity* or *resonator*.

3.1.2 Definition: Modes and Resonance Frequencies

The resolution of Maxwell's equations (1.2) to (1.5), with boundary conditions (1.8) and (1.9) imposed on the surface of a conducting enclosure, leads to an ***eigenvalue problem***. This problem possesses a set of different solutions, called the *resonant modes,* which are the eigenfunctions. To each eigenfunction is associated an eigenvalue, which is the *frequency of resonance.* An infinite number of modes do exist, forming a discrete set.

3.1.3 Particular Case: Cavity Formed of a Waveguide Section

When a waveguide section is terminated at both ends by metal planes perpendicular to its axis, it forms a particular kind of cavity, which is uniform in at least one direction (§ 2.1.1). This cavity resonates when its length is equal to a ***full number of half-guide wavelengths*** (Sec. 3.3). The field distribution may be determined in terms of the expressions obtained for the waveguides (Ch. 2).

3.1.4 Extension of the Definition: Open Resonator

Similar resonance properties are also found in structures which are not entirely surrounded by a conducting enclosure. This happens for resonators made up of sections of open lines (microstrip, slot line, etc.). Non-radiating apertures may be opened in the walls of certain cavities (for instance, to measure properties of materials). Apertures are also required to excite the fields within a cavity. For a cavity with openings, certain resonance modes may produce radiation towards the outside (cavity-antenna).

3.1.5 Application: Resonant Circuit

When the cavity dimensions are on the order of a few centimeters, the lowest resonant frequencies appear in the microwave range (§ 1.1.2). Consequently, cavities play the role of ***resonant circuits*** and are utilized in the following devices:

1. Generators (Ch. 4). The frequency of oscillation of most microwave generators is defined by a resonant cavity, placed either within a feedback loop, or actually containing the active element;
2. Frequency meters or wave meters (Ch. 5). A cavity of adjustable dimensions is used to measure the frequency or frequencies of a microwave signal;
3. Filters. The frequency-selective properties of resonant cavities are used in the design of filters, most often band-pass or band-stop.

3.1.6 Application: Measurement of Materials

The introduction of a material sample within a resonant cavity produces a variation of its resonance characteristic. The measurement of the resulting perturbation allows one to determine the properties of the material (§ 3.4.9 and Sec. 8.5).

3.1.7 Application: Heating

In a resonant cavity (microwave oven), the electromagnetic energy can be contained and used to heat up, cook, or dry a material placed inside (Sec. 8.3).

3.2 STUDY OF A CLOSED HOMOGENEOUS CAVITY

3.2.1 Definition of the Model

The general characteristics of a resonant cavity are first determined for the ideal structure of Figure 3.1, which has the following specifications:

1. its shape is arbitrary,
2. the material filling the cavity is homogeneous, linear, and isotropic: its properties $\underline{\epsilon}$, $\underline{\mu}$ and σ are independent of the position within the cavity, and also of the signal level. Isotropy means that electric and displacement phasor-vectors are colinear, and that magnetic and induction phasor-vectors are also colinear (§ 1.4.6);

3. the wall, which entirely surrounds the cavity, is formed of perfect conductors, either electric (pec, approximation of a metal wall, short circuit), or magnetic (pmc, corresponding to a plane of symmetry, open circuit) (§ 1.4.3);
4. there is no electric charge density within the cavity ($\underline{\rho} = 0$).

3.2.2 Similarity between Cavity and Waveguide

The definitions of the problem (§ 3.2.1) are very similar to those made for the homogeneous waveguide (§ 2.2.1), except that the cavity is three-dimensional, whereas the waveguide is studied in terms of its two transverse dimensions. Due to the uniformity along the direction of propagation, the longitudinal field dependence is considered separately. The solution of Maxwell's equations is obtained using similar approaches for these two problems.

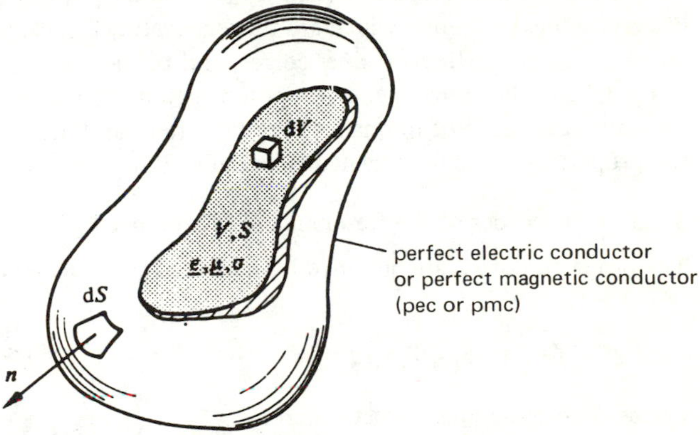

Fig. 3.1 Homogeneous cavity of arbitrary shape.

3.2.3 Helmholtz's Equation, Wave Equation

In the medium filling the cavity, the curl Maxwell's equations (1.2) and (1.3) are combined:

$$\nabla \times \nabla \times \underline{E} = -\mathrm{j}\omega\underline{\mu}\ \nabla \times \underline{H} = -\mathrm{j}\omega\underline{\mu}(\mathrm{j}\omega\underline{\epsilon} + \sigma)\underline{E} = \underline{k}^2\underline{E} \qquad \mathrm{V/m^3} \qquad (3.1)$$

where the wavenumber $\underline{k}$ (2.28) was defined as:

$$\underline{k}^2 = -\mathrm{j}\omega\underline{\mu}(\mathrm{j}\omega\underline{\epsilon} + \sigma) \qquad \mathrm{m^{-2}} \qquad (3.2)$$

Using vector relations, the left-hand term of (3.1) is developed (10.21), yielding

$$\nabla \times \nabla \times \underline{E} = \nabla(\nabla \cdot \underline{E}) - \nabla^2\underline{E} \qquad \mathrm{V/m^3} \qquad (3.3)$$

As there are no free charges within the medium, the divergence of the electric field vanishes, so that by combining (3.1) and (3.3) one obtains:

$$\nabla^2 \underline{E} + \underline{k}^2 \underline{E} = 0 \qquad \text{V/m}^3 \qquad (3.4)$$

Similarly, for the magnetic field, one takes the curl of (1.3) and gets:

$$\nabla^2 \underline{H} + \underline{k}^2 \underline{H} = 0 \qquad \text{A/m}^3 \qquad (3.5)$$

The two expressions (3.4) and (3.5) are the three-dimensional *Helmholtz's equations* or *wave equations.* The similarity with the two-dimensional equations (2.30) and (2.59) for waveguides is self-evident.

3.2.4 Definition: Resonance Wavenumber $\underline{k}_p$

When the boundary conditions upon the conducting walls of the cavity (1.8) and (1.9) are taken into account, equations (3.4) and (3.5) have solutions for certain discrete values of $\underline{k}$ only, which are the eigenvalues $\underline{k}_p$, called the *resonance wavenumbers.* These particular values correspond to the resonance modes of the cavity (§ 3.1.2). They are dependent upon its geometry (shape and size) and upon the particular mode of resonance. On the other hand, they are independent of the properties of the homogeneous medium filling the cavity.

■ 3.2.5 Properties of the Resonance Wavenumber $\underline{k}_p$

The properties of $\underline{k}_p$ are obtained by means of the divergence theorem, (§ 10.2.2):

$$\int_V \nabla \cdot \underline{A} \, \mathrm{d}V = \oint_S \boldsymbol{n} \cdot \underline{A} \, \mathrm{d}A \qquad \text{V}^2/\text{m} \qquad (3.6)$$

Let us choose the phasor-vector here:

$$\underline{A} = \underline{E} \times (\nabla \times \underline{E}^*) \qquad \text{V}^2/\text{m}^3 \qquad (3.7)$$

Its divergence becomes

$$\nabla \cdot \underline{A} = (\nabla \times \underline{E}) \cdot (\nabla \times \underline{E}^*) - \underline{E} \cdot \nabla \times (\nabla \times \underline{E}^*) \qquad \text{V}^2/\text{m}^4 \qquad (3.8)$$

Utilizing (3.3) and substituting these values into (3.6) yields:

$$\int_V |\nabla \times \underline{E}|^2 \mathrm{d}V + \int_V \underline{E} \cdot \nabla^2 \underline{E}^* \mathrm{d}V = \oint_S \underline{E} \times (\nabla \times \underline{E}^*) \cdot \boldsymbol{n} \mathrm{d}A \qquad \text{V}^2/\text{m} \qquad (3.9)$$

The right-hand term, integrated over the cavity boundary, always vanishes, due to the boundary conditions (1.8) or (1.9). The second integral is manipulated, making use of Helmholtz's equation (3.4), and the value of $\underline{k}_p$ is then given by:

$$\underline{k}_p^2 = \frac{\int_V |\nabla \times \underline{E}|^2 \mathrm{d}V}{\int_V |\underline{E}|^2 \mathrm{d}V} \qquad \text{m}^{-2} \qquad (3.10)$$

As both integrands are square moduli, the eigenvalues $\underline{k}_p^2$ are all positive real numbers (an identical result was obtained for p^2 in waveguides, in § 2.2.23). The positive root will be taken from now on for k_p.

3.2.6 Complex Resonance Angular Frequency $\underline{\omega}_p$

The angular frequency for the resonant mode is obtained by taking the *complex eigen-angular frequency* $\underline{\omega}_p$ out of (3.2), taking into account the real nature of $\underline{k}_p^2$

$$\underline{\omega}_p = \frac{j\sigma}{2\underline{\epsilon}} + \sqrt{\frac{\underline{k}_p^2}{\underline{\epsilon}\underline{\mu}} - \left(\frac{\sigma}{2\underline{\epsilon}}\right)^2} \qquad s^{-1} \qquad (3.11)$$

The positive square root is taken because the angular frequency is a positive quantity in the lossless case (as is the frequency itself).

When losses, either magnetic, dielectric, or of conduction, are present in the material filling the cavity, the angular frequency is complex.

3.2.7 Meaning of a Complex Angular Frequency

The complex nature of $\underline{\omega}_p$ means that the angular frequency has both a real and an imaginary part:

$$\underline{\omega}_p = \omega_{pr} + j(1/\tau) \qquad s^{-1} \qquad \textbf{(3.12)}$$

where ω_{pr} is the real part, or *eigen-angular frequency,* while τ is the ***relaxation time*** at the resonance.

Utilizing (1.1), the time-dependence of the electric field is given by:

$$\begin{aligned} \underline{E}(r,t) &= \mathrm{Re}[\sqrt{2}\underline{E}(r)\exp(j\underline{\omega}_p t)] \\ &= \sqrt{2}E_0(r)\exp(-t/\tau)\cos[\omega_{pr}t + \varphi(r)] \qquad \mathrm{V/m} \qquad (3.13) \end{aligned}$$

letting:

$$\underline{E}(r) = E_0(r)\exp[j\varphi(r)] \qquad \mathrm{V/m} \qquad (3.14)$$

The curve representing (3.13) as a function of time is a damped sine wave (Fig. 3.2): inside a closed cavity with losses, the signal amplitude decreases as a function of time.

3.2.8 Particular Case: Volume Conduction Losses

When the losses within the resonant cavity are entirely due to conduction, i.e., when $\epsilon'' = 0$ and $\mu'' = 0$, and the losses remain small, one obtains from (3.11) and (3.12):

$$\tau = 2\epsilon/\sigma \qquad s \qquad (3.15)$$

$$\omega_{pr} = \sqrt{\omega_0^2 - (1/\tau)^2} \qquad \text{rad/s} \qquad (3.16)$$

where one further defines:

$$\omega_0 = k_p/\sqrt{\epsilon\mu} \qquad \text{rad/s} \qquad (3.17)$$

as long as the condition $\tau\omega_0 > 1$ is met.

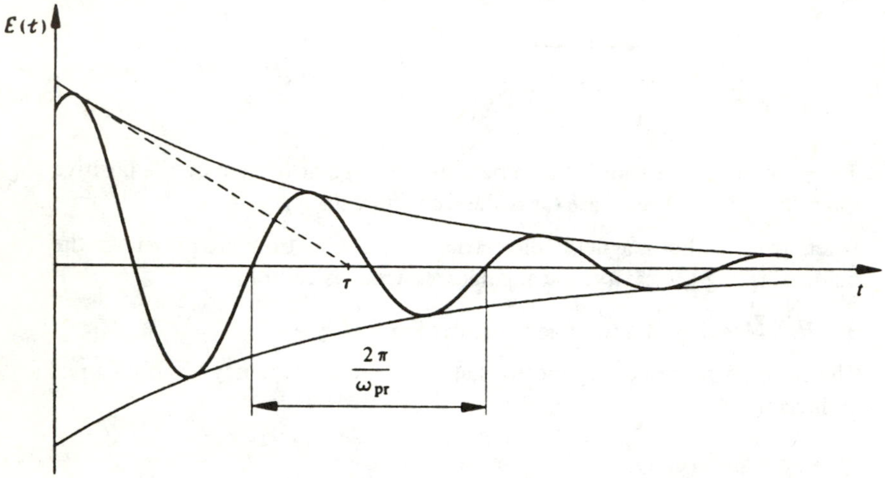

Fig. 3.2 Time-dependence of the electric field for a lossy closed cavity.

The existence of two different characteristic angular frequencies must be noted:

1. ω_{pr} is the ***eigen-angular frequency of the cavity***, which appears when the fields decay after the excitation has been removed (Fig. 3.2);
2. ω_0 is the ***resonance angular frequency***, which corresponds to the extremum of the cavity response under continuous wave operation.

For usual microwave cavities, the difference between the two angular frequencies is quite small, as the product $\tau\omega_0$ is very much greater than unity.

3.2.9 Definition: Unloaded Quality Factor Q_0

A high-grade resonant circuit must have losses as small as possible, and thus a relaxation time much greater than the period of the signal. The product $\tau\omega_0$ appearing in the previous paragraph is a measure of this quality. It is proportional to the *unloaded quality factor* Q_0, defined as:

$$Q_0 = \omega_0\tau/2 \qquad 1 \qquad (3.18)$$

The complex resonant angular frequency may further be expressed, in terms of ω_0 and of Q_0:

$$\underline{\omega}_p = \omega_{pr} + j\,\frac{\omega_0}{2Q_0} = \omega_0\left(\sqrt{1 - (1/2\,Q_0)^2} + j/2Q_0\right) \qquad s^{-1} \qquad (3.19)$$

3.2.10 Effect of Volume Conduction Losses

Replacing τ by its value given in (3.15), the unloaded quality factor Q_{0v} is obtained: it results from the ***losses within the medium filling the cavity***:

$$Q_0 = Q_{0v} = \omega_0\,\epsilon/\sigma \qquad 1 \qquad (3.20)$$

3.2.11 Note

Within real-life cavities (non-ideal), other loss mechanisms exist, in particular on the walls. They are considered in Section 3.4, which defines the global unloaded quality factor, covering all dissipative effects (§ 3.4.12).

3.2.12 Equivalent Circuit

The damped sine wave of Figure 3.2 is analogous to the step response of an *RLC* bipole. Letting a voltage correspond to the electric field in the cavity, a current to the magnetic field, one obtains the circuit of Figure 3.3, which provides the frequency response of a resonant mode.

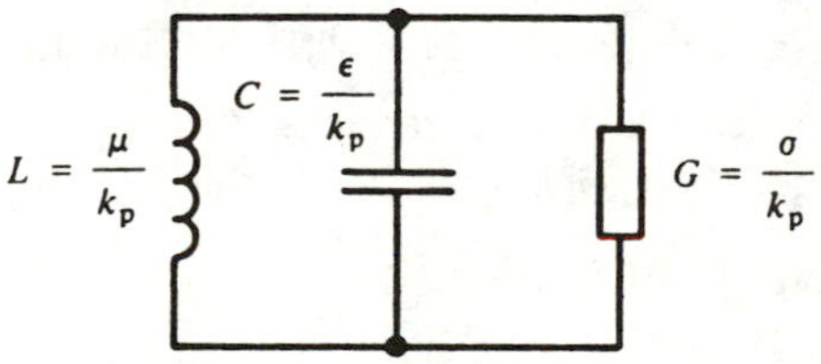

Fig. 3.3 Equivalent circuit of a resonant mode of the unloaded cavity.

The energies stored within the circuit are evaluated:

$$W_m = \frac{1}{2}\,L|\underline{I}|^2 = \frac{1}{2}\,|\underline{U}|^2/\omega_0^2 L = \frac{1}{2}\,C|\underline{U}|^2 = W_e \qquad J \qquad (3.21)$$

The dissipated power is given by:

$$P_d = G|\underline{U}|^2 \qquad W \qquad (3.22)$$

The quality factor is defined in circuit theory as

$$Q_0 = \frac{\omega_0 \cdot \text{maximum stored energy}}{\text{average power dissipated}}$$

$$= \frac{\omega_0 C|U|^2}{G|U|^2} = \frac{\omega_0 C}{G} = \frac{\omega_0 \epsilon}{\sigma} \qquad 1 \qquad (3.23)$$

One obtains in this manner the value given in (3.18).

It must be noted that all the elements of the equivalent circuit are defined within a multiplicative constant: C and G may be multiplied by any arbitrary factor, while L is divided by that same factor, without affecting either the resonant frequency or the unloaded quality factor.

3.2.13 Note: Multiple Solutions

The modes of resonance, or eigenfunctions of a cavity, are infinite in number. Each one is associated with one equivalent circuit, as in Figure 3.3. The resulting equivalent circuit of the cavity is, therefore, not one single parallel RLC circuit, but an infinity of such circuits. As the modes of a closed cavity with lossless walls are ***orthogonal***, these circuits ***are not coupled to each other***.

3.3 CAVITY MADE OF A SECTION OF TRANSMISSION LINE

3.3.1 Definition

For practical purposes, most cavities consist of a section of transmission line or of waveguide, terminated at both ends by short-circuit planes (or other reactive terminations, § 3.3.15).

3.3.2 Section of Transmission Line with Two Short Circuits

The input impedance of a lossless transmission line of length d terminated by a short circuit (Fig. 3.4) is given by (9.20), letting $\underline{Z}_L = 0$:

$$\underline{Z}_{in} = jZ_c \tan(\beta d) \qquad \Omega \qquad (3.24)$$

When a second short-circuit is placed at the input of the line (Fig. 3.5), it actually imposes the condition $Z_{in} = 0$, and by (3.24) leads to:

$$\tan(\beta d) = 0 \qquad 1 \qquad (3.25)$$

This condition is met when

$$\beta d = l\pi \qquad \text{integer } l \qquad 1 \qquad (3.26)$$

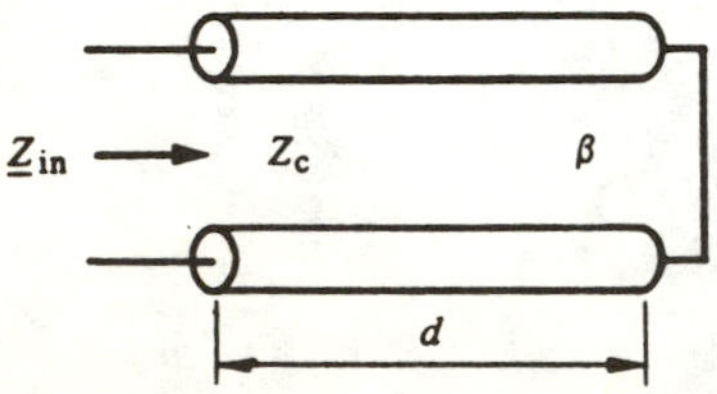

Fig. 3.4 Transmission line terminated by a short-circuit.

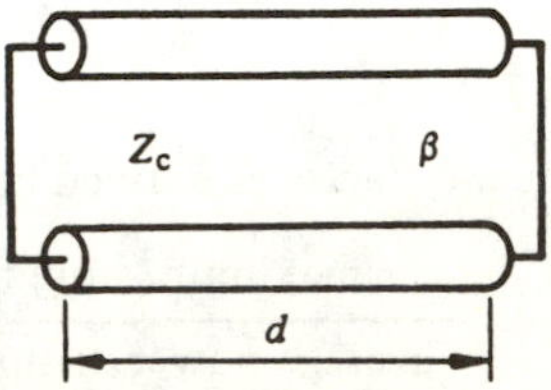

Fig. 3.5 Transmission line terminated by two short-circuits.

which in turn specifies the length d:

$$d = l\pi/\beta = l\lambda_g/2 \qquad \text{m} \tag{3.27}$$

At the resonant frequency, the length d must be a full number of half guide wavelengths $\lambda_g/2$ within the line or waveguide.

3.3.3 Waveguide with Two Short Circuits

The dispersion relation in a lossless waveguide above cutoff is given by (2.32):

$$\beta^2 = -\gamma^2 = k^2 - p^2 \qquad \text{m}^{-2} \tag{3.28}$$

The value of k_p^2 can be inferred:

$$k_p^2 = p^2 + (l\pi/d)^2 \qquad \text{m}^{-2} \tag{3.29}$$

3.3.4 Example: Rectangular Waveguide Cavity

For a cavity made up of a rectangular waveguide, of cross section $a \times b$ and of length d (Fig. 3.6), the values of p are given by (2.109), yielding in turn for k_p:

$$k_p^2 = k_{mnl}^2 = \left(\frac{m\pi}{a}\right)^2 + \left(\frac{n\pi}{b}\right)^2 + \left(\frac{l\pi}{d}\right)^2 \qquad \text{m}^{-2} \tag{3.30}$$

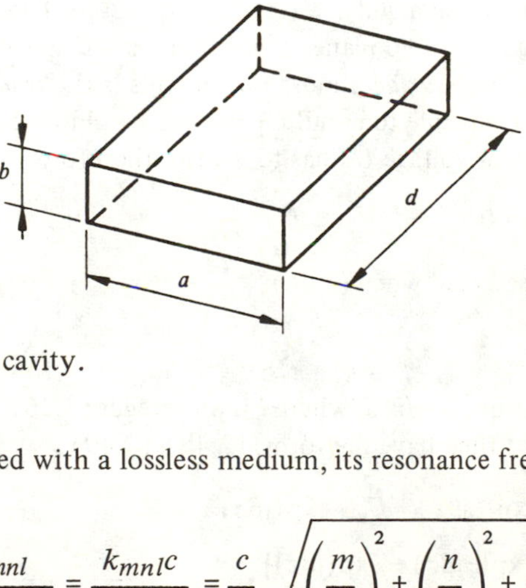

Fig. 3.6 Rectangular cavity.

When the cavity is filled with a lossless medium, its resonance frequencies are given by:

$$f_{mnl} = \frac{\omega_0}{2\pi} = \frac{k_{mnl}}{2\pi\sqrt{\epsilon\mu}} = \frac{k_{mnl}c}{2\pi} = \frac{c}{2}\sqrt{\left(\frac{m}{a}\right)^2 + \left(\frac{n}{b}\right)^2 + \left(\frac{l}{d}\right)^2} \qquad \text{Hz} \tag{3.31}$$

These frequencies correspond to the distance from the origin to points of coordinates $(cm/2a, cn/2b, cl/2d)$ in frequency space (Fig. 3.7).

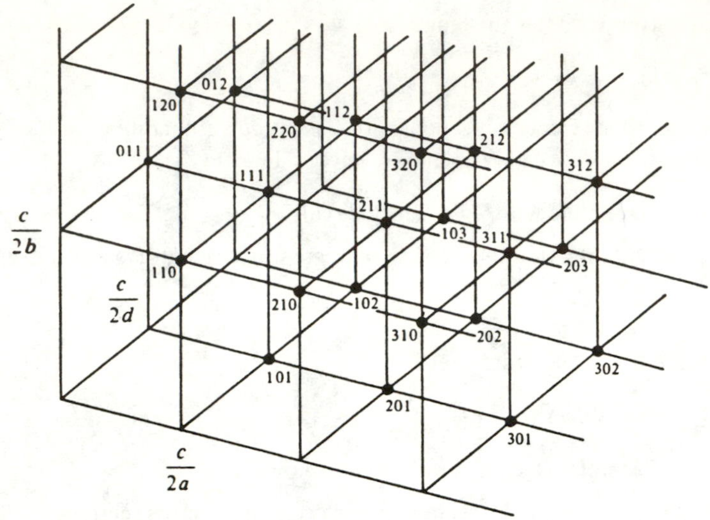

Fig. 3.7 Resonance modes in frequency space.

▪ 3.3.5 TE_{mnl} Mode in a Lossless Rectangular Cavity

The electromagnetic fields of the TE_{mnl} mode in the cavity are related to those of the TE_{mn} mode in rectangular waveguide (§ 2.3.7). The presence of a perfect electric conductor in the two planes at $z = 0$ and $z = d$ specifies that the electric field ***tangential to these surfaces*** must vanish: this is the ***transverse electric field*** $\underline{E}_t$. It is clearly not possible to annul $\underline{E}_T$, as this would lead to a trivial solution (2.38); the equivalent voltage $\underline{U}_e$ has to vanish, therefore, at both ends:

$$\underline{U}_e(0) = \underline{U}_e(d) = 0 \qquad \text{V} \tag{3.32}$$

This condition is satisfied when (2.48):

$$\underline{U}_{e-} = -\underline{U}_{e+} \qquad \text{V} \tag{3.33}$$

and when, in addition, $\beta = l\pi/\mathrm{d}$, where l is an integer (3.26). The equivalent voltage and current then have the forms (2.48), (2.49):

$$\begin{aligned}\underline{U}_e(z) &= \underline{U}_{e+} \exp(-\mathrm{j}\beta z) + \underline{U}_{e-} \exp(\mathrm{j}\beta z) \\ &= \underline{U}_{e+} [\exp(-\mathrm{j}\beta z) - \exp(\mathrm{j}\beta z)] \\ &= -2\mathrm{j}\,\underline{U}_{e+} \sin \beta z = -2\mathrm{j}\,\underline{U}_{e+} \sin(l\pi z/d) \qquad \text{V}\end{aligned} \tag{3.34}$$

$$\begin{aligned}\underline{I}_e(z) &= Y_e [\underline{U}_{e+} \exp(-\mathrm{j}\beta z) - \underline{U}_{e-} \exp(\mathrm{j}\beta z)] \\ &= Y_e \underline{U}_{e+} [\exp(-\mathrm{j}\beta z) + \exp(\mathrm{j}\beta z)] \\ &= 2\,Y_e \underline{U}_{e+} \cos\beta z = 2\,Y_e \underline{U}_{e+} \cos(l\pi z/d) \qquad \text{A}\end{aligned} \tag{3.35}$$

The fields are obtained by multiplying their transverse dependencies (2.115), (2.119), and (2.120) by the longitudinal dependence (3.34), (3.35). This is done using (2.38), (2.39), and (2.58):

$$\underline{E} = \mathrm{j}\omega\mu \underline{A}_{mn} \left(-\boldsymbol{e}_x \frac{n\pi}{b} \cos\frac{m\pi x}{a} \sin\frac{n\pi y}{b} \sin\frac{l\pi z}{d} + \boldsymbol{e}_y \frac{m\pi}{a} \sin\frac{m\pi x}{a} \cos\frac{n\pi y}{b} \sin\frac{l\pi z}{d} \right) \quad \text{V/m} \tag{3.36}$$

$$\underline{H} = \underline{A}_{mn} \left\{ \boldsymbol{e}_x \frac{l\pi}{d} \frac{m\pi}{a} \sin\frac{m\pi x}{a} \cos\frac{n\pi y}{b} \cos\frac{l\pi z}{d} + \boldsymbol{e}_y \frac{l\pi}{d} \frac{n\pi}{b} \cos\frac{m\pi x}{a} \sin\frac{n\pi y}{b} \cos\frac{l\pi z}{d} - \boldsymbol{e}_z \left[\left(\frac{m\pi}{a}\right)^2 + \left(\frac{n\pi}{b}\right)^2 \right] \cos\frac{m\pi x}{a} \cos\frac{n\pi y}{b} \sin\frac{l\pi z}{d} \right\} \quad \text{A/m} \tag{3.37}$$

where

$$\underline{A}_{mn} = 2\underline{U}_{e+} C_{mn}/\omega\mu \quad \text{Am} \tag{3.38}$$

one of the subscripts m or n may vanish, but the subscript l must always be non-zero (otherwise, the trivial solution $\underline{\boldsymbol{E}} \equiv 0, \underline{\boldsymbol{H}} \equiv 0$ is obtained).

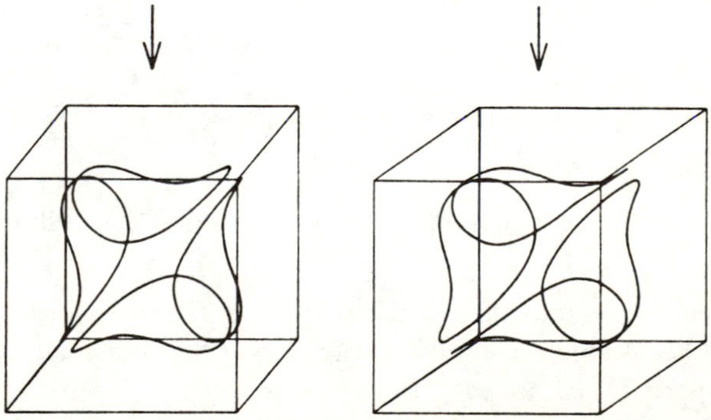

Fig. 3.8 Lines of magnetic field for the TE_{111} mode in a cubic cavity. This is a stereoscopic pair: look at the figure from a distance of about 25 cm, placing a sheet of paper so that one eye sees only one of the two images. After some time, the two images converge and one single three-dimensional image is perceived. It is recommended to let the arrows converge first, then to look at the image underneath.

One must therefore have:

$$(m+n)l \neq 0 \qquad 1 \qquad (3.39)$$

Magnetic field lines for the TE_{11} mode in a cubic cavity are represented in Figure 3.8.

3.3.6 TM_{mnl} Mode in a Lossless Rectangular Cavity

In an analogous manner, the fields of the TM_{mnl} mode are obtained, starting with the TM_{mn} mode in rectangular waveguide (2.107), (2.111) and (2.112):

$$\underline{E} = \underline{B}_{mn} \left\{ -e_x \frac{l\pi}{d} \frac{m\pi}{a} \cos\frac{m\pi x}{a} \sin\frac{n\pi y}{b} \sin\frac{l\pi z}{d} - e_y \frac{l\pi}{d} \frac{n\pi}{b} \sin\frac{m\pi x}{a} \cos\frac{n\pi y}{b} \sin\frac{l\pi z}{d} + e_z \left[\left(\frac{m\pi}{a}\right)^2 + \left(\frac{n\pi}{b}\right)^2\right] \sin\frac{m\pi x}{a} \sin\frac{n\pi y}{b} \cos\frac{l\pi z}{d} \right\} \quad \text{V/m} \qquad (3.40)$$

$$\underline{H} = \mathrm{j}\omega\epsilon \underline{B}_{mn} \left(e_x \frac{n\pi}{b} \sin\frac{m\pi x}{a} \cos\frac{n\pi y}{b} \cos\frac{l\pi z}{d} - e_y \frac{m\pi}{a} \cos\frac{m\pi x}{a} \sin\frac{n\pi y}{b} \cos\frac{l\pi z}{d} \right) \quad \text{A/m} \qquad (3.41)$$

letting

$$\underline{B}_{mn} = 2\underline{U}_{e+} C_{mn}/\mathrm{j}(l\pi/d) \qquad \text{Vm} \qquad (3.42)$$

The two subscripts m and n must here both be non-zero, as in the waveguide (2.108).

The subscript l, on the other hand, may vanish: the resonant frequency of the mode TM_{mn0} is independent of the cavity length, it is actually the cutoff frequency of the TM_{mn} mode in the waveguide.

3.3.7 Degenerate Modes of the Rectangular Cavity

When neither one of the integers m, n, and l vanishes, the same resonant frequency, f_{mnl}, is obtained for modes TE_{mnl} and TM_{mnl}, which are then degenerate. Two modes are called *degenerate* when they have the ***same eigenvalue*** (resonant frequency) but ***different eigenfunctions*** (field distributions). A similar situation is encountered in waveguide (§ 2.2.36). Any linear combination of two degenerate modes is also a resonant degenerate mode of the cavity.

□ 3.3.8 Hertz Potentials

Since the fields are solenoidal (divergenceless), Hertz potentials may be introduced: the magnetic one for a TE mode, the electric one for a TM mode [16].

Since its divergence vanishes, the $\underline{\boldsymbol{E}}$ field derives from a potential:

$$\underline{\boldsymbol{E}} = -\mathrm{j}\omega\mu \, \nabla \times \underline{\Lambda}_{mnl} \qquad \text{V/m} \qquad (3.43)$$

For a TE mode, this potential is given by the expression:

$$\underline{\Lambda}_{mnl} = \boldsymbol{e}_z \underline{A}_{mn} \cos \frac{m\pi x}{a} \cos \frac{n\pi y}{b} \sin \frac{l\pi z}{d} \qquad \text{Am} \qquad (3.44)$$

The magnetic field is obtained with (3.43) and (1.2):

$$\underline{\boldsymbol{H}} = \nabla \times \nabla \times \underline{\Lambda}_{mnl} \qquad \text{A/m} \qquad (3.45)$$

Similarly, the $\underline{\boldsymbol{H}}$ field derives from a potential:

$$\underline{\boldsymbol{H}} = \mathrm{j}\omega\epsilon \, \nabla \times \underline{\Pi}_{mnl} \qquad \text{A/m} \qquad (3.46)$$

For a TM mode, this potential is given by:

$$\underline{\Pi}_{mnl} = \boldsymbol{e}_z \underline{B}_{mn} \sin \frac{m\pi x}{a} \sin \frac{n\pi y}{b} \cos \frac{l\pi z}{d} \qquad \text{Vm} \qquad (3.47)$$

and the electric field is obtained with (3.46) and (1.3):

$$\underline{\boldsymbol{E}} = \nabla \times \nabla \times \underline{\Pi}_{mnl} \qquad \text{V/m} \qquad (3.48)$$

The expressions of Sections 3.3.5 and 3.3.6 are then obtained by derivation.

3.3.9 Comment on the Triple Symmetry

The rectangular cavity is a quite particular case, presenting uniformity along its three coordinate axes. There are thus three possible waveguide choices, for the three possible directions of propagation (section $a \times b$, length d; section $a \times d$, length b; section $b \times d$, length a). For each one of the three choices, the resonant modes may be designated TE or TM, and six different types of modes are defined in this manner; they are not linearly independent (degenerate).

The direction of propagation z is most often taken along the longest edge. In the present situation (Fig. 3.6), the sequence $d \geqslant a \geqslant b$ was selected. The field distributions obtained when choosing the direction of propagation along x or y are linear combinations of the TE and TM modes with respect to z, when modes are degenerate (§ 3.3.7).

3.3.10 Example: Cavity in Circular Waveguide

For a cavity formed of a circular waveguide section of radius a and of length d (Fig. 3.9), the values of p correspond to the zeros and extrema of Bessel functions (2.141) and (2.148):

$$k_{mnl}^{2\,(\mathrm{TE,TM})} = p_{mn}^{2\,(\mathrm{TE,TM})} + \left(\frac{l\pi}{d}\right)^2 \qquad \mathrm{m}^{-2} \qquad (3.49)$$

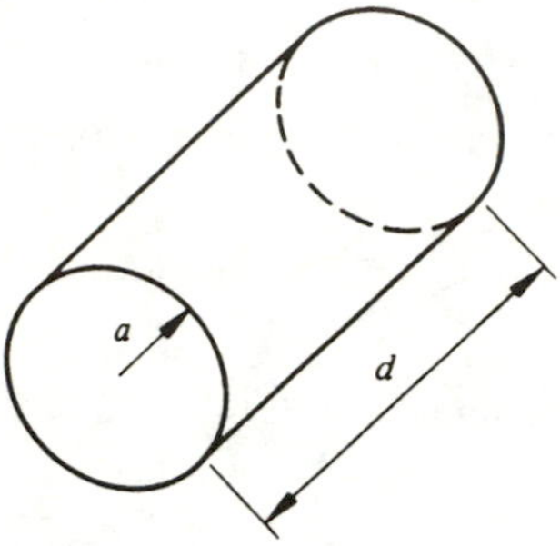

Fig. 3.9 Circular waveguide cavity.

For a cavity filled with a lossless medium, the resonant frequencies are given by:

$$f_{mnl} = \frac{k_{mnl}c}{2\pi} = \frac{c}{2}\sqrt{\left(\frac{p_{mn}^{\mathrm{TE,TM}}}{\pi}\right)^2 + \left(\frac{l}{d}\right)^2} \qquad \mathrm{Hz} \qquad \mathbf{(3.50)}$$

Figure 3.10 permits the quick determination of the resonant frequencies of the first modes when the dimensions a and d are known.

3.3.11 TE$_{mnl}$ Mode in a Lossless Circular Cavity

Here also, the field distribution is obtained from the solution for circular waveguide, given by (2.149), (2.151), and (2.152). Proceeding exactly as in Section 3.3.5, one obtains:

$$\underline{E} = \mathrm{j}\omega\mu\underline{A}_{mn}\left(e_\rho \frac{m}{\rho} \mathrm{J}_m(p_{mn}^{\mathrm{TE}}\rho)\begin{Bmatrix}-\sin m\varphi\\ \cos m\varphi\end{Bmatrix} \sin\frac{l\pi z}{d} - e_\varphi p_{mn}^{\mathrm{TE}} \mathrm{J}'_m(p_{mn}^{\mathrm{TE}}\rho)\begin{Bmatrix}\cos m\varphi\\ \sin m\varphi\end{Bmatrix} \sin\frac{l\pi z}{d}\right) \qquad \mathrm{V/m} \qquad (3.51)$$

$$\underline{H} = -\underline{A}_{mn}\left(e_\rho \frac{l\pi}{d} p_{mn}^{\mathrm{TE}} \mathrm{J}'_m(p_{mn}^{\mathrm{TE}}\rho) \begin{Bmatrix}\cos m\varphi\\ \sin m\varphi\end{Bmatrix} \cos\frac{l\pi z}{d}\right.$$

$$+e_\varphi \frac{l\pi}{d}\frac{m}{\rho} J_m(p_{mn}^{TE}\rho) \begin{Bmatrix} -\sin m\varphi \\ \cos m\varphi \end{Bmatrix} \cos\frac{l\pi z}{d}$$

$$+e_z (p_{mn}^{TE})^2 J_m(p_{mn}^{TE}\rho) \begin{Bmatrix} \cos m\varphi \\ \sin m\varphi \end{Bmatrix} \sin\frac{l\pi z}{d}\Bigg) \quad \text{A/m} \qquad (3.52)$$

where the subscript l cannot vanish.

3.3.12 TM$_{mnl}$ Mode in a Lossless Circular Cavity

Proceeding as in the previous sections, making use of expressions (2.143), (2.145), and (2.146) one obtains the field distributions:

$$\underline{E} = \underline{B}_{mn} \Bigg(-e_\rho \frac{l\pi}{d} p_{mn}^{TM} J'_m(p_{mn}^{TM}\rho) \begin{Bmatrix} \cos m\varphi \\ \sin m\varphi \end{Bmatrix} \sin\frac{l\pi z}{d}$$

$$-e_\varphi \frac{l\pi}{d}\frac{m}{\rho} J_m(p_{mn}^{TM}\rho) \begin{Bmatrix} -\sin m\varphi \\ \cos m\varphi \end{Bmatrix} \sin\frac{l\pi z}{d}$$

$$+e_z (p_{mn}^{TM})^2 J_m(p_{mn}^{TM}\rho) \begin{Bmatrix} \cos m\varphi \\ \sin m\varphi \end{Bmatrix} \cos\frac{l\pi z}{d}\Bigg) \quad \text{V/m} \qquad (3.53)$$

$$\underline{H} = j\omega\epsilon\, \underline{B}_{mn} \Bigg(e_\rho \frac{m}{\rho} J_m(p_{mn}^{TM}\rho) \begin{Bmatrix} -\sin m\varphi \\ \cos m\varphi \end{Bmatrix} \cos\frac{l\pi z}{d}$$

$$-e_\varphi p_{mn}^{TM} J'_m(p_{mn}^{TM}\rho) \begin{Bmatrix} \cos m\varphi \\ \sin m\varphi \end{Bmatrix} \cos\frac{l\pi z}{d}\Bigg) \quad \text{A/m} \qquad (3.54)$$

The subscript l may vanish in this case, the resonant frequency then being independent of the length d of the cavity (horizontal lines in Figure 3.10).

3.3.13 Example: Coaxial Cavity, TEM Resonant Modes

For the TEM resonant modes, $p = 0$ and then (3.29) becomes:

$$k_l^{TEM} = l\pi/d \quad \text{m}^{-1} \qquad (3.55)$$

$$f_l^{TEM} = cl/2d \quad \text{Hz} \qquad \textbf{(3.56)}$$

The eigenvalue is only related to the length of the cavity and to the filling medium, but not to the shape and size of the cross section.

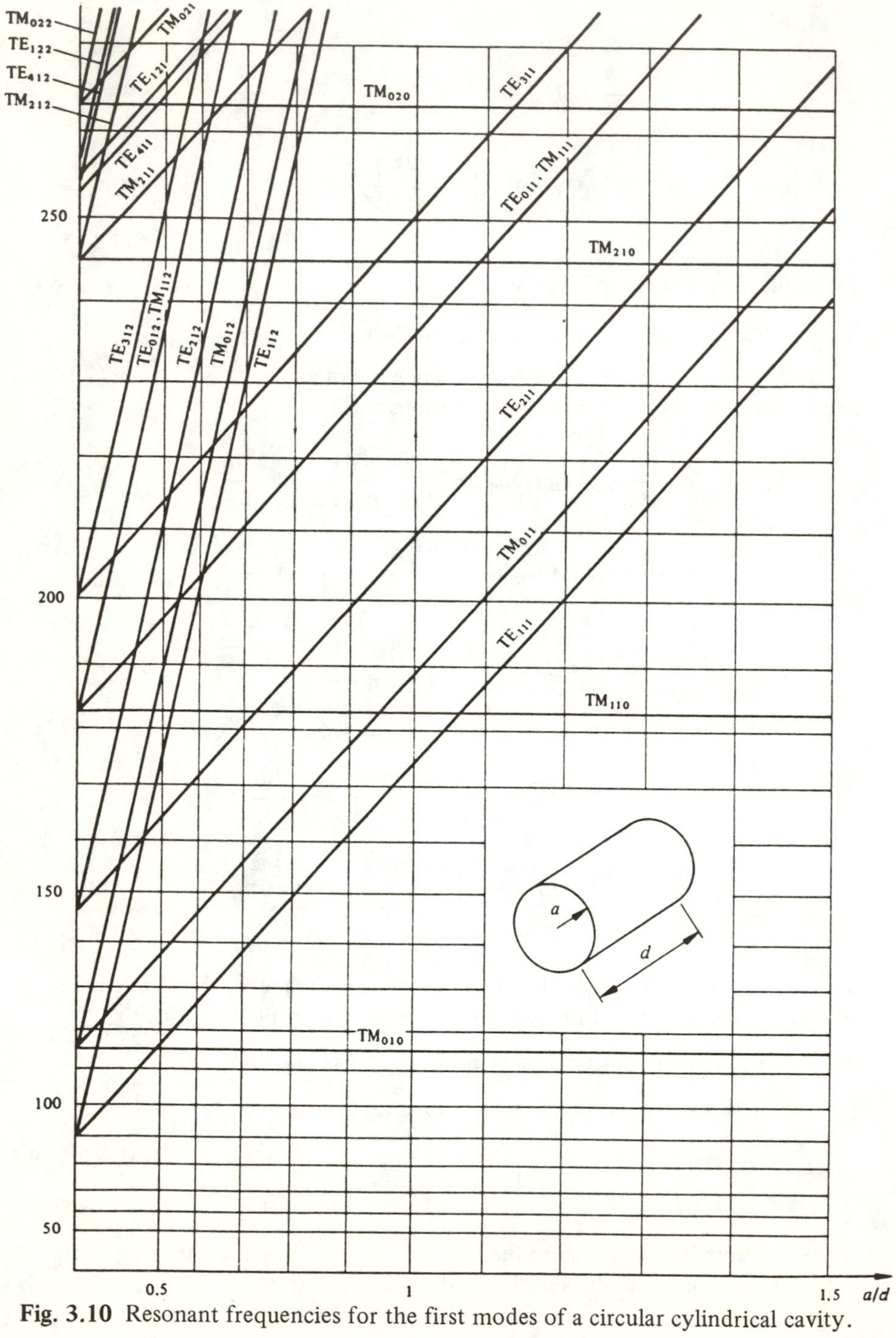

Fig. 3.10 Resonant frequencies for the first modes of a circular cylindrical cavity.

3.3.14 Note: Higher Order Modes

TE and TM modes can propagate on a coaxial line when the signal frequency exceeds their respective cutoff frequencies [49] : a coaxial cavity therefore has, in addition to its TEM resonant modes (§ 3.3.13), resonant TE_{mnl} and TM_{mnl} modes.

3.3.15 Cavity Made Up of a Line Section Terminated by an Impedance

A transmission line terminated at one end by a short-circuit (Fig. 3.11) may also form a resonator when the load impedance $\underline{Z}_L$ is very different from the characteristic impedance of the line $\underline{Z}_c$. For resonance to occur, the losses in the system must remain small.

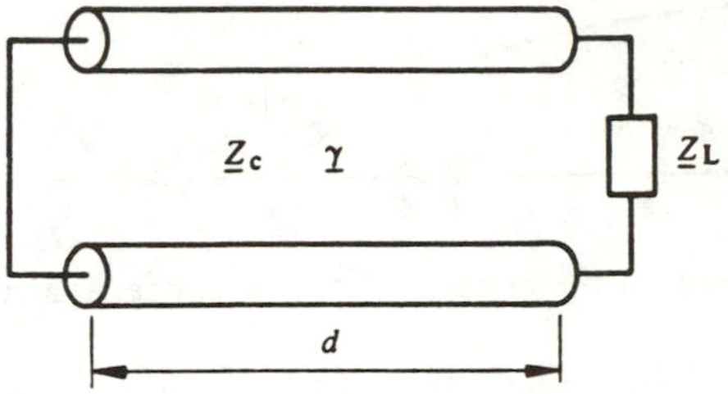

Fig. 3.11 Definition of the equivalent model.

The boundary conditions in the short-circuit plane are met when the impedance in this plane vanishes; this occurs for (9.18):

$$\underline{Z}_L + \underline{Z}_c \tan \underline{\gamma} d = 0 \qquad \Omega \qquad (3.57)$$

In the ideal lossless condition, $\underline{\gamma} = j\beta$, $\underline{Z}_L = jX_L$, and Z_c is real:

$$X_L + Z_c \tan\beta d = 0 \qquad \Omega \qquad (3.58)$$

one then deduces:

$$d = \frac{l\lambda_g}{2} - \frac{1}{\beta} \arctan(X_L/Z_c) = \frac{\lambda_g}{2} \left[l - \frac{1}{\pi} \arctan(X_L/Z_c) \right] \qquad \text{m} \qquad (3.59)$$

This dependence is shown in Figure 3.12.

The particular case of Section 3.3.2 is obtained with a lossless line and a load $\underline{Z}_L = 0$. The short-circuit resonances are found in the figure for $d = l\lambda_g/2$. In addition, a line terminated by an open circuit ($X_L = \infty$) resonates for $d = \lambda_g/4 + l\lambda_g/2$.

When a short circuit is replaced by an inductive load, the relative length (d/λ_g) decreases (Fig. 3.12). For a fixed value of d, the wavelength increases and, therefore, the resonant frequency is reduced. Replacing a short-circuit by a capaci-

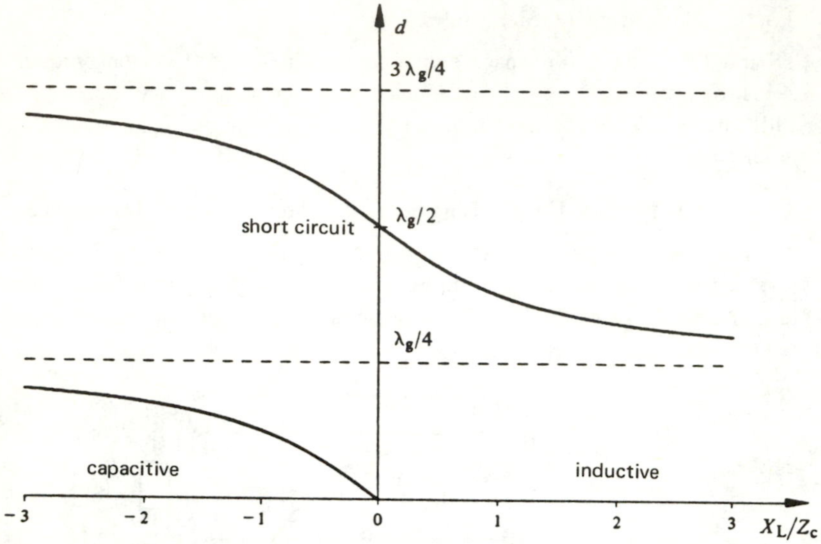

Fig. 3.12 Length of the resonator for resonance, as a function of load reactance.

tance leads to an increase in frequency. By duality, the contrary occurs when the starting point is an open circuit.

■ 3.3.16 Resonance of a Microstrip Line Section: Fringing Capacitance

A section of microstrip line having an upper conductor of length d (Fig. 3.13) is also a resonator. As the fields extend slightly beyond the ends of the line on both sides, these are not ideal open circuits, but may be represented by fringing capacitive reactances (§ 6.3.23), when one neglects radiation losses (§ 2.11.7). The equivalent circuit of Figure 3.14 is then obtained: a transmission line of length d terminated at both ends by *fringing capacitances* C_0. The value of the latter depends on dimensions and substrate permittivity (Fig. 3.15) [50].

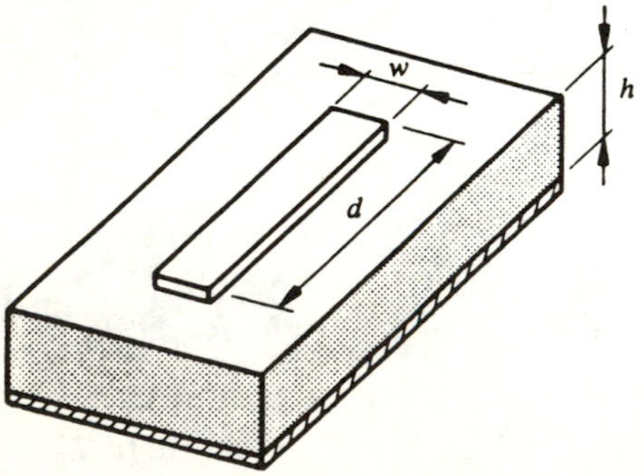

Fig. 3.13 Microstrip resonator.

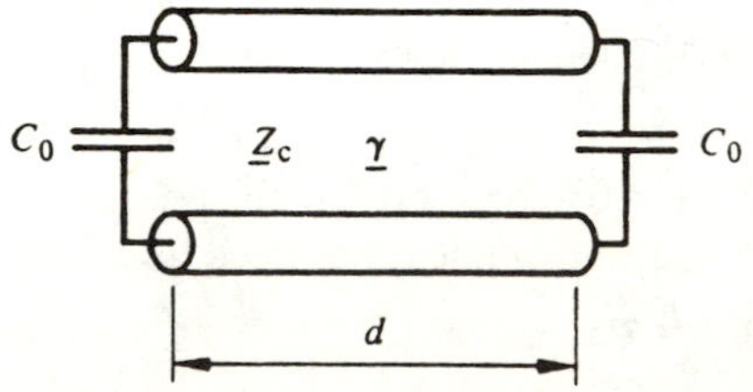

Fig. 3.14 Equivalent circuit.

The resonant frequencies are obtained with equation (9.18)

$$\underline{Z}_L + \underline{Z}_c \frac{\underline{Z}_L + \underline{Z}_c \tanh \underline{\gamma} d}{\underline{Z}_c + \underline{Z}_L \tanh \underline{\gamma} d} = 0 \qquad \Omega \qquad (3.60)$$

Solving for tanh $\underline{\gamma} d$ yields:

$$\tanh \underline{\gamma} d = - \frac{2\underline{Z}_c \underline{Z}_L}{\underline{Z}_c^2 + \underline{Z}_L^2} = - \frac{2\underline{Z}_L/\underline{Z}_c}{1 + (\underline{Z}_L/\underline{Z}_c)^2} = - \frac{2\underline{Z}_c/\underline{Z}_L}{1 + (\underline{Z}_c/\underline{Z}_L)^2} \qquad 1 \qquad (3.61)$$

When both the line and the load are lossless, $\underline{Z}_L$ is purely imaginary, $\underline{Z}_c$ real, and $\underline{\gamma} = \mathrm{j}\beta$. Introducing the impedance of the capacitive load $\underline{Z}_L = 1/\mathrm{j}\omega C_0$ one then finds:

$$\tan \beta d = \frac{2\omega C_0 Z_c}{(\omega C_0 Z_c)^2 - 1} \qquad 1 \qquad (3.62)$$

The resonant frequencies are smaller than those of a line ending with two open circuits ($C_0 = 0$).

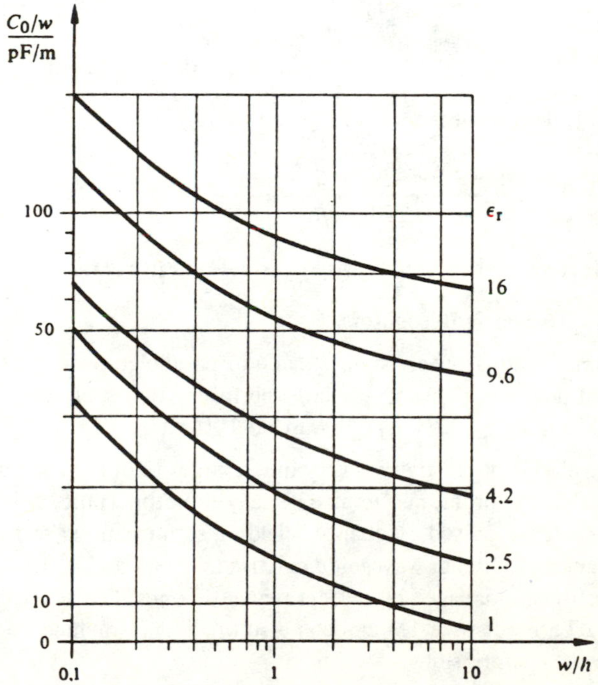

Fig. 3.15 Equivalent capacitance of an open-ended microstrip.

3.3.17 Resonance of a Microstrip Line Section: Effective Lengthening

The end effects, represented by fringing capacitances in Section 3.3.16, decrease the resonant frequency. This result may also be obtained by an *equivalent lengthening* Δd of the line at both ends, as shown in Figure 3.16.

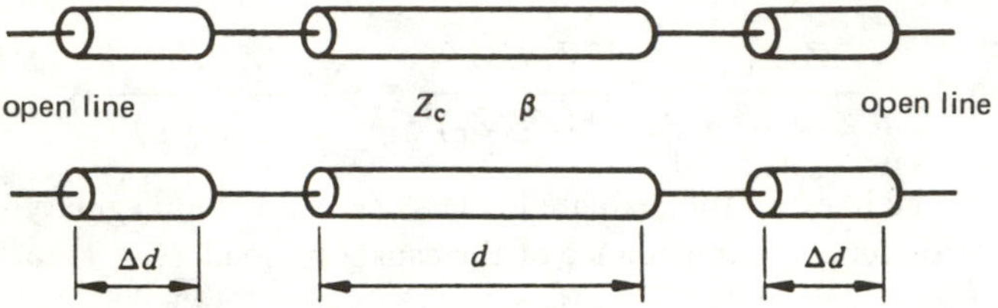

Fig. 3.16 Equivalent circuit for a microstrip resonator.

Here, the capacitances are replaced by sections of transmission line (having the same characteristic Z_c and β). Resonances then appear for:

$$\beta(d + 2\,\Delta d) = \pi \qquad l \text{ integer} \qquad 1 \qquad (3.63)$$

that is, when

$$d + 2\,\Delta d = l\lambda_g/2 \qquad \text{m} \qquad (3.64)$$

The equivalent lengthening Δd is given by [34]:

$$\Delta d = 0.412\,h\,\frac{\epsilon_e + 0.300}{\epsilon_e - 0.258}\,\frac{w/h + 0.262}{w/h + 0.813} \qquad \text{m} \qquad (3.65)$$

The effective relative permittivity ϵ_e is defined in Section 2.11.5.

3.3.18 Dielectric Resonators

Very small resonators are simply made of parallelepipeds or of cylinders (Fig. 3.17) of dielectrics having very large permittivities, such as rutile (titanium dioxide, TiO_2, $\epsilon_r = 100(1 - j\,0.00025)$ at 100 MHz).

As the relative permittivity is very much larger than the surrounding air, the walls are, to the first order, perfect magnetic conductors (pmc, open circuit § 1.4.3). This problem is solved by duality, yielding results similar to those obtained for rectangular or circular waveguide cavities (§ 3.3.4 and § 3.3.10). Perfect electric conductors are replaced by perfect magnetic ones, $\underline{\boldsymbol{E}}$ fields by $\underline{\boldsymbol{H}}$ fields and vice versa, a TE mode by a TM mode and so on. The resonance frequencies are given by the same expressions.

For a cylindrical cavity, a more thorough study considers a dielectric waveguide of length d terminated by two open circuits, utilizing the Section 2.10.

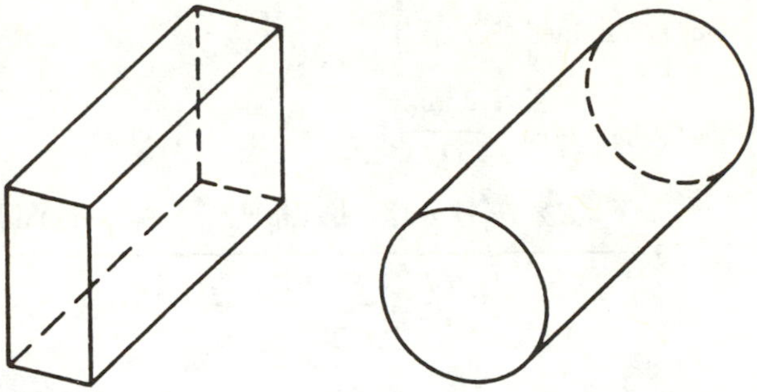

Fig. 3.17 Dielectric resonators.

A more detailed study of the problem is only possible using computer techniques [51].

3.4 PERTURBATION METHODS

3.4.1 Introduction

The change in the complex angular frequency produced by a small modification of the cavity is determined with the ***perturbation method***, which is analogous to the one developed for waveguides in Section 2.7. It is used in particular to evaluate the effects of wall losses, of the introduction of samples of material, or of a change in shape. The development closely parallels the one for the cutoff frequency in a waveguide (§ 2.7.9). In the present situation the system is three-dimensional and may be lossy.

■ 3.4.2 Mathematical Development

An unperturbed ***lossless*** cavity, filled with a homogeneous medium of properties ϵ and μ is first considered. In this medium, the fields satisfy expressions:

$$\nabla \times \underline{E}_0 + \mathrm{j}\, \omega_{\mathrm{p}0}\, \mu \underline{H}_0 = 0 \qquad \mathrm{V/m^2} \qquad (3.66)$$

$$\nabla \times \underline{H}_0 - \mathrm{j}\, \omega_{\mathrm{p}0}\, \epsilon \underline{E}_0 = 0 \qquad \mathrm{A/m^2} \qquad (3.67)$$

When no losses are present, the unperturbed resonant angular frequency $\omega_{\mathrm{p}0}$ is real.

In the cavity perturbed by ***source currents*** $\underline{J}_{\mathrm{e}}$ and $\underline{J}_{\mathrm{m}}$ the expressions become:

$$\nabla \times \underline{E} + \mathrm{j}\, \underline{\omega}_{\mathrm{p}}\, \mu \underline{H} + \underline{J}_{\mathrm{m}} = 0 \qquad \mathrm{V/m^2} \qquad (3.68)$$

$$\nabla \times \underline{H} - \mathrm{j}\, \underline{\omega}_{\mathrm{p}}\, \epsilon \underline{E} - \underline{J}_{\mathrm{e}} = 0 \qquad \mathrm{A/m^2} \qquad (3.69)$$

where the resonant angular frequency $\underline{\omega}_{\mathrm{p}}$ becomes complex when losses are involved. Proceeding as in Section 2.7, a perturbation expression for the difference in frequencies is obtained (3.19):

$$\underline{\omega}_p - \omega_{po} = \omega_{pr} - \omega_{po} + \frac{2\,j\omega_0}{Q_0}$$

$$= j\,\frac{\oint_S (\underline{E} \times \underline{H}_0^*) \cdot \boldsymbol{n}\,dl + \int_V (\underline{J}_e \cdot \underline{E}_0^* + \underline{J}_m \cdot \underline{H}_0^*)\,dV}{\int_V (\epsilon \underline{E}_0^* \cdot \underline{E} + \mu \underline{H}_0^* \cdot \underline{H})\,dV} \qquad s^{-1} \qquad \textbf{(3.70)}$$

3.4.3 Note

The unloaded quality factor Q_0 which appears in (3.70) considers the losses produced by the wall surfaces or by samples placed in the cavity.

3.4.4 Effect of Wall Losses, Metal Quality Factor Q_{0m}

The same assumptions that were made for waveguides (§ 2.7.3) are made here, namely:

1. no source currents, $\underline{J}_e = 0$, $\underline{J}_m = 0$;
2. the perturbation does not affect to first order the magnetic field, $\underline{H} \cong \underline{H}_0$;
3. the perturbation does not affect the overall electric field, so that $\underline{E} \cong \underline{E}_0$ in the denominator of (3.70), where the integration is carried out over the whole cavity volume;
4. on the other hand, the electric field tangential to the cavity wall no longer vanishes, but is proportional to the surface current, which is in turn proportional to the tangential magnetic field (the normal vector $\boldsymbol{n}$ is pointed towards the outside).

$$\underline{E}_{tan} = \underline{Z}_m \underline{A} = -\underline{Z}_m(\boldsymbol{n} \times \underline{H}) \cong -\underline{Z}_m(\boldsymbol{n} \times \underline{H}_0) \qquad \text{V/m} \qquad (3.71)$$

where $\underline{Z}_m \cong (1 + j)\sqrt{\omega\mu_m/2\sigma}$ is the wave impedance in the metal (2.182).

Utilizing expressions (3.70) and (3.71), an expression is obtained for the *metallic quality factor* Q_{0m} and the change of resonance angular frequency:

$$\frac{2\omega_0}{Q_{0m}} - j(\omega_{pr} - \omega_{po}) \cong \underline{Z}_m \frac{\oint_S |\underline{H}_0|^2\,dA}{2\int_V \mu |\underline{H}_0|^2\,dV} \qquad s^{-1} \qquad \textbf{(3.72)}$$

The change in resonant angular frequency due to the wall losses is most often negligible. As the real and imaginary parts of (3.72) have the same value, one deduces that:

$$Q_{\mathrm{om}} = \frac{2\omega_0}{\omega_{\mathrm{po}} - \omega_{\mathrm{pr}}} \qquad 1 \qquad (3.73)$$

3.4.5 Application: Rectangular Waveguide Cavity

Introducing into (3.72) the values of the magnetic field determined in (3.37) and in (3.41), one obtains, following integration [52]:

1. for a TE_{mnl} mode

$$Q_{\mathrm{om}} = \frac{\pi Z_0}{R_{\mathrm{m}}} \frac{abd}{4} \cdot$$

$$\frac{(u^2 + v^2)(u^2 + v^2 + w^2)^{3/2}}{\zeta ad\,[u^2 w^2 + (u^2 + v^2)^2] + \eta bd\,[v^2 w^2 + (u^2 + v^2)^2] + abw^2(u^2 + v^2)} \qquad 1 \qquad (3.74)$$

with

$$u = m/a, \; v = n/b, \; w = l/d, \; \zeta = \begin{cases} 1 & \text{if } n \neq 0 \\ 1/2 & \text{if } n = 0 \end{cases}, \; \eta = \begin{cases} 1 & \text{if } m \neq 0 \\ 1/2 & \text{if } m = 0 \end{cases}$$

2. for a TM_{mnl} mode

$$Q_{\mathrm{om}} = \frac{\pi Z_0}{R_{\mathrm{m}}} \frac{abd}{4} \frac{(u^2 + v^2)(u^2 + v^2 + w^2)^{1/2}}{u^2 b(\eta a + d) + v^2 a(\eta b + d)} \qquad 1 \qquad (3.75)$$

$$\text{with } \eta = \begin{cases} 1 & \text{if } l \neq 0 \\ 1/2 & \text{if } l = 0 \end{cases}$$

The change of the quality factor, as a function of cavity size, is represented in Figure 3.18 for a particular geometry.

3.4.6 Application: Circular Waveguide Cavity

The expressions for the magnetic field (3.52) and (3.54) are introduced into (3.72), yielding after integration:

1. for a TE_{mnl} mode

$$Q_{\mathrm{om}} = \frac{Z_0 a}{2R_{\mathrm{m}}} \frac{[1 - (m/p_{mn}^{\mathrm{TE}} a)^2]\,[(p_{mn}^{\mathrm{TE}})^2 + (l\pi/d)^2]^{3/2}}{(p_{mn}^{\mathrm{TE}})^2 + (l\pi/d)^2(2a/d) + (1 - 2a/d)(mn\pi/d)^2(1/ap_{mn}^{\mathrm{TE}})^2} \qquad 1 \qquad (3.76)$$

2. for a TM_{mnl} mode

$$Q_{0m} = \frac{Z_0 a}{2R_m} \frac{\sqrt{(p_{mn}^{TM})^2 + (l\pi/d)^2}}{1 + \eta 2a/d} \qquad 1 \tag{3.77}$$

The geometrical dependence of the quality factor is shown in Figure 3.18 for two cavities.

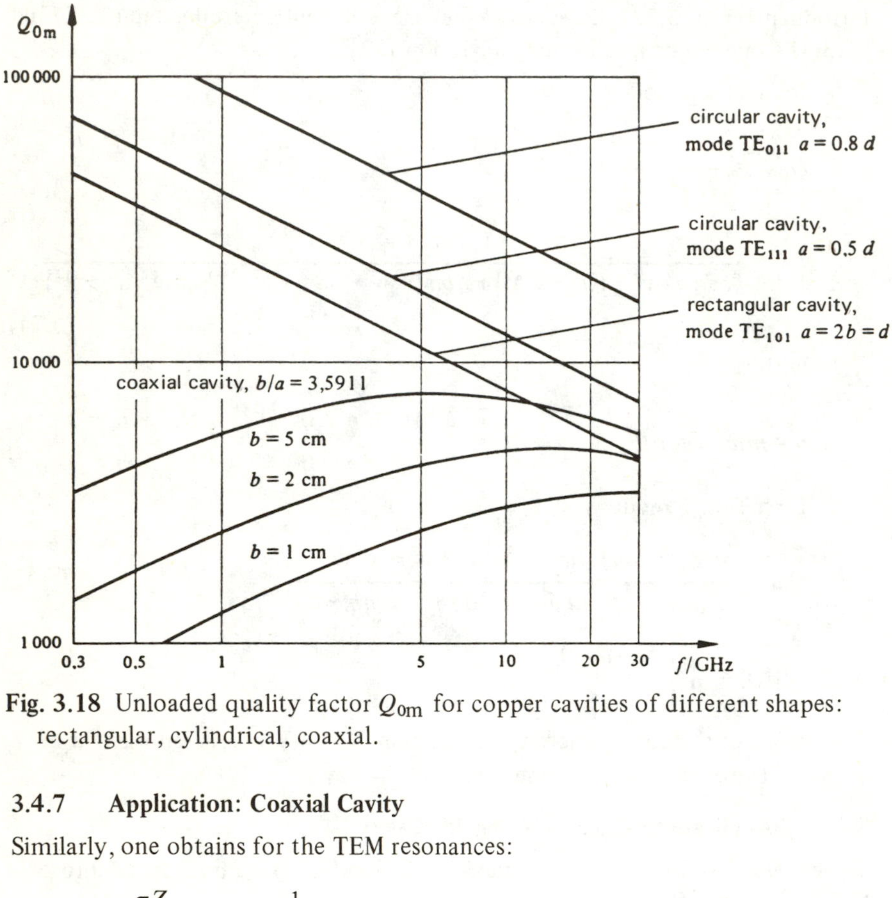

Fig. 3.18 Unloaded quality factor Q_{0m} for copper cavities of different shapes: rectangular, cylindrical, coaxial.

3.4.7 Application: Coaxial Cavity

Similarly, one obtains for the TEM resonances:

$$Q_{0m} = \frac{\pi Z_0}{R_m} \frac{1}{4 + \dfrac{2d}{b} \dfrac{1 + b/a}{\ln(b/a)}} \qquad 1 \tag{3.78}$$

The effect of dimensions on the quality factors is represented in Figure 3.18 for three cavities having the optimum diameter ratio $b/a = 3.5911$ (§ 2.7.7).

3.4.8 Application: Microstrip Resonator

For the resonant structure formed of a microstrip line section, represented in

Figure 3.13, the quality factor for the metal losses is approximately given by (2.280):

$$Q_{om} = \frac{\beta}{2\alpha_c} = \frac{\pi f \sqrt{\epsilon_e}}{\alpha_c c_0} \cong \frac{\pi f \sqrt{\epsilon_e}}{c_0} \frac{wZ_c}{R_m} \qquad 1 \qquad (3.79)$$

3.4.9 Dielectric Sample in a Cavity, Sample Quality Factor Q_{oe}

An ***air-filled*** cavity with lossless walls is considered; a dielectric sample of relative complex permittivity $\underline{\epsilon}_r$ is then introduced (Fig. 3.19). The source current density $\boldsymbol{J}_e$ in the volume ΔV of the sample is given by (2.197):

$$\underline{J}_e = j\omega_{pr}\,\epsilon_0(\underline{\epsilon}_r - 1)\underline{E} \qquad \text{A/m}^2 \qquad (3.80)$$

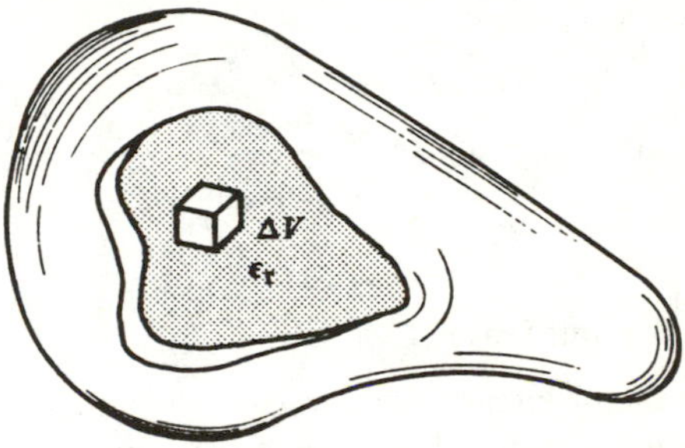

Fig. 3.19 Dielectric sample within a cavity.

The introduction of the dielectric material (small sample) is assumed to leave the cavity fields more or less as they were in the unperturbed situation, so that an approximate expression is obtained, defining the *sample quality factor* Q_{oe}:

$$\underline{\omega}_p - \omega_{po} = \omega_{pr} - \omega_{po} + 2j\frac{\omega_{po}}{Q_{oe}} \cong -\frac{\omega_{pr}}{2} \frac{\int_{\Delta V}(\underline{\epsilon}_r - 1)\,|\underline{E}_0|^2 \,\mathrm{d}V}{\int |\underline{E}_0|^2 \,\mathrm{d}V} \qquad \mathrm{s}^{-1} \qquad \textbf{(3.81)}$$

The denominator integral covers the whole cavity volume, the one in the numerator the volume of the dielectric sample only. Making use of this relation, it is possible to determine the permittivity of a material, by introducing a small sample of it into the cavity (§ 8.5.11). The change in resonant frequency is used to determine ϵ_r', that of the quality factor yields ϵ_r'' (§ 1.4.2). This expression can also be used to experimentally determine the amplitude distribution of the electric field inside the cavity: a sample of material of known properties is moved in the cavity, and the shift in resonant frequency is determined as a function of the sample position.

□ 3.4.10 Ferrite Sample in a Cavity

In a cavity filled with a non-magnetic medium (air, dielectric), a piece of magnetized ferrite is introduced (§ 6.7.6). In addition to the perturbation due to its permittivity, a change due to its magnetic properties appears, represented by a source magnetic current within the ferrite volume ΔV:

$$\underline{J}_m = j\omega_{pr}\mu_0(\bar{\bar{\underline{\mu}}}_r - 1)\underline{H} \qquad \text{V/m}^2 \qquad (3.82)$$

The relative permeability $\bar{\bar{\underline{\mu}}}_r$ is a ***tensor***, meaning that the material is ***anisotropic***, *i.e.*, the induction $\underline{\boldsymbol{B}}$ and the magnetic field $\underline{\boldsymbol{H}}$ are not colinear (Sec. 6.7, § 1.4.6).

To distinguish the dielectric perturbation from the magnetic one, a small ferrite sample is placed at a location in the cavity where the electric field vanishes. The measured variation is then related only to the magnetic properties of the ferrite by the expression:

$$\underline{\omega}_p - \omega_{po} = \omega_{pr} - \omega_{po} + 2j\frac{\omega_0}{Q_{oe}} \cong -\frac{\omega_{pr}}{2}\frac{\int_{\Delta V}\underline{\boldsymbol{H}}_0^* \cdot (\bar{\bar{\underline{\mu}}}_r - \bar{\bar{1}}) \cdot \underline{\boldsymbol{H}}_0 \, dV}{\int_V |\underline{\boldsymbol{H}}_0|^2 \, dV} \qquad \text{s}^{-1} \qquad (3.83)$$

where $\bar{\bar{1}}$ is the unit tensor.

The denominator integral extends over the whole volume of the cavity, the one in the numerator over the volume of the sample. The ***hermitian*** part of the relative permeability tensor $\bar{\bar{\underline{\mu}}}_r$ modifies the resonant frequency, the ***antihermitian*** part affects the sample quality factor Q_{oe} [53].

3.4.11 Modification of the Shape of a Cavity

The volume of the perturbed cavity is smaller by an amount ΔV than the volume of the unperturbed one. This perturbation is localized on the cavity envelope, in both instances, the cavity walls are perfectly conducting (Fig. 3.20).

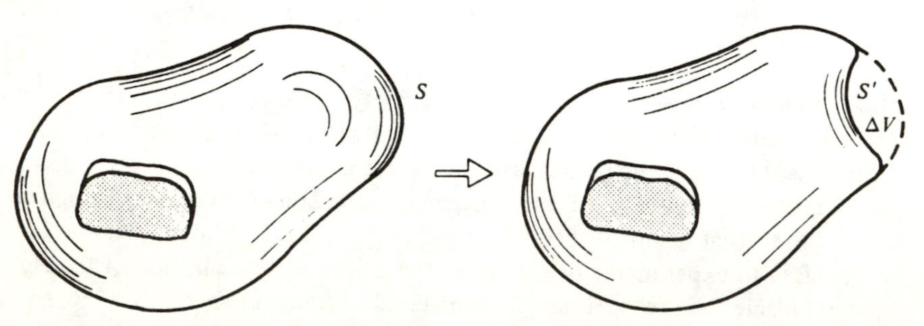

Fig. 3.20 Alteration of the shape of a cavity.

The treatment for a small perturbation parallels the one previously given for a waveguide in Section 2.7.11. The same arguments are used, giving:

$$\omega_{pr} - \omega_{po} = -\omega_{pr} \frac{\int_{\Delta V} (\epsilon \, |\underline{E}_0|^2 - \mu \, |\underline{H}_0|^2) \mathrm{d}V}{\int_V (\epsilon \, |\underline{E}_0|^2 + \mu \, |\underline{H}_0|^2) \mathrm{d}V} \qquad \text{rad/s} \qquad (3.84)$$

When the reduction in size occurs at a place where the electric energy is predominant, the resonant frequency decreases; it increases when the change in volume takes place in a position of large magnetic field. Results in the opposite direction are obtained when enlarging the cavity.

3.4.12 Global Unloaded Quality Factor Q_0, Overvoltage Factor

All losses reduce the quality factor, which would become infinite for an absolutely lossless cavity. Several loss-producing mechanisms were considered in the previous paragraphs:

1. The homogeneous medium which fills the cavity is (slightly) lossy. The losses determine the ***volume*** quality factor Q_{ov} in Section 3.2.10.
2. The metal walls enclosing the cavity always present ohmic losses, as the conductor never is perfectly conducting. The ***metallic*** quality factor Q_{om} takes this source of damping into account; its expression is given in Section 3.4.4.
3. The introduction of a sample of lossy material in a cavity produces additional losses, represented by the ***sample*** quality factor Q_{oe} (§ 3.4.9 and § 3.4.10).

In practical situations, with the exception of microwave ovens, cavity losses remain small and there is no significant interaction between the different loss mechanisms. The *global unloaded quality factor* Q_0 or *overvoltage factor* is then obtained by adding the three contributions, losses being inversely proportional to the quality factors:

$$\frac{1}{Q_0} = \frac{1}{Q_{ov}} + \frac{1}{Q_{om}} + \frac{1}{Q_{oe}} \qquad 1 \qquad (3.85)$$

The second term in the right-hand side is non-zero for any resonator having metal walls. Very high quality factors may be achieved when using supraconductors: values of 10^7 [54] and even of 10^9 [55] have been reported in the literature. At the ambient temperature, quality factors of microwave cavities are located between several hundreds and several thousands. Lower values are obtained with microstrip resonators, for which $Q_{ov} \cong \epsilon_r'/\epsilon_r'' = 1/\tan\delta$.

The global unloaded quality factor will be used to consider the effect of losses in the following developments.

3.5 OPEN CAVITIES

3.5.1 General Remarks

In every practical application where the resonant properties of a cavity are to be used, some form of ***coupling*** has to be established between the cavity and the outside world. An opening has to be cut through the wall, permitting the connection of the cavity to a waveguide or, using probes, to a two-conductor line. The opening must be sufficiently small so that the fields within the cavity are not significantly perturbed: the developments of the previous sections should remain valid. In addition, the opening (or the probe) must be located where it can provide a transfer of signal from the line to the resonant circuit: the field distribution in the cavity and on the element used for the excitation must be compatible, in which case coupling occurs (Sec. 3.6). The transmission line is assumed to be lossless.

3.5.2 Schematic Description of a Cavity

Following the norm IEC 1129/1150.1.1/1150.1.2 [56], a cavity coupled to one or to two transmission lines is schematically represented by the symbols of Figure 3.21.

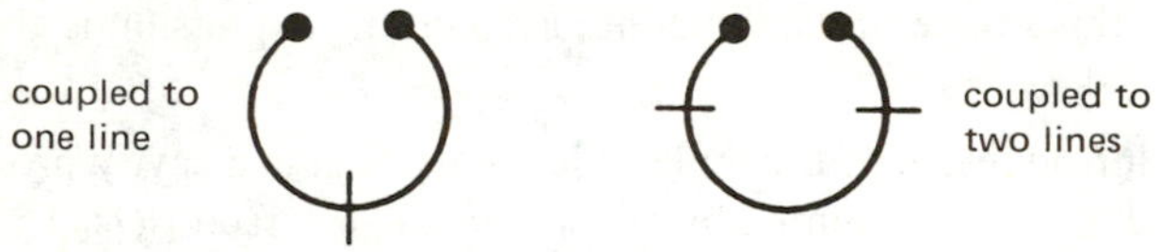

Fig. 3.21 Schematic representation of a cavity coupled to one or several lines.

3.5.3 Cavity Coupled to a Transmission Line

A cavity possesses an infinity of resonant modes (§ 3.1.2); when an opening is made to couple it with a line, the latter couples not to a single resonant mode, but ***to all the modes*** which satisfy the compatibility conditions for the fields. The resulting equivalent circuit is thus formed of an infinite number of series-connected resonant circuits (Fig. 3.22). Each circuit represents a resonant mode.

3.5.4 Simplified Study: Coupling to a Single Mode, Plane of the Detuned Short-Circuit

When the resonant frequencies of the different modes are spaced sufficiently far apart from one another, and the quality factors are high, the interaction between adjacent circuits becomes negligibly small. Every mode can then be considered separately, as long as it is not degenerate (§ 3.3.7).

The reference plane is defined at a voltage minimum on the line ***when the cavity is detuned*** (detuned short-circuit plane). Next to the resonant frequency of the

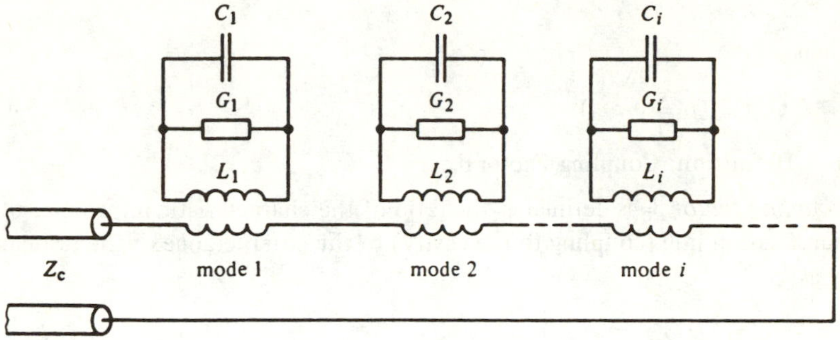

Fig. 3.22 Equivalent circuit of a cavity coupled to a transmission line.

mode considered, the cavity is then represented, approximately, by the simplified circuit of Figure 3.23. A shift of the reference plane can be made following the procedure of Section 6.1.26.

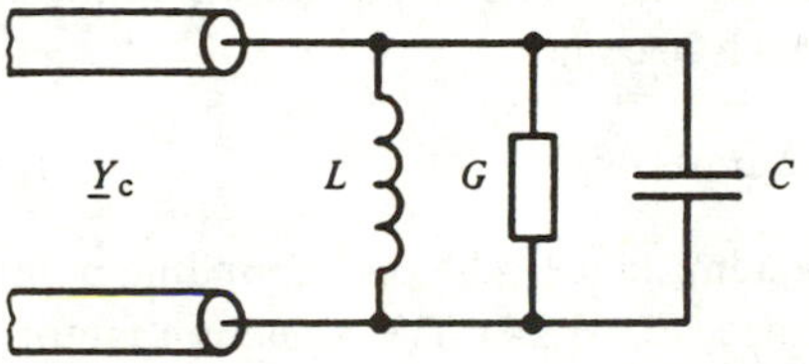

Fig. 3.23 Equivalent circuit of a cavity in the vicinity of a resonant frequency, in the plane of the detuned short circuit.

3.5.5 Input Admittance of a Cavity

The input admittance of the circuit, in the plane of the detuned short, is given by:

$$\underline{Y} = G + j\omega C + 1/(j\omega L) \qquad \text{S} \qquad (3.86)$$

or, introducing the resonance angular frequency ω_0 and the unloaded quality factor Q_0, respectively defined in (3.17) and in (3.85).

$$\underline{Y} = G\left[1 + jQ_0\left(\frac{\omega}{\omega_0} - \frac{\omega_0}{\omega}\right)\right] \qquad \text{S} \qquad (3.87)$$

When $\omega = \omega_0$, the admittance is purely real. Since only the frequency range close to the resonance has to be considered (due to the assumptions of § 3.5.4), an approximation for the term between brackets in (3.87) is taken, letting $\Delta\omega = \omega - \omega_0$ with $\Delta\omega \ll \omega_0$:

$$\frac{\omega}{\omega_0} - \frac{\omega_0}{\omega} = \frac{\omega_0 + \Delta\omega}{\omega_0} - \frac{\omega_0}{\omega_0 + \Delta\omega} \cong \frac{2\Delta\omega}{\omega_0} \qquad 1 \qquad (3.88)$$

and then

$$\underline{Y} \cong G(1 + 2jQ_0 \Delta\omega/\omega_0) \qquad \text{S} \qquad (3.89)$$

3.5.6 Definition: Coupling Factor β_c

The *coupling factor* β_c is defined as the ratio of the characteristic impedance of the transmission line (coupling to the cavity) to the conductance of the resonant cavity as:

$$\beta_c = Y_c/G = 1/GZ_c \qquad 1 \qquad (3.90)$$

To avoid possible confusion with the phaseshift per unit length β, the subscript c was added.

3.5.7 Reflection from the Cavity

The reflection depends on the ratio of input to line admittances, and making use of (3.89) and (3.90) it becomes:

$$\underline{Y}/Y_c \cong (1 + 2jQ_0 \Delta\omega/\omega_0)/\beta_c \qquad 1 \qquad (3.91)$$

In the Smith chart for admittances, the corresponding point moves on a circle for constant conductance (Fig. 3.24). The reflection factor is then defined by (9.17):

$$\underline{\rho} = \frac{1 - \underline{Y}/Y_c}{1 + \underline{Y}/Y_c} \cong \frac{\omega_0(\beta_c - 1) - 2jQ_0\Delta\omega}{\omega_0(\beta_c + 1) + 2jQ_0\Delta\omega} \qquad 1 \qquad (3.92)$$

The ratio of absorbed power P to incident power P_{in} is given by:

$$\frac{P}{P_{in}} = 1 - |\underline{\rho}|^2 \cong \frac{4\beta_c}{(\beta_c + 1)^2 + (2Q_0\Delta\omega/\omega_0)^2} \qquad 1 \qquad (3.93)$$

3.5.8 Definition: External Quality Factor Q_e

The external quality factor is defined by the ratio of the capacitive susceptance of the cavity to the characteristic line admittance:

$$Q_e = \omega_0 C/Y_c \qquad 1 \qquad \mathbf{(3.94)}$$

It is related to the unloaded quality factor by:

$$Q_e = (\omega_0 C/G)(G/Y_c) = Q_0/\beta_c \qquad 1 \qquad (3.95)$$

3.5.9 Definition: Loaded Quality Factor Q_c

The loaded quality factor is defined in terms of the sum of admittances at resonance (line and cavity) as follows:

$$Q_c = \omega_0 C/(Y_c + G) \qquad 1 \qquad \textbf{(3.96)}$$

it is related to the two other quality factors Q_0 and Q_e by the relation:

$$Q_c = Q_0/(\beta_c + 1) = Q_e\beta_c/(\beta_c + 1) \qquad 1 \qquad (3.97)$$

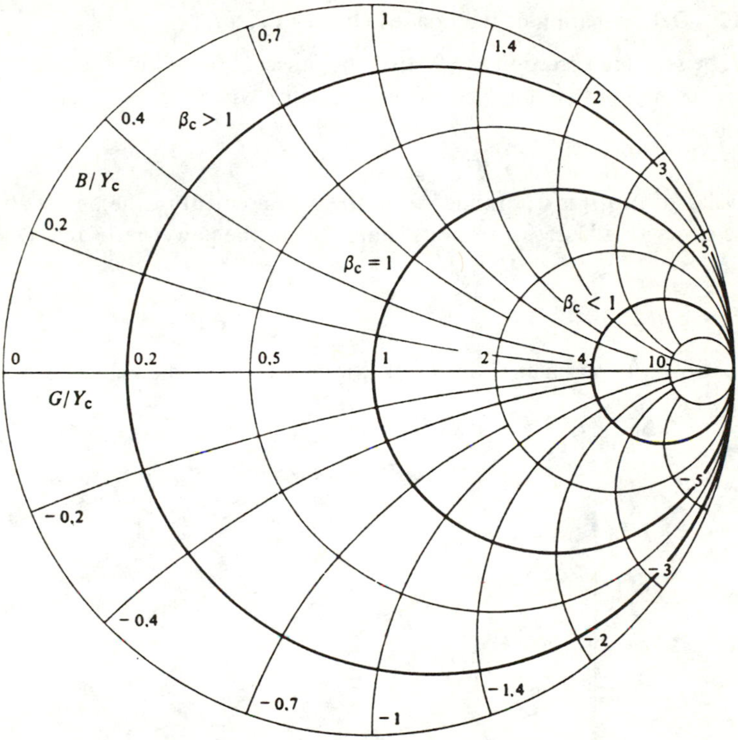

Fig. 3.24 Cavity admittance in the Smith chart.

3.5.10 Relation between the Three Quality Factors

The loaded quality factor is the inverse of the sum of inverses of the two others:

$$\frac{1}{Q_c} = \frac{1}{Q_0} + \frac{1}{Q_e} \qquad 1 \qquad (3.98)$$

3.5.11 Power Absorbed by the Cavity

The relation (3.93) may be expressed in terms of the loaded quality factor (3.97). In this manner one obtains:

$$\frac{P}{P_{\text{in}}} \cong \frac{4\beta_c}{(\beta_c + 1)^2} \frac{1}{1 + (2Q_c \Delta\omega/\omega_0)^2} \tag{3.99}$$

This expression is illustrated in Figure 3.25. It will be noted that the two curves for β_c and for $1/\beta_c$ coincide (§ 3.5.13).

3.5.12 Determination of the Loaded Quality Factor

It can be seen in expression (3.99) that the power absorbed by the cavity becomes one half of that absorbed at the resonance when:

$$2Q_c\Delta\omega_{1/2}/\omega_0 = 1 \tag{3.100}$$

The value of $\Delta\omega_{1/2}$ is determined from the measurement of the power absorbed in the cavity (Fig. 3.25) or from its complement, the power reflected. One then obtains:

$$Q_c = \omega_0/(2\Delta\omega_{1/2}) \tag{3.101}$$

The narrower the resonance line on the frequency scale, the higher the loaded quality factor.

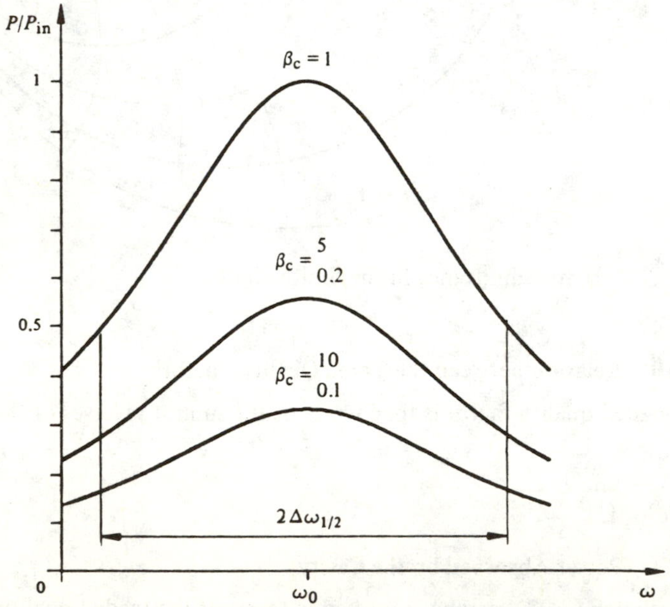

Fig. 3.25 Power absorbed by a resonant cavity.

3.5.13 Determination of the Coupling Factor

The coupling factor is determined by comparison of the power levels at resonance η_r; one actually has (3.99):

$$\left.\frac{P}{P_{in}}\right|_{\Delta\omega=0} = \eta_r = \frac{4\beta_c}{(\beta_c + 1)^2} \qquad 1 \qquad (3.102)$$

from which β_c is expressed as:

$$\beta_c = (2/\eta_r) - 1 \pm \sqrt{(2/\eta_r - 1)^2 - 1} \qquad 1 \qquad (3.103)$$

This expression has two solutions, which correspond respectively to β_c and to $1/\beta_c$. The measurement of the absorbed power does not indicate whether β_c is larger or smaller than unity.

3.5.14 Phase of the Reflected Signal: Overcoupled Cavity, Undercoupled Cavity, Critical Coupling

In order to determine whether β_c is larger or smaller than unity, the phases of the signals reflected at resonance and off-resonance are compared (Fig. 3.24). Considering (3.92), it is clear that when the signal frequency moves away from the resonance, the reflection factor tends to $\rho = -1$ (short-circuit), its phase is then 180°. At the resonance, the reflection factor is given by:

$$\left.\underline{\rho}\right|_{\Delta\omega=0} = \frac{\beta_c - 1}{\beta_c + 1} \qquad 1 \qquad (3.104)$$

Three different situations are encountered:

1. When $\beta_c > 1$, ρ is real positive at the resonance; its phase is equal to zero. Detuning the cavity then causes a phase shift of 180°. In this case, the cavity is *overcoupled.*
2. When $\beta_c < 1$, ρ is real negative at the resonance: its phase is then 180°. Detuning the cavity does not modify the phase of the reflection factor. The cavity is then *undercoupled.*
3. In the intermediate case $\beta_c = 1$ or *critical coupling*, the reflection factor vanishes at resonance, so that all the power is absorbed by the cavity ($\eta_r = 1$ in (3.102).

The techniques available to measure the phase are presented in Chapter 7.

3.5.15 Cavity Connected to Two Transmission Lines

When coupling takes place with a single mode of the cavity, as considered in Section 3.5.4, the equivalent circuit has only one resonant circuit, coupled to two transmission lines by ideal transformers [57] (Fig. 3.26).

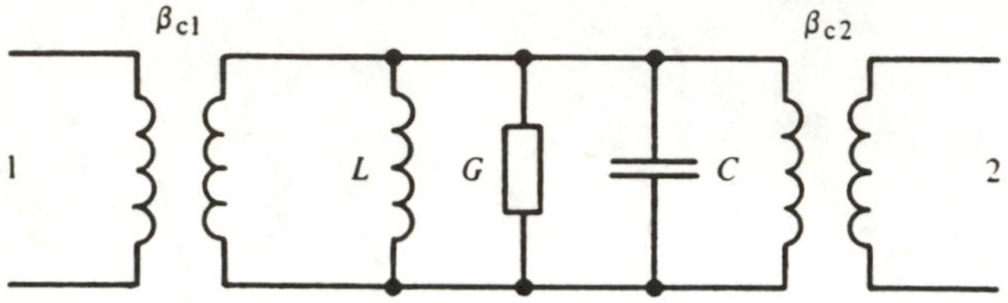

Fig. 3.26 Equivalent circuit of a cavity coupled to two transmission lines.

This circuit is studied in a similar manner as the single access cavity. Two coupling factors are defined here: one, called β_{c1}, with respect to line 1, the other, β_{c2}, with respect to line 2. The ratio of transmitted to incident power, under matched load and generator conditions, is given by:

$$\frac{P_{tr}}{P_{in}} \cong \frac{4\beta_{c1}\beta_{c2}}{(1+\beta_{c1}+\beta_{c2})^2\,[1+(2Q_c\Delta\omega/\omega_0)^2]} \qquad (3.105)$$

In this situation, the loaded quality factor is defined in terms of the two lines coupled to the cavity by:

$$Q_c = Q_0/(1+\beta_{c1}+\beta_{c2}) \qquad (3.106)$$

The power versus frequency curves are similar to those of Figure 3.24. The power transmitted at resonance is represented in Figure 3.27 as a function of both β_{c1} and β_{c2}.

The power transfer increases with both coupling factors, tending towards unity as the two factors tend to infinity. Such a condition corresponds to a heavily overcoupled cavity, which at the limit does not resonate ($Q_c \to 0$).

3.5.16 Comment

When applying a continuous signal to the cavity, the resonance appears at the frequency $f_0 = \omega_0/2\pi$. At that frequency, the ***power absorbed*** by a single port cavity is the largest (Fig. 3.25). With a two-port cavity, the ***transmitted power*** has a maximum at the frequency f_0.

On the other hand, as soon as the excitation is interrupted, the signal decreases as a damped sine wave (Fig. 3.2). The sine wave frequency (natural response of the cavity) is $f_{pr} = \omega_{pr}/2\pi$. It is always smaller than f_0; however, the difference is negligible in practice, except for cavities with very large losses.

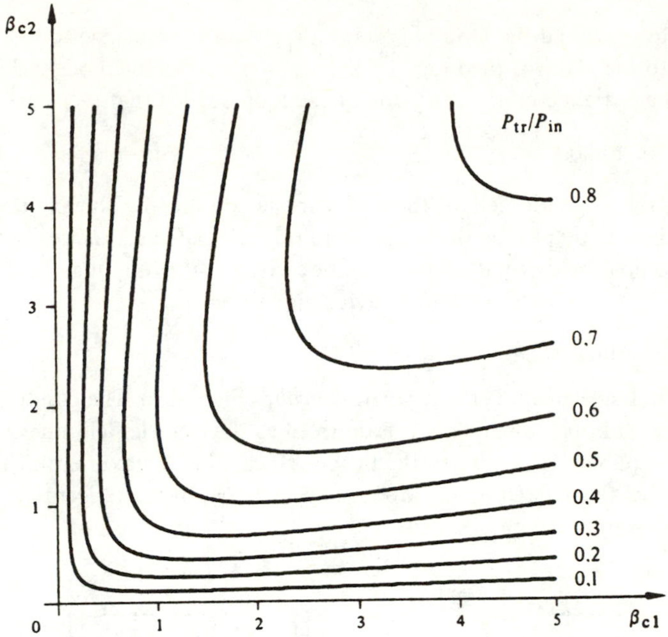

Fig. 3.27 Power transfer through a cavity at resonance, as a function of the two coupling factors. The lines correspond to constant power transfer.

3.6 NOTIONS OF EXCITATION

3.6.1 General Remark

An electromagnetic field can exist inside of a cavity when an electromagnetic signal is injected by an outside circuit. The excitation elements may be simple apertures, probes or still electron streams.

3.6.2 Inductive Loop

When a cavity is coupled to a transmission line (generally a coaxial line), an *inductive loop* is most often utilized (Fig. 2.38).

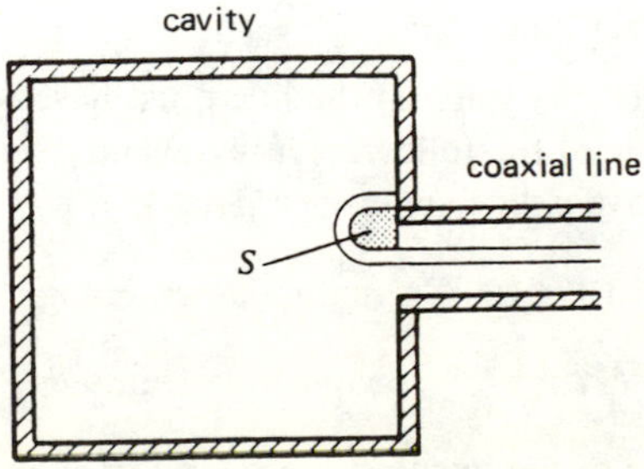

Fig. 3.28 Cavity coupled with an inductive loop.

A current flows around the loop of conductor, generating a magnetic field perpendicular to the plane of the loop (1.3). This loop excites the modes which have a magnetic field crossing the plane of the loop S, such that:

$$\int_S \underline{H}_C \cdot \boldsymbol{n}\, dA \neq 0 \qquad \text{Am} \qquad (3.107)$$

where $\underline{H}_c$ is the magnetic field of the cavity mode considered. The excitation may be optimized by placing the loop where this integral has its ***largest value***. Modes for which this integral vanishes are not excited: the loop may be positioned so as to excite only particular cavity modes.

3.6.3 Capacitive Probe

Another coupling element is the *capacitive probe* (Fig. 3.29). The conductor is here terminated into an open line; it is coupled to the electric field lines parallel to it. The coupling is largest when the probe is located at a maximum of the electric field of the excited mode. Modes having zeroes of the tangential electric field at this location are not excited.

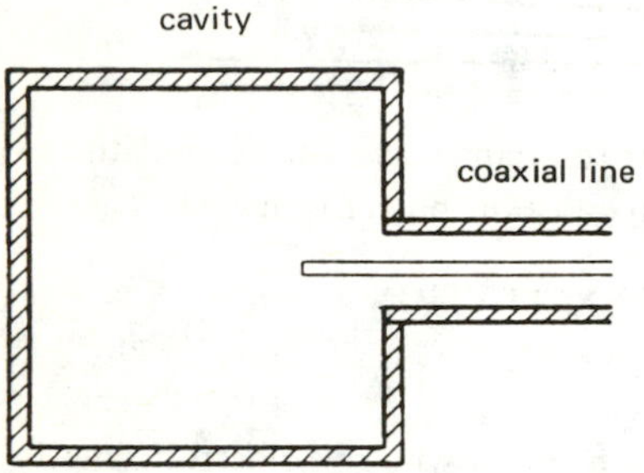

Fig. 3.29 Cavity coupled by a capacitive probe.

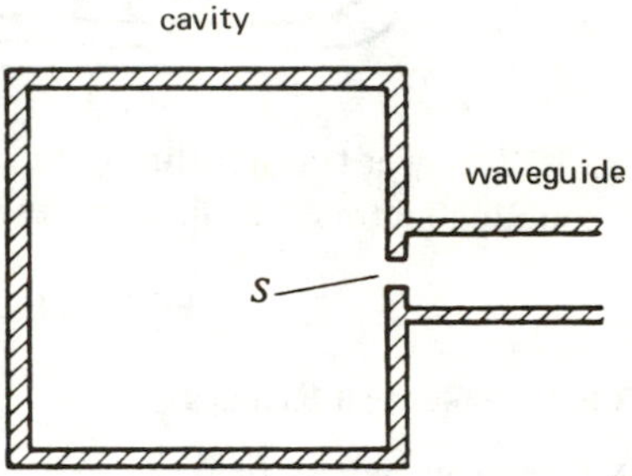

Fig. 3.30 Cavity coupled through an aperture.

3.6.4 Radiating Aperture

A slot, an iris, or an opening of arbitrary shape are cut through the wall between a cavity and a transmission line (waveguide, Fig. 3.30), allowing radiation of the signal from one to the other.

The fields in the cavity and in the line must have components in the same direction; at least one of the following integrals must be non-zero (subscripts L and C denote, respectively, line and cavity fields):

$$\int_S \underline{H}_C \cdot \underline{H}_L\, dA \neq 0 \qquad \text{inductive coupling} \qquad A^2 \qquad (3.108)$$

$$\int_S \underline{E}_C \cdot \underline{E}_L\, dA \neq 0 \qquad \text{capacitive coupling} \qquad V^2 \qquad (3.109)$$

The positioning of the opening allows one to excite selected cavity modes.

3.6.5 Electron Beam

A stream of loaded particles is equivalent to a current flowing along a conductor. When an ***electron beam*** travels across a cavity, it creates a magnetic field surrounding the beam. The resulting excitation is similar to the one produced by the inductive loop (§ 3.6.2). The coupling between an electron beam and the fields of the cavity is utilized to generate microwave signals in electron tubes like the ***magnetrons*** (Sec. 4.2) and ***klystrons*** (Sec. 4.3).

3.6.6 Comment

Coupling structures of the types mentioned here also permit, by reciprocity (§ 6.1.21), to transfer a microwave signal from a cavity towards an outside circuit, and also to couple a cavity to another cavity, or a line to another line (for instance, a waveguide to a coaxial line) [58].

3.7 PROBLEMS

3.7.1 Determine the resonant frequency, the quality factor and the relaxation time of a cavity for which $\underline{\omega}_p = 1.8 \cdot 10^{10} + j\,10^8\ s^{-1}$.

3.7.2 A cavity is formed of a waveguide between two short circuits. The length of the waveguide section is much larger than its transverse dimensions. Three resonant frequencies are measured at 8.965 GHz, 10.538 GHz, and 12.403 GHz. Determine:

1. the cutoff frequency of the waveguide,
2. the length of the cavity (filled with air).

3.7.3 For a parallelepiped shaped cavity (Fig. 3.6) having a 15 cm length and a cross section of 7.5 cm x 5 cm, determine all the resonant frequencies between 3 and 4 GHz. Indicate all the corresponding modes and the possible degeneracies.

3.7.4 A cubic cavity with 10 cm wide sides is fed by a rectangular waveguide having a 2.5 cm x 5 cm cross section. Determine the lowest possible frequency at which a mode can be excited in the cavity.

3.7.5 A parallelepiped shaped cavity having sides with three different lengths *a, b,* and *d* has its three first resonant frequencies at 5.196 GHz, 6 GHz, and 6.708 GHz. Determine the dimensions of the three sides.

3.7.6 Demonstrate that, starting with the Hertz potentials defined in Section 3.3.8, the electromagnetic fields of Sections 3.3.5 and 3.3.6 are obtained.

3.7.7 Determine the equations of the magnetic field lines and of the lines of surface current of the mode TE_{101} in a rectangular cavity of sides *a, b,* and *d.* Show that, on the cavity wall, the two families of curves are orthogonal.

3.7.8 A circular cavity having a 5 cm diameter must resonate at 6 GHz. How long should it be?

3.7.9 Determine the resonant frequency of the TM_{112} mode in a 5 cm long cavity having a 2 cm radius.

3.7.10 Starting from the expressions of Section 3.3.11, determine the Hertz potential of the TE_{mnl} mode in a circular cavity.

3.7.11 Determine the first resonant frequency of a microstrip line having $h = 0.6$ mm, $w = 2$ mm, $d = 14$ mm and $\epsilon_r = 9.5$.

3.7.12 Which are the resonant frequencies of a ring in microstrip on a substrate of relative permittivity $\epsilon_r = 4.5$ and of thickness 0.8 mm? The average diameter of the ring is 25 mm and its width $w = 0.1$ mm.

3.7.13 Determine, in terms of the resonant frequency, the quality factor of the TE_{101} mode in a cubical aluminum cavity.

3.7.14 Calculate the quality factor of a copper circular cavity resonating at 6 GHz in the TE_{112} mode as a function of the diameter-to-length ratio.

3.7.15 How does the resonant frequency of a cubical cavity vary when a dielectric sphere of permittivity $\epsilon_r = 100$ crosses the cavity? The latter resonates (when empty) in the TM_{110} mode at 3 GHz. The sphere moves along a straight line, parallel to the y-axis, at $x = a/2$ and $z = d/2$. The volume of the sphere is one ten-thousandth (10^{-4}) of that of the cavity.

3.7.16 A cavity loosely coupled to a transmission line is considered. At the two frequencies of 7.924 GHz and 7.946 GHz, the power absorbed by the cavity is exactly one half of the one absorbed at the resonance. Determine the resonant frequency, the three quality factors, the coupling factor, and the relaxation time, knowing that the power reflected at resonance is the 29% of the incident power.

3.7.17 Repeat the previous problem, with the same numerical values, for a cavity strongly coupled to the transmission line.

CHAPTER 4

GENERATORS AND AMPLIFIERS

4.1 BASIC PRINCIPLES

4.1.1 Low Frequency Amplifiers

An active electron device (tube, transistor) allows one to control the power fed to a load by adjusting the bias of a component. When the power required for biasing is smaller than the one dissipated in the load, the device behaves as an ***amplifier***. It may to some extent be compared to a water faucet, in which a flow of liquid having a large mechanical power is controlled by a slight shift of a mechanical valve.

The flow of carriers (electrons or holes) which crosses the device can be adjusted by means of the bias voltage. This effect may be called the ***modulation of the flow***.

4.1.2 Oscillator

An oscillator is an electronic circuit which generates a periodic signal at a steady frequency. Its principle of operation is most often ***positive feedback*** (§ 6.1.18): a part of the signal at the output of an amplifier is sampled and injected at the amplifier's input (Fig. 4.1). Oscillations build up when the gain of the feedback

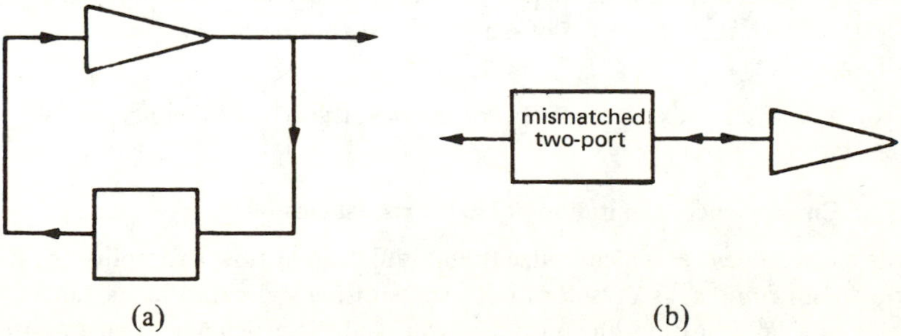

Fig. 4.1 Basic schematics of an oscillator: (a) feedback loop, (b) negative resistance.

loop is equal to or larger than unity. As the power level increases, the gain decreases and an equilibrium state is reached, at which the device generates a sine wave at a constant level.

The frequency of oscillation is determined by the electrical length of the feedback loop. The latter generally contains a resonant circuit (filter) which, at microwave frequencies, is most often a ***resonant cavity*** (Ch. 3).

A ***negative resistance amplifier***, for which a single port is simultaneously the input and output (§ 6.2.5) may also be used. In this case, the feedback loop is completed by a partially reflecting (mismatched) two-port connected to the amplifier.

In the microwave range, signals are also generated by an amplifier placed within a feedback loop. The two functions are often very closely overlapping within a single device.

4.1.3 Definition: Transit Time

The charge carriers (electrons, holes) require a certain amount of time, called the ***transit time***, to cross the device. The transit time is directly proportional to the distance covered by the carriers, it may also depend on the bias voltage, when the speed of the carriers is field-dependent. As frequency goes up, the period of the signal decreases; when it is of the same order as the transit time, the phase shift between the control and the amplified signal becomes significant.

In addition, the trajectories of the particles within the electron device do not all have the same length. At high frequencies, the signal may tend to spread out, and the amplification is reduced.

4.1.4 Parasitic Reactances

Every electron device is made of a system of metal electrodes for the injection, the control, and the collection of the charge carriers in the interacting region (vacuum or crystal). The metal electrodes are part of a ***capacitor system***, having an effect which increases with frequency: the displacement current density $\partial \boldsymbol{D}/\partial t$ may reach the same order of magnitude as the conduction current density $\boldsymbol{J}$.

The connecting wires exhibit a ***series inductance***, the effect of which also increases with frequency.

4.1.5 Consequence: Limitation at High Frequencies

All the electron devices which utilize the modulation of flow to amplify (§ 4.1.1) are ***frequency-limited***, as a result of both transit time and parasitic reactances. In practice, the amplification, the output power, and the efficiency of an amplifier (§ 4.11.4) all decrease with frequency. Beyond a certain frequency boundary, an amplifier no longer has any gain.

4.1.6 Size Reduction

Since the transit time is proportional to the distance covered by the carriers, it may be reduced by shrinking the device size. However, interelectrode capacitances then increase, so that the electrode area must also be reduced, which in turn reduces the power available from the device, as well as its amplification and efficiency.

For conventional vacuum tubes, the operating range of ***triodes*** could be extended up to a few GHz.

In the semiconductor field, the reduction in size was much more successful, as evidenced by the spectacular developments of MESFET ***transistors***, operating up to 10 or even 20 GHz. The upper frequency limit could be pushed upwards thanks to the technological developments by which electrodes can be realized with dimensions smaller than one micrometer. The amplification and the output power remain nevertheless rather modest (Sec. 4.8).

4.1.7 Other Amplification Principles

When the first radars were developed around 1940 (§ 1.3.4, Sec. 8.1), the transistor was yet to be invented. As for available tubes, their performances at high frequencies were too limited to ensure satisfactory operation of radars. New techniques for amplification, not limited by size effects, had to be found. The transit time itself was in fact utilized to produce amplification.

When an electromagnetic wave and mobile charge carriers are allowed to interact over distances several wavelengths long, energy may be transferred from the charges to the field, producing an amplification of the latter. Within crossed-field tubes (Sec. 4.2: magnetron) electrons are submitted to the conjugated effects of static electric and magnetic fields perpendicular to each other. They follow complex spinning trajectories, and in the process excite the resonances of a cavity. In some other tubes, the speed of the electrons is modulated (Sec. 4.3: klystron; Sec. 4.4: BWO, TWT). Within solid state devices, the transit time is adjusted in such a way that a negative resistance effect is obtained (Sec. 4.7).

Very low level signals are amplified in nonlinear reactive components (Sec. 4.10). These also permit frequency multiplication: very high stability signals are produced by frequency multiplier chains fed by an oscillator driven from a quartz crystal (Sec. 4.9).

4.2 CROSSED FIELD TUBES: MAGNETRONS

4.2.1 Definition: Cyclotron Frequency

An electron, circulating at a speed $\boldsymbol{v}$ within a magnetic induction field $\boldsymbol{B}$ is submitted to a Lorentz force given by:

$$F_m = q(\boldsymbol{v} \times \boldsymbol{B}) = m\frac{d\boldsymbol{v}}{dt} \qquad \text{N} \qquad (4.1)$$

where q and m are the (negative) charge, and the mass of an electron, respectively.

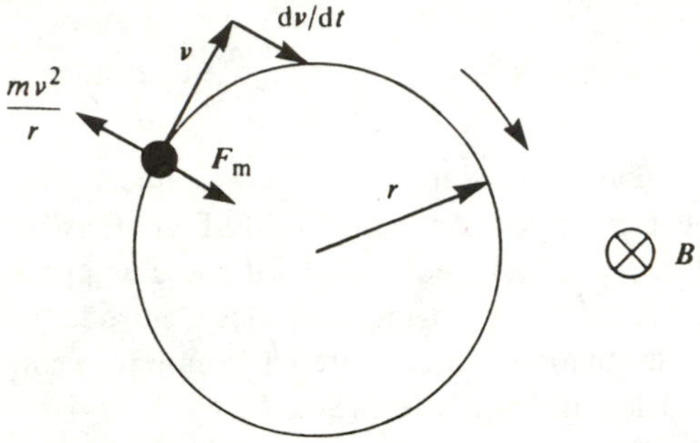

Fig. 4.2 Electron trajectory in a homogeneous induction field.

This force acts in a direction which is perpendicular to both the electron speed $\boldsymbol{v}$ and the induction field $\boldsymbol{B}$, as long as these two directions are not parallel. It produces a directional change of the speed vector and the electron trajectory, when located in a plane perpendicular to the induction field, becomes a circle of radius r (Fig. 4.2). The speed and the magnetic induction determine the radius of the circular trajectory. The balance of the force and the centrifuge acceleration permits the three quantities to be linked.

$$-qvB = mv^2/r \qquad \text{N} \qquad (4.2)$$

where B is the modulus of the induction $\boldsymbol{B}$. The angular velocity of rotation $\omega_B = v/r$ is defined, in terms of which:

$$-q\,\omega_B B r = m\omega_B^2 r \qquad \text{N} \qquad (4.3)$$

The frequency f_B at which the electron revolves on its trajectory is called the *cyclotron frequency.*

$$f_B = \omega_B/2\pi = -(q/m)B/2\pi \qquad \text{Hz} \qquad (4.4)$$

This frequency is independent of the radius of the trajectory. It is proportional to the induction field $\boldsymbol{B}$, the proportionality factor being:

$$f_B/B = -q/2\pi m = 2.8 \cdot 10^{10} \qquad \text{Hz/T} \qquad (4.5)$$

An electron current circulating within an induction field thus moves along a circular trajectory with a frequency f_B, which is within the microwave range for values of induction between about 0.1 T and 10 T. This effect is utilized in crossed field tubes, the magnetron being the main element of this family.

4.2.2 Description of the Magnetron, Diode Operation

The magnetron is basically made of a coaxial structure: a cylindrical cathode of radius a, inside a tubular anode with an inside radius b. Electrons are emitted in the vacuum by the cathode, which is heated by an internal filament. They are accelerated towards the anode, which is biased at a positive voltage U, producing a radial electric field $\boldsymbol{E}$.

$$\boldsymbol{E}(\rho) = -\boldsymbol{e}_\rho \frac{U}{\ln(b/a)} \frac{1}{\rho} \qquad \text{V/m} \tag{4.6}$$

When ***no other field*** is present, the electrons move radially, accelerated by the electric field, and reach the anode. This is the *vacuum diode operation.* The energy $-qU$ provided by the electric field to each accelerating electron is dissipated in the form of heat as the electron impinges upon the anode.

4.2.3 Crossed Fields, Cutoff Parabola

An additional static induction field is applied, most often produced by a permanent magnet, and perpendicular to the electric field. The force acting on an electron is then the combined effect of the two crossed fields:

$$\boldsymbol{F} = q(\boldsymbol{E} + \boldsymbol{v} \times \boldsymbol{B}) = m \frac{\mathrm{d}\boldsymbol{v}}{\mathrm{d}t} \qquad \text{N} \tag{4.7}$$

Replacing $\boldsymbol{E}$ and $\boldsymbol{B}$ by their values, one obtains:

$$\boldsymbol{F} = m \frac{\mathrm{d}\boldsymbol{v}}{\mathrm{d}t} = q\left(-\boldsymbol{e}_\rho \frac{U}{\ln(b/a)} \frac{1}{\rho} + \boldsymbol{v} \times \boldsymbol{e}_z B\right) \qquad \text{N} \tag{4.8}$$

Under the effect of the induction $\boldsymbol{B}$, the electron trajectory curves and takes a cycloidal shape (Fig. 4.3). When the induction exceeds a threshold, the electrons are even pushed back towards the cathode by the magnetic force, and can no longer reach the anode. This *cutoff* condition depends upon the magnetron dimensions and the amplitudes of the biasing fields [59]:

$$U_c = \frac{m}{8(-q)} \left(\frac{qB_c}{m}\right)^2 b^2 \left(1 - \frac{a^2}{b^2}\right)^2 \qquad \text{V} \tag{4.9}$$

The cutoff voltage U_c is proportional to the square modulus of the cutoff induction B_c. In a voltage U versus induction B diagram, this expression defines the *cutoff parabola* (Fig. 4.4). A point located above the parabola represents a situa-

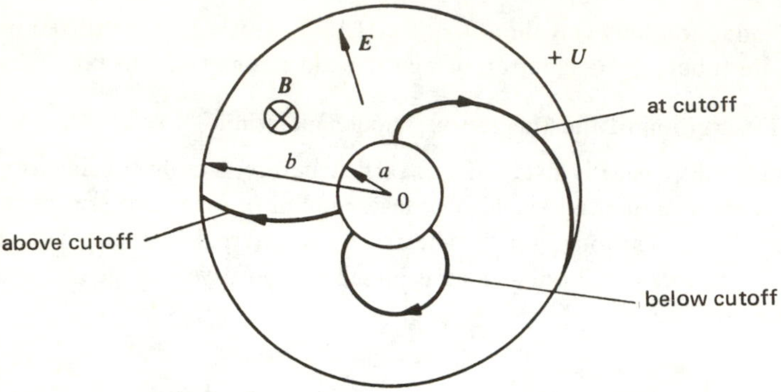

Fig. 4.3 Electron trajectories in the central structure of a magnetron.

tion at which the electrons reach the anode, producing a current in the bias circuit. The magnetron then operates in a ***diode*** mode, in the same way as when no induction is present (§ 4.2.2). The points located just upon the parabola correspond to trajectories ending tangentially to the anode. When the point is placed below the parabola, electrons return to the cathode. In this last situation, a double stream of electrons flows within the tube: from the cathode, part way towards the anode and back. The region close to the anode is void of carriers and no current flows in the outside circuit.

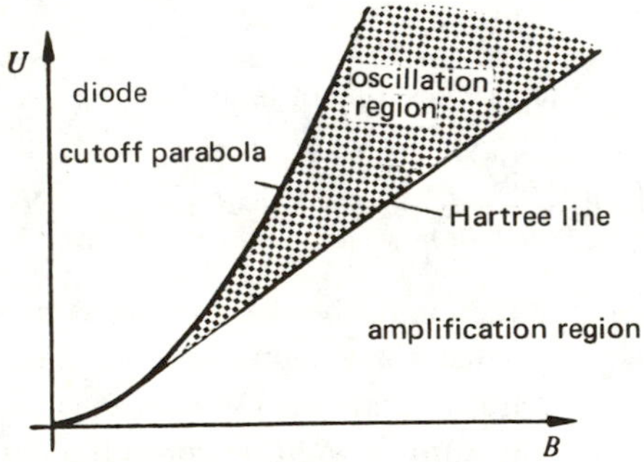

Fig. 4.4 *U-B* characteristics of a magnetron.

A more thorough study of the interactions shows that, when operating below cutoff, the radial velocity of the electrons slows down in the vicinity of certain particular radii [60]. An ***accumulation of space charges*** build up and striations appear. Since the cylindrical cathode possesses a large emitting surface, the charge densities and the electron currents may become quite large.

4.2.4 Spontaneous Oscillation Region, Hartree Line

Below the cutoff parabola, the electrons cannot reach the anode, not due to lack of energy, but because their velocity vector progressively moves away from the direction of the anode, under the effect of the static induction field. When certain conditions are satisfied, it only takes a small rotation of the vector to let the electrons reach the anode. Such a rotation may be produced by an azimuthal component of the electric field, for instance the one of an electromagnetic wave.

A small additional drive signal may then give rise to an important current. Within a region of the *U-B* diagram, *spontaneous oscillations* may appear, grow up, and remain steady. This region is limited by the *Hartree line* which, for an oscillation of angular frequency ω, is given by the relation

$$U_{\mathrm{H}} = \frac{1}{2} B_{\mathrm{H}}\omega(b^2 - a^2) - \frac{m}{2(-q)} \omega^2 b^2 \qquad \mathrm{V} \qquad (4.10)$$

The straight line is tangent to the cutoff parabola (Fig. 4.4).

4.2.5 Operating Principle

The oscillation phenomenon described can be taken advantage of by coupling the magnetron cavity (cathode and anode) to an electromagnetic circuit. For a large transfer of power to the circuit to take place, the rotating velocity of the electrons should coincide with the electromagnetic wave rotating velocity (synchronism, § 4.4.8). This can be done by reducing the velocity of the electromagnetic wave, an effect achieved by means of ***periodically loaded transmission lines*** [67]. The anode is cut by slots or alveoli, which make up a set of secondary cavities.

The change in direction produced by the electromagnetic signal allows most electrons to reach the anode after a whirling movement (Fig. 4.5). The electron

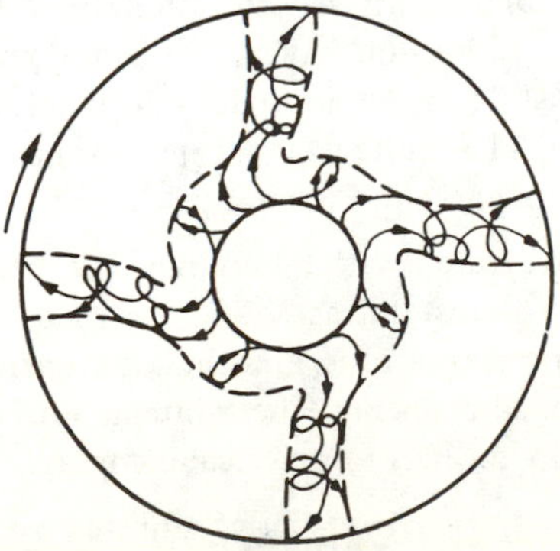

Fig. 4.5 Electron trajectories in a magnetron.

trajectories actually group together, forming the ***spokes of a wheel*** which rotates with the speed of the resonant mode. The current resulting from the electron flow within the spokes is coupled to the field of the rotating mode, to which an important part of the energy is then transferred. The cavity is in turn coupled to the outside circuit, to a waveguide, or a coaxial line (Sec. 3.6).

Some of the electrons are slowed down by the electromagnetic field and fall back on the cathode, as when no such field was present. In this situation however, the electrons draw some energy from the microwave field, reducing the efficiency slightly. The additional heating up of the cathode produced by the electron bombardment permits the heating provided by the filament to be reduced, or even be suppressed.

4.2.6 Instabilities

The slots, alveoli, and secondary cavities are loosely coupled to each other across the interaction region of the magnetron. As a result, the resonant mode structure of the assembly is dominated by the secondary cavities. Several resonant frequencies are obtained for the whole, they are rather close to each other and correspond to different phase shifts between adjacent slots. If no precautions are taken, the magnetron may well oscillate on one mode during a pulse of the signal (in pulsed operation, § 5.8.3), on another mode during the following pulse, and so on. The magnetron may even jump modes during a single pulse. Such instabilities produce an important ***phase noise***, which severely curtails the use of the device.

4.2.7 Stabilization of the Oscillation, Straps, Rising Sun, Coaxial Magnetron

Several techniques are available to suppress the instabilities due to mode jumping [61]. The coupling between adjacent slots can be strengthened by adding *straps*. The separation between resonant frequencies then becomes broader; in particular, one of them is clearly set aside from the others. A similar result is obtained by cutting successive slots having different depths: this is called the *rising sun* structure. In still another approach, the whole anode is placed inside of a *coaxial cavity*, which has quite well-defined resonant frequencies. These three structures are represented in Figure 4.6.

Since the magnetron structure is rather complex, the interactions between electrons and fields become almost untractable. Simplified models were elaborated upon to develop the first magnetrons, and the experimental study played an important part in the final adjustments. These interactions can nowadays be studied in more detail by means of numerical computer techniques.

The signal frequency of a magnetron can be adjusted over a relative range of about 15% by mechanical means: plungers are moved within the secondary

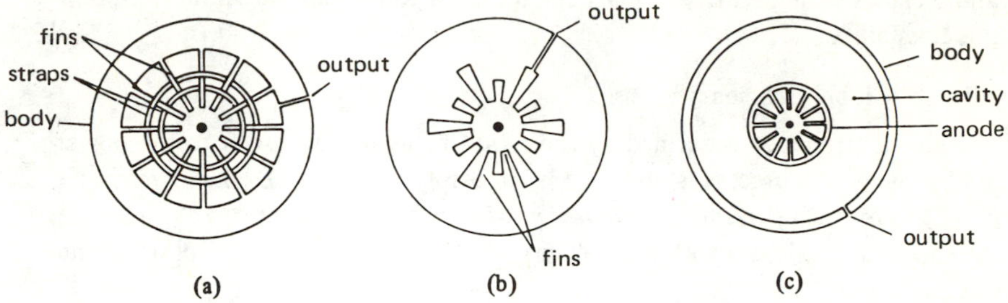

Fig. 4.6 Magnetron geometries: (a) with fins and straps, (b) rising sun, (c) coaxial.

cavities. Since a hard vacuum must exist within the tube, this control, which is carried out from the outside, requires vacuum-tightness; this is usually realized by means of ***metal bellows***.

Certain low power magnetrons are electrically tuned over a narrow range (VTM: Voltage Tuned Magnetron) [62].

4.2.8 Electronic Efficiency

The voltage source furnishes each electron which reaches the anode with an energy $-qU$. The kinetic energy lost by the electron when reaching the anode is in most cases much smaller than this value. The difference yields the energy transferred to the microwave circuit. The *electronic efficiency* η_e is defined as the ratio of the power fed to the microwave circuit to the power provided by the voltage bias. An approximate estimate of the kinetic energy of the electrons permits the evaluation of this efficiency as [60]

$$\eta_e \cong 1 - \frac{U}{U_c}\left(\frac{B_c}{B}\right)^2 \qquad 1 \qquad (4.11)$$

where U_c and B_c are the voltage and the induction on the cutoff parabola (4.9), respectively. As may be seen in (4.11), this parabola is the locus of points with ***zero efficiency***. The efficiency increases when moving away from the parabola. The largest values are obtained for operating points located upon the Hartree line, for which one obtains

$$\eta_e \cong \left[1 - \frac{2}{B}\,\frac{m}{-q}\,\omega\,\frac{1}{1-(a/b)^2}\right]^2 = \left[1 - \frac{f}{f_B}\,\frac{2}{1-(a/b)^2}\right]^2 \qquad 1 \qquad (4.12)$$

The electronic efficiency may apparently be increased arbitrarily by operating at large values of the induction *B*, *i.e.*, with $f \ll f_B$. In practice, however, the bulk and weight of the permanent magnets, required to provide the induction, limit this possibility.

4.2.9 Global Efficiency

The global efficiency is defined by the ratio of the output power to the power fed by the bias source. It is obtained by multiplying the ***electronic efficiency*** (§ 4.2.8) by the ***efficiency of the microwave circuit*** (attenuation), which takes into account the losses within the cavity (§ 3.4.12) and in the coupling to the outside circuit.

4.2.10 Magnetron Applications

Magnetrons can produce significant output powers, with global efficiencies up to 90% [63]. They can be manufactured in large series at moderate costs. Pulse powers (§ 5.8.3) of several hundred kW up to several MW are achieved in pulsed operation. In continuous wave operation, tens to hundreds of kW are supplied by the most powerful magnetrons. Magnetrons are, however, rather ***noisy*** generators: their output signal does not exhibit a pure spectrum, due to instabilities and spurious oscillations. Magnetrons are mostly utilized within pulsed radar systems and for microwave heating, two applications which provide a very large slice of the generator market. Figure 4.7 shows the magnetron of an airborne radar, and a magnetron for heating.

Fig. 4.7 Magnetrons: (a) for aircraft radar; (b) for heating applications.

4.2.11 Crossed Field Amplifiers

A magnetron structure may also be utilized to amplify a signal. In this case, spontaneous oscillations should not occur, so that the operating point is placed below the Hartree line (Fig. 4.4). The firm Raytheon manufactures amplifiers of this type under the trade name ***Amplitron***. Laboratory prototypes yielded global efficiencies in the 90% range [64]. Power levels of the same order as those achieved with magnetrons are possible. The gain of these amplifiers remains rather modest, in the 10 dB to 20 dB range.

4.3 KLYSTRONS

4.3.1 Principles of Velocity Modulation, Applegate's Diagram

A simple klystron amplifier, having two cavities, is shown in Figure 4.8. The principle of operation is the following:

1. An electron beam is emitted by a heated cathode, focused by electrodes and accelerated by a static bias voltage (beam voltage). In the first section of the tube, the electrons all travel at the same velocity, independent of time.

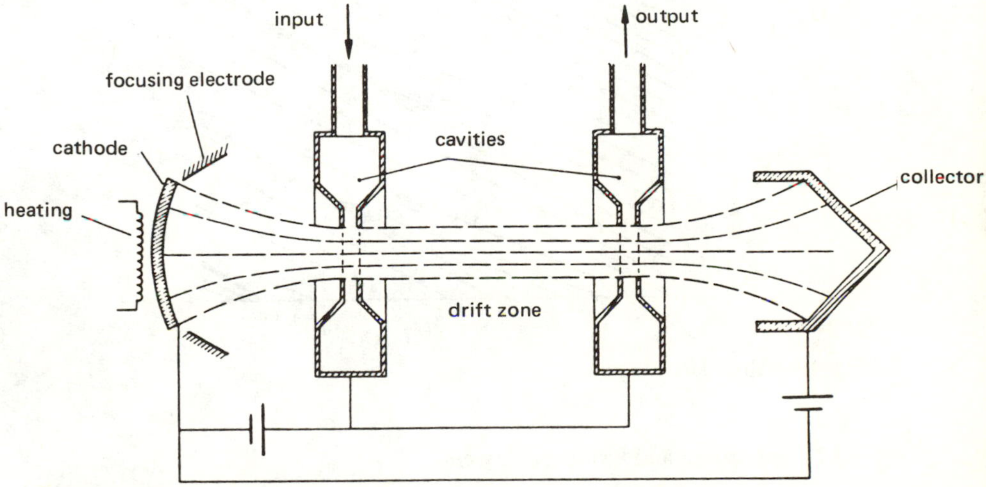

Fig. 4.8 Two-cavity klystron amplifier.

2. The beam then crosses a ring-shaped resonant cavity. The resonant mode of the cavity has an electric field component along the electron beam. Depending upon the time at which an electron crosses the aperture of the cavity, it may be accelerated or slowed down (half a period later). The *speed of the electrons is modulated* while they move across the cavity.

3. The beam then enters the ***drift region***. This is a region where no significant fields are applied, so that electrons keep moving at the same speed as at the cavity output. The fast electrons catch up to the slower electrons of the previous half-period: the beam ***bunches***, as shown in *Applegate's diagram* (Fig. 4.9). This diagram shows the location of particular electrons as a function of time. Within the drift region, the ***speed modulation*** is gradually converted into an ***amplitude modulation*** of the electron beam.
4. At a distance d from the modulating cavity, the bunching reaches its maximum value. A second cavity, tuned to the same frequency as the first one, is excited by the electron beam that is amplitude modulated at this point.
5. Electrons are then gathered by a collector.
6. An axial magnetic field generally stabilizes the electron beam. It does not play a basic role in the amplifying process.

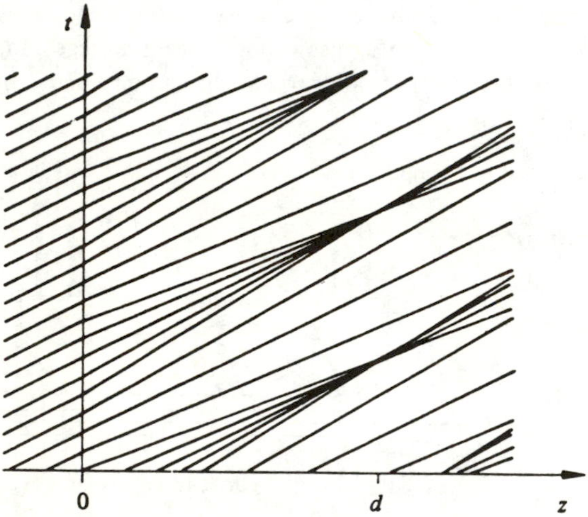

Fig. 4.9 Applegate diagram.

4.3.2 Received and Produced Powers

The modulating cavity, at $z = 0$, does not transfer any average energy to the electrons. While some electrons are accelerated, and thus receive additional energy, others are slowed down, returning the same amount of energy to the cavity. Only a ***small amount*** of energy is required to balance the losses and sustain the oscillation of the cavity.

The cavity excited by bunches of electrons at $z = d$ can withdraw an ***important portion of the energy*** of the electron beam, which was provided by the voltage bias. The power coupled to the output circuit is clearly larger than the one fed to the first cavity: the device is an ***amplifier***.

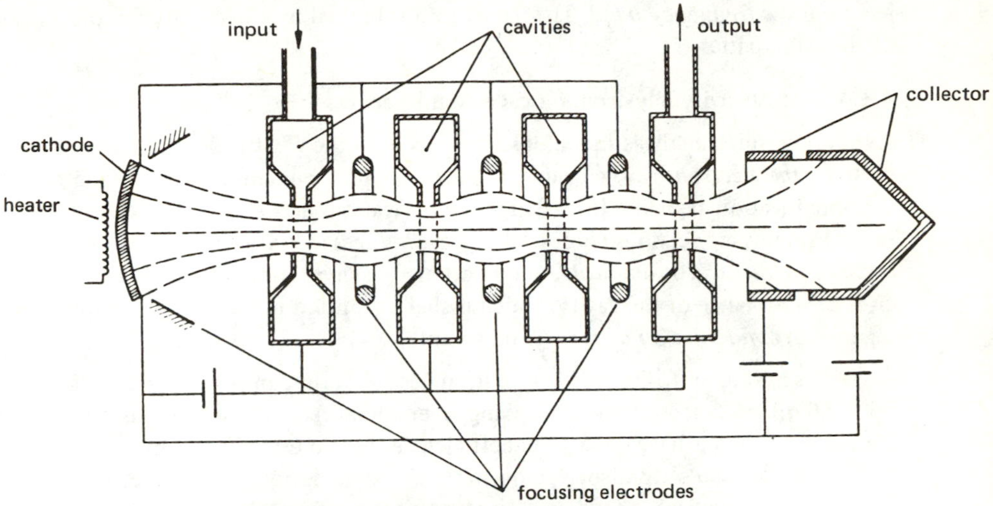

Fig. 4.10 Multicavity klystron.

4.3.3 Comment: Narrow Band

The amplification principle utilizes resonant cavities, and the resulting amplifier has a ***narrow frequency band***. The bandwidth is limited by the quality factors of the two cavities.

4.3.4 Multicavity Klystron, Cascade Amplifier

A more thorough analysis should take into account the electrical repulsion ***between the electrons*** of the beam: it tends to oppose bunching, and thus limits the amplification available [65]. Larger amplifications are obtained by cascading several amplifying sections (Fig. 4.10). This produces a *multicavity klystron* or *cascade amplifier.* The ***amplitude*** modulation at each intermediate resonator excites the resonance of the cavity, which is slightly shifted in frequency, producing in turn a ***modulation of speed***, and so on.

Klystrons having up to 7 cavities have been designed; they provide amplifications up to 60 dB and output powers reaching several hundred kW, continuous wave. Amplifier klystrons have good global efficiencies (up to 60%). They are quite stable, with a clean output spectrum (no spurious oscillations). They are, however, expensive devices. They have found many applications in communications and research: radar astronomy (§ 8.1.18), accelerators (Sec. 8.8), and so on.

4.3.5 Klystron Oscillator

If a portion of the amplified signal is fed to the modulating cavity with an adequately adjusted phaseshift, a feedback oscillator is realized, operating as described in Section 4.1.2 (Fig. 4.11).

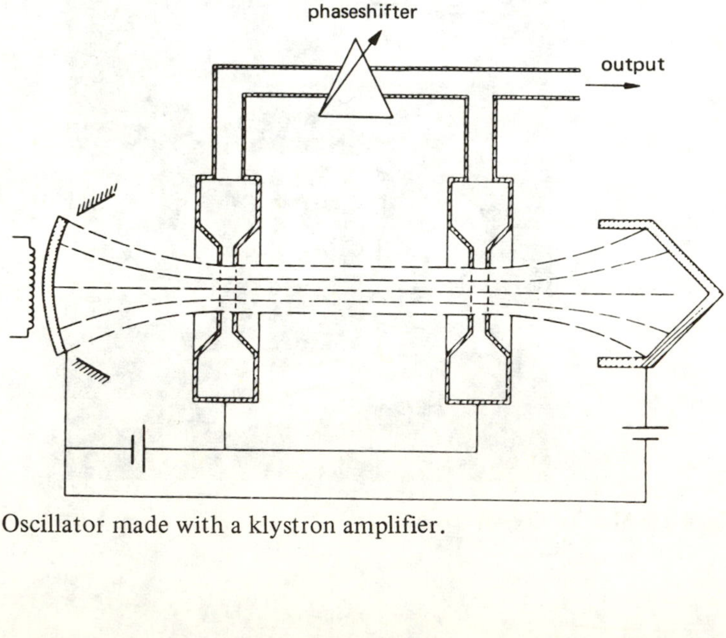

Fig. 4.11 Oscillator made with a klystron amplifier.

This oscillator would be difficult to operate in practice. For any change in frequency, the two cavities must be individually tuned to the desired frequency, and the phase shift of the feedback loop must be adjusted to obtain oscillation. For this reason, this setup is seldom utilized.

4.3.6 Reflex Klystron

One of the two cavities in Figure 4.11 may be removed by adding a negatively biased electrode, the ***reflector*** or ***repeller***, which acts as a mirror (Fig. 4.12).

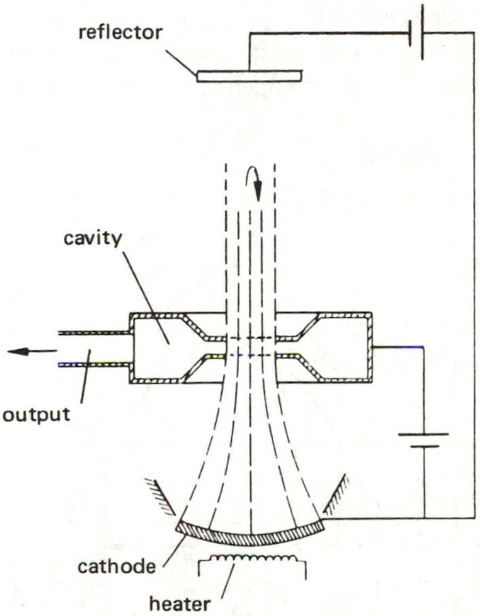

Fig. 4.12 Reflex klystron.

A single cavity then modulates the velocity of the electron beam and collects the amplitude-modulated beam after reflection. The reflector is negatively biased with respect to the cathode, so that electrons cannot reach it. In Applegate's diagram, their trajectories assume a parabolic shape (Fig. 4.13).

When they cross the cavity on the way out, electrons are accelerated by the electric field during half a period, then slowed down during the next one. The accelerated electrons move closer to the reflector and return later than those which were slowed down. ***Bunches are formed***, in a similar way as in the amplifier klystron (§ 4.3.1). Electrons transfer their energy to the oscillation when they return during an accelerating half period. The transfer of energy obtained is largest when the bunch comes back when the signal is maximum. The average transit time of the electrons is then $(n + 3/4)T$, where n is an integer and T is the period of the signal (Fig. 4.13).

To tune a reflex klystron, two adjustments are required:

1. tuning of the resonant frequency of the cavity,
2. adjustment of the bias voltage for the reflector.

Reflex klystrons generate signals from a few milliwatts up to several watts. They are tunable over frequency bands up to 40% (usable band of a waveguide). Their main applications are as local oscillators in radars, as pump sources in parametric amplifiers (§ 4.10.4), or as laboratory generators. They are progressively being replaced by solid state oscillators, which provide similar characteristics while being less bulky and much simpler to operate (Sections 4.6 and 4.7).

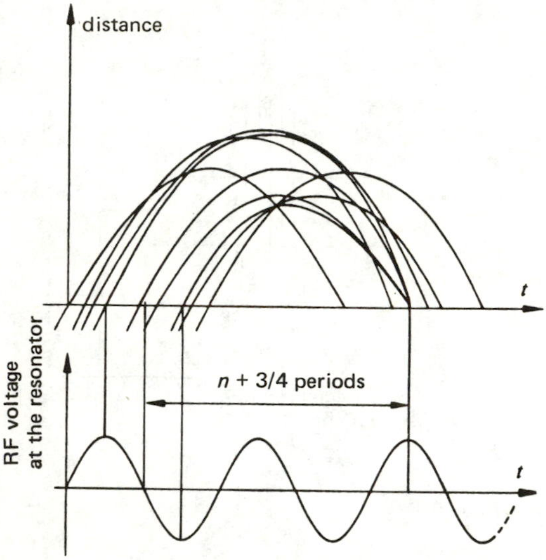

Fig. 4.13 Applegate diagram for a reflex klystron.

4.3.7 Small Signal Theory, Equivalent Beam Admittance

The electric field within the cavity at the time t at which an electron moves across it acts on its trajectory and determines the transit time. The electron current flowing through the cavity at the time $t + t_r$ is therefore a function of the voltage at time t. This function describes the effect of the beam in the operation of the reflex klystron. When nonlinear effects are neglected and only ***small signals*** considered (§ 4.4.3), the ratio of the electron current to the voltage across the cavity defines the *beam equivalent admittance* $\underline{Y}_f$ [66]:

$$\underline{Y}_f = \frac{I_0}{V_a}\frac{M^2}{2}\,j\theta\exp(-j\theta) \qquad \text{S} \qquad (4.13)$$

where I_0 is the dc current in the beam, V_a the beam voltage, M a beam coupling factor and θ the phase shift produced by the transit time.

$$\theta = 2\pi t_r/T = 2\pi f t_r = \omega t_r \qquad \text{rad} \qquad (4.14)$$

The beam admittance $\underline{Y}_f$ is connected in shunt with the resonant cavity of the klystron (Fig. 4.14). The components of its equivalent circuit take into account the resonant mode of the cavity, including the ***external circuit*** (load) connected to it. Admittances are all defined in the same reference plane (§ 6.1.5).

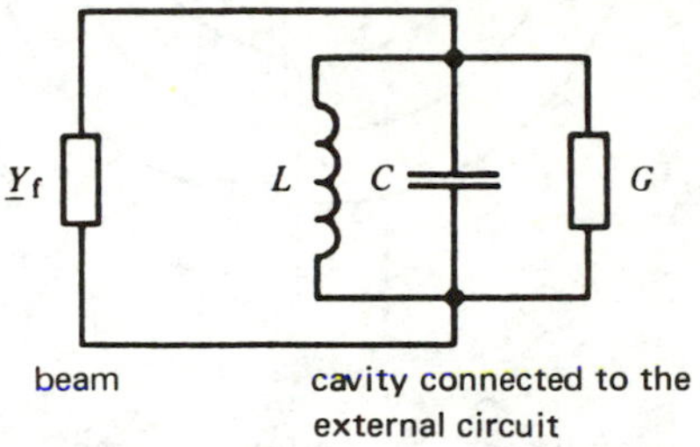

Fig. 4.14 Equivalent circuit of a reflex klystron.

The admittance of the cavity and of the external circuit is given by (3.89)

$$\underline{Y} \cong G\left(1 + 2jQ_0\frac{\Delta\omega}{\omega_0}\right) \qquad \text{S} \qquad (4.15)$$

For the klystron to oscillate, one must have:

$$\text{Re}[\underline{Y}_f + \underline{Y}] < 0 \qquad \text{S} \qquad (4.16)$$

The region over which this requirement is satisfied is located in the left-hand part of Figure 4.15.

4.3.8 Modes of Oscillation of the Klystron

When the frequency or the reflector voltage (which affects t_r) varies, the admittance $\underline{Y}_f$ describes a spiral in the complex plane. The klystron oscillates over certain ranges of the reflector voltage, which correspond to the portions of the spiral located on the left of the vertical line going through $-G$ producing the different *modes of oscillation of the klystron.* For each mode, the output power and the frequency depend upon the reflector voltage (Fig. 4.16). The maximum power is delivered at the cavity resonance.

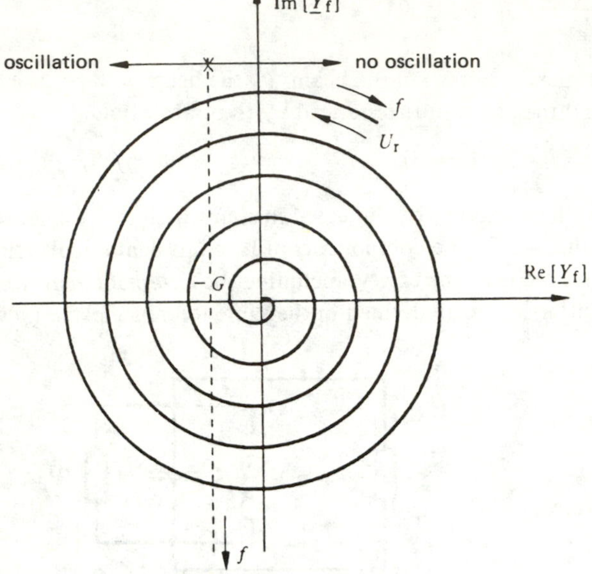

Fig. 4.15 Oscillation conditions for a reflex klystron.

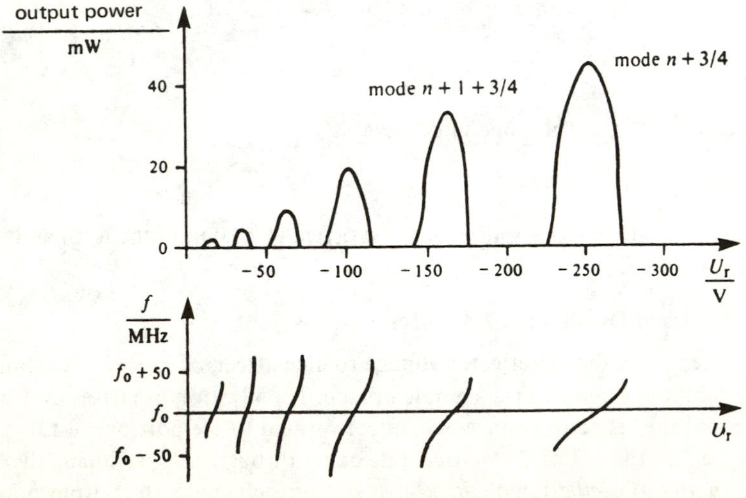

Fig. 4.16 Signal produced by a reflex klystron, in terms of reflector voltage.

4.3.9 Note: Effect of the External Circuit

The admittance (4.15) represents the cavity ***and the external circuit***. Any modification of this circuit produces a variation of the admittance $\underline{Y}$ and, therefore, of the signal at the generator's output (frequency and power). This cause of possible instability is avoided in practice by matching the load for minimum reflection (Sec. 9.6). This means that there should be no signal reflected by the load on the transmission line.

4.4 TUBES WITH DISTRIBUTED COUPLING

4.4.1 Comment: Limited Frequency Bandwidth

The microwave tubes of the ***first generation, i.e.,*** the magnetrons (Sec. 4.2) and the klystrons (Sec. 4.3), all utilize resonant cavities, which implies that their operating frequency range will be narrow (§ 3.1.5). A signal may only be amplified when its frequency is close to the resonant frequency of the amplifier's cavity. A generator provides a signal at a single frequency, determined entirely by the geometry of the tube. Tuning by mechanical means is always slow; these tubes are therefore not capable of providing rapid changes of frequency.

However, radar and communications require increasingly wide frequency bands for operation. One wishes on the one side to amplify wide band signals, on the other to generate signals easily tuned by electrical means. Such functions may be achieved by tubes of the ***second generation***, in which the interaction of an electron beam with an electromagnetic field is spread out over several wavelengths. The basic interactions are the same ones as in a klystron, but the functions of speed modulation, drift, and output coupling are simultaneously carried out over the complete tube length.

4.4.2 Longitudinal Waves on an Electron Beam

A simple one-dimensional model of the electron beam is introduced to show the ***principle of the interaction***. The beam is represented by an electron density $\rho(t,z)$ moving in the z-direction with a velocity $v_z(t,z)$ under the effect of an electric field, also directed along z. The current density is then

$$J_z(t,z) = \rho(t,z)\, v_z(t,z) \qquad \text{A/m}^2 \tag{4.17}$$

This is a ***nonlinear*** expression, as the right-hand side term is the product of two time-dependent terms.

■ 4.4.3 Definition: Linearization

Since the exact resolution of a nonlinear system usually becomes very mathematically complex, this study is restricted to small signals. This is done by *linearizing* the nonlinear equation(s). Each function of time is separated into a static and a time-variable term. For a sine wave excitation at the angular frequency ω, the variable term is represented by its complex notation (phasor-vectors, § 1.4.1). The higher order terms appearing within products of functions are neglected (angular frequencies $n\omega$).

The functions appearing in (4.17) are developed:

$$\rho(t,z) = \rho_0 + \mathrm{Re}[\sqrt{2}\,\underline{\rho}(z)\exp(\mathrm{j}\omega t)] \qquad \mathrm{As/m^3} \qquad (4.18)$$

$$v_z(t,z) = v_0 + \mathrm{Re}[\sqrt{2}\,\underline{v}(z)\exp(\mathrm{j}\omega t)] \qquad \mathrm{m/s} \qquad (4.19)$$

$$J_z(t,z) = J_0 + \mathrm{Re}[\sqrt{2}\,\underline{J}(z)\exp(\mathrm{j}\omega t)] \qquad \mathrm{A/m^2} \qquad (4.20)$$

Introducing these three expressions into (4.17) and identifying the ω-dependent terms, the linearized relationship is obtained:

$$\underline{J} = \rho_0 \underline{v} + v_0 \underline{\rho} \qquad \mathrm{A/m^2} \qquad (4.21)$$

For a particular beam, the terms v_0 and ρ_0 are known: they depend upon the voltage applied to accelerate the beam and the emissivity of the heated cathode. The two phasors $\underline{\rho}$ and $\underline{v}$ are functions of the microwave electric phasor $\underline{E}$. The continuity equation is utilized to determine $\underline{\rho}$ [(1.3) and (1.4)]:

$$\mathrm{j}\omega\underline{\rho} = -\nabla \cdot \underline{J} = -\frac{\partial \underline{J}}{\partial z} \qquad \mathrm{A/m^3} \qquad (4.22)$$

The balance of the forces acting upon the electron provides an expression for its speed; in the time domain it is written as

$$m\frac{\mathrm{d}\boldsymbol{v}}{\mathrm{d}t} = m\left(\frac{\partial \boldsymbol{v}}{\partial t} + (\boldsymbol{v}\cdot\nabla)\boldsymbol{v}\right) = q\boldsymbol{E} \qquad \mathrm{N} \qquad (4.23)$$

This expression is also nonlinear. It is linearized and the one-dimensional assumption is made, yielding:

$$\mathrm{j}\omega\underline{v} + v_0\frac{\partial \underline{v}}{\partial z} = (q/m)\underline{E}_z \qquad \mathrm{m/s^2} \qquad (4.24)$$

■ 4.4.4 Equivalent Beam Conductivity, Plasma Angular Frequency

Phasors are all assumed to have only a lossless forward longitudinal dependence, so that (§ 9.3.4)

$$\underline{X}(z) = \underline{X}(0)\exp(-\mathrm{j}\beta z) \qquad (4.25)$$

The differentiation with respect to z is replaced by $-\mathrm{j}\beta$, expressions (4.22) and (4.24) becoming, respectively,

$$\underline{\rho} = (\beta/\omega)\underline{J} \qquad \text{As/m}^3 \qquad (4.26)$$

$$\mathrm{j}(\omega - \beta v_0)\underline{v} = (q/m)\underline{E}_z \qquad \text{m/s}^2 \qquad (4.27)$$

Eliminating $\underline{\rho}$ and $\underline{v}$ with (4.21), (4.26) and (4.27), Ohm's law for the electron beam is obtained and the *equivalent beam conductivity* $\underline{\sigma}_f$ is defined

$$\underline{J} = \frac{\rho_0(q/m)\omega}{\mathrm{j}(\omega - \beta v_0)^2}\,\underline{E}_z = \underline{\sigma}_f\underline{E}_z \qquad \text{A/m}^2 \qquad (4.28)$$

For a longitudinal wave, the curl of the phasors vanishes, since the direction of propagation coincides with their own orientation. Maxwell's equation (1.3) then becomes

$$\nabla \times \underline{H} = (\mathrm{j}\omega\underline{\epsilon} + \sigma)\underline{E} = 0 \qquad \text{A/m}^2 \qquad (4.29)$$

Therefore,

$$(\mathrm{j}\omega\epsilon_0 + \underline{\sigma}_f)\underline{E}_z = 0 \qquad \text{A/m}^2 \qquad (4.30)$$

for the beam.

For a non-vanishing solution to exist, the bracketed term within (4.30) must vanish. With (4.28), the dispersion relation for the longitudinal waves is obtained as

$$\beta = \frac{\omega \pm \omega_p}{v_0} \qquad \text{rad/m} \qquad \mathbf{(4.31)}$$

where the *plasma angular frequency* is defined as

$$\omega_p = \left(\frac{\rho_0 q}{m\epsilon_0}\right)^{1/2} = \left(\frac{q^2 n_0}{m\epsilon_0}\right)^{1/2} = 56.4\sqrt{n_0(\mathrm{m}^{-3})} \qquad \text{rad/s} \qquad (4.32)$$

n_0 is the electron density in the beam at equilibrium. The dispersion diagram is represented in Figure 4.17.

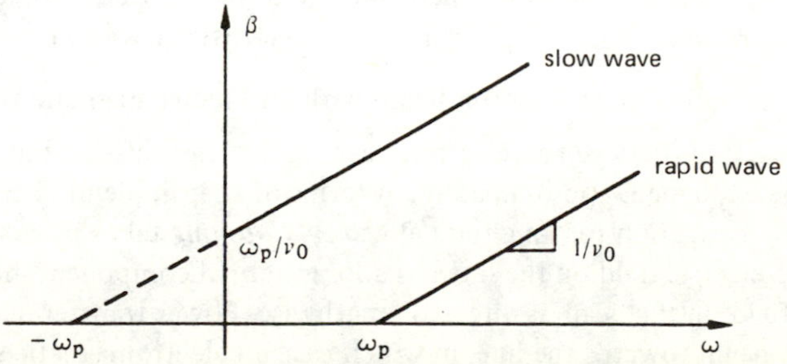

Fig. 4.17 Dispersion diagram for longitudinal waves on an electron beam.

4.4.5 Definitions: Fast and Slow Waves

Two waves can propagate on the beam, corresponding to the two signs within (4.31). Their characteristics are straight lines having the same slope. For both waves, the ***group velocity*** is constant and equal to the average electron velocity

$$v_g = (\partial\beta/\partial\omega)^{-1} = v_0 \qquad \text{m/s} \qquad (4.33)$$

On the other hand, the ***phase velocities*** of the two waves are different. For the upper line, one has

$$v_{\varphi-} = \omega/\beta_- = v_0 \frac{\omega}{\omega + \omega_p} < v_0 \qquad \text{m/s} \qquad (4.34)$$

This velocity is always smaller than that of the beam, it corresponds to the *slow wave.* For an observer moving along the beam at the average electron speed, it looks like a backward wave. Therefore, it transfers a ***negative power*** [67]. The lower line on the diagram corresponds to a wave with a phase velocity larger than the average speed of the beam:

$$v_{\varphi+} = \omega/\beta_+ = v_0 \frac{\omega}{\omega - \omega_p} > v_0 \qquad \text{m/s} \qquad (4.35)$$

It is a *fast wave*; moving faster than the beam, it transports a ***positive power***.

4.4.6 Definition: Electrokinetic Waves

For longitudinal waves moving along the beam, the magnetic field is identically zero. Indeed, for a forward wave moving along z, $\nabla \rightarrow -j\beta \mathbf{e}_z$, and thus using (1.2)

$$\underline{\mathbf{H}} = \frac{-1}{j\omega\mu_0} \nabla \times \underline{\mathbf{E}} = \frac{-1}{j\omega\mu_0} (-j\beta \mathbf{e}_z \times \mathbf{e}_z \underline{E}_z) \equiv 0 \qquad \text{A/m} \qquad (4.36)$$

The longitudinal waves are ***not electromagnetic*** ones. As the energies involved are the electric energy and the kinetic energy of the particles, these waves are called *electrokinetic* (or still plasma waves, electrostatic waves).

4.4.7 Coupling of an Electron Beam with an Electromagnetic Wave

The interactions between an electron beam and a transmission line placed near by can be explained, approximately, in terms of the simple model used for the study of propagation for longitudinal waves. ***Coupling*** takes place (Sec. 3.6) when the electric field on the line has a longitudinal component, because the electric field on the beam is directed lengthwise. Power transfer may then occur from the beam towards the line, in which case an electromagnetic signal is amplified (microwave tube). Transfer of energy from the line to the beam is used in particle accelerators (Sec. 8.8).

The two structures (beam and line) can guide a number of different modes, and a complete study of all possible interactions would be tedious and useless. The principle of operations can be described by considering only one of the waves on the beam and one on the line, both having positive phase velocities v_ϕ. For the resulting coupled modes, the propagation factors are given by [68]

$$\underline{\gamma} \cong \mathrm{j}\,\frac{\beta_1 + \beta_2}{2} \pm \mathrm{j}\,\sqrt{\left(\frac{\beta_1 - \beta_2}{2}\right)^2 + \frac{p_2}{p_1}\,K^2} \qquad \mathrm{m}^{-1} \qquad \textbf{(4.37)}$$

where β_1 and β_2 are, respectively, the phase shifts per unit length on the two lines in the absence of coupling; K is the distributed coupling factor; and p_1 and p_2 are the signs for the power transfer. For a usual $L'C'$ line, the power is transmitted in the same direction as the wave, and $p = +1$. For an inverse wave the power flows in the opposite direction, so that $p = -1$: this situation is encountered on helical lines and on lines periodically loaded with obstacles. For longitudinal waves on an electron beam, one has $p = +1$ for the fast wave, and $p = -1$ for the slow wave.

4.4.8 Definition: Synchronism

When, at a given signal frequency, the phase shifts per unit length of the two waves are identical ($\beta_1 = \beta_2$), their phase velocities v_φ (2.9) are also equal. The two waves are then *synchronous.* The propagation factor becomes then (4.37)

$$\underline{\gamma} \cong \mathrm{j}\beta_1 \pm \mathrm{j}K\sqrt{p_2/p_1} = \mathrm{j}\beta_1 \pm \Delta\underline{\gamma} \qquad \mathrm{m}^{-1} \qquad (4.38)$$

where $\Delta\underline{\gamma} = \mathrm{j}K\sqrt{p_2/p_1}$

4.4.9 Dispersion Diagram for Coupled Lines Close to the Synchronism

The two possible conditions are sketched in Figure 4.18.

In the first case, the coupling suppresses the crossing of the two curves; in the second one, the propagation factor gets a real part, in the vicinity of the synchronism. Since the system is lossless, this does not represent an attenuation, but a transfer of power between the two lines.

The coupling effects become most important at synchronism: the smallest separation of the two characteristics, or the maximum value of α occurs there. The transfer of energy between the two lines also takes its largest value at synchronism [68]. A complete transfer of power from one line to the other then becomes possible.

4.4.10 How to Reach Synchronism

The efficiency of a tube can only be large when a large transfer of power from the beam to the line is realized in practice. This happens when the tube operates at the synchronism or very close to it. However, the phase velocities of electro-

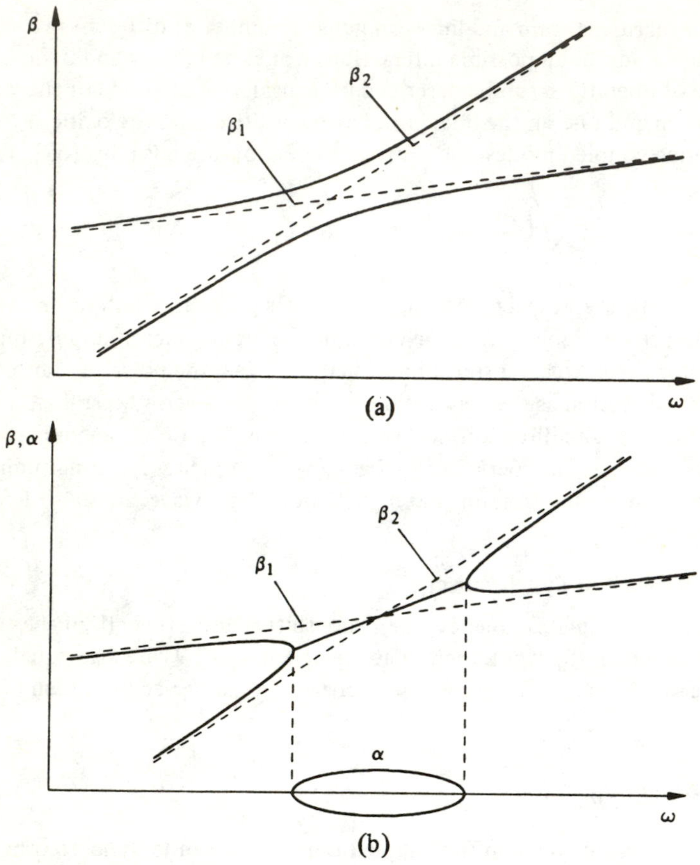

Fig. 4.18 Dispersion diagram close to synchronism: (a) for $p_2/p_1 = +1$, (b) for $p_2/p_1 = -1$.

kinetic waves (§ 4.4.5) are of the order of v_0, the speed of the electrons in the beam. To obtain synchronism, two possibilities exist:

1. accelerate electrons to a relativistic speed ($v_0 \cong c_0$),
2. reduce the velocity of the electromagnetic wave by means of slow wave structures; for instance, periodically loaded lines [67], helical lines, or still dielectrically loaded waveguides.

The second alternative is the only practical one. Figure 4.19 shows a structure for the interaction of the beam (line 1) with a wave on a helical line (line 2). In the latter one, the phase velocity is reduced approximately by the factor $\sin\psi$, where ψ is the pitch of the helix [69]. The interaction takes place over the region $0 < z < d$.

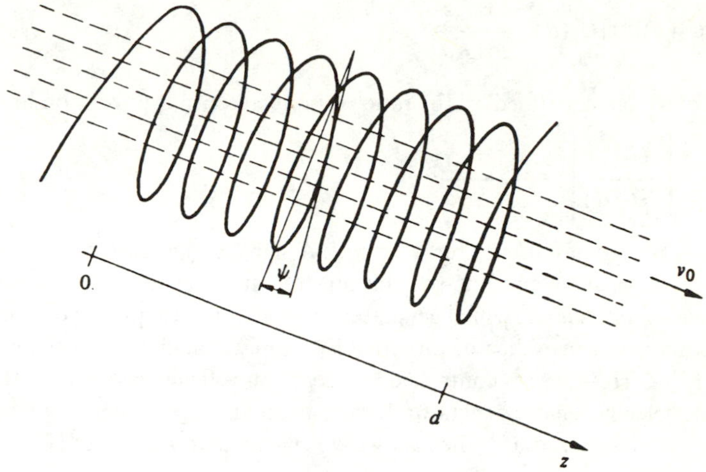

Fig. 4.19 Electron beam and helix. The longitudinal phase velocity of the wave on the helix is approximately $c_0 \sin \psi$.

4.4.11 Equivalent Voltages on the Two Lines at Synchronism

If both lines are assumed to have the same equivalent characteristic impedance, the voltages on the two lines are

$$\underline{U}_1(z) = [\underline{A} \exp(-\Delta\underline{\gamma} z) + \underline{B} \exp(+\Delta\underline{\gamma} z)] \exp(-\mathrm{j}\beta_1 z) \quad \mathrm{V} \tag{4.39}$$

$$\underline{U}_2(z) = [\underline{A} \exp(-\Delta\underline{\gamma} z) - \underline{B} \exp(+\Delta\underline{\gamma} z)] \exp(-\mathrm{j}\beta_1 z + \varphi) \tag{4.40}$$

where $\underline{A}$ and $\underline{B}$ are constants to be determined from the boundary conditions and φ is a phase shift whose value could be obtained from a more complete study, but which is not significant as far as powers are concerned.

Considering the particular situation $\underline{U}_1 = 0$, in which the electrons reaching the interaction region are not modulated, the voltages (4.39) and (4.40) become

$$\underline{U}_1(z) = -2\underline{A} \sinh(\Delta\underline{\gamma} z) \exp(-\mathrm{j}\beta_1 z) \quad \mathrm{V} \tag{4.41}$$

$$\underline{U}_2(z) = 2\underline{A} \cosh(\Delta\underline{\gamma} z) \exp(-\mathrm{j}\beta_1 z) \quad \mathrm{V} \tag{4.42}$$

4.4.12 Backward Wave Oscillator, BWO

The slow wave on the beam ($p_1 = -1$) couples to an inverse wave on the line ($p_2 = -1$). In this case, $p_2/p_1 = +1$, and therefore $\Delta\underline{\gamma} = \mathrm{j}K$. Specifying the output voltage on the microwave line $\underline{U}_2(d)$, the voltage amplitudes along the lines are:

$$|\underline{U}_1(z)| = |\underline{U}_2(d)| \frac{\sin Kz}{\cos Kd} \quad \mathrm{V} \tag{4.43}$$

$$|\underline{U}_2(z)| = |\underline{U}_2(d)| \frac{\cos Kz}{\cos Kd} \qquad \text{V} \qquad (4.44)$$

The power gain, defined by the ratio of input to output powers on line 2 is

$$G = \left| \frac{\underline{U}_2(0)}{\underline{U}_2(d)} \right|^2 = \frac{1}{\cos^2 Kd} \qquad 1 \qquad (4.45)$$

This gain becomes infinite for $Kd = \pi/2$. A signal comes out of the interaction region, on the line, when no signal is injected at the other end. The device obtained is an ***oscillator***, which generates a signal at the frequency of synchronism. This frequency may then be modified by simply changing the speed v_0 of the electrons. This speed is controlled by the beam voltage bias only. In this manner, an electrically tunable generator, having a tuning range which may reach an octave, can be realized; it is the *backward wave oscillator* or *BWO* [70].

The BWO has military applications, where its frequency agility is an important asset for countermeasures (creation of spurious responses in radar systems, (§ 8.1.19). It is also the center of swept frequency generators for laboratory work, an important element in spectrum analyzers (Sec. 5.4). BWOs are, however, increasingly being replaced by solid state devices.

Figure 4.20 shows how the power levels vary along the length of the lines; they are proportional to $p_i|\underline{U}_i|^2$ on the two lines. The power $|\underline{U}_1(z)|^2$ is progressively coupled from the beam to the line, and extracted at its beginning in $z = 0$.

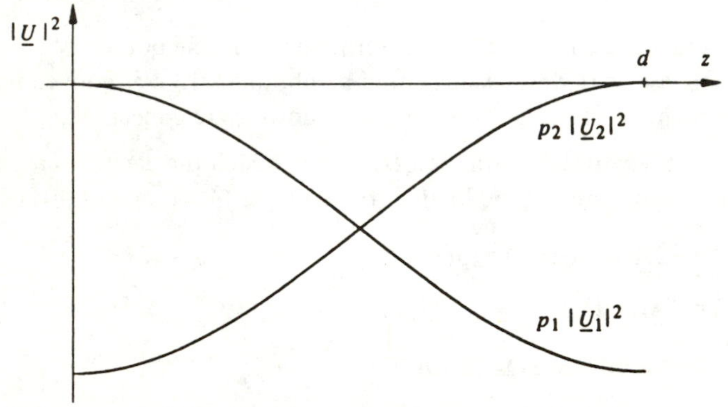

Fig. 4.20 Power diagram in a backward wave oscillator.

4.4.13 Traveling Wave Tube, TWT

The amplification of a signal is realized by the interaction of the slow wave, having $p_1 = -1$, with a fast wave upon the line, for which $p_2 = +1$. A *traveling wave tube* or *TWT* is realized in this manner [71]. The relation (4.38) becomes $\Delta\gamma = K$, so that (4.41) and (4.42) yield

$$|\underline{U}_1(z)| = |\underline{U}_2(0)| \sinh(Kz) \qquad \text{V} \qquad (4.46)$$

$$|\underline{U}_2(z)| = |\underline{U}_2(0)| \cosh(Kz) \qquad \text{V} \qquad (4.47)$$

The power gain is here given by

$$G = \left| \frac{\underline{U}_2(d)}{\underline{U}_2(0)} \right|^2 = \cosh^2(Kd) \qquad 1 \qquad (4.48)$$

The power diagram for a TWT is sketched in Figure 4.21.

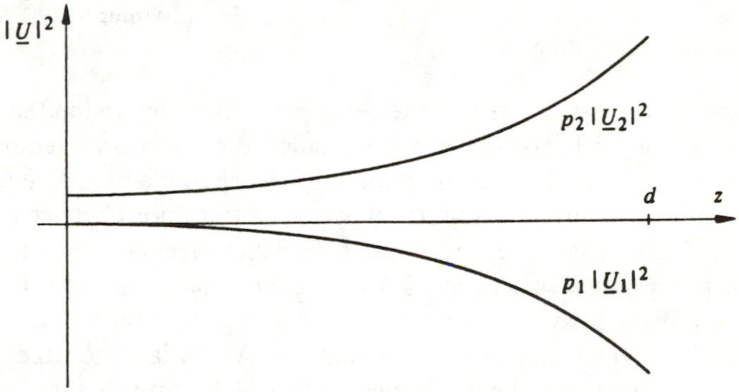

Fig. 4.21 Power diagram in a traveling wave amplifier.

When the two characteristics $\beta_1(\omega)$ and $\beta_2(\omega)$ can be made to coincide, a broadband amplifier is obtained (ranges up to 10:1 have been realized). It is mostly used for communications, for satellites in particular.

4.4.14 Comment

The model considered in the previous paragraphs is very much simplified. It describes the interaction mechanism between an electron beam and a slow wave transmission line when both the coupling and the electron density within the beam are very small.

4.4.15 Practical Implementation

The sketch of a coupled wave tube is given in Figure 4.22. The heated cathode emits an electron beam along the axis of the helical line. The beam is shaped as a hollow cylinder, with a concentration of the electrons near the helix. As the device may be rather long, the beam must be focused, most often by a longitudinal induction field.

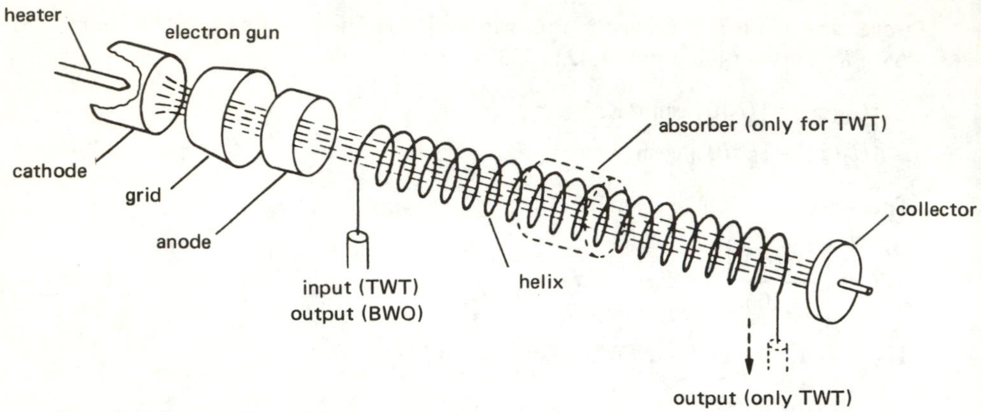

Fig. 4.22 Coupled line tube.

The dimensions of the structure and the beam parameters determine the type of operation (BWO or TWT) and the frequency range. Precautions are required to avoid spurious oscillations. Since operation takes place over a broad frequency range, no cavities are allowed within the device, and transitions between the helical line and the outside lines must have broadband capability, which prevents the filtering of unwanted signals. In a TWT, where one obviously does not want to operate as a BWO, lossy material is deposited over part of the helix. This cuts the feedback loop which may produce oscillations. A very large accuracy is required when manufacturing the mechanical parts which comprise these tubes. They are therefore quite expensive.

Coupled line tubes may also utilize a circular geometry, labelled M (for magnetron), in which the beam is bent by a static induction field (§ 4.2.1).

4.5 GYROTRONS

4.5.1 Power-Frequency Limitations

The current density of an electron beam is limited by the cathode. The electron speed results from the accelerating voltage, which is also physically limited by the electrode geometry. The beam energy thus cannot be increased arbitrarily, it is specified by the technology available. Furthermore, as frequency increases, conductor losses also increase and efficiency goes down [72].

As a result, the power that classical microwave tubes (magnetrons, klystrons, BWOs, TWTs) can deliver decreases approximately as f^{-5} [73]. A new frequency boundary appears therefore in the millimeter wave range (Fig. 4.45), available power decreasing very steeply beyond 30 GHz.

In order to overcome this limitation, a new amplification principle is presently under development.

4.5.2 Cyclotron Resonance Masers

Electrons moving within a static induction field follow a circular trajectory (§ 4.2.1) at the cyclotron frequency (4.4). Some electrons may ***excite*** a microwave signal on an external circuit. On the other hand, other electrons ***absorb*** part of the signal. To amplify or to generate a signal, the number of generating electrons should exceed that of absorbing ones. In the magnetron, this is done by means of a crossed-field structure, in which complex interactions between the electron stream and a periodically loaded transmission line (secondary cavities) take place. Absorbing electrons fall back on the cathode: the resulting loss of efficiency becomes prohibitive at large frequencies.

This energy loss may be reduced in another manner, utilizing relativity and inhomogeneous fields. As the speed of the electron approaches the velocity of light, its mass increases and the cyclotron frequency is reduced (4.4). This effect may be utilized to form ***bunches of electrons***, similar to those observed in klystrons, in a device called the *cyclotron resonance maser* [74]. Several different geometries have been considered, and one of them, the ***gyrotron***, generates large power levels in the millimeter range.

4.5.3 Principle of Operation of the Gyrotron

The *gyrotron* structure, developed in Gorki in the Soviet Union [75], is represented schematically in Figure 4.23.

The geometry presents an axial symmetry. Electrons are emitted by the lateral faces of the cathode, which yield a large current density with small variations of speed. Electrons are accelerated by an electric field. Due to the presence of a non-uniform induction field, their rotation speed increases and the beam is constricted. The structure is an inverted magnetic mirror (developed for fusion machines). The rotation energy of the electrons increases, at the expense of the longitudinal motion energy provided by the electrostatic field. The interactions between the particles and the microwave field take place in a circular waveguide, whose cross section varies slightly with position. The frequency of the signal is close to the cutoff of the mode (high dispersion region, Fig. 2.8). The electrons are guided by a static magnetic field which remains almost constant: they follow helical trajectories. The interaction with a rotating TE mode produces bunching: electrons yield a large portion of their energy to the field of the mode. Further along, the induction decreases, it is the decompression zone, in which electrons settle on the surface of the collector.

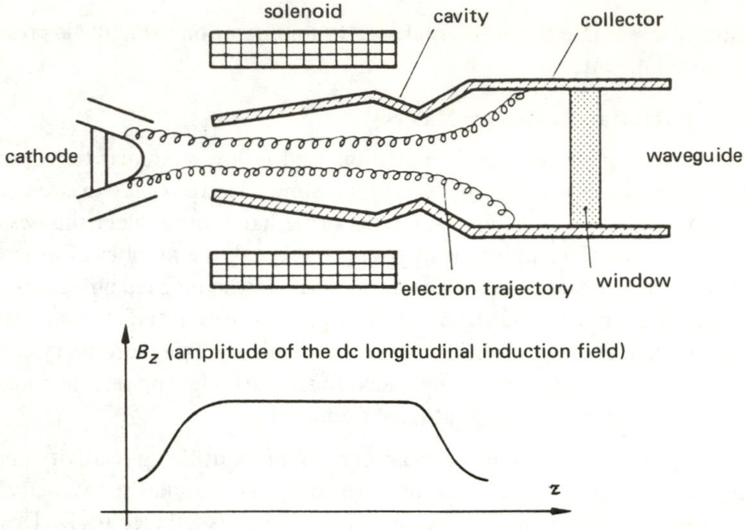

Fig. 4.23 Gyrotron geometry and distribution of the longitudinal induction field.

In contrast with the classical microwave tubes (§ 4.5.1), the gyrotron structure has a ***regular*** shape: no sharp changes in the cross section, no secondary cavities, no helical line, etc. The field distributions and the electron trajectories are therefore more uniform, which reduces losses. Also, a large electron current density is available. The different functions (emission, bunching, collection) are well separated in space.

4.5.4 Realizations

Gyrotrons have produced, under continuous wave operation, 12 kW at 107 GHz, 2.4 kW at 157 GHz and 1.5 kW at 330 GHz [76]. The corresponding efficiencies are 31%, 9.5%, and 6%. These tubes require large static inductions of 4.3 and 6 Tesla, provided by supraconducting coils.

The powers produced by gyrotrons are 100 to 1000 times larger than those available at the same frequencies from classical microwave tubes. Gyrotrons and other cyclotron resonance masers open the way to high power applications of millimeter waves.

4.6 TRANSFERRED ELECTRON GENERATORS

4.6.1 Multiple Valleys in Semiconductors, Negative Differential Resistivity

The study of conduction processes (quantum physics) within semiconductors shows that there is generally not a single conduction band. In most semiconductors, this band is formed of ***several valleys*** within the energy-momentum diagram [77]. In most materials however, the electrons possess the same properties in all the valleys, and the distinction has no observable effect. The valleys are then called equivalent.

Some semiconductors of the III-V group, among them n-type ***Gallium Arsenide*** (***GaAs***) and ***Indium Phosphide*** (***InP***) possess two valleys having different shapes (Fig. 4.24).

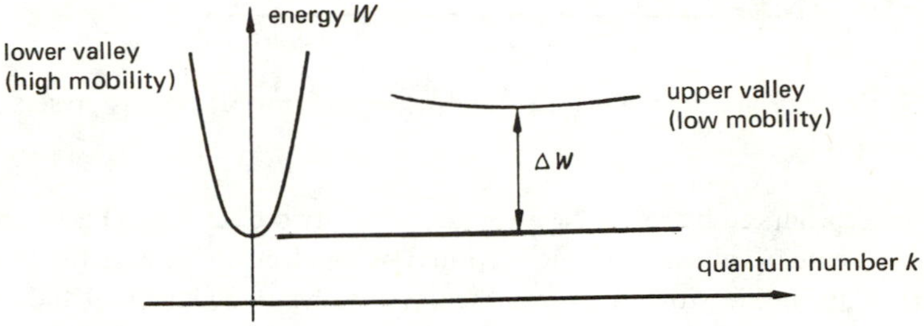

Fig. 4.24 Diagram energy/momentum for GaAs.

The diagram of Figure 4.24 shows the energy W of an electron in terms of its quantum wave number k. The speed of the electron is determined from

$$v = \frac{2\pi}{h} \frac{\mathrm{d}W}{\mathrm{d}k} \qquad \mathrm{m} \qquad (4.49)$$

where h is Planck's constant.

The speed of an electron is directly proportional to the slope of the curves. The lower valley is steep, which implies that there are rapid changes of speed with voltage, or in other terms, that the electron has a high mobility and a small effective mass. The upper valley is broader, and electrons there are slower, having low mobility and a large effective mass.

For a small applied voltage, all electrons lie within the lower valley. A voltage rise produces an increase in the electric field, electrons move faster and get more energy. For an increase of energy larger than ΔW, electrons jump into the upper valley: this is called ***electron transfer***. A sudden reduction of speed is associated with this jump (Fig. 4.25).

When the electric field exceeds a threshold $|E_g|$, the average speed of the electrons and consequently the current decrease as the field increases. This is called the *negative differential resistivity.*

4.6.2 Gunn Effect, High Intensity Field Domain

An electron moves within a III-V semiconductor (GaAs) under the effect of the

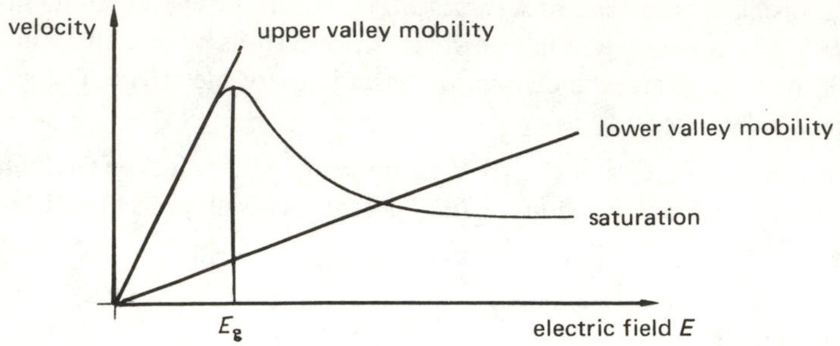

Fig. 4.25 Velocity in terms of electric field when electron transfer is present.

electric field produced by the voltage applied across two electrodes (Fig. 4.26). After an increase in voltage, the field surrounding an electron exceeds the threshold value $|E_g|$. The electron jumps into the upper valley of Figure 4.24 and its speed decreases drastically. The result is analogous to what happens on a highway when a car slows down, a ***bunch*** starts forming. Electrons slowed by the large electric field tend to bunch together. The resulting concentration of charge carriers further increases the local field, and a *high intensity field domain* is formed; it moves at the speed of the electrons in the upper valley.

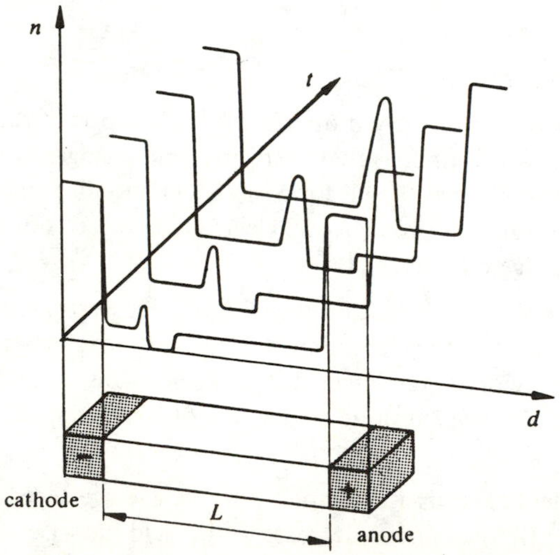

Fig. 4.26 Gunn domains (n = electron density).

This local concentration of electrons and of the field in time and space implies an equivalent reduction within the remainder of the semiconductor. A large portion of the voltage applied across the semiconductor actually appears across the domain. The electric field elsewhere in the semiconductor then remains well below the threshold. The existence of a domain actually inhibits the formation of other ones. Only when the domain has crossed the whole semiconductor and reached the anode can another domain start to grow. The electron flow, injected continuously by the cathode, reaches the anode in the form of bunches. The result is analogous to the one observed in the klystron (Sec. 4.3). In the present situation, it is produced within a homogeneous semiconductor by quantum interactions.

This effect was discovered fortuitously by J.B. Gunn in 1963 [78] while studying conduction in GaAs submitted to an intense electric field. An 0.1 mm long GaAs sample then produced current oscillations in the vicinity of 1 GHz.

4.6.3 Oscillation Conditions

The usual connotation ***"Gunn diode"*** is improper, as there is no junction between two different materials within this device. The Gunn effect is a volume phenomenon, which takes place within a slab of homogeneous semiconductor.

Most often, the domain starts growing close to the cathode, where the change in doping level produces a somewhat inhomogeneous region. The domain gradually drifts towards the anode, the transit time t_t being the ratio of the slab length L by the speed v of the electrons (upper valley)

$$t_t = L/v \qquad \text{s} \qquad (4.50)$$

A certain building-up time is also required for the domain to reach its maximum dimension. The building-up time is smaller than the transit time when

$$nL \geqslant 10^{16} \qquad \text{m}^{-2} \qquad (4.51)$$

where n is the carrier density.

Furthermore, for the building-up time to be shorter than a period of the signal, it is necessary that

$$n/f \geqslant 5 \cdot 10^{10} \qquad \text{s/m}^3 \qquad (4.52)$$

4.6.4 Modes of Oscillation

The characteristics of a Gunn generator depend not only upon the semiconductor material, but also upon the microwave circuit to which it is connected, which may more or less deeply influence the operation of the assembly. A large number of modes of operation thus exists.

4.6.5 Transit Mode

The domain may cross entirely the semiconductor and then disappear; this is called the *transit mode.* The signal frequency is imposed by the material and $f_t = 1/t_t$ (4.50). This mode is the one observed by Gunn in his original experiments. The output power and the efficiency can be increased by coupling the device to a resonant circuit (cavity), which stabilizes and permits small adjustments of frequency.

4.6.6 Delayed Mode

For the *delayed mode,* the transit time t_t is shorter than the period T of the signal in the resonant circuit. The building up of a new domain is delayed by applying a voltage below the threshold. For maximum power to appear at the output, the circuit is adjusted for the domain to reach the anode when the applied voltage is the smallest. This is a procedure similar to the one utilized in class C amplifiers: the current maximum coincides with the voltage minimum, increasing the efficiency [79]. The oscillation frequency is controlled by the resonant circuit, it may be tuned over a frequency band up to one octave (the output power does not remain constant over the whole band).

4.6.7 Quenched Mode

When the voltage drops below a specific value, the domain disappears while still in transit, this is the *quenched mode.* The semiconductor is then in the ohmic part of its characteristic for a short period, and a current of fast electrons flows until a new domain can start growing. Here, too, the current flows while the voltage is minimum (class C). This mode is tunable over a broad frequency range.

4.6.8 Limited Space Charge Accumulation (LSA) Mode

When the microwave signal has a large enough amplitude for the voltage to drop below the threshold at each cycle and, in addition, its frequency is large enough, domain formation may be prevented. *Limited space charge accumulation modes* are then achieved. The oscillating frequency of these modes is independent of transit time, and thicker semiconductors may then be utilized [80].

In theory, LSA modes yield better efficiencies at higher power outputs than the other modes [81].

In practice, however, it is difficult to actually prevent domain formation (domains appear at all material inhomogeneities), so that these devices hardly passed the prototype stage.

4.6.9 Hybrid Modes

A whole collection of *hybrid modes* is scattered between the domain and the LSA modes. They exhibit intermediate properties.

4.6.10 Gunn Oscillator

In practice, generators are made by placing a Gunn diode within a waveguide, a coaxial line or a microstrip (Fig. 4.27), and a continuous bias voltage is applied (in the 6 to 15 volt range, depending upon the particular diode used). Frequency tuning is done by using a piston or a tuning screw. Available power levels vary from 10 to 100 mW typically. Efficiencies are low, on the order of a few percent.

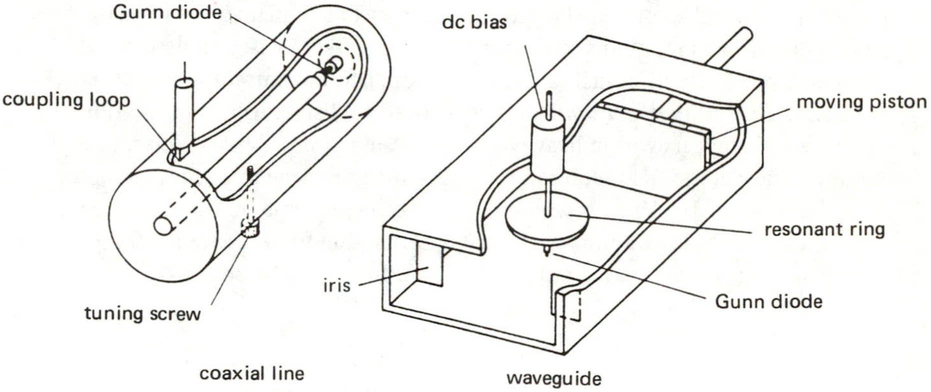

Fig. 4.27 Gunn diode oscillators.

The power obtained is of the same order as that produced by a reflex klystron (§ 4.3.6), the operation of the device being much simpler. Gunn generators thus made possible new microwave applications, for which limited power is adequate, but which must be easy to use.

4.6.11 Gunn Amplifier

When a Gunn oscillator is biased slightly below the threshold voltage, its input impedance exhibits a negative resistance component at frequencies close to the oscillation frequency. Amplification may then be realized. This amplifier only has one port (at microwaves), through which pass both the input and the output signals. Particular care is then required to prevent a spurious feedback, which would produce instabilities (§ 6.2.5).

4.6.12 Note

The Gunn diode is ***not a negative resistance element***: it cannot be compared to the tunnel diode. The basic mechanisms of the Gunn effect are the build-up and the evanescence of the domains, which recur with a period related to the size of the semiconductor. The Gunn diode amplifier only has a narrow operating frequency. In contrast, negative resistance amplifiers allow amplification to occur over a broad frequency range.

4.7 AVALANCHE AND TRANSIT TIME

4.7.1 Avalanche Mechanism

The avalanche effect within a semiconductor is similar to the arc discharge within a gas (electrical arc). An intense electric field accelerates charge carriers (electrons and holes) until their energy is sufficient to ionize an atom of the semiconductor and liberate an electron-ion pair, which in turn is accelerated by the field, producing other pairs and so on. As the process is somewhat similar to the formation of ***snow slides*** in the mountains, the term *avalanche effect* is used to describe the phenomenon. There is, in fact, a chain reaction, and the current increases exponentially. The term ***impact ionization*** is utilized to distinguish from the tunnel effect, which appears within heavily doped p-n junctions (Zener diodes, tunnel diodes). The electric field required to trigger the avalanche in a semiconductor is of the order of 10^7 V/m. It would clearly not be possible to establish such a field within the bulk of a homogeneous material (due to excessive heating). Therefore an inversely biased p-n junction is utilized (Fig. 4.28) [140].

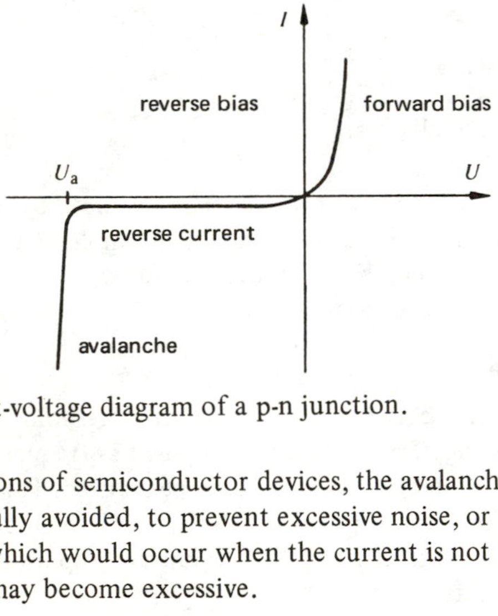

Fig. 4.28 Current-voltage diagram of a p-n junction.

In usual applications of semiconductor devices, the avalanche zone and its close vicinity are carefully avoided, to prevent excessive noise, or even the destruction of the junction, which would occur when the current is not limited, in which case the heating may become excessive.

4.7.2 Avalanche under Sine Wave Operation

Once the avalanche voltage is applied to a p-n junction, the current increases ***exponentially***, due to the impact ionization. A certain time is required for the current to build up, as carriers must move within the avalanche region, creating pairs by ionizing collisions as they move. For a junction biased just below the avalanche voltage, to which a small sine wave voltage is added, the avalanche current starts flowing just as soon as the voltage exceeds the threshold. This current grows exponentially. When the voltage drops below the threshold, the avalanche effect stops and the current decreases. Consequently, the current flowing through a junction is delayed by roughly a quarter of a period with respect to the voltage (Fig. 4.29) [82].

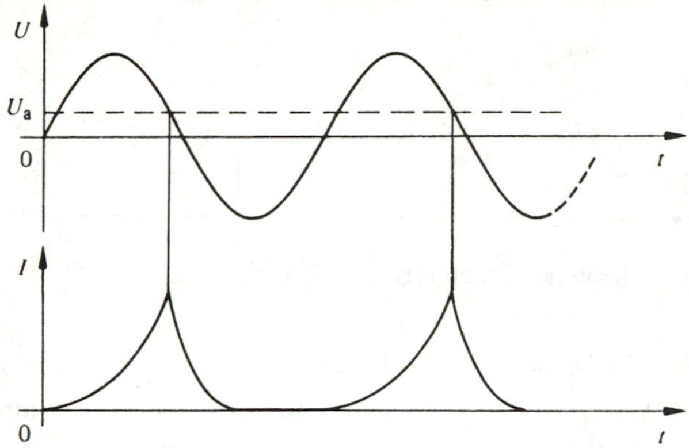

Fig. 4.29 Delay between current and voltage in the avalanche region.

4.7.3 Drift Region

A negative resistance corresponds to a shift of 180° between voltage and current. The charge carriers are then made to cross a *drift zone*, free of charges and subjected to a large electric field. The carriers cross this region at their saturated drift velocity (which is constant for high field values). Electrons injected at time t within the drift region reach the anode at $t + t_t$, where t_t is the time required for the transit. A current flows within the external circuit during this period. By adjusting the length of the drift zone, an additional phase shift of 90° is obtained for the current, which is added to the 90° already produced by the avalanche (§ 4.7.2).

4.7.4 IMPATT Diode

The combined effects of avalanche and transit time are utilized in the *IMPATT* diode (*IMPact Avalanche and Transit Time*). The profile originally proposed by Read [83] has an abrupt n^+p junction, followed by a drift zone of intrinsic semiconductor (i) and a p^+ zone for connections. Only the holes participate to the delayed current within this structure. Several simpler profiles were realized in practice: n^+pp^+, $n^+p_1p_2p^+$, and n^+npp^+. The last profile possesses two drift zones, one of them for holes, the other for electrons, on either side of the junction. Figure 4.30 represents the four profiles mentioned together with the corresponding electric fields. Seen from the outside, IMPATT diodes are similar to Gunn diodes, but with a larger impedance. They require larger operating voltages (70-100 V) and produce larger power levels (up to several watts) with a higher efficiency. On the negative side, they have a higher noise level.

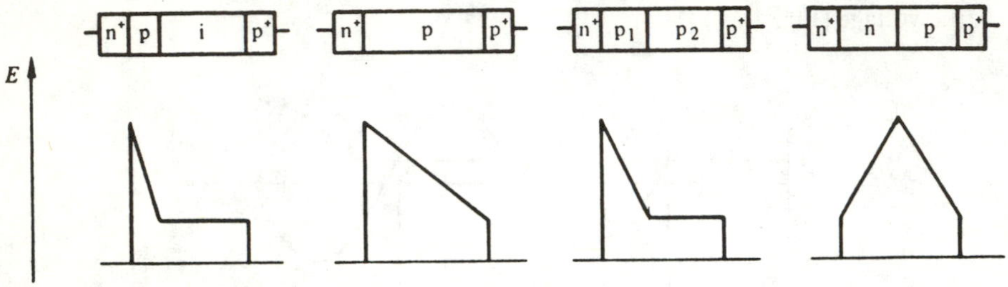

Fig. 4.30 IMPATT diodes and electric field profile.

4.7.5 TRAPATT Diode

The *TRApped Plasma Avalanche Triggered Transit* takes place within an n^+pp^+ structure in silicon or germanium (Fig. 4.31). The p region is lightly doped and behaves as a drift region.

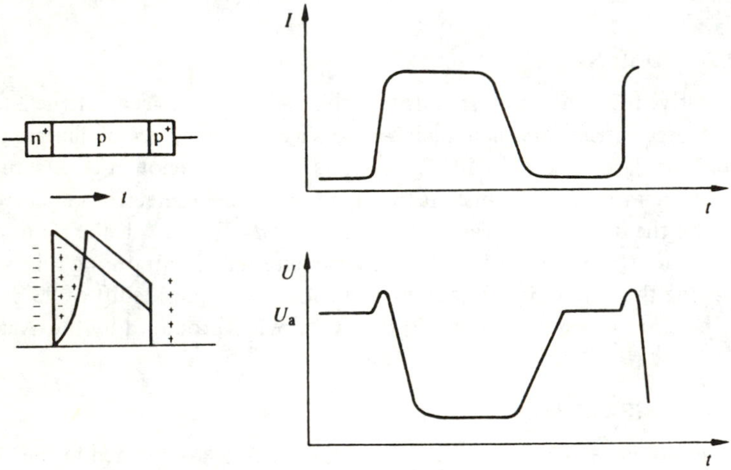

Fig. 4.31 TRAPATT diode.

When this diode is inversely polarized, the electric field in the n^+p junction is large enough to sustain the avalanche and majority carriers move at the saturation velocity across the drift region. A short voltage pulse, produced by the outside circuit, triggers a fast expansion of the avalanche across the whole drift zone. The diode is then filled with a dense plasma and becomes highly conductive at low voltage. When the carriers have been evacuated, the voltage increases again, up to the avalanche level. The diode thus switches, when triggered by an outside pulse, from a high current/low voltage state to a low current/high voltage state [84]. This negative resistance effect can be utilized in the design of oscillators and amplifiers. In this way, one can obtain high powers in pulsed operation (§ 5.8.2) with a very good efficiency, of the order of 60%. The operation is ***strongly nonlinear***, however, and harmonics must be filtered.

4.7.6 BARITT Diode

In the *BARrier Injected Transit Time* diode, carriers are injected into the drift zone by a forward-biased junction (Fig. 4.32).

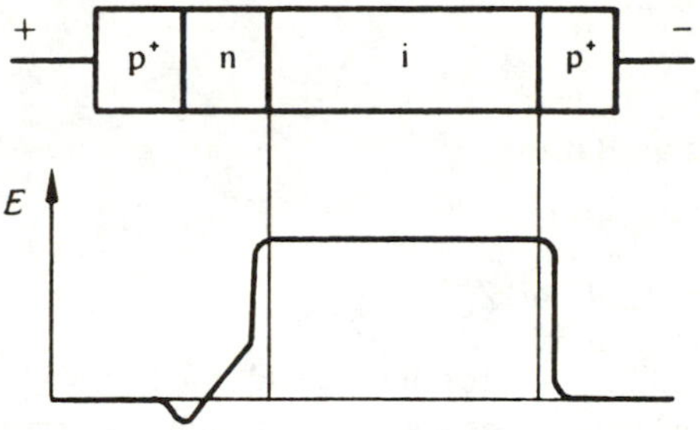

Fig. 4.32 BARITT diode.

The p^+n junction is in its conducting state, emitting holes into the intrinsic region, which they cross at their saturation velocity to reach the p^+ contact. A negative resistance with a phase shift of π radians, is obtained in this manner. BARITT diodes operate at a low bias voltage and are much less noisy than IMPATT diodes. Their output power, however, does not exceed a few milliwatts [85].

4.8 TRANSISTORS

4.8.1 Introduction

The utilization of transistors at microwave frequencies is relatively recent. It follows the quite spectacular development of the solid state technology, which led to very accurate positioning of the masks during manufacturing. Electrode dimensions could be brought down into the ***sub-micrometer*** region, transit times then being on the order of picoseconds. As for parasitic reactances, their effects are reduced by the introduction of matching elements on the very semiconductor substrate [86]. The frequency limit of bipolar transistors is located around 5-6 GHz, MESFETs being used at higher frequencies (§ 4.8 4).

Transistors are ***three-poles*** mostly used for amplification. They may be used in oscillators, by coupling them by means of a feedback circuit (§ 4 1.2). The block diagram of a microwave transistor amplifier is given in Figure 4.33. Particularly worth noting are the ***matching circuits*** [87].

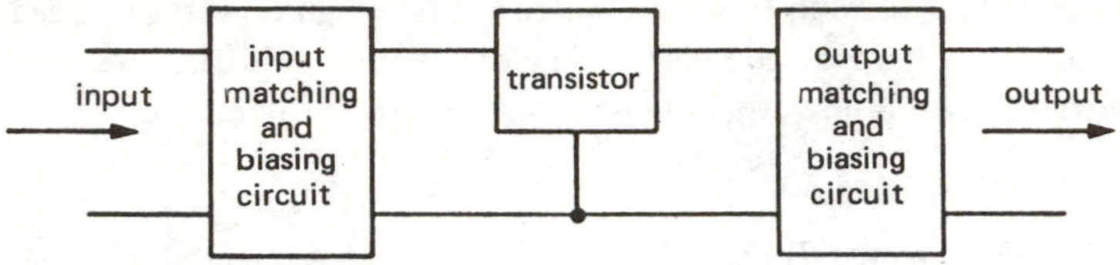

Fig. 4.33 Transistor amplifier.

4.8.2 Amplification Ratios

The *power amplification* is given by

$$P_s/P_e = (i_s/i_e)^2 \operatorname{Re}[\underline{Z}_s]/\operatorname{Re}[\underline{Z}_e] \qquad 1 \qquad (4.53)$$

where the indices s and e stand for output and input respectively. The ratio i_s/i_e is the current gain, which decreases with frequency as f_t/f, the gain becoming unity at the ***cutoff frequency*** f_t

$$P_s/P_e = (f_t/f)^2 \operatorname{Re}[\underline{Z}_s]/\operatorname{Re}[\underline{Z}_e] \qquad 1 \qquad (4.54)$$

For power amplification to be significant at microwaves, one must not only have a high cutoff frequency f_t, but also a large ratio between the real parts of the output impedance $\underline{Z}_s$ and the input impedance $\underline{Z}_e$. The device geometry and the doping must be optimized to obtain the best possible compromise.

4.8.3 Bipolar Transistors

Transistors for microwaves are most often realized in planar ***n-p-n silicon technology*** [88]. They are connected with a common emitter for large gain, or in a common base for power. The cutoff frequency is inversely proportional to the transit time τ, the sum of four contributions:

1. charging time of the emitter-base junction
2. transit time across the base
3. transit time of the depletion region of the base-collector junction
4. charging time of the base-collector junction.

It follows from (4.54) that a low input impedance and a large output impedance are desired. These must be matched to the lines, which most often have a 50 Ω characteristic impedance. Metal oxide capacitors (MOS) and discrete components, placed within the encapsulation, allow matching to be performed over a wide frequency range.

Several geometries are utilized to widen as much as possible the junctions without increasing the surface of the electrodes (Fig. 4.34).

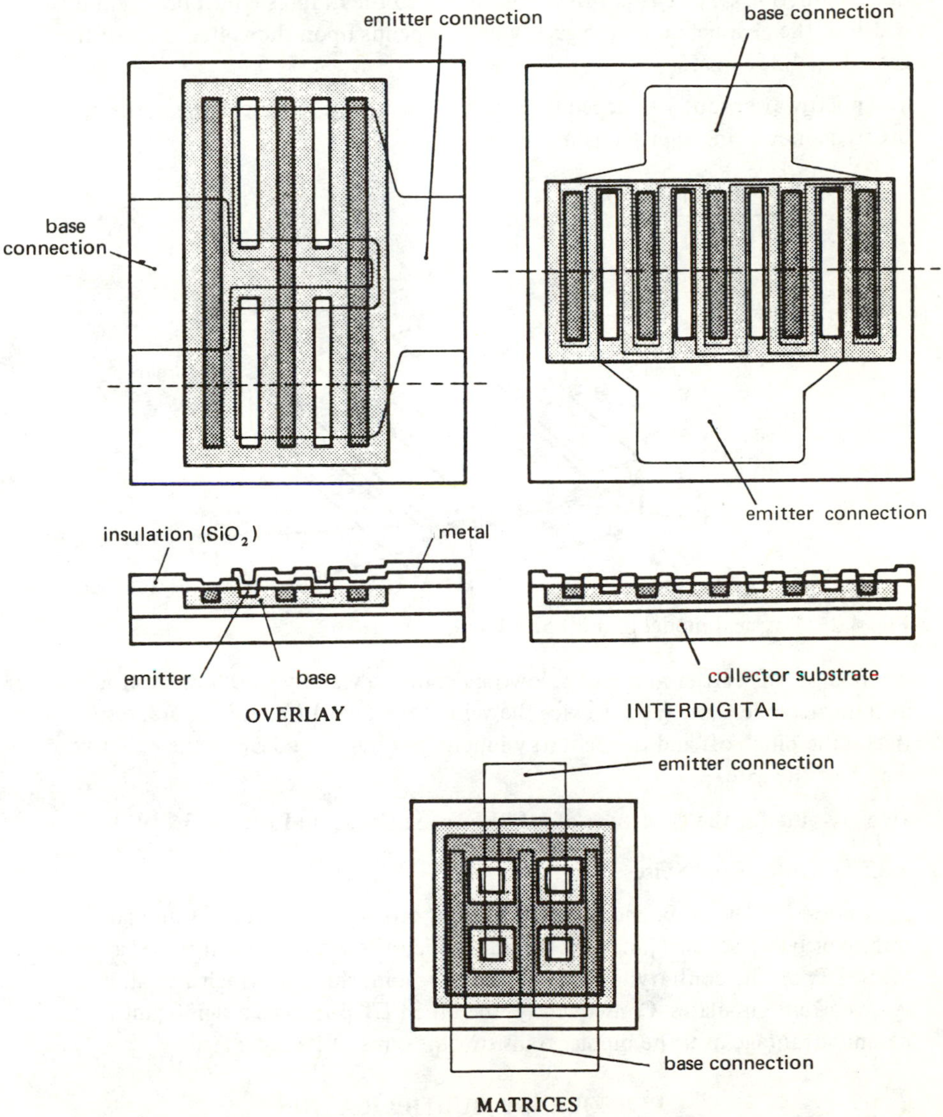

Fig. 4.34 Geometries of bipolar transistors.

4.8.4 MESFET

A *MEtal Semiconductor Field Effect Transistor* is a particular kind of field effect transistor in which the gate junction is a Schottky junction (metal-semiconductor junction). The current from the source to the drain is controlled by the width of the channel under the gate, which depends upon the voltage bias of the gate (Fig. 4.35) [89].

The microwave properties depend upon the transit time for the carriers, the cutoff frequency being approximately given by [90]:

$$f_t \cong \frac{\mu_P(U_P + U_B)}{2\pi L_G^2} \quad \text{Hz} \tag{4.55}$$

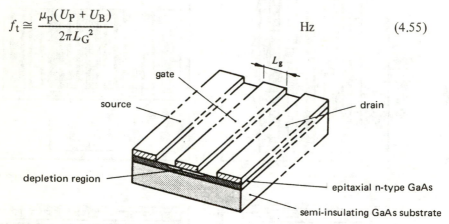

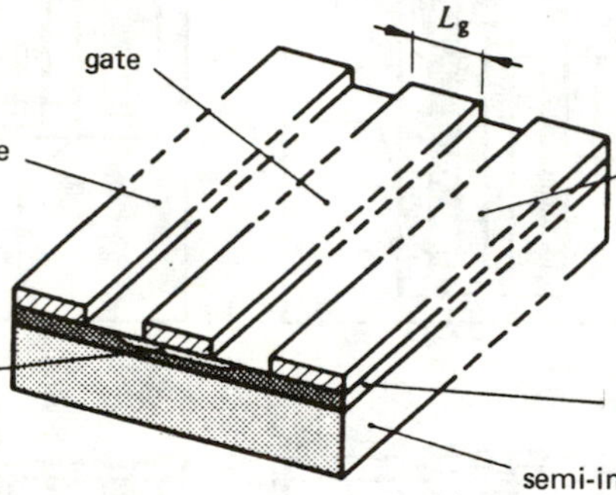

Fig. 4.35 Physical model of a MESFET.

where μ_P is the carrier mobility at low fields, in the vicinity of 0.5 m^2/Vs for gallium arsenide GaAs (about twice the value for silicon) U_P and U_B are, respectively, the pinch off and the Schottky junction voltages, and L_G is the effective length of the gate.

Two lay-outs for the electrodes of MESFETs are shown in Figure 4.36 [91].

4.8.5 Note: Shot Noise

Shot noise [92] is proportional to the current crossing a junction. A bipolar transistor possesses two junctions through which the whole current flows; a MESFET, on the contrary, only has a gate junction, through which a small inverse current circulates. Consequently, the MESFET possesses a significant inherent advantage over the bipolar transistor in terms of its noise figure.

4.9 FREQUENCY MULTIPLICATION

4.9.1 Harmonic Generation

A sine wave signal whose frequency is a fraction of the desired frequency is ap-

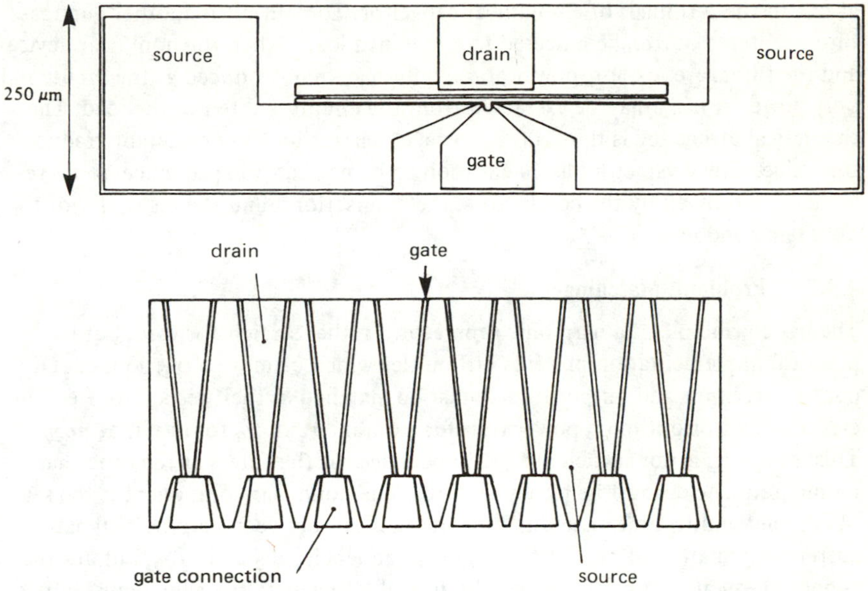

Fig. 4.36 Geometries of MESFET transistors.

plied to a nonlinear device [93]. ***Harmonics*** of the input signal are generated in the device and a signal at the desired frequency is extracted through a filter [94].

Nonlinear resistors or nonlinear capacitors may be utilized, consisting of semiconductor p-n junctions. However, due to its inherent losses, the efficiency of a nonlinear resistor cannot exceed $1/n^2$, where n is the harmonic number [95]. On the other hand, a nonlinear capacitor (varactor) may generate harmonics with a theoretical efficiency close to 100% [96].

The block diagram of a frequency multiplier is shown in Figure 4.37.

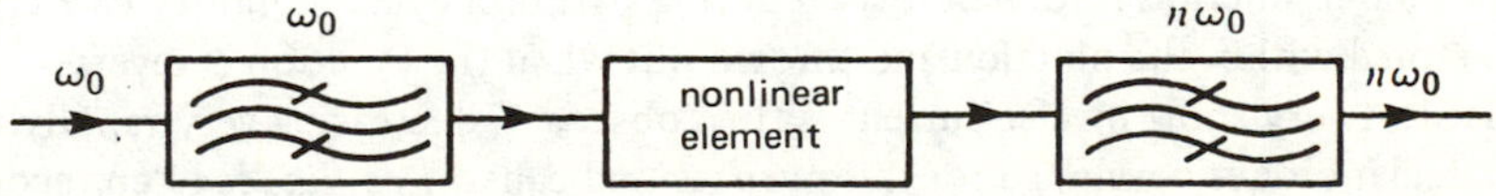

Fig. 4.37 Basic schematic of a frequency multiplier.

A signal of angular frequency ω_0 is applied, through a bandpass filter centered at ω_0, to the terminals of a nonlinear capacitor. The output is another bandpass filter, centered at $n\omega_0$, connected to a matched load. When the nonlinear device and the filters are lossless, power at ω_0 only may be introduced at the input, and only power at $n\omega_0$ may be extracted from the circuit and fed to the load. The theoretical efficiency is thus 100%, however, circuit losses considerably reduce this value. Every varactor has, in addition to its nonlinear capacitance, series resistances produced by the conductors, the connections, and the resistivity of the bulk semiconductor.

4.9.2 Problem: Matching

The basic principle of a harmonic generator is rather straightforward, but its practical implementation presents difficulties which cannot be overlooked. Of course, the input and output filters must be matched at their respective frequencies to allow for optimum power transfer from input at ω_0 to output at $n\omega_0$. This, however, is not sufficient. The impedances of the filters at the other harmonic frequencies must be purely reactive. The nonlinear component is the seat of complex multiplication and mixing interactions between harmonic signals: currents circulate, not only at the angular frequencies ω_0 and $n\omega_0$, but at other angular frequencies $k\omega_0$ [94]. To obtain high efficiency, the filters must also possess the correct impedance at intermediate frequencies. The impact of higher order harmonics is less significant. It is therefore difficult to make multipliers with nonlinear capacitors having large values of n. Generally, the largest value encountered is $n = 4$.

4.9.3 Varactors

The capacitance of a p-n junction varies with voltage as:

$$C = C_0(1 - V/\phi)^{-\gamma} \qquad \text{for } V < \phi \qquad \text{F} \qquad (4.56)$$

where C_0 is the capacitance at zero bias, ϕ the barrier potential of the junction and γ is a constant, which is 1/2 for an ***abrupt junction*** and 1/3 for a ***graded junction*** The change in capacitance is produced by the widening of the depletion region when the inverse voltage increases.

4.9.4 Stored Charges, Step Diode

When a p-n junction is forward-biased during part of a cycle, minority carriers are stored within the junction and thrown out when the junction is reverse-biased. A very short inverse current is then observed, decreasing very rapidly in particular diodes developed for this purpose and called *step diodes* (step recovery, snap back). The current which circulates is shown in Figure 4.38 As the transition period may be very short (on the order of a few picoseconds), this current has harmonic components of very high orders.

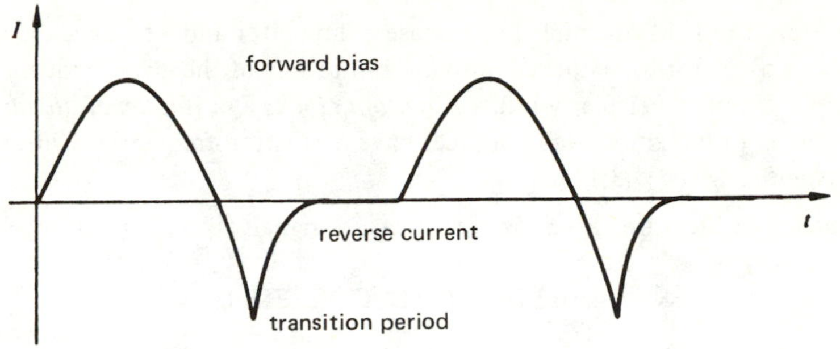

Fig. 4.38 Current in a snap-back diode.

The charge storage phenomenon may be considered as an increase of the varactor capacitance variation. However, the principle of multiplication is fairly different from the one of a simple nonlinear capacitor. Step diodes are used as generators of higher order harmonics ($n > 8$) [97].

4.9.5 Active Multipliers

Harmonics (Fig. 4.37) may also be generated by an active nonlinear device, for instance a class B or class C transistor amplifier. One then has amplification and frequency multiplication at the same time.

4.9.6 Application

Frequency multipliers are most often utilized to realize sources of very ***high stability***. Starting with a stable oscillator controlled by a quartz crystal, the signal is amplified and its frequency multiplied within a chain of multipliers. An example is shown in Figure 4.39.

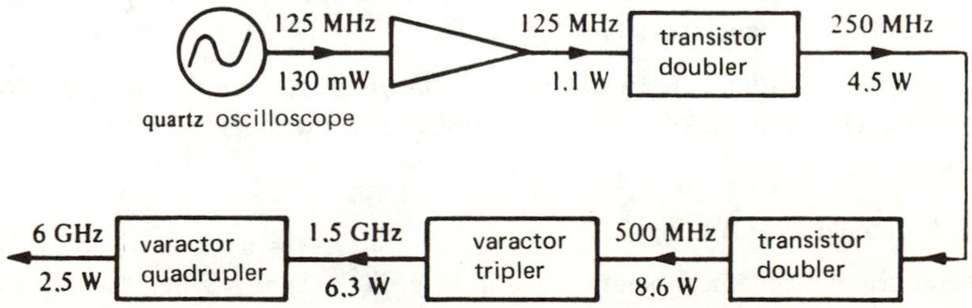

Fig. 4.39 Block diagram of a generator with frequency multipliers.

4.9.7 Application: Frequency Divider

A multiplication chain, coupled to a phase comparator and a feedback loop, also allows an oscillator to be locked onto a sub-multiple of the given frequency. This principle is more particularly utilized in ***atom clocks*** and ***frequency standards***. As a matter of fact, most stable sources have oscillation frequencies within the microwave range (§ 5.3.11).

Frequency dividers are also utilized to measure signals (§ 5.3.8).

4.10 LOW NOISE AMPLIFIERS

4.10.1 Introduction

The electromagnetic background noise level happens to have its lowest values within the microwave range (§ 7.6.4), the sensitivity of a receiver is therefore highest in this range. This sensitivity is, however, strongly reduced when transistor or tube amplifiers are utilized, because these amplifiers also add a noise which is larger, by several orders of magnitude, than the background noise itself (§ 7.6.5). In order to fully exploit the very high sensitivity capability of microwaves, devices which ***do not utilize free electrons*** (noise sources) have been implemented. Interactions between signals at ***different frequencies*** within nonlinear capacitances of parametric amplifiers and paramagnetic crystals of MASERs are utilized.

4.10.2 Principle of Parametric Amplification

By varying *one parameter* in a system, it is possible to add energy to a signal. Let us assume that in an LC resonant circuit, the distance between the capacitor plates may be instantaneously changed. If the plates are pulled apart when the voltage is maximum, the capacitance decreases and the voltage increases, since the charge $Q = CU$ is not affected. The energy $1/2\ CU^2$ increases, and some work must be provided to pull the plates apart. The plates are brought back to their initial location when the voltage goes through zero: no energy is then returned to the mechanical system (Fig. 4.40).

The signal is amplified here by ***mechanical pumping*** at twice the frequency of the signal. The principle of parametric amplification was discovered by Faraday in 1831.

4.10.3 Mechanical Analogy

A well-known mechanical parametric amplifier is the ***swing***. When the deviation goes through its maximum, a parameter is modified, the position of the center of gravity. Energy is thus added to the oscillation. The initial center of gravity is restored when moving through the equilibrium position. For anatomical reasons, it is not possible to amplify during the second alternation.

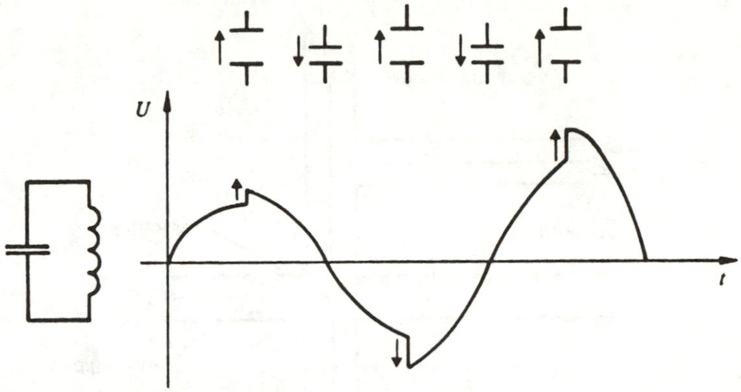

Fig. 4.40 Pumping of a capacitor.

4.10.4 Practical Implementation of the Parametric Amplifier, Pump, Idler

The capacitor plates of Figure 4.40 cannot be just mechanically pulled back and forth at twice the frequency of the microwave signal. However, the same effect may be provided by electrical means, in a solid-state p-n junction, in a variable capacitance or varactor (§ 4.9.3). The varactor is placed in a resonant circuit (cavity), fed by a signal at frequency f, which is to be amplified, and a large amplitude *pump signal* at a frequency $f_p > f$. In the example of Figure 4.40, the pump frequency is exactly twice the signal frequency. This is not a basic requirement, as amplification also takes place when this ratio is not maintained. In the latter situation, an additional signal at the intermediate, or *idler*, frequency $f_i = f_p - f$ is obtained. The circuit must therefore resonate at the three frequencies. A simple representation is given in Figure 4.41. The three cavity modes must, of course, each produce a voltage across the varactor.

Very complex interactions take place within the cavity. The mixing of the signal and pump yields difference and sum signals. These signals in turn interact with the pump, producing an amplified signal coming out of the cavity. Of course, phase consistency between all the signals is required for amplification to occur [98].

A low noise level is obtained by cooling the amplifier (for instance, down to the temperature of liquid nitrogen, 77 K), or by utilizing a pump frequency much larger than the signal frequency (in the millimeter wave region).

At the signal frequency, the parametric amplifier only has one port, which is, at the same time, the input and the output (§ 6.2.5). Parametric amplifiers have gains between 10 and 20 dB, with bandwidths up to 20%. They are sensitive, but expensive devices.

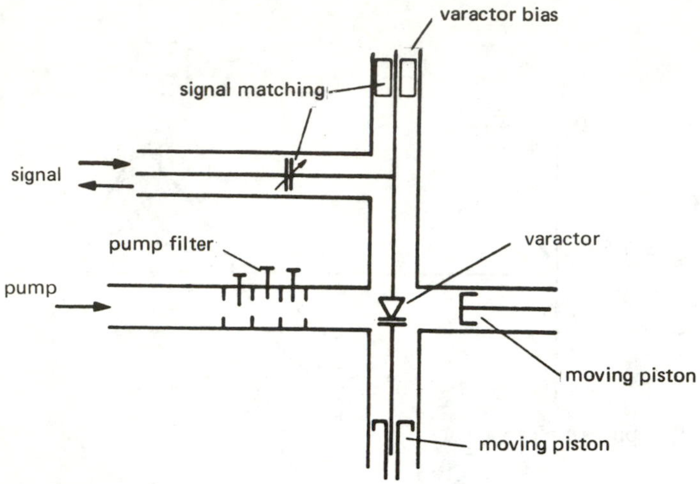

Fig. 4.41 Parametric amplifier.

4.10.5 MASER

The *MASER (Microwave Amplification by Stimulated Emission of Radiation)* utilizes quantum properties within certain materials, among them ruby crystals. Low energy electrons are ***pumped*** into an upper energy level. In the presence of an input signal of adequate frequency, they fall down to a lower level, releasing a quantum of energy to the signal. The basic principle is sketched in Figure 4.42.

A high intensity pump signal of frequency $f_{13} = (W_3 - W_1)/h$ (where $h = 6.624 \cdot 10^{-34}$ J/s is Planck's constant) transfers electrons from level 1 up to level 3 and back. The probabilities for transition from 1 to 3 and from 3 to 1 are equal, so the resulting populations tend to balance when the pump signal has a very large power.

The population of level 3, in the example shown in Figure 4.42, then becomes larger than that of level 2. A signal of frequency $f_{23} = (W_3 - W_2)/h$ gets electrons to jump from 2 to 3, absorbing energy, and from 3 to 2, releasing energy to the signal. Since more electrons are at level 3 than at level 2 the signal is amplified. Electrons then fall from level 2 to 1 by relaxation processes and the cycle is completed.

Distances between energy levels in atoms correspond to frequencies in the visible spectrum. For transitions to take place in the microwave range, the Zeeman effect (paramagnetic resonance) in a ruby crystal (Al_2O_3 + Cr^{3+} ions) is utilized. In the presence of a magnetic field, levels are subdivided (a hyperfine structure), and the spacing between levels can thus be adjusted, providing frequency tuning. This phenomenon is observed in a ***very limited number*** of materials.

The population difference must also be sufficiently large. At equilibrium, Boltzmann's equation yields:

$$N_3/N_2 = \exp[(E_2 - E_3)/kT] \cong 1 - \frac{1}{20}\,\frac{f_{23}\,(\mathrm{GHz})}{T\,(\mathrm{K})} \qquad 1 \qquad (4.57)$$

At the ambient temperature, the difference is too small for amplification. This difference is increased by cooling the maser to the temperature of liquid helium (4.2 K), or of liquid nitrogen (77 K)[99].

4.10.6 Cavity Maser

The basic structure is the same as the one of the parametric amplifier, but with the nonlinear capacitance replaced by a ruby crystal. The cavity must simultaneously resonate at the three frequencies f_{13} (pump), f_{23} (signal) and f_{12} (idler). Amplifications up to 20-25 dB can be achieved over narrow bands, on the order of MHz. The input and output signals appear at the same port (§ 6.2.5).

4.10.7 Traveling Wave Maser

Utilizing a hybrid structure, obtained by crossing a maser with a TWT (§ 4. 4.13), one may realize a wider bandwidth, on the order of tens of MHz. The ruby crystal, in which the pump signal propagates, is surrounded by a low phase velocity line (helix, periodically loaded line, § 4.4.10). Interactions produce an amplification when the frequency and field excitation conditions

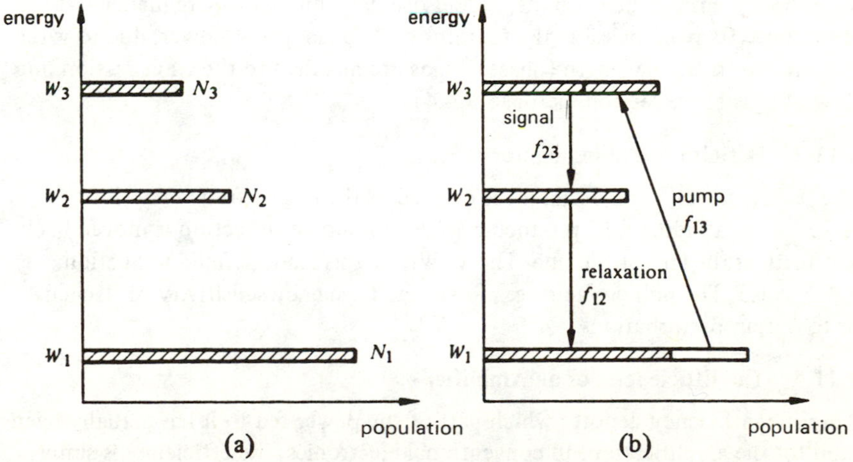

Fig. 4.42 Population of the energy levels (3-level maser): (a) at equilibrium, (b) under maser operation.

are satisfied. Even more than in the TWT, precautions must be taken to avoid bidirectional amplification, which is a cause of instability.

4.10.8 Note

Masers provide amplification with extremely low noise added, their equivalent noise temperature even dropping below 10 K (§ 7.6.5). They are the ***most sensitive*** of all amplifiers, but also are the most expensive ones to operate, as they require ***cryogenic conditions***. They are therefore utilized only in applications where the utmost in sensitivity is an absolute must.

4.11 PRACTICAL NOTES

4.11.1 Effect of the Load

Every oscillator is influenced by the load connected to it, the effect being more or less severe for different devices. The load has a direct effect on the cavity or the electrical circuit of the oscillator, modifying not only the power (9.26), but also the signal frequency at the oscillator's output.

4.11.2 Rieke's Diagram

The effect of the load on an oscillator is represented by plotting, on the Smith chart for the load (§ 9.5.3), the loci of constant power and of constant frequency for the signal coming out of the generator. As an example, the *Rieke diagram* for a klystron is shown in Figure 4.43. Over a region where the impedance has a large reactive component, the klystron actually cannot oscillate. In other highly mismatched regions, it does oscillate, but at two frequencies at the same time. To avoid possible fluctuations of frequency and power, due to variations in electrical line length (phase), loads are matched to the transmission line (lowest possible reflection factor, § 9.4.4).

4.11.3 Definition: Pulling Factor

The *pulling factor* of an oscillator is defined as the largest frequency shift observed when a sliding load producing a 20% amplitude reflection is moved back and forth at the end of the line. The VSWR of this load, defined in Section 7.2.2, is 1.5. The pulling factor expresses the frequency sensitivity of oscillator to output perturbations.

4.11.4 The Efficiencies of an Amplifier

Amplifier efficiency denotes which part of the power fed to it has actually been used for the amplification. In conventional electronics, the efficiency is simply defined as the ratio of the output power to the power supplied by the bias. The

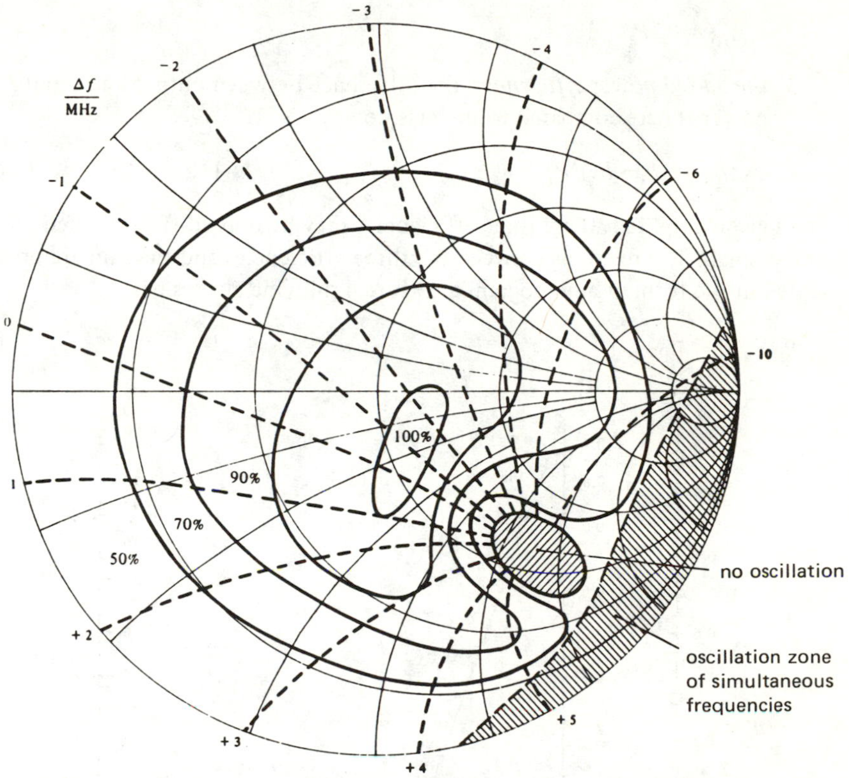

Fig. 4.43 Rieke diagram for a klystron under unfavorable operating conditions. (long line effect) —— constant power lines -----constant frequency lines.

input power is generally quite small, and not considered. At microwave frequencies, the gain of an amplifier may be quite low, so that ***three different definitions*** of efficiency are encountered, which should not be used indiscriminately:

1. *the total efficiency*: it is the quotient of the output power to all the power supplied to the amplifier

$$\eta_t = P_s/(P_e + P_a) \qquad 1 \qquad (4.58)$$

where P_s is the output power, P_e the input signal power and P_a the power coming from the bias supply;

2. *the partial efficiency*: the power of the input signal is not considered

$$\eta_p = P_s / P_a \qquad 1 \qquad (4.59)$$

3. *the added power efficiency*: the difference between output and input power is here compared to the bias power

$$\eta_{pa} = (P_s - P_e)/P_a \qquad 1 \qquad (4.60)$$

For a given amplifier, these three efficiencies may be quite different when the gain is small. The relations between the three efficiencies and the gain are presented in the form of a nomogram in Figure 4.44. One always has:

$$\eta_p \geqslant \eta_t \geqslant \eta_{pa} \qquad 1 \qquad (4.61)$$

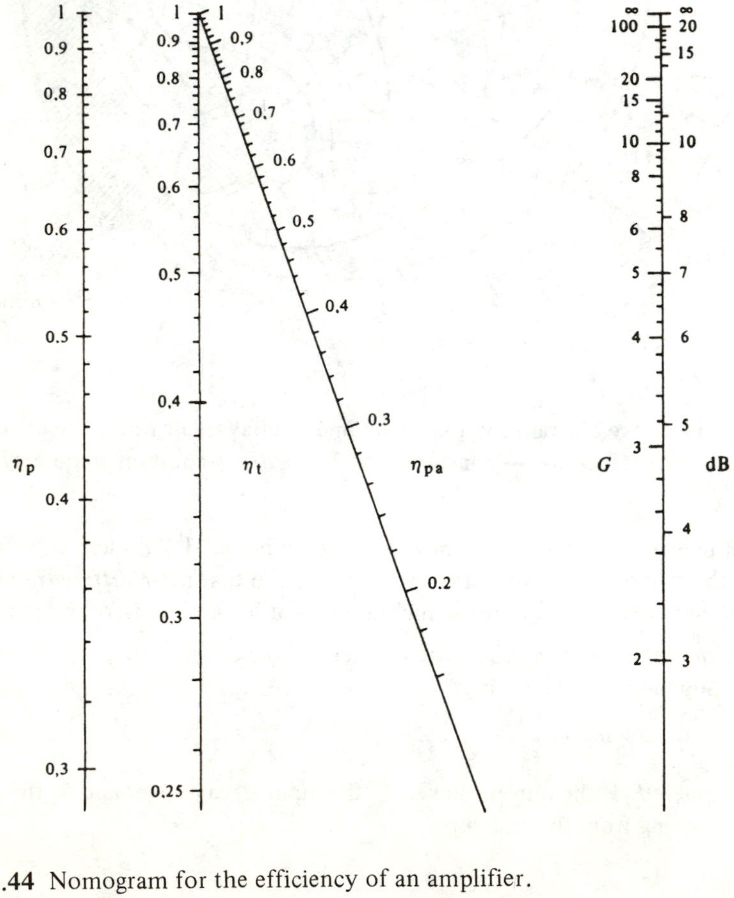

Fig. 4.44 Nomogram for the efficiency of an amplifier.

4.11.5 Available Output Powers

For indicative purposes, Figure 4.45 represents the highest power levels presently achieved using available sources described in Chapter 4. As technology is in constant evolution, these curves may well move upwards and towards the right with time. It is therefore recommended to keep up to date by consulting specialized technical publications and manufacturer's catalogs.

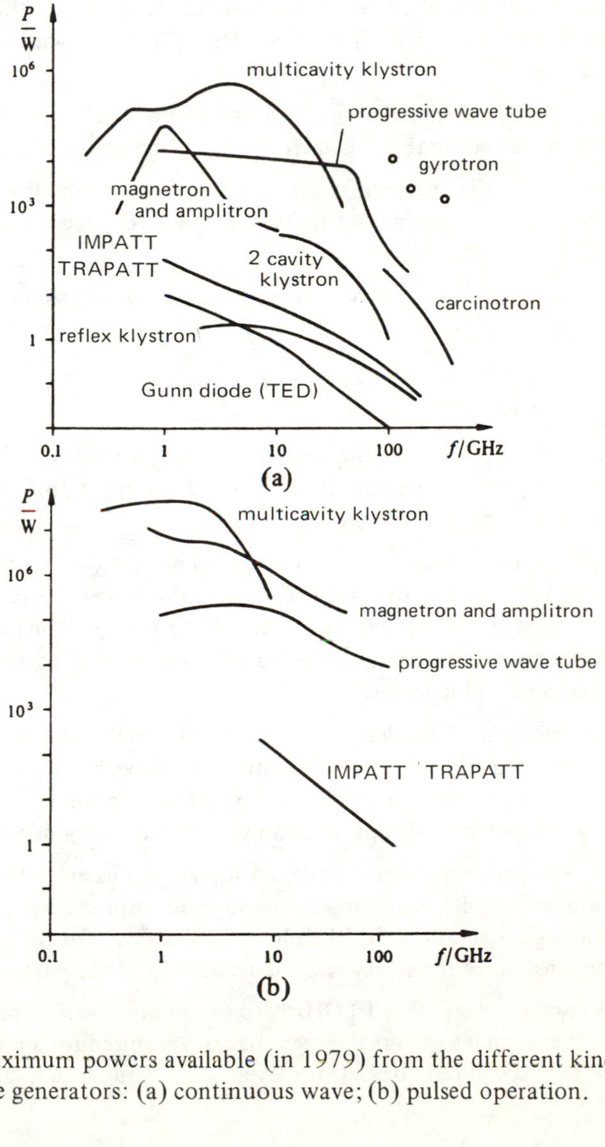

Fig. 4.45 Maximum powers available (in 1979) from the different kinds of microwave generators: (a) continuous wave; (b) pulsed operation.

4.12 PROBLEMS

4.12.1 The speed of charge carriers in a transistor is 10^5 m/s. Determine the time required for a carrier to cross the base, which is 10 micrometers (10 μm) thick. At which frequency does this transit time correspond to the quarter of a period?

4.12.2 In the characteristic diagram of the magnetron (Fig. 4.4), determine the position of the point at which the Hartree line is tangent to the parabola. How does this point move with frequency? Determine the locus of constant electron efficiency.

4.12.3 Determine the transit time of electrons in a reflex klystron for which the power is the largest at a specified frequency f (maximum of a mode).

4.12.4 Determine the phase shifts per unit length and the propagation velocities for slow and fast waves in the situation $n_0 = 4 \cdot 10^6$ electrons/m^3, $v_0 =$ 100 m/s and $f = 2$ GHz.

4.12.5 Two transmission lines have linear propagation characteristics, respectively given by:

$$\beta_1 = a_1 \omega + b_1$$

$$\beta_2 = a_2 \omega + b_2$$

The two lines are coupled, the coupling factor being K (§ 4.4.7). Determine the characteristics of the coupled lines for the two situations $p_2/p_1 = 1$ and $p_2/p_1 = -1$. What kind of curves does one obtain?

4.12.6 A Gunn diode produces a 10 mW signal when fed by a 150 mA direct current under a voltage bias of 7 volts. Calculate the power dissipated and the efficiency. Knowing that the active part of the diode is 10 μm long and that its cross section is circular with a radius of 50 μm, determine the power density dissipated in watts per cubic meter.

4.12.7 A varactor has a barrier potential $\phi = 0.6$ V and a peak inverse voltage of -100 V. Determine its maximum and minimum capacitances and their ratios, for bias voltages going from the peak inverse voltage to a forward bias of $\phi/2$. Compare the values for an abrupt junction with those for a graded junction.

4.12.8 A parametric amplifier is designed for a signal having a frequency of 3 GHz. It utilizes a cavity in rectangular waveguide, with the TE_{101} mode resonance for the signal, the TE_{103} for the idler and the TE_{105} for the pump. Determine the dimensions of the cavity and the frequency of the pump.

4.12.9 A microwave signal at 10 GHz is to be amplified with a maser. What is the distance between levels in the energy diagram required for amplification? What is the population ratio of the two levels at the equilibrium: at the ambient temperature of 20° Celsius, at the temperature of liquid nitrogen and at that of liquid helium?

4.12.10 The power fed to an amplifier by its dc bias is 20 watts. The total efficiency is 45.5%, the power added efficiency 40%. Determine the input power, the output power, and the gain in dB.

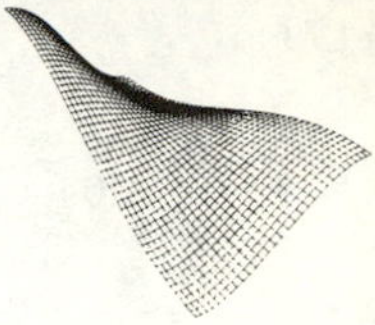

CHAPTER 5

SIGNAL MEASUREMENTS

5.1 QUANTITIES CHARACTERIZING THE SIGNAL

5.1.1 Definitions, Spectrum

A microwave signal, generated by one of the oscillators presented in Chapter 4, is characterized by two quantities:

1. its frequency (in GHz);
2. its power (in watts).

Any modulated signal may be represented by a combination of monochromatic signals (single frequency), obtained by Fourier series or transforms. These elementary signals, also called ***spectral lines***, form the *spectrum* of the signal, which is defined in terms of the respective frequencies and amplitudes of each of the component lines.

5.1.2 Comparison with Standards

The measurement of the signal characteristics (frequency, power) requires a comparison with similar quantities outside of the system being measured, which are called *standards.* In contrast, when characterizing components (Ch. 7), one compares similar quantities, which are, however, all measured within the system under study (comparison of amplitudes and phases of signals).

5.1.3 Note

Within a waveguide, the voltage, the current, and the characteristic impedance are quantities derived from transverse field components and from their ratio (§ 2.2.25). They are not uniquely defined and acknowledged by all. In addition, they generally cannot be measured directly. As a result, they are seldom utilized at the microwave level: power and associated quantities are more often utilized to define the level of a signal (Sec. 6.1).

5.1.4 Frequency Measurement

Two very different principles are utilized to determine the frequency of a signal. One of them is based on the ***spatial distribution of the fields*** within cavities and waveguides (Sec. 5.2), the other on the ***counting of periods*** by means of digital techniques (Sec. 5.3). Another approach, utilized in the spectrum analyzer, ***shifts the signal towards lower frequencies***, where it can be studied by classical electronics methods (Sec. 5.4).

5.1.5 Power Measurement

At frequencies below the microwave range, the power of a periodical signal is generally measured after rectification: a semiconductor diode rectifier yields a rectified signal, which has a direct current component proportional to the ***average power*** of the periodical signal. This approach is commonly used in frequency bands below 1 GHz. At higher frequencies, several practical difficulties are encountered, such as matching a nonlinear element (the diode impedance varies with the power level). As a result, rectification was seldom used in the past when measuring power [100]. Other mechanisms for detection are often preferred, such as:

1. mechanical measurement of the radiation pressure (Sec. 5.5);
2. thermal conversion of the microwave signal (Sec. 5.6).

5.1.6 Note

While diode detectors are seldom used to measure signal powers, they are on the other hand very widely utilized to compare signal levels (Ch. 7).

5.2 MECHANICAL MEASUREMENTS OF FREQUENCY

5.2.1 Cavity Frequency Meter or Wavemeter

Cavities are the microwave equivalents of tuned circuits (Ch. 3). The resonant frequency of certain modes may be modified by varying cavity dimensions, most often the length. Coaxial or cylindrical cavities containing a sliding piston (short circuit, Fig. 5.1) are generally used. The cavity is coupled to a line or a waveguide by means of one of the three schemes in Figure 5.2.

5.2.2 Principle of the Measurement

The position of the piston is mechanically shifted until the resonant frequency of the cavity coincides with the signal frequency. For the reactive cavity, the amplitude of the transmitted signal moves through a minimum. For the transmission cavity, it passes a maximum (§ 3.5.15). The reflected signal of a one-port cavity has a minimum at resonance (§ 3.5.11). The position of the piston which produces these extrema is determined, yielding the signal frequency

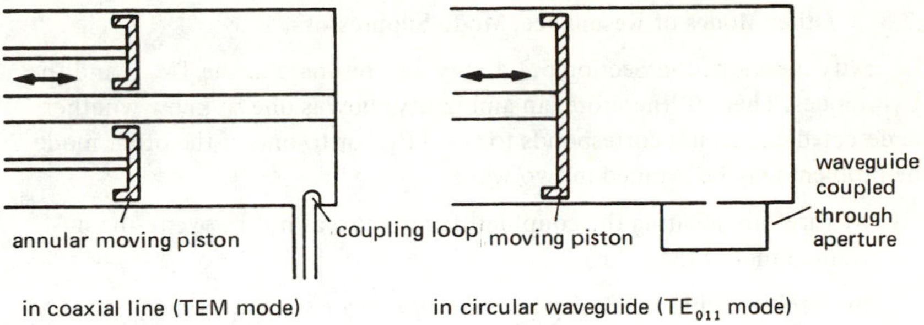

Fig. 5.1 Cavity wavemeters.

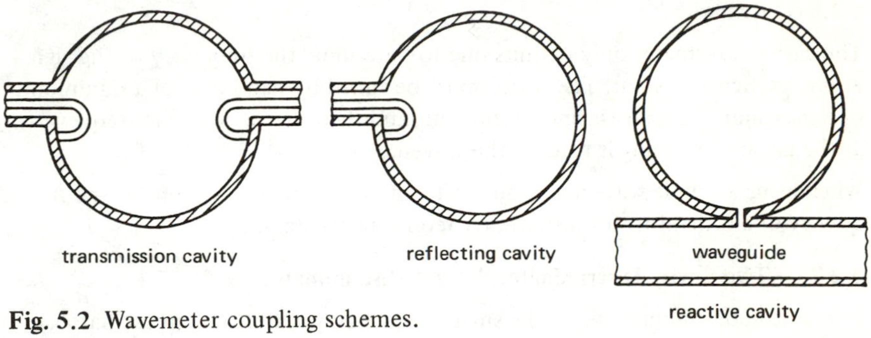

Fig. 5.2 Wavemeter coupling schemes.

[101]. Most high-grade wavemeters presently on the market have a scale directly graduated in frequency.

5.2.3 Accuracy

The resolution of a wavemeter depends upon the width of its resonance curve. To obtain a narrow resonance, a cavity with a large loaded quality factor Q_C is required (§ 3.5.12). For this reason, the modes selected are most often the TE_{011} and TE_{111} in the circular cylindrical cavity, and the first TEM mode in the coaxial cavity (Fig. 3.18). Typically, the loaded quality factor of a wavemeter should be equal or larger than 10,000 for good resolution. The measurement accuracy of the wavemeter is on the order of 0.1%. It is subject to a thermal drift around 10 ppm/°C.

5.2.4 Example: Selection of Wavemeter Dimensions

A wavemeter for the 8.2 to 12.4 GHz band utilizes the TE_{111} mode in a circular cavity. To determine the dimensions, Figure 3.10 is considered. After some calculations, a radius $a = 14.3$ mm and a length d, adjustable from 13.9 to 17.2 mm are selected.

5.2.5 Other Modes of Resonance, Mode Suppressors

The cavity mentioned in Section 5.2.4 may also resonate in the TM_{011} and the TE_{211} modes. There is, therefore, an ambiguity: how is one to know whether the detected extremum corresponds to the TE_{111} or to one of the other modes? The problem may be avoided in two ways:

1. by carefully locating the coupling structure so as not to excite the unwanted modes (Sec. 3.6);
2. by placing within the cavity a *mode suppressor* structure which perturbs the unwanted modes, generally by lowering their Q-factor. This structure must however not modify the mode utilized for the measurement.

5.2.6 Remark

The cavity wavemeter only permits one to determine the frequency of the signal at one particular instant. The cavity must be tuned by hand to read a minimum or a maximum on a measuring instrument, and then the value of the frequency at the instant of tuning is read on the wavemeter.

When using a simple wavemeter, one just cannot continuously monitor the frequency of a generator, for instance to determine its stability.

5.2.7 Two-Cavity Discriminator, Pound Discriminator

For continuous monitoring of the small frequency variations of an oscillator, a *discriminator* may be utilized. It is a microwave circuit having one or more cavities (Fig. 5.3).

The voltage appearing at the output of the differential amplifier is ***approximately a linear function*** of frequency over a limited range of frequencies. For the *two-cavity discriminator* (Fig. 5.3a), this range is located between the two resonant frequencies. For the *Pound discriminator* (Fig. 5.3c), the two detectors are similarly connected to a differential amplifier. Its output voltage has a frequency-dependence similar to the two-cavity discriminator [102]. In both situations, a voltage directly proportional to a frequency deviation is obtained; this voltage may be recorded as a function of time, to determine the stability of the oscillator.

5.2.8 Frequency Stabilizer

When an oscillator with electronic tuning is used, the output signal from the discriminator (§ 5.2.7) may be utilized to lock the oscillator frequency onto the central frequency f_0 (Fig. 5.3b).

5.2.9 Wavelength Measurement

In a transmission line terminated into a reflecting load, the superposition of an incident and a reflected wave yields standing waves (§ 9.4.6). Two successive

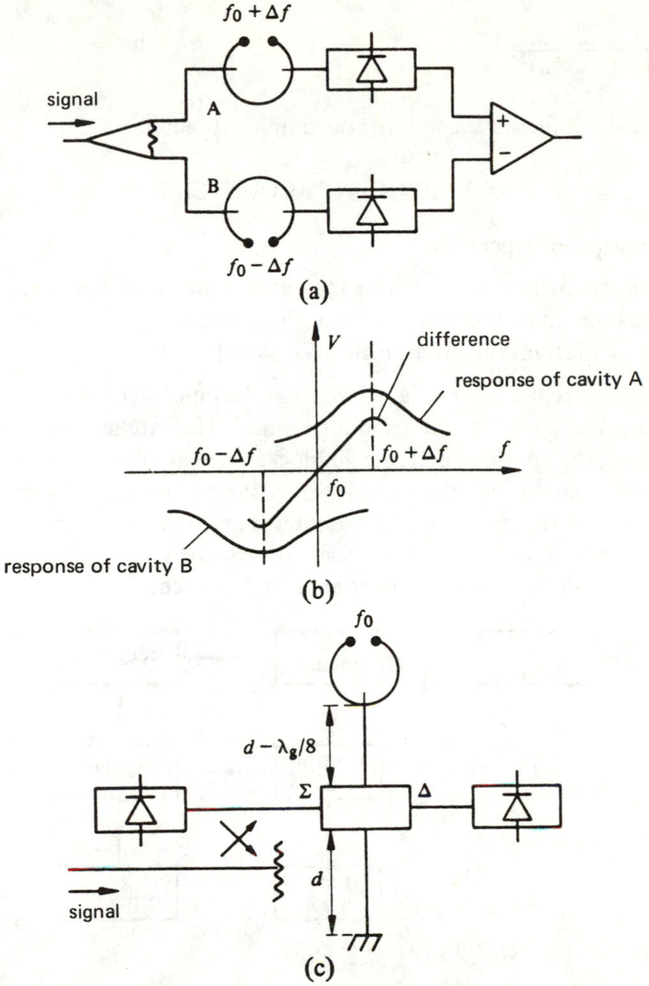

Fig. 5.3 Discriminators: (a) two-cavity discriminator; (b) frequency response of the two-cavity discriminator; (c) Pound discriminator.

similar extrema of voltage or of current are then separated by a half-guide wavelength, $\lambda_g/2$. This distance may be measured by using a slotted line (Sec. 7.2). In an air-filled TEM line (Sec. 2.6), the measured distance is equal to $\lambda/2$, and then:

$$f = c_0/\lambda \qquad \text{Hz} \tag{5.1}$$

For a rectangular waveguide propagating the dominant TE_{10} mode, the open-space wavelength is obtained with (2.123), yielding:

$$\lambda = \frac{\lambda_{10}}{\sqrt{1 + (\lambda_{10}/2a)^2}} \quad \text{m} \tag{5.2}$$

where λ_{10} is the guide wavelength for the dominant mode.

5.3 FREQUENCY COUNTERS

5.3.1 Principle of Operation

An electronic frequency counter is an apparatus made up of digital circuits, which counts the number of periods during a rigorously specified period of time. Its block diagram is given in Figure 5.4 [103].

The input circuit replaces the original signal at the input (which is assumed to be a sine wave) by a sequence of calibrated pulses. The gate lets pulses go through during the counting period, and blocks them at all other times. The opening and closing of the gate are ordered by the control unit, driven in turn by the signal from the time base. The counter is made up of a sequence of logic triggers, and adds up the number of pulses crossing the gate while it is open. The number obtained in this way is stored and displayed.

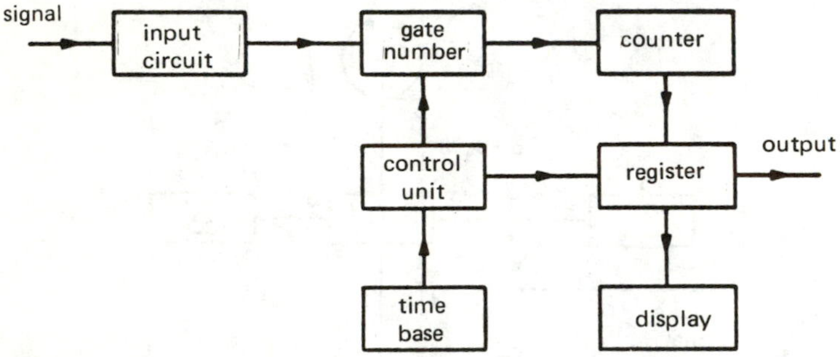

Fig. 5.4 Direct measurement of frequency.

5.3.2 Inaccuracies

Two sources of errors do exist:

1. The opening and the closing of the gate takes place at arbitrary times, as far as the signal period is concerned. Therefore, the counting is accurate to within ± 1 pulse (quantization effect). This error is negligible when the duration of the opening is much greater than one period.

2. The precision is directly related to the duration of counting, specified by the time base. High stability quartz sources provide a 10^{-8} to 10^{-9} frequency accuracy. The oscillator may further be locked to a frequency standard (§ 5.3.12), whose accuracy may reach 10^{-12} to 10^{-13}.

5.3.3 Frequency Limitations

The time required to carry out the logical operations in the counter is ***quite small, but non-zero***. When the period of the signal falls below a given minimal value, the pulse rate becomes too fast to be counted reliably by the instrument. A standard laboratory counter can measure frequencies up to about 500 MHz, while more sophisticated instruments reach about 1.5 GHz. As high-speed logic keeps developing, these frequency limits may be expected to shift upward. Nevertheless, most microwave signals remain well beyond the reach of ***direct*** electronic counting.

5.3.4 Frequency Shift, Heterodyning

By *shifting down the frequency* of the signal (*heterodyning*), the measurement range of digital counters is greatly extended at microwaves [104]. The signal to be measured, which has a frequency f, is applied to a mixer (§ 5.3.14), also connected to a reference source at frequency f_0. The mixer produces an output signal at the difference frequency $|f - f_0|$. When the reference source is correctly selected, the frequency of the output signal is within the range of the counter. This technique however does not permit one to determine which one of the two frequencies is the higher one.

This approach yields the same resolution as the direct measurement of the original signal. Frequency shift techniques have allowed measurement of laser frequencies up to 2.54 THz (2540 GHz) [105].

5.3.5 Automatic Heterodyne Counter

In practice, the stable source of the counter itself is utilized as a reference, and its frequency is multiplied (Sec. 4.9). One of the harmonics is mixed with the measured signal (Fig. 5.5).

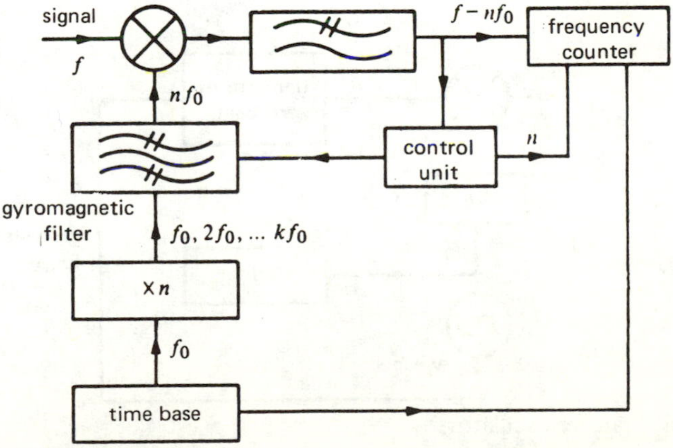

Fig. 5.5 Heterodyne converter.

A tunable gyromagnetic filter (§ 6.7.23) sequentially selects the harmonics. When a signal appears at the output of the low-pass filter, the current driving the tunable filter is maintained constant and the counting operation takes place. Systematically repeating this operation, moving up the frequency range avoids the ambiguity in sign produced by the mixer. When several spectral lines are present at the same time, their frequencies may be determined one after the other. After measuring the lowest frequency, the measurement procedure is resumed, starting from a slightly higher frequency, and so on.

5.3.6 Transfer Oscillator

The input signal is mixed with another signal, in this case the harmonics of a tunable auxiliary source (Fig. 5.6). The auxiliary source is tuned so that one of its harmonics is synchronized with the input signal (zero frequency beat) for a frequency $f_0 = f_{01}$ of the auxiliary source. This frequency is, in turn, measured on a counter, and the signal frequency is known to be

$$f = nf_{01} \qquad \text{Hz} \qquad (5.3)$$

However, the order n of the harmonic is unknown. To determine it, the operation is repeated, increasing the frequency of the auxiliary source until the next zero beat is obtained, for $f_0 = f_{02}$, in terms of which the signal frequency corresponds to:

$$f = (n - 1)f_{02} \qquad \text{Hz} \qquad (5.4)$$

The frequency f is then the smallest common multiple of the two measured frequencies f_{01} and f_{02}. The value of n is obtained from (5.3) and (5.4):

$$n = f_{02}/(f_{02} - f_{01}) \qquad 1 \qquad (5.5)$$

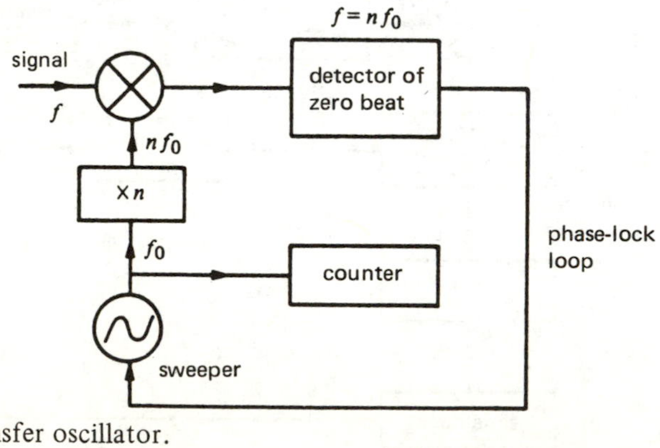

Fig. 5.6 Transfer oscillator.

By definition, n is an integer: the value obtained from the measurements may be slightly inaccurate, it is rounded off before calculating f with either (5.3) or (5.4).

The resolution of the frequency measurement with a transfer oscillator, as compared to the direct measurement, is ***reduced*** by the factor n.

5.3.7 Phase-Locking of the Auxiliary Oscillator

Most counters include a phase-locking feedback loop (PLL), which locks the auxiliary source on a subharmonic of the signal measured, provided that its frequency drift is not too important. The multiplication by n may also be carried out directly by the counter, which then displays f. One can continuously monitor the signal frequency in this manner.

5.3.8 Frequency Divider

Another technique to increase the range of a counter is to use a frequency divider, described in Section 4.9.7. The principle of operation is similar to that of the transfer oscillator, except that the multiplication factor n is kept constant. The signal to be measured is directly connected to the input of the divider, while the counter measures the signal provided at the output. The resolution is then also reduced by a factor n.

5.3.9 Note: Digital Divider

Frequency division may be carried out equally well with digital circuits, which are the building blocks of electronic counters. Such dividers are subject to the limitations already outlined in Section 5.3.3. Dividers with an upper frequency limit of 1.2 GHz and division factors up to 128 are available.

5.3.10 Note

In contrast to cavities, counters may be utilized to continuously monitor the frequency of the signal produced by an oscillator. The measurement is, however, not instantaneous, it represents an average taken over the counting period.

5.3.11 Atomic Frequency Standards

When an atom jumps from one energy level to another (quantum physics), one photon is either absorbed or emitted, having a frequency directly proportional to the energy difference between the two levels (§ 4.10.5)

$$f_{pq} = (W_q - W_p)/h \qquad \text{Hz} \qquad (5.6)$$

where h is Planck's constant.

Certain materials possess very narrow (hyperfine) energy levels, which provide extremely stable interaction frequencies [103]. This happens in particular for hydrogen, rubidium, cesium and thallium, which all exhibit a transition between two hyperfine levels in the microwave range:

1. hydrogen maser 1.420405751 GHz
2. rubidium cell 6.834682608 GHz
3. cesium beam 9.192631770 GHz
4. thallium beam 21.310833945 GHz

5.3.12 Cesium Beam Frequency Standard

The standard most generally used nowadays to provide high accuracy makes use of a cesium beam (Fig. 5.7).

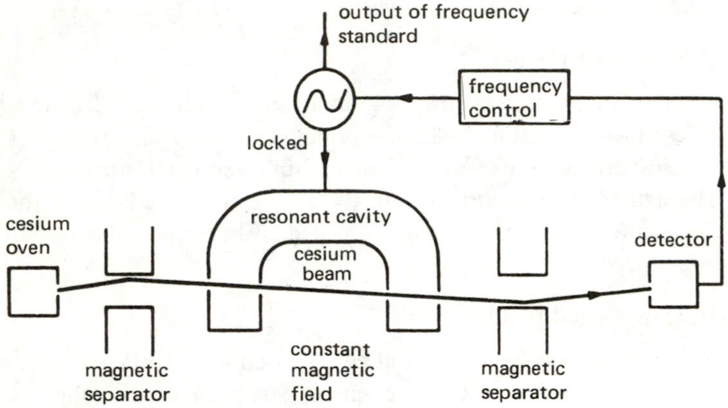

Fig. 5.7 Cesium frequency standard.

A beam of cesium atoms is emitted by an oven containing liquid cesium. A magnetic state selector directs those atoms which are in one particular hyperfine state towards the interaction regions of the resonant cavity A small dc magnetic field separates the different sublevels of the hyperfine state (Zeeman effect), so that only transitions between two well-defined levels may be stimulated by the electromagnetic field in the cavity. A second magnetic state selector directs towards the detector the atoms that underwent the transition. The current in the detector exhibits a maximum when the signal in the cavity is exactly at the transition frequency, *i.e.,* at 9.192631770 GHz. A feedback loop keeps the oscillator tuned at precisely that frequency.

The accuracy provided by the cesium beam is on the order of 10^{-12}, whereas the most accurate measurements of length and mass yield only a 10^{-9} accuracy.

5.3.13 Atomic Clock

Frequency is the inverse of time, and a frequency standard is also a standard for the ***measurement of time***. In October 1967, the general conference on weights and measures adopted the definition of the second [106]:

> *La seconde est la durée de* 9 192 631 770 *périodes de la radiation correspondant à la transition entre les deux niveaux hyperfins de l'état fondamental de l'atome de césium 133.*

Since the second is defined in terms of the hyperfine transition in cesium, the cesium source is called a ***primary standard***.

An atomic cesium clock is made up of a frequency standard and a counter.

5.3.14 Mixer, Intermodulation Products

A *mixer* consists of a nonlinear device, most often a semiconductor diode (Schottky diode, metal-semiconductor junction) [140] and connecting circuits. At low voltages, the current-to-voltage relationship is:

$$I = I_s[\exp(\alpha U) - 1] = I_s \left[\alpha U + \frac{(\alpha U)^2}{2!} + \frac{(\alpha U)^3}{3!} + \ldots\right] \quad \text{A} \qquad (5.7)$$

where the constant α is about 40 V^{-1}.

The sum of two sine wave voltages is applied to the diode:

$$U = U_1 \sin(\omega_1 t) + U_2 \sin(\omega_2 t) \quad \text{V} \qquad (5.8)$$

The current I then possesses components at all frequencies $|mf_1 + nf_2|$, with $m = \pm\, 0, 1, 2 \ldots$ and $n = \pm\, 0, 1, 2 \ldots$, which are called the *intermodulation products.* Considering the square term of (5.7), one can develop it as follows:

$$I_s \frac{(\alpha U)^2}{2} = \frac{I_s \alpha^2}{2} [U_1^2 \sin^2(\omega_1 t) + U_2^2 \sin^2(\omega_2 t) + 2 U_1 U_2 \sin(\omega_1 t) \sin(\omega_2 t)]$$

$$= \frac{I_s \alpha^2}{4} [U_1^2 + U_2^2 - U_1^2 \cos(2\omega_1 t) - U_2^2 \cos(2\omega_2 t)$$

$$+ 2 U_1 U_2 \cos(\omega_1 - \omega_2)\, t - 2 U_1 U_2 \cos(\omega_1 + \omega_2)\, t] \quad \text{A} \qquad (5.9)$$

Among the various components of the current flowing through the diode, one is at the difference frequency $|f_1 - f_2|$. It is not possible to determine whether f_1 is larger or smaller than f_2.

The mixer is the ***fundamental device for shifting frequency***. In general, it permits a signal to slide from one band to another, and in particular, to transfer a microwave signal down to lower frequencies. The resulting signal may then be processed with classical electronics methods. The symbol of Figure 5.8 schematically represents a mixer.

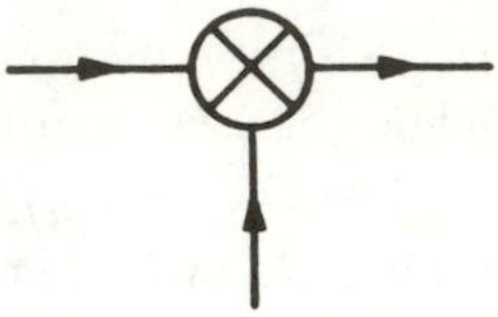

Fig. 5.8 Symbol for the mixer.

5.4 SPECTRUM ANALYZER

5.4.1 Remark

All frequency measurements considered in the two previous sections assume a purely sinusoidal signal. A cavity, as well as a counter, associates one figure to each measurement of the frequency; this is an average value for the frequency. These instruments do not provide any insight into the ***fine structure*** of the signal (the modulation). Only the discriminators provide a truly instantaneous measurement of frequency, allowing one to demodulate a frequency-modulated signal (§ 5.2.7). For a more detailed description of the signal, a spectrum analyzer is required.

5.4.2 Definition

A *spectrum analyzer* is an instrument used to study a signal in the frequency domain by displaying its spectrum (Fourier transform [215]), most often on the screen of an oscilloscope.

5.4.3 Simple Design: Swept Tunable Filter

An elementary spectrum analyzer is simply a tunable filter with a very narrow bandwidth, for instance, a gyromagnetic filter (YIG, § 6.7.23) connected to a variable current source, which in turn is connected to a display unit (Fig. 5.9).

The filter window, controlled by a sawtooth shaped current, periodically sweeps the measurement range selected in the frequency domain. A signal of frequency f may only pass through the filter and reach the detector at the particular time t at which the filter's window is centered on the frequency f. One obtains in this manner at the detector's output a transposition of the frequency spectrum into the time domain. The oscilloscope, driven by the same sweep current, permits the display of the spectrum on the screen.

5.4.4 Problem: Filter Bandwidth

The simple spectrum analyzer consisting of a tunable filter presents a drawback: both the bandwidth and the attenuation of the filter window vary when the center frequency is moved. As a matter of fact, the curve displayed on the oscilloscope is not the true spectrum of the signal itself, but the ***convolution*** of the signal by the transfer function of the filter. The quality of a network analyzer is

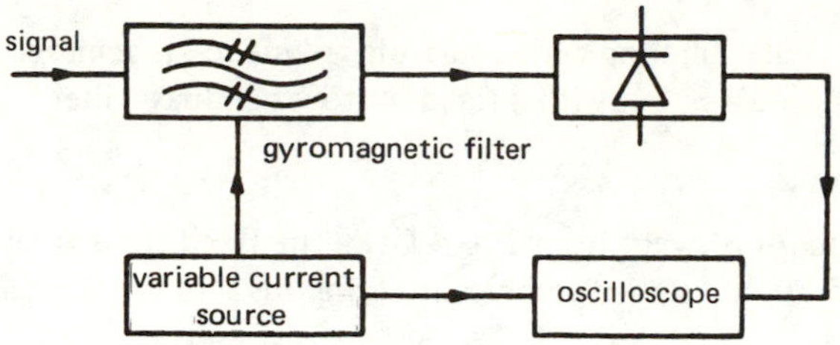

Fig. 5.9 Simple spectrum analyzer.

thus determined by the filter. The simple spectrum analyzer of Figure 5.9 does not permit an accurate study of the signal, it is therefore seldom used in practice.

5.4.5 Variable Frequency Shift

A more elaborate technique to determine the spectrum of a signal is to shift its frequency by a time-dependent amount, and then feed the shifted signal through a filter having a ***fixed narrow bandwidth***. The general schematic design is represented in Figure 5.10.

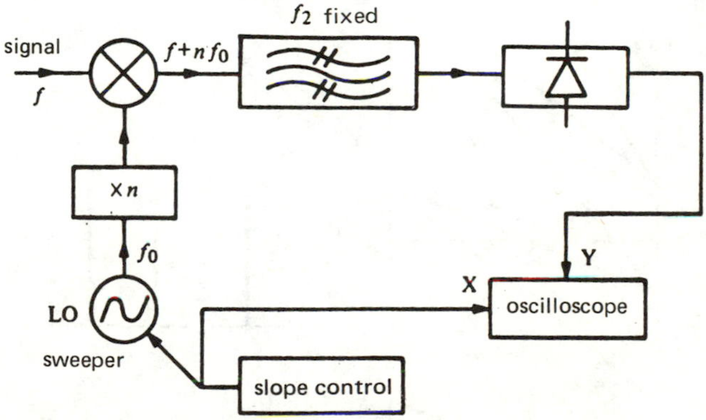

Fig. 5.10 Spectrum analyzer with swept frequency transfer.

5.4.6 Operating Principle

The signal at frequency f enters a mixer (§ 5.3.14), where it is mixed with the harmonics of the local oscillator signal (LO) of frequency f_0. At the output of the mixer, components at frequencies $|f + nf_0|$ are obtained, where n is an integer, positive or negative. For one of the components to pass through the fixed bandpass filter, one must have:

$$|f + nf_0| = f_2 \qquad \text{Hz} \qquad (5.10)$$

When this condition is satisfied, a signal appears on the screen of the oscilloscope. The local oscillator is swept in frequency, so that f_0 varies linearly with time. In this manner filtering with a variable frequency is achieved, in a similar way as in Section 5.4.3, but with a fixed narrow bandpass filter.

5.4.7 Example

The local oscillator is swept from 2 to 4 GHz, the fixed filter is set at 2 GHz. The solutions of (5.10) are shown graphically in Figure 5.11 for this particular set of values.

A 5 GHz input signal produces three different outputs in this case, corresponding to three solutions of (5.10):

1. with the −3 characteristic for $f_0 = 2.333$ GHz
2. with the −1 characteristic for $f_0 = 3.000$ GHz
3. with the −2 characteristic for $f_0 = 3.500$ GHz

The oscilloscope display (Fig. 5.11) shows three different lines. The three are, however, all produced by the same signal at 5 GHz.

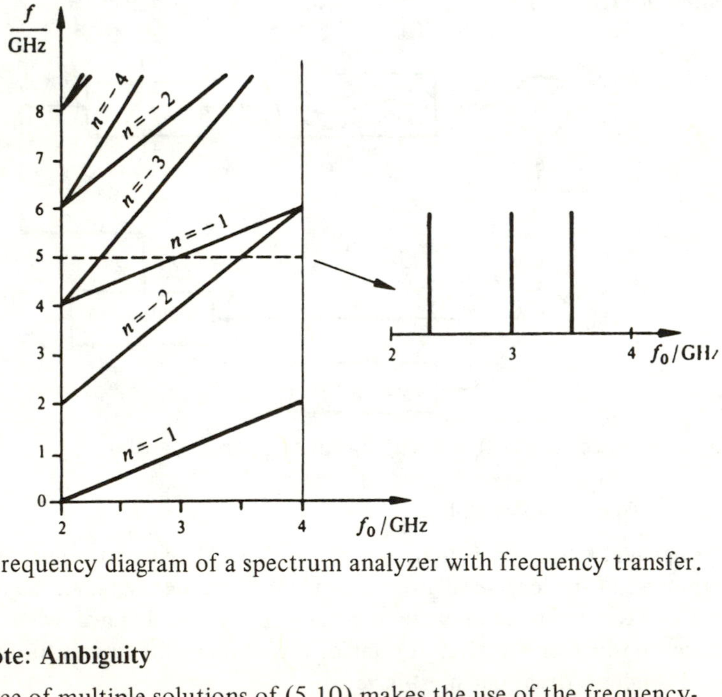

Fig. 5.11 Frequency diagram of a spectrum analyzer with frequency transfer.

5.4.8 Note: Ambiguity

The existence of multiple solutions of (5.10) makes the use of the frequency-shift analyzer more difficult. When a line is observed on the screen, one must first determine to which harmonic it corresponds (value and sign of n). This ambiguity may be lifted by the shifting of the limits of the sweep, for which an additional operation is required. It is also possible to impose a value for $|n|$, by placing a filter after the multiplier. This reduces the measurement range, but does not completely suppress the indetermination. A more favorable approach utilizes a preselector.

5.4.9 Preselector Analyzer, Image Frequencies

The frequency-shift filtering (§ 5.4.5) is completed by prefiltering, carried out by a gyromagnetic filter (§ 6.7.23). The latter is only utilized to eliminate the unwanted solutions of (5.10), called *image frequencies.* It may have a fairly broad bandwidth, as it does not carry the fine analysis of the spectrum.

5.4.10 Comment: Signal Multiplication

The mixer's response (§ 5.3.14) contains, in fact, all the intermodulation products, at frequencies $|mf + nf_0|$. However, when the amplitude of the input signal remains sufficiently small, the terms for which $m \neq 1$ are of negligible amplitude. The input signal must therefore be kept at a low enough level so that this condition is satisfied.

5.5 MECHANICAL MEASUREMENT OF POWER

□ 5.5.1 Radiation Pressure

The pressure exerted by an electromagnetic signal upon an obstacle may be measured by mechanical methods. When a waveguide is terminated into a perfectly reflecting piston, the force F produced by the signal on the piston is [108]:

$$F = \frac{2P}{c_0} \frac{\lambda}{\lambda_g} \qquad \text{N} \qquad (5.11)$$

where c_0 is the velocity of light in vacuum, λ and λ_g are, respectively, the wavelengths in space and in the guide. The force F can be measured and the power P directly determined from (5.11). The proportionality factor only depends upon waveguide size and frequency.

□ 5.5.2 Practical Implementation

Connecting directly a microwave generator to a total reflection would produce a large variation of the signal measured (Sec. 4.11). Precautions must therefore be taken to suppress the reflected wave, for instance by placing an isolator within the line (§ 6.3.10), or still an adequately terminated hybrid coupler (§ 6.5.18). Measurements carried out at power levels in the 10 to 50 watt range determine the power within 1 to 2 watts.

□ 5.5.3 Torque Pendulum Wattmeter

The technique described in Section 5.5.1 has the significant drawback of reflecting all the power; it is therefore difficult to use in practical applications. Another approach is based on the use of a torque pendulum having one or two vanes (Fig. 5.12).

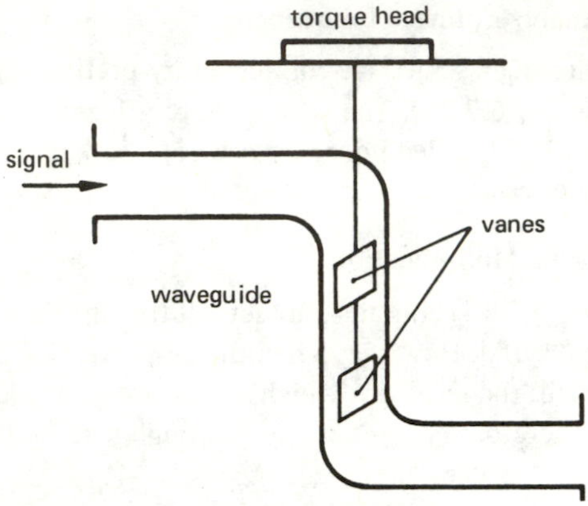

Fig. 5.12 Powermeter with torque pendulum.

The signal traveling through the guide exerts a torque on the vanes, producing an observable rotation. The initial orientation is restored by turning the torque head: the power is then proportional to the angle of rotation. The calibration of the instrument is straightforward. Matching elements compensate the reflection produced by the vanes, when they are in their equilibrium position. Since the resulting torque always remains quite small, this approach may only be used to measure rather large powers, those on the order of hundreds of watts (like the previous one). Measuring instruments based on this principle were realized. They are used mostly in calibration laboratories.

□ 5.5.4 Mechanical Wattmeters for Small Powers

By concentrating the fields within a cavity, it is possible to increase the sensitivity of torque pendulum devices, and therefore to use them to measure lower power levels. Instruments realized with this approach are, however, delicate and difficult to use.

5.5.5 Pros and Cons

The interest of mechanical measurement techniques for power lies mainly in their ease of calibration: a force or a torque is measured, which may easily be compared with standards. However, this force is most often extremely small, so that very sensitive instruments are required. They must be operated within a vibration-free environment. Mechanical methods are therefore seldom used in practice to measure power.

5.6 MEASUREMENT OF POWER BY THERMAL CONVERSION

5.6.1 Basic Principle

The microwave power is dissipated in an absorbing material, producing heat. The resulting temperature increase of the element is determined, most often by electrical means (thermocouples, bolometers, etc.). It may come as a surprise that one would transform the so much nobler power of a microwave signal into thermal energy to measure it. Still, in practice, most of the usual microwave wattmeters operate on this principle [109].

Two alternative approaches commonly used are:

1. for high accuracy measurements (standards laboratories), a calorimeter, either dry or using a fluid (water). Only rather high power levels (1 W or more) may be measured (§ 5.6.2 to § 5.6.4);
2. for routine measurements, the power is dissipated within a temperature-sensitive resistor or in a bimetallic thermocouple junction, source of a variable voltage (§ 5.6.5 to § 5.6.10).

5.6.2 Static Dry Calorimeter

Two elements are placed within a thermally insulated enclosure. One of them absorbs the microwave signal, the other one, which is as identical as possible to the first one, acts as a reference (Fig. 5.13).

The temperature difference between the two elements is measured with a set of

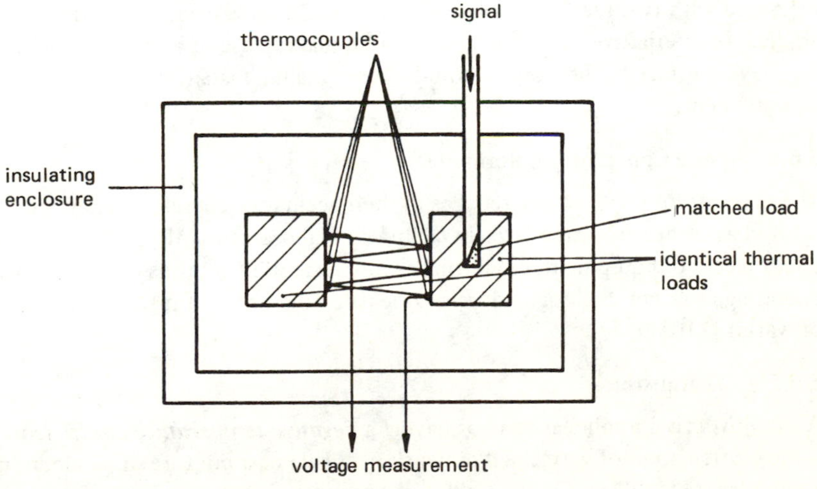

Fig. 5.13 Schematic of a steady state calorimeter.

thermocouples (a thermopile). The voltage developed is proportional to the dissipated power. Precautions must be taken to prevent heat leakage along the input line or waveguide, which would warp the measurement.

5.6.3 Continuous Flow Calorimeter

The absorbing material in this case is a liquid (most often water), which absorbs the microwave signal and transforms it into heat. The temperature difference between input and output is then proportional to the dissipated power:

$$P = qc_p \Delta T \qquad \text{W} \qquad (5.12)$$

where q is the flow of the liquid in kg/s, c_p its specific heat content in J/(kg K), and ΔT the temperature difference in Kelvin. A continuous flow calorimeter is shown in Figure 5.14.

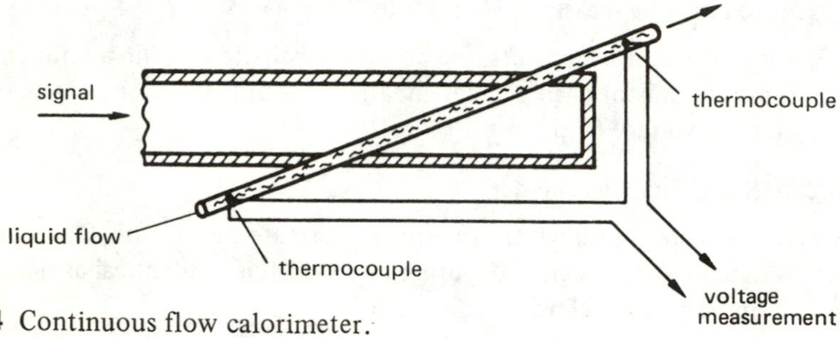

Fig. 5.14 Continuous flow calorimeter.

The signal is absorbed within a water load (§ 6.2.2), the temperature variation is measured with two thermocouples. The flow of liquid is measured and (5.12) is applied, or alternatively, a calibration system making use of a substitution heating may be utilized. The flow of liquid must remain constant during the whole measurement.

5.6.4 Note Concerning Calorimeters

The main interest of calorimeters lies in their accuracy. Therefore, they are utilized as comparison standards in calibration laboratories. Most often, they are bulky pieces of equipment, exhibiting a slow response (in terms of minutes), and operating at rather large signal levels (at least of the order of mW, and often even of watts) [110].

5.6.5 Thermistor

A *thermistor* is a nonlinear resistor having a ***negative temperature coefficient***. It is most often a pill of very small dimensions, made of a mixture of semiconducting oxides (Mn, Ni, Co, Cu, etc.), skillfully proportioned by the manufacturer. The resistance of a thermistor not only depends upon the power absorbed, but also on the ambient temperature.

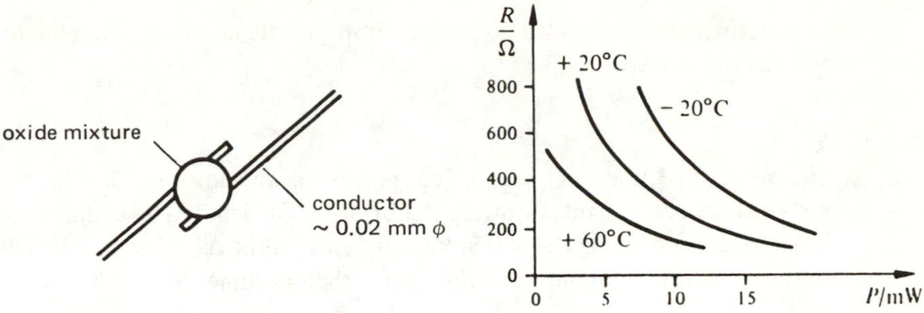

Fig. 5.15 Thermistor: geometry and electrical characteristics.

Thermistor response times are of the order of 0.5 to 1 second, allowing one to perform much faster measurements than with calorimeters.

5.6.6 Bolometer or Barretter

Another sensitive device is the *bolometer* or *barretter*, made with a metal conductor: a thin wire or a metal film deposited onto an insulating substrate. The power absorbed heats up the conductor, whose resistance increases: the resistance change is measured, and the power of the signal deduced. The response time of a bolometer is less than a millisecond.

5.6.7 Milliwattmeter with Nonlinear Resistor

The sensitive device (thermistor or bolometer) terminates the transmission line, absorbing the incoming microwave signal (Fig. 5.16). At the same time, it is inserted within a Wheatstone bridge, fed by either a direct or a low-frequency source [109].

The measurement procedure follows two steps:

1. In the absence of the microwave signal, the measurement bridge is balanced, *i.e.,* the bias voltage is adjusted until the nonlinear resistor presents the same resistance R as the precision resistors in the other branches of the bridge. The voltage at which balance is attained (no current flows across

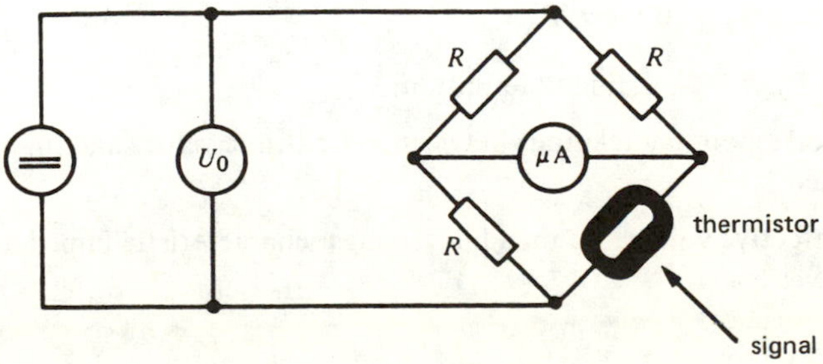

Fig. 5.16 Power measurement bridge.

the microammeter) is called U_0. At equilibrium the nonlinear component absorbs the power:

$$P_0 = U_0^2/4R \qquad \text{W} \qquad (5.13)$$

2. The microwave power is then applied, producing an additional heating of the sensitive component, throwing the bridge off-balance. The equilibrium is restored by reducing the bias voltage by an amount ΔU. The power fed to the nonlinear component by the circuit then becomes:

$$P_1 = (U_0 - \Delta U)^2/4R \qquad \text{W} \qquad (5.14)$$

The decrease in the power supplied by the circuit compensates the microwave power P absorbed by the sensitive element; it is given by:

$$P = P_0 - P_1 = \frac{\Delta U}{4R}(2U_0 - \Delta U) \qquad \text{W} \qquad (5.15)$$

5.6.8 Automatic Balancing

In most general purpose milliwattmeters, a feedback loop automatically balances the bridge. A measurement scale directly indicates the value of P in milliwatts or in the power level.

Generally, the instrument must be adjusted when starting a measurement: one must check that it actually indicates zero when no microwave signal is applied. This adjustment must be checked periodically during the measurements, to avoid a possible drift of the measurement circuits [111].

5.6.9 Power Level

Logarithmic values are often utilized to define the power level, comparing it to a standard level. The *power level* LP is defined by:

$$LP(\text{dBm}) = 10 \log(P/P_m) \qquad \text{dBm} \qquad (5.16)$$

where, by definition, P_m = 1 mW.

$$LP(\text{dBW}) = 10 \log(P/P_w) \qquad \text{dBW} \qquad (5.17)$$

where P_w = 1 W, again by definition.

The corresponding relations between logarithmic values and linear ones are given in Table 5.17.

The effective voltage U_e on a line having a characteristic impedance Z_c is related to power by:

$$U = \sqrt{PZ_c} \qquad \text{V} \qquad (5.18)$$

Table 5.17 Relationship between power and power levels.

P	W	*LP*/dBm	*LP*/dBW	Voltage over 50 Ω line (V)
1 GW	10^{9}	120	90	2.236 10^{5}
1 MW	10^{6}	90	60	7.071 10^{3}
1 kW	10^{3}	60	30	2.236 10^{2}
1 W	10^{0}	30	0	7.071
1 mW	10^{-3}	0	−30	2.236 10^{-1}
1 μW	10^{-6}	−30	−60	7.071 10^{-3}
1 nW	10^{-9}	−60	−90	2.236 10^{-4}
1 pW	10^{-12}	−90	−120	7.071 10^{-6}
1 fW	10^{-15}	−120	−150	2.236 10^{-7}
1 aW	10^{-18}	−150	−180	7.071 10^{-9}

The power level may be defined with respect to the voltage on a 50 Ω line:

$$LP(\text{dBm}) = 10 \log [(U^2/Z_c)/P_m] = 20 \log(U/U_m) + 13.0103 \quad \text{dBm} \qquad (5.19)$$

where $U_m = 1$ V.

5.6.10 Thermocouple

When two different conducting materials are in contact, a voltage appears across the junction. This voltage is proportional to temperature (Seebeck effect). Connecting in series one junction subjected to heating (hot junction) and one protected from the microwave signal (cold junction), a voltage difference proportional to the incident power develops. This voltage difference is measured directly with a high sensitivity voltmeter. While the operating principle of the thermocouple is straightforward, its practical implementation encountered severe difficulties, related in particular to the matching of the power-sensitive element. These problems have been solved using thin-film circuits deposited on a semiconductor substrate, an example of which is shown in Figure 5.18 [112].

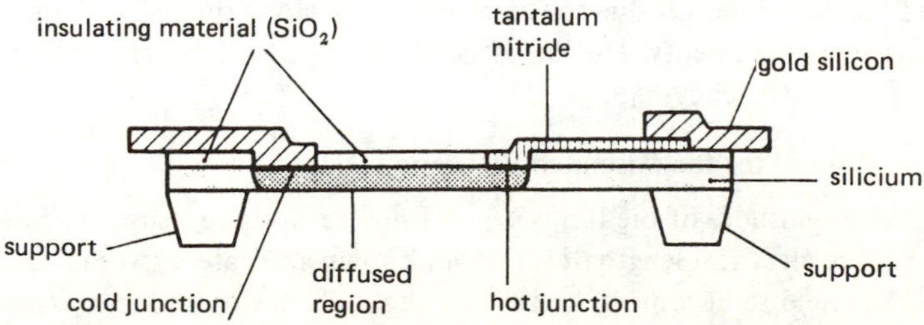

Fig. 5.18 Thermocouple for the measurement of microwave power.

5.7 SOURCES OF ERRORS

5.7.1 Introduction

The measurement of power is plagued by errors due to various causes, some of a very general nature (matching), others specifically related to a particular power-sensitive device. The main sources of errors are reviewed in this section.

5.7.2 Mismatch Efficiency

When a power measuring device of any kind is ***not matched to the line***, a part of the signal is reflected and does not contribute to the measurement: there is a systematic error ***by default***. When in addition the generator is not matched to the line, which is often the case, the signal returned by the load is in turn partially reflected by the generator, producing ***multiple reflections***. This situation is schematically represented in Figure 5.19 by a ***signal-flow graph***, defined in Section 6.1.14.

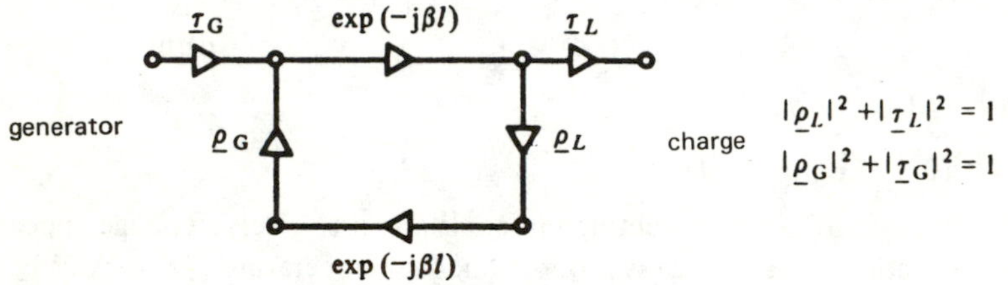

Fig. 5.19 Flow graph for a mismatched measuring instrument.

The power P_m measured by the mismatched instrument is different from the power P_a, which a matched load would absorb. In the latter situation $\underline{\rho}_L = 0$ and $|\underline{\tau}_L| = 1$. The ratio of the two powers is given by (9.26):

$$\frac{P_m}{P_a} = \frac{1 - |\underline{\rho}_L|^2}{|1 - \underline{\rho}_G \underline{\rho}_L \exp(-2j\beta l)|^2} = \eta_d \qquad 1 \qquad \textbf{(5.20)}$$

where βl is the phaseshift due to the propagation along the line of length l which joins the two instruments. The line is assumed to be lossless. This ratio is called the mismatch efficiency η_d.

5.7.3 Bounds for the Mismatch Efficiency

While the magnitudes of the terms $\underline{\rho}_G$ and $\underline{\rho}_L$ are easily measurable (Sec. 7.2 and 7.3), the electrical length βl is seldom known accurately. At microwaves, phaseshifts tend to become quite large: a slight change in frequency (modulation) or a change in length (thermal dilatation) may suffice to produce a significant drift of the phaseshift. Generally, one then only knows the ***bounds*** of η_d:

$$\frac{1 - |\underline{\rho}_L|^2}{(1 + |\underline{\rho}_G \underline{\rho}_L|)^2} \leqslant \eta_d \leqslant \frac{1 - |\underline{\rho}_L|^2}{(1 - |\underline{\rho}_G \underline{\rho}_L|)^2} \quad 1 \tag{5.21}$$

A nomogram yielding directly these two bounds is provided in Figure 5.20. In certain situations, the upper bound may exceed unity.

5.7.4 Practical Counsels

The numerator of (5.20) produces a systematic error, easily corrected when $|\underline{\rho}_L|$ has been determined. The denominator, on the other hand, creates an

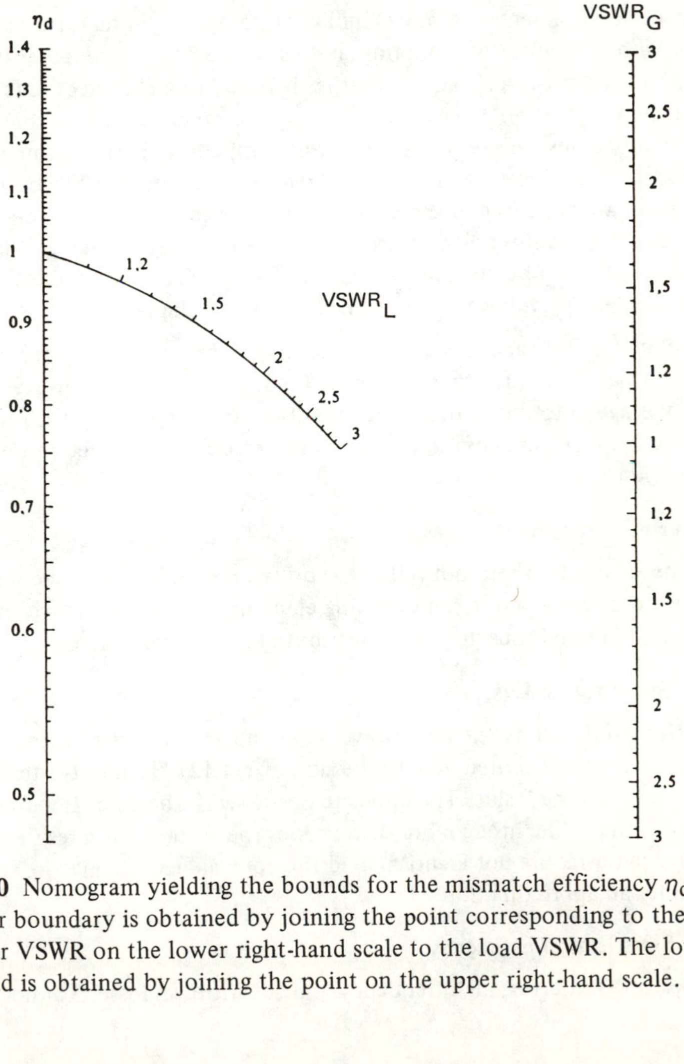

Fig. 5.20 Nomogram yielding the bounds for the mismatch efficiency η_d. The upper boundary is obtained by joining the point corresponding to the generator VSWR on the lower right-hand scale to the load VSWR. The lower bound is obtained by joining the point on the upper right-hand scale.

uncertainty, which may only be reduced by decreasing the amplitudes of the reflections:

1. $\underline{\rho}_L$, by matching the measuring instrument,
2. $\underline{\rho}_G$, by matching the generator to the transmission line. It is difficult, in practice, to reach a good match, in particular when the generator must operate over a broad frequency band. An isolator may be inserted (§ 6.3.10) between the generator and the measuring instrument. The isolator output must then be well matched to the line.

5.7.5 Ambient Temperature

Thermistor and bolometer (§ 5.6.7) milliwattmeters are, in fact, thermometers. They measure a temperature variation, but cannot determine whether it is produced by heating from a signal, or whether it is due to a change of ambient temperature (Fig. 5.15). A temperature change produces a ***drift***: it is necessary to continually adjust the zero level on old uncompensated milliwattmeters. More recent instruments possess an ***internal automatic compensation***. Two measurement bridges are used, one of them absorbing the signal, the other one isolated from it. The two sensitive elements are placed next to each other, so that they are similarly affected by ambient conditions. The power is determined by comparing the balancing voltages on both measurement bridges.

Calorimeters (§ 5.6.2 and § 5.6.3) are self-compensated by their very principle. The same is partly true for thermocouples (§ 5.6.10). It must, however, be noted that the voltage output of a thermocouple does not vary linearly with temperature: an additional compensation circuit is therefore incorporated in the measuring instrument.

5.7.6 High Frequency Losses

Part of the signal absorbed (not reflected) by the measuring instrument is dissipated within its walls and in the matching elements. It does not reach the power-sensitive element and thus does not contribute to the measurement.

5.7.7 Substitution Losses

The heating produced by the microwave signal in the sensitive element (thermistor) is not evenly distributed, due to the skin effect [214]: surfaces tend to be more heated than the inside. The opposite occurs with the dc or low frequency signal, which heats the inside more than it does the surface. As a result, the two heating mechanisms are not identical, and the equivalence assumed in Section 5.6.7 is only an approximation.

5.7.8 Substitution Efficiency

The two sources of error, high frequency and substitution losses, cannot be distinguished in practice. Both produce a systematic error, which can be compensated for by ***calibration*** of the instrument. The *substitution efficiency* η_s of a thermistor mount is defined by the ratio:

$$\eta_s = \frac{P_0 - P_1}{P} = \frac{\text{dc substituted power}}{\text{absorbed, microwave power}} \quad 1 \tag{5.22}$$

the different powers involved were defined in Section 5.6.7 This efficiency is situated between $0.92 < \eta_s < 1$ [113].

5.7.9 Paired Coaxial Mounts

In coaxial line systems, precautions are required to prevent the dc or low frequency signal from leaking into the transmission line. Usually a setup involving two sensitive elements and a bypass capacitor is utilized (Fig. 5.21).

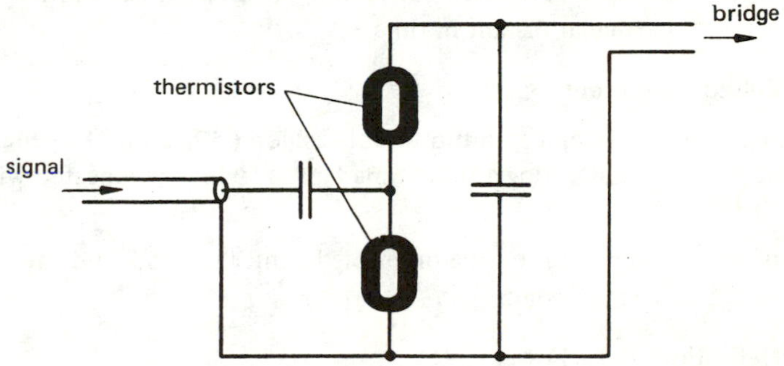

Fig. 5.21 Coaxial measurement head.

For the dc or low frequency substituted signal, the two sensitive elements are series-connected. From the microwave point of view, on the other hand, the capacitors practically become short-circuits, so that the power-sensitive elements appear to be shunt-connected. If the two thermistors are not rigorously identical, the power distribution will be different for the microwave and for the dc signal. The pairing unbalance may thus reach a few percent. This is also a substitution error (§ 5.7.7), which can be corrected by calibration.

5.7.10 Thermo-Electric Effect

The very effect on which thermocouple measurements are based becomes a source of error in thermistor bridges. When the metal-semiconductor junctions on both sides of the thermistor (Fig. 5.15) are not at the same temperature, a direct voltage appears. This voltage affects the equilibrium of the Wheatstone bridge, and it may produce a significant error at low power levels. Its amplitude may be evaluated by inverting the bias voltage across the measurement circuit.

This source of error is avoided by using an alternating low frequency signal for the substitution.

5.7.11 Measuring Instruments

Electrical and electronic devices utilized within the measurement bridge, self-balancing circuits and measuring instruments are not ideal. Resulting inaccuracies may reach a few percent.

5.8 MEASUREMENT OF THE PULSE POWER

5.8.1 Note

All the methods described in Section 5.6 actually measure the average value of power over a relatively long period of time, as compared to the period of the signal. The response time is on the order of seconds for thermistors. When utilized in pulsed operation, these devices do not permit the direct measurement of the actual power at a particular instant in time.

5.8.2 Pulsed Operation

Radars most often operate in an intermittent fashion (§ 8.1.7). The generator emits a microwave signal during a short time lapse τ, which is repeated with a period τ_r (Fig. 5.22).

The duration τ of the pulse is on the order of the microsecond, and the repetition period τ_r a few milliseconds.

5.8.3 Definition: Pulse Power

The average signal power during the pulse is called *pulse power* P_i. When the

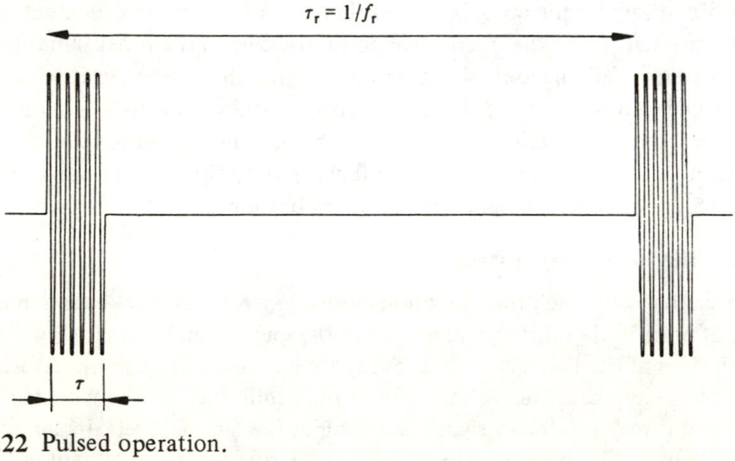

Fig. 5.22 Pulsed operation.

pulse shape (envelope) is rectangular, a simple relationship exists between the pulse power and the measured average power P_m:

$$P_i = P_m \tau_r / \tau \qquad \text{W} \qquad (5.23)$$

5.8.4 Note: Peak Power

The *peak power, i.e.,* the maximum instantaneous value taken by the power, is twice the pulse power P_i, which is an effective power.

5.8.5 Actual Pulses

In practice, pulses are seldom rectangular. Their rise and fall times are never infinitely short. In addition, the signal amplitude varies within the interval between rise and fall (Fig. 5.23).

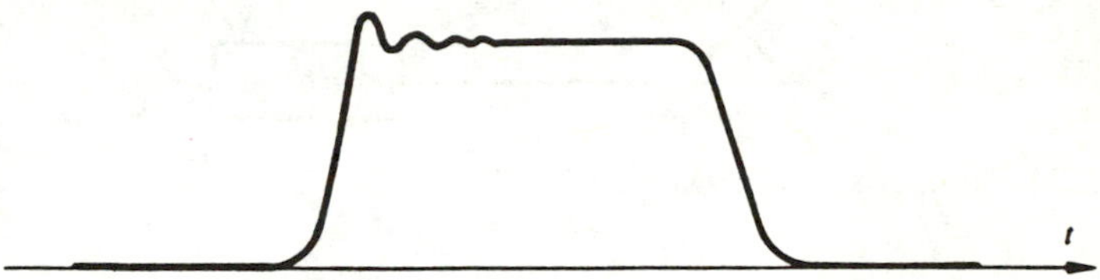

Fig. 5.23 Actual pulse (envelope).

The maximum average power of the ***actual pulse*** may produce a breakdown in the system. The expression (5.23) for the pulse power yields an approximation which may be unreliable, so more elaborate techniques are required to determine the true shape of the pulse and to measure its maximum.

5.8.6 Comparison Method

The amplitudes of the pulses may be reduced, either by using a directional coupler (§ 6.5.2), or a calibrated attenuator (Sec. 6.3). The lower amplitude pulses are then rectified with a crystal detector (§ 7.1.5), and their shape is displayed on the screen of a low frequency oscilloscope. The same detector is alternately fed from a continuous wave source, the level adjusted to coincide on the screen (Fig. 5.24).

When the pulse maximum just coincides with the level from the cw signal, the pulse level is given by:

$$\mathrm{L}P_i = \mathrm{L}P_m + \mathrm{L}C_1(\mathrm{dB}) - \mathrm{L}A_2(\mathrm{dB}) + \mathrm{L}C_2(\mathrm{dB}) \qquad \text{dBm} \qquad (5.24)$$

where $\mathrm{L}C_1$, $\mathrm{L}C_2$ and $\mathrm{L}A_2$ are, respectively, the couplings of the two directional couplers and the attenuation of the second coupler, all expressed in dB. This method allows one to determine the shape of the pulse, *i.e.,* the power distribution as a function of time.

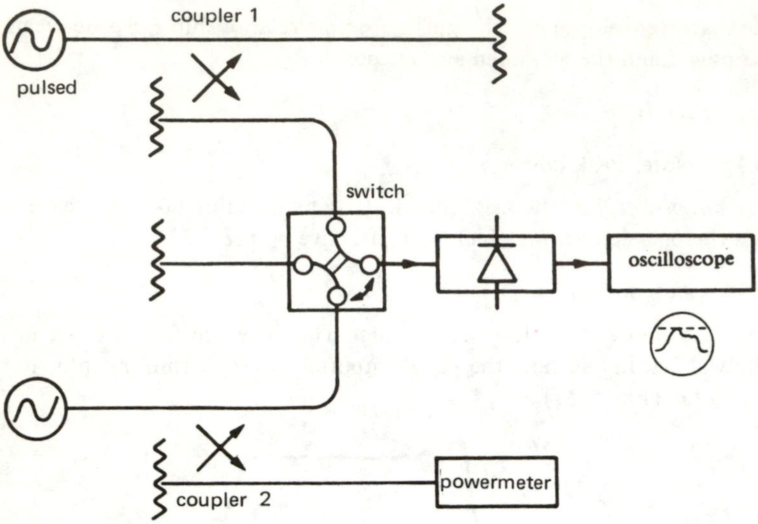

Fig. 5.24 Comparison of power levels.

5.8.7 Integrating Bolometer Measurement

The time constant of a bolometer or barretter (§ 5.6.6) is smaller than a millisecond, it is less than the pulse repetition period. This element may thus be used as an integrator. The signal produced by the bolometer is then differentiated, its positive part approximately reproduces the original pulse shape (Fig. 5.25).

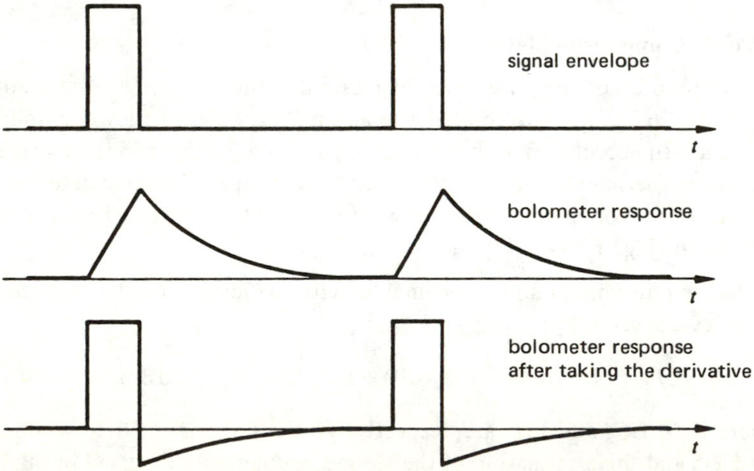

Fig. 5.25 Principle of the measurement of peak power with a bolometer.

A peak voltmeter determines the maximum value, which corresponds to P_i. This simple approach only provides a limited accuracy [114].

5.9 CATHODE RAY TUBES

5.9.1 General Comment

Much information may be gathered by watching a signal displayed on the screen of an oscilloscope; most of it would be difficult to obtain otherwise. For instance, a frequency counter provides an average value of the frequency, but it does not detect short term instabilities. The same happens with power measurements. Oscilloscopes do not actually allow one to directly view the behavior of signals above roughly 1 GHz. Two principles are utilized for particular measurements:

1. rectification,
2. sampling.

5.9.2 Rectification and Oscilloscope

By rectifying the microwave signal with a crystal detector (§ 7.1.5), one obtains a continuous or low frequency signal having an amplitude proportional to the envelope of the microwave signal. This envelope is then observed on a low frequency oscilloscope. An example of application is the determination of the pulse shape (§ 5.8.6). Other ones are swept frequency measurements (§ 7.3.6).

5.9.3 Sampling Scope

When studying repetitive events, the frequency range of a cathode ray tube may be extended by sampling the signal at periodical time intervals. This provides a transfer towards a lower frequency, at which the signal can be observed on a fast response oscilloscope. Signals of frequencies up to 18 GHz can be observed indirectly in this manner.

5.9.4 Caution

Oscilloscopes permit the observation of the shape of a signal and thus are very useful instruments in practice. However, the accuracy provided by the measurements is always limited (beam width, parallax, etc.). The oscilloscope is a most valuable complementary device, which does not, however, replace high precision measuring instruments.

5.10 PROBLEMS

5.10.1 Determine the most favorable dimensions for the design of a circular guide wavemeter for the 9 to 10 GHz band. Provide the calibration of the instrument, the resonant frequency as a function of piston location.

5.10.2 A discriminator is to be made with two resonant cavities having the same loaded quality factor Q_c. Which frequency spacing Δf should one select to yield a frequency response as linear as possible (Fig. 5.3)?

5.10.3 In a rectangular waveguide of width $a = 22.86$ mm, the frequency is measured with a slotted line. Two successive minima are detected at $x = 6.74$ cm and $x = 8.27$ cm. What is the signal frequency?

5.10.4 To measure its frequency, a signal is mixed with the output of a 10.085 GHz stable source. A resultant frequency of 168 MHz is measured. Repeating the process with a 9.938 GHz source provides an intermediate signal at 315 MHz. What is the signal frequency?

5.10.5 A transfer oscillator is utilized to measure the frequency of a signal. Two successive zero beats are observed with local oscillator frequencies of 161.74 and 165.42 MHz. Which are the harmonic numbers in both cases, and what is the signal frequency?

5.10.6 A spectrum analyzer without preselector has an IF filter tuned at 3 GHz and a local oscillator sweeping the 2-4 GHz band. For which frequencies of the local oscillator does one obtain a signal at the filter's output when the signal frequency is 7 GHz?

5.10.7 In a spectrum analyzer without preselector, signals appear at the filter's output for several frequencies of the local oscillator: 1566, 1740, 1827, 2192, 2740, 3480 and 3653 MHz. Is it possible that all these frequencies correspond to a single frequency signal? If the answer is yes, determine which is the signal frequency and the frequency of the selective bandpass filter.

5.10.8 Determine the force produced by a 10 watt, 8.5 GHz signal on a perfectly reflecting piston terminating a rectangular waveguide having a width $a =$ 22.86 mm.

5.10.9 What is the temperature increase produced by a 100 mW signal absorbed by a continuous flow calorimeter, the water flowing at 10 liters per minute?

5.10.10 A thermistor bridge has three precision resistors, each one of resistance $R = 450\ \Omega$. The circuit is balanced, when no microwave signal is applied, for $U_0 = 18$ V. When the microwave signal is present, the balance is restored by decreasing the voltage to 14.5 V. What is the power of the measured signal, when the substitution efficiency of the thermistor mount is 94%? The measuring device is matched.

5.10.11 A measuring instrument having a VSWR = 2 at its input measures a 5 mW power. The generator is known to have a voltage reflection factor of 0.3. What is the signal power? If the instrument is substituted by a matched load, how much power should this load be able to dissipate?

5.10 12 A batch of Gunn diodes is tested in quality control: their output power is specified as 10 mW minimum into a matched load. A thermistor bridge available to measure them has a substitution efficiency of 0.95 and a VSWR of 1.5. It is further known that the output reflection factor of the oscillators is 0.35. For which values of the measured power can one be certain that an oscillator fulfills the specification? When is one sure that it is defective?

5.10.13 Power is measured with a wattmeter having a substitution efficiency of 0.93 and a VSWR of 1.8. With different line lengths between the generator and the load, the detected powers range from 15.6 to 22.5 watts. How much power would a matched load absorb? What is the VSWR of the generator?

5.10.14 A 250 watt power must be measured, and a milliwattmeter with an upper range value of 10 mW is available. What devices are needed to carry out the measurement?

5.10.15 A pulsed operation is under study. The pulse length of the signal is 0.6 μs, the pulse repetition rate is 0.7 kHz. The average power measured is 420 watts. What may be said of the pulse power?

5.10.16 The circuit of Figure 5.24 is set up to determine the pulse power of a radar. The first coupler has 40 dB coupling, the second one 3 dB coupling and 3 dB attenuation. When the oscilloscope traces coincide, the milliwattmeter indicates 8 mW. What is the signal's pulse power?

CHAPTER 6

MICROWAVE DEVICES

6.1 SCATTERING MATRIX

6.1.1 Definition: Microwave Device

The general term *microwave device* will be used here to define a structure connected to n uniform transmission lines (Ch. 2): waveguides, coaxial lines, microstrip, optical fibers, etc. (Fig. 6.1).

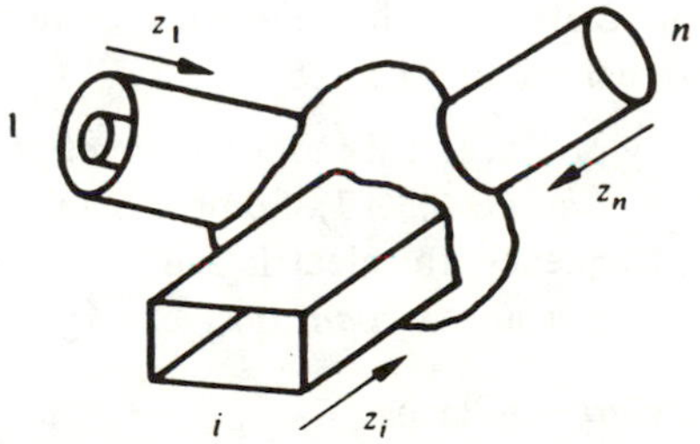

Fig. 6.1 Microwave device.

6.1.2 Basic Assumption

All connecting lines are assumed to be ***lossless***. ***A single mode only***, the dominant mode, can propagate on each line, implicating directly a constraint on the ***size of the line*** or on the ***frequency of the signal***.

6.1.3 Comment

The rigorous study of the electromagnetic fields within a device, solving Maxwell's equations and then applying the boundary conditions, may only be carried out for a very limited number of simple geometries. Nevertheless, the application of physical properties such as reciprocity, losslessness and symmetry permit one to define the main characteristics of a device.

6.1.4 Onset of Higher Order Modes

When any kind of discontinuity is placed inside the waveguide, the uniformity along the direction of propagation (§ 2.1.1) is destroyed. On the surface of the obstacle, additional boundary conditions have to be satisfied by the electromagnetic fields. The modes of propagation, which were studied in chapter 2 for a uniform waveguide, do not satisfy – ***separately*** – these additional boundary conditions. However, the modes form part of a complete set [115], so that any distribution of the fields which satisfies Maxwell's equations may be expressed as a development over the waveguide modes. The boundary conditions on any obstacle may then be satisfied, not by a single mode alone, but by a combination of several modes.

When a signal propagating in the dominant mode reaches an obstacle, it is partly transmitted and partly reflected. In addition, to satisfy the boundary conditions, higher order modes are excited at the discontinuity. However, since operating conditions were selected so that only one mode propagates (§ 6.1.2), all higher order modes are evanescent (§ 2.2.30). When one moves away from the discontinuity, in its far field region, the signals of the dominant mode remain alone.

Evanescent higher order modes exist only next to the discontinuities. They store locally magnetic and electric energy; one of the two is predominant, depending on the field components. Since a TE mode below cutoff stores an excess amount of ***magnetic energy*** (§ 2.2.30), any obstacle that preferentially excites TE modes has an ***inductive*** behavior (§ 6.3.16).

When an obstacle mainly produces TM modes, more ***electric energy*** is stored and the discontinuity is ***capacitive*** (§ 6.3.17). Some obstacles excite both kinds of modes: at a particular frequency, the electric and the magnetic energy have equal total amplitudes. These are ***resonant*** obstacles (§ 6.3.18).

6.1.5 Definition: Reference Planes

On every access line *i* connecting a device to the outer world, a coordinate axis z_i is defined. The origin of the coordinates ($z_i = 0$) defines the *reference plane* at port *i.* This plane must be located far enough from the component, so that all the evanescent higher order modes excited at the discontinuity between line and device are sufficiently damped down in this plane. About one guide wavelength is generally found to be an adequate distance.

6.1.6 Limit of Validity

When two devices are connected to each other by transmission lines shorter than about one wavelength, coupling between the ***evanescent modes*** excited at both discontinuities may take place. The properties described in this chapter, considering only interactions by the dominant modes, are then no longer valid.

6.1.7 Definition: Complex Normalized Waves

The *complex normalized waves* $\underline{a}_i$ and $\underline{b}_i$ [116] are defined in terms of the voltage $\underline{U}_i$, the current $\underline{I}_i$ and the characteristic impedance Z_{ci} of the equivalent line i

$$\underline{a}_i = \frac{\underline{U}_i + Z_{ci}\underline{I}_i}{2\sqrt{Z_{ci}}} \qquad \mathrm{W}^{1/2} \qquad \textbf{(6.1)}$$

$$\underline{b}_i = \frac{\underline{U}_i - Z_{ci}\underline{I}_i}{2\sqrt{Z_{ci}}} \qquad \mathrm{W}^{1/2} \qquad \textbf{(6.2)}$$

The dimension of complex normalized waves is always the square root of a power.

6.1.8 Reverse Relation

The voltage $\underline{U}_i$ and the current $\underline{I}_i$ along line i are in turn obtained in terms of the complex normalized waves $\underline{a}_i$ and $\underline{b}_i$ by respectively adding or subtracting the two expressions (6.1) and (6.2), yielding

$$\underline{U}_i = \sqrt{Z_{ci}}\;(\underline{a}_i + \underline{b}_i) \qquad \mathrm{V} \qquad (6.3)$$

$$\underline{I}_i = (\underline{a}_i - \underline{b}_i)/\sqrt{Z_{ci}} \qquad \mathrm{A} \qquad (6.4)$$

6.1.9 Correspondence with Line Theory

On the lossless transmission line i, the general solution of the telegraphist's equation is given by (2.48) and (2.49)

$$\underline{U}_i = \underline{U}_{i+}\exp(-\mathrm{j}\beta_i z_i) + \underline{U}_{i-}\exp(\mathrm{j}\beta_i z_i) \qquad \mathrm{V} \qquad (6.5)$$

$$\underline{I}_i = (1/Z_{ci})[\underline{U}_{i+}\exp(-\mathrm{j}\beta_i z_i) - \underline{U}_{i-}\exp(\mathrm{j}\beta_i z_i)] \qquad \mathrm{A} \qquad (6.6)$$

Introducing these expressions into (6.1) and (6.2) yields:

$$\underline{a}_i = \frac{\underline{U}_{i+}}{\sqrt{Z_{ci}}}\exp(-\mathrm{j}\beta_i z_i) \qquad \mathrm{W}^{1/2} \qquad (6.7)$$

$$\underline{b}_i = \frac{\underline{U}_{i-}}{\sqrt{Z_{ci}}}\exp(\mathrm{j}\beta_i z_i) \qquad \mathrm{W}^{1/2} \qquad (6.8)$$

The term $\underline{a}_i$ corresponds to the ***forward wave***. It moves towards increasing values of z_i, and thus enters the device (incident wave). The $\underline{b}_i$ term corresponds to the ***backward wave***, representing the signal coming out of the component.

6.1.10 Power at a Port of the Device

The active power at port i is given by

$$P_i = \mathrm{Re}[\underline{U}_i \underline{I}_i^{\,*}] = \mathrm{Re}[(\underline{a}_i + \underline{b}_i)(\underline{a}_i^{*} - \underline{b}_i^{*})] = |\underline{a}_i|^2 - |\underline{b}_i|^2 \qquad \mathrm{W} \qquad (6.9)$$

The total active power is the difference between the power $|\underline{a}_i|^2$ entering the device at port i and the power $|\underline{b}_i|^2$ coming out of the device through the same port.

6.1.11 Comment

It is impractical to use voltages in a waveguide, because they are not uniquely defined (§ 5.1.3). On the other hand, since the complex normalized waves $\underline{a}_i$ and $\underline{b}_i$ are directly determined in terms of power measurements (Ch. 5), they permit one to study devices connected by waveguides, and are therefore consistently utilized at the microwave level.

6.1.12 Scattering Matrix

For any linear device, the incoming and outgoing signals are related to each other by linear algebraic relations. Then one has a system of n equations with n unknowns, which may be written in matrix form, defining the *scattering matrix* $(\underline{s})$:

$$\begin{pmatrix} \underline{b}_1 \\ \underline{b}_2 \\ \vdots \\ \underline{b}_n \end{pmatrix} = \begin{pmatrix} \underline{s}_{11} & \underline{s}_{12} & \cdots & \underline{s}_{1n} \\ \underline{s}_{21} & \underline{s}_{22} & \cdots & \underline{s}_{2n} \\ \vdots & \vdots & & \vdots \\ \underline{s}_{n1} & \underline{s}_{n2} & \cdots & \underline{s}_{nn} \end{pmatrix} \begin{pmatrix} \underline{a}_1 \\ \underline{a}_2 \\ \vdots \\ \underline{a}_n \end{pmatrix} \qquad \mathrm{W}^{1/2} \qquad (6.10)$$

In abbreviated symbolic form, this expression becomes

$$(\underline{b}) = (\underline{s})(\underline{a}) \qquad \mathrm{W}^{1/2} \qquad (6.11)$$

6.1.13 Transfer Function, Intrinsic Reflection

Every term in the scattering matrix is a *transfer function.* When $i \neq j$, the $\underline{s}_{ij}$ term is the transfer function from port j to port i (input j, output i, $\underline{s}_{i \leftarrow j}$). As for the diagonal term $\underline{s}_{ii}$, it represents the *intrinsic reflection* of the device at its port i. In both situations, only one single port is excited (j or i). All other ports are terminated on matched loads

$$\underline{s}_{ij} = \frac{\underline{b}_i}{\underline{a}_j} \qquad \text{with } \underline{a}_k = 0,\ k \neq j \qquad 1 \qquad (6.12)$$

6.1.14 Signal Flow Graph

The physical meaning of the terms in the scattering matrix can be illustrated by means of *signal flow graphs* [117]. Every port is represented by ***two nodes***, one at which the input wave $\underline{a}_i$ arrives, the other from which the output wave $\underline{b}_i$ departs. Every one of the $\underline{s}_{ij}$ terms is associated to an arrow pointing from input node j to output node i. The signal flow graph of a two-port device is sketched in Figure 6.2.

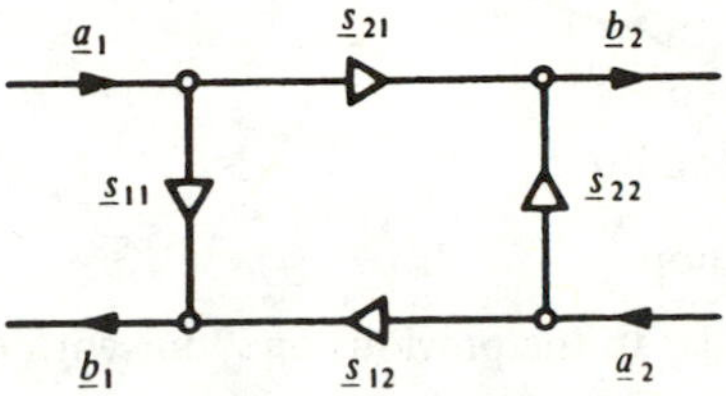

Fig. 6.2 Flow graph of a two-port or quadrupole.

The value taken by the signal at any given node is obtained by summing all the contributions for every arrow reaching this node. A ***full*** arrow designates an incoming or an outgoing signal. The contribution of ***white*** arrows is the product of the transfer function $\underline{s}_{ij}$, associated to the arrow, by the value of the signal at node j. The latter is itself the sum of all the contributions at that node.

In this manner, for a two-port device (Fig. 6.2), one obtains

$$\underline{b}_1 = \underline{s}_{11}\underline{a}_1 + \underline{s}_{12}\underline{a}_2 \qquad \mathrm{W}^{1/2} \qquad (6.13)$$

$$\underline{b}_2 = \underline{s}_{21}\underline{a}_1 + \underline{s}_{22}\underline{a}_2 \qquad \mathrm{W}^{1/2} \qquad (6.14)$$

This actually yields (6.10) with $n = 2$.

6.1.15 Reduction Rules

Signal flow graphs are of particular interest in studying the interconnection of several devices. Long mathematical derivations may be avoided by using a few simple combination rules.

6.1.16 Multiplication

Two arrows connected in series represent the multiplication of two transfer functions (Fig. 6.3).

Fig. 6.3 Flow graph multiplication.

6.1.17 Addition

Two arrows connected in shunt are replaced by the sum of their transfer functions (Fig. 6.4). The two arrows are pointed in the same direction.

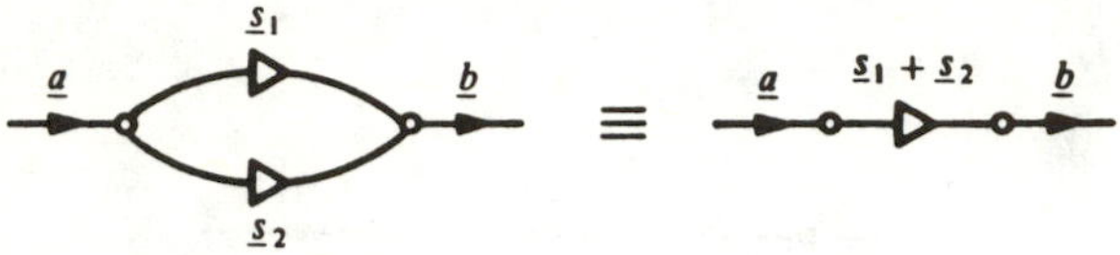

Fig. 6.4 Flow graph addition.

6.1.18 Feedback Loop

The schematic is similar to the previous one, but with one arrow pointing in the opposite direction (Fig. 6.5).

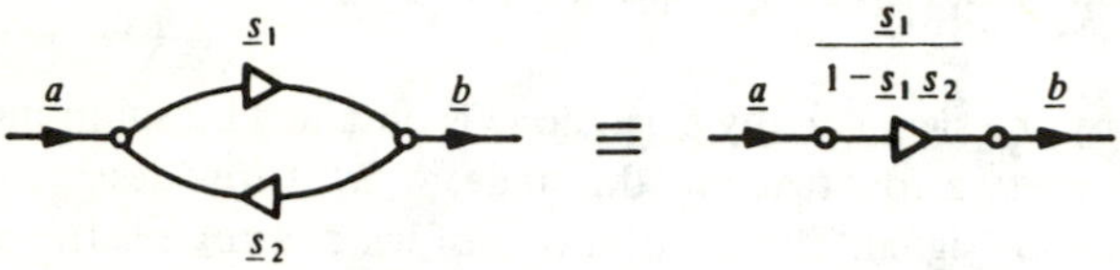

Fig. 6.5 Feedback loop flow graph.

Here, the application of the rule stated under 6.1.14 yields

$$\underline{b} = \underline{s}_1(\underline{a} + \underline{s}_2\underline{b}) \qquad \text{W}^{1/2} \qquad (6.15)$$

or, in other terms

$$\underline{b}(1 - \underline{s}_1\underline{s}_2) = \underline{s}_1\underline{a} \qquad \text{W}^{1/2} \qquad (6.16)$$

and therefore

$$\underline{b} = \frac{\underline{s}_1}{1 - \underline{s}_1\underline{s}_2}\,\underline{a} \qquad \text{W}^{1/2} \qquad (6.17)$$

The transfer function of Figure 6.5 is then obtained.

6.1.19 Examples

The effect of multiple reflections on the measurement of power produces a feedback loop (Fig. 5.19). In the particular case where $\underline{s}_1\underline{s}_2 = 1$, a singularity is encountered. This means that a signal $\underline{b}$ may come out of the device in the absence of any applied signal ($\underline{a} = 0$). The device realized in this manner is an oscillator (Fig. 4.1).

6.1.20 Properties of Devices

The scattering matrix of an n-port device contains n^2 complex terms. These are generally not independent from each other, so that some simplifications may be carried out by considering the ***physical properties of the device***. The number of measurements required to completely characterize a device may then be reduced.

6.1.21 Reciprocity

When the basic assumptions of the ***reciprocity theorem*** are satisfied, the transfer function linking two different ports does not depend upon the direction of transfer [118], meaning that

$$\underline{s}_{ji} = \underline{s}_{ij} \qquad i \neq j \qquad 1 \qquad (6.18)$$

The number of independent terms for a reciprocal device is thus reduced to $n(n + 1)/2$.

6.1.22 Nonreciprocal Devices

The reciprocity theorem does not apply when devices contain gyrotropic materials, characterized by a dyadic permeability or permittivity. This happens in particular in magnetized ferrites (Sec. 6.7), plasmas [119], or semiconductors (Hall effect). In general, active devices are also nonreciprocal.

▪ 6.1.23 Energy Conservation

In a lossless passive circuit, where no energy is dissipated nor produced, the sum of all incoming powers must equal that of all outgoing powers, which is expressed by

$$\sum_{i=1}^{n} |\underline{a}_i|^2 = \sum_{i=1}^{n} |\underline{b}_i|^2 \qquad \text{W} \qquad (6.19)$$

In abbreviated symbolic notation, this expression becomes

$$(\tilde{\underline{a}})(\underline{a}) - (\tilde{\underline{b}})(\underline{b}) = 0 \qquad \text{W} \qquad (6.20)$$

where the tilde (~) denotes the conjugate transposed matrix.

$$(\tilde{\underline{a}}) = (\underline{a}_1^* \, \underline{a}_2^* \ldots \underline{a}_n^*) \qquad \text{W}^{1/2} \qquad (6.21)$$

Relation (6.11) and its transpose are then utilized as

$$(\tilde{\underline{b}}) = (\tilde{\underline{a}})(\tilde{\underline{s}}) \qquad \text{W}^{1/2} \qquad (6.22)$$

Replacing within (6.20) and grouping the terms yields

$$(\tilde{\underline{a}})\,[(1) - (\tilde{\underline{s}})(\underline{s})]\,(\underline{a}) = 0 \qquad \text{W} \qquad (6.23)$$

where (1) is the unit matrix of order n.

This relation must be satisfied for any excitation $(\underline{a})$, yielding the condition for *energy conservation*

$$(\tilde{\underline{s}})(\underline{s}) = (1) \qquad 1 \qquad \textbf{(6.24)}$$

The condition may be further developed for the terms of the scattering matrix

$$\sum_{i=1}^{n} \underline{s}_{ij}^{*} \underline{s}_{ik} = \delta_{jk} \qquad 1 \qquad (6.25)$$

where δ_{jk} is Kronecker's delta symbol, $\delta_{jk} = 1$ for $j = k$, $\delta_{jk} = 0$ for $j \neq k$.

A device which neither absorbs nor yields energy has a scattering matrix whose terms must satisfy n^2 complex quadratic equations (6.25).

6.1.24 Reflectionless Match

A device is said to be *matched at its port i* when no signal comes out of it when this port is the only one excited (*i.e.*, for $\underline{a}_j = 0, j \neq i$). One then has

$$\underline{s}_{ii} = 0 \qquad 1 \qquad (6.26)$$

A device is matched at all its ports when ***all the terms on the diagonal*** of the scattering matrix ***vanish***.

6.1.25 Symmetry

When a reciprocal device (§ 6.1.21) possesses one or several planes of geometrical symmetry, and when, in addition, the reference planes are located symmetrically, the terms of the scattering matrix related to the symmetrically located ports are either equal or of opposite sign, depending upon the orientation of the electric field. Examples of symmetrical devices are the T junctions (§ 6.4 8 and § 6.4.9) and the Y junctions (§ 6.4.10).

6.1.26 Translation of the Reference Plane

The origin of the z_i coordinate on line i is defined arbitrarily, subject to negligible amplitudes for higher order modes (§ 6.1.5). The reference plane may be shifted along the transmission line by a distance Δz_i (Fig. 6.6).

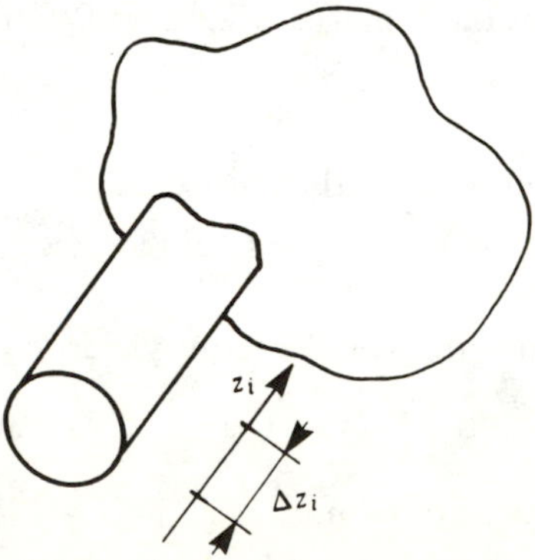

Fig. 6.6 Translation of the reference plane at port *i*.

Within the new system of the displaced coordinate $z_i^d = z_i - \Delta z_i$, the forward $\underline{a}_i$ and the backward $\underline{b}_i$ waves keep their respective amplitudes, but their phases are shifted in proportion to the distance Δz_i (6.7) and (6.8).

$$\underline{a}_i^d = \underline{a}_i \exp(-j\varphi_i) \qquad W^{1/2} \qquad (6.27)$$

$$\underline{b}_i^d = \underline{b}_i \exp(j\varphi_i) \qquad W^{1/2} \qquad (6.28)$$

with

$$\varphi_i = -\beta_i \Delta z_i \qquad \text{rad} \qquad (6.29)$$

■ 6.1.27 Translation of the Reference Planes

Shifts like the one considered in Section 6.1.26 may be carried out at the same time within all n ports of the device. The column vectors of the incoming signals $(\underline{a})$ and of the outgoing signals $(\underline{b})$ become, by using (6.27) and (6.28),

$$(\underline{a}) = (\text{diag} \exp(j\varphi)) (\underline{a}^d) \qquad W^{1/2} \qquad (6.30)$$

$$(\underline{b}^d) = (\text{diag} \exp(j\varphi)) (\underline{b}) \qquad W^{1/2} \qquad (6.31)$$

where the diagonal matrix was defined as

$$(\text{diag} \exp(j\varphi)) = \begin{pmatrix} \exp(j\varphi_1) & 0 & \dots 0 \\ 0 & \exp(j\varphi_2) & \dots 0 \\ \vdots & \vdots & \vdots \\ 0 & 0 & \dots \exp(j\varphi_n) \end{pmatrix} \qquad 1 \qquad (6.32)$$

The scattering matrix within the system of the displaced reference planes $(\underline{s}^d)$ is obtained by applying (6.11), (6.30) and (6.31)

$$(\underline{b}^d) = (\text{diag} \exp(j\varphi)) (\underline{s}) (\text{diag} \exp(j\varphi)) (\underline{a}^d) = (\underline{s}^d) (\underline{a}^d) \qquad W^{1/2} \qquad (6.33)$$

The expression for each component of the matrix then becomes

$$\underline{s}_{ij}^d = \underline{s}_{ij} \exp[j(\varphi_i + \varphi_j)] \qquad 1 \qquad (6.34)$$

A shift of the reference planes produces a change in the ***argument*** of the terms of the scattering matrix. Their modulus is, however, not modified. The phases of n terms may be arbitrarily set in this manner, *i.e.*, for one term within each line and each column, by selecting particular reference planes at the n ports of the device. As an example, n terms could be specified to be positive real numbers.

6.1.28 Transformation: From the Scattering Matrix to the Impedance Matrix

The impedance matrix $(\underline{Z})$ of a device is obtained by using the definition of the complex normalized waves (6.1) and (6.2), and the one of the scattering matrix (6.11). Two auxiliary diagonal matrices are defined in the process

$$(G) = (\text{diag}\, Z_{ci}) = \begin{pmatrix} Z_{c1} & 0 & \dots & 0 \\ 0 & Z_{c2} & \dots & 0 \\ \vdots & \vdots & & \vdots \\ 0 & 0 & \dots & Z_{cn} \end{pmatrix} \quad \Omega \qquad (6.35)$$

$$(F) = \left(\text{diag}\, \frac{1}{2\sqrt{Z_{ci}}} \right) = \begin{pmatrix} \frac{1}{2\sqrt{Z_{c1}}} & 0 & \dots & 0 \\ 0 & \frac{1}{2\sqrt{Z_{c2}}} & \dots & 0 \\ \vdots & \vdots & & \vdots \\ 0 & 0 & \dots & \frac{1}{2\sqrt{Z_{cn}}} \end{pmatrix} \quad \Omega^{-1/2} \qquad (6.36)$$

After some matrix manipulations, one obtains

$$\begin{aligned} (\underline{Z}) &= (F)^{-1}[(1) + (\underline{s})]\,[(1) - (\underline{s})]^{-1}(F)\,(G) \\ &= (F)^{-1}[(1) - (\underline{s})]^{-1}[(1) + (\underline{s})](F)\,(G) \end{aligned} \quad \Omega \qquad (6.37)$$

6.1.29 Inverse Transformation

In a quite similar manner, the scattering matrix is obtained from the impedance matrix

$$(\underline{s}) = (F)[(\underline{Z}) - (G)]\,[(\underline{Z}) + (G)]^{-1}(F)^{-1} \quad 1 \qquad (6.38)$$

6.1.30 Remark

In circuit theory, the scattering matrix formalism is also utilized. However, no transmission lines and characteristic impedances are defined there. Normalization is usually carried out with respect to the load impedances, which may be set equal to each other by insertion of ideal transformers. In microwave devices, these simplifications are not possible.

6.2 ONE-PORT DEVICES

6.2.1 Introduction

The scattering matrix of a device having only *one port* reduces to a single term: its reflection factor (Fig. 6.7). It is obviously not possible to match a lossless one-port device; the two requirements are contradictory.

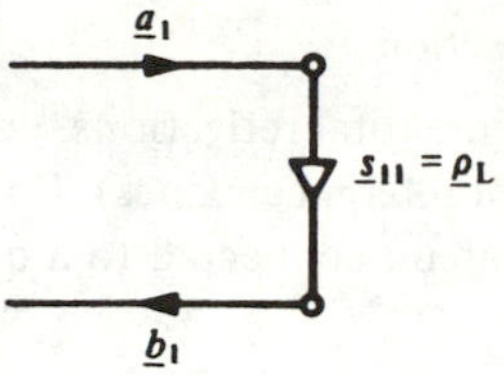

Fig. 6.7 Flow graph of a one-port.

6.2.2 Matched Load

The one-port device most often encountered is the *load,* or *matched termination* (matched to the transmission line), for which $|s_{11}| \cong 0$. In practice, a matched load consists of a gradual transition from an empty waveguide to a waveguide loaded with a lossy material, often some thermoplastic loaded with finely divided carbon or iron powder, or also a thin metal layer deposited onto a dielectric substrate. Several structures are sketched in Figure 6.8.

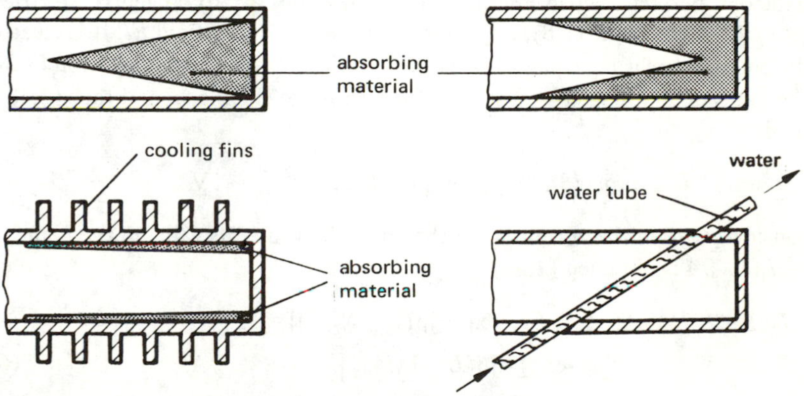

Fig. 6.8 Matched loads (sections).

Since all the power of the incoming signal must be dissipated in the absorbing material, a matched load should be able to transfer to the outside the thermal energy produced. In a high power load, the absorbing material is deposited on the metal walls, in intimate thermal contact with them; the walls are then cooled (air or water, as needed). Water may also be used directly as the absorbing material (§ 5.6.3). The liquid must flow continuously whenever the microwave signal is on, otherwise boiling may take place.

Practically speaking, a rigorously matched load cannot be realized. A very good waveguide load has a VSWR (§ 7.2.2) smaller than about 1.005, while average quality loads have VSWRs between 1.01 and 1.05. Somewhat lower performances are achieved in coaxial loads: at best, on the order of 1.02.

6.2.3 Fixed Total Reflection

Several measurements require total reflections, either open- or short-circuits (depending upon the choice of reference planes). For a reflection factor as close to unity as possible, a short-circuit connected to a quarter-wave section of line is used (Fig. 6.9).

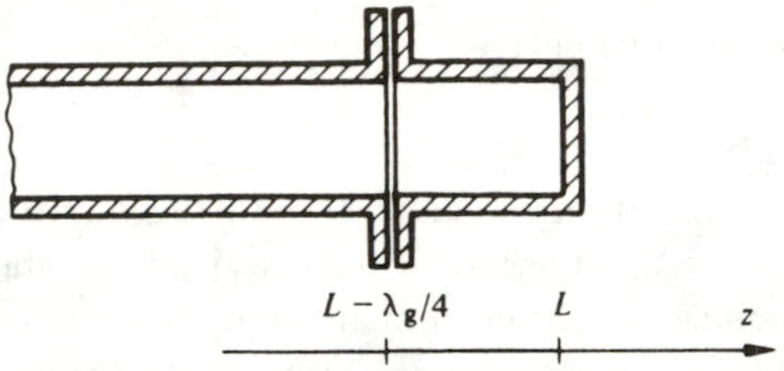

Fig. 6.9 Fixed short-circuit with large reflection (section).

At the end of the waveguide (at $z = L$), the short-circuit plane, assumed to be made of a perfect electric conductor (pec), imposes a zero tangential electric field within the plane (1.8), *i.e.*, a zero transverse electric field $\underline{E}_t$. It is clearly not possible to let $\underline{E}_T$ equal zero, as this would impose a trivial solution (2.38). The equivalent voltage $\underline{U}_e$ must therefore vanish in the plane, and thus (2.48) yields

$$\underline{U}_e(L) = \underline{U}_{e+} \exp(-j\beta L) + \underline{U}_{e-} \exp(+j\beta L) = 0 \qquad \text{V} \qquad (6.39)$$

Considering then the current $\underline{I}_e$ on the equivalent line, its value in the plane $z = L - \lambda_g/4$ is obtained (2.49)

$$\underline{I}_e(L - \lambda_g/4) = (1/\underline{Z}_e)\left\{\underline{U}_{e+} \exp[-j\beta(L - \lambda_g/4)] - \underline{U}_{e-} \exp[+j\beta(L - \lambda_g/4)]\right\} \qquad \text{A} \qquad (6.40)$$

In addition, it is known that

$$\beta\lambda_g/4 = (\beta \cdot 2\pi/\beta)/4 = \pi/2 \qquad 1 \qquad (6.41)$$

Making use of this expression to develop the terms in $\beta\lambda_g$ of (6.40), one finds with (6.39)

$$\underline{I}_e(L - \lambda_g/4) = (j/Z_e)[\underline{U}_{e+} \exp(-j\beta L) + \underline{U}_{e-} \exp(+j\beta L)] = (j/Z_e)\,\underline{U}_e(L) = 0 \qquad \text{A} \qquad (6.42)$$

The current $\underline{I}_e$ on the equivalent line vanishes in the plane $z = L - \lambda_g/4$. This means that the transverse magnetic field is zero in this plane (2.39); the same is true for the longitudinal component of the surface current (1.12). Since no line of current crosses the plane $z = L - \lambda_g/4$, a separation in this plane does not perturb the electromagnetic fields within the waveguide. The possible contact resistances between waveguide and short-circuit flanges do not have any effect.

6.2.4 Sliding Short or Piston

Some measurements require variable reflection phases, as do oscillators (§ 4.6.10) and amplifiers (§ 4.10.4) for frequency tuning. A good reflection factor ($|\underline{\rho}| \cong 1$) should be maintained for all piston locations. Spring loaded contacts may be used, and also quarter-wave buckets or more sophisticated filter structures (Fig. 6.10).

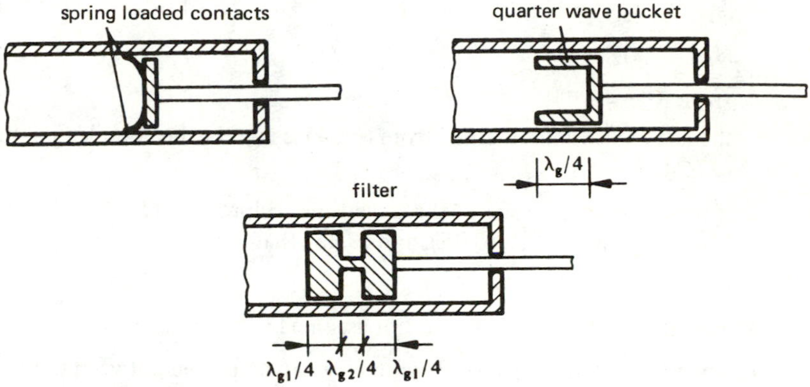

Fig. 6.10 Sliding shorts or pistons.

6.2.5 Reflection Amplifiers

Several kinds of amplifiers utilized at microwave only have a single port, through which both the incident and the amplified signals pass; these are particular one-port devices for which $|\underline{s}_{11}| > 1$. Their load must necessarily be located within the input circuit, and any mismatch may produce oscillations.

The availability of circulators (§ 6.7.13) has greatly simplified the use of reflection amplifiers, separating their input from their output.

The most common reflection amplifiers are parametric amplifiers (§ 4.10.4), cavity masers (§ 4.10.6), tunnel diode, Gunn diode (§ 4.6.11) and avalanche diode amplifiers (Sec. 4.7).

6.3 TWO-PORT DEVICES

6.3.1 General Remarks

In a *two-port device* or ***quadrupole***, the scattering matrix has four terms

$$(\underline{s}) = \begin{pmatrix} \underline{s}_{11} & \underline{s}_{12} \\ \underline{s}_{21} & \underline{s}_{22} \end{pmatrix} \qquad (6.43)$$

The corresponding signal flow graph is sketched in Figure 6.2.

6.3.2 Particular Properties

Applying the expressions developed in Section 6.1 to the scattering matrix (6.43), it appears that one has:

1. for a reciprocal two-port (§ 6.1.21):

   $$\underline{s}_{12} = \underline{s}_{21} \qquad 1 \qquad (6.44)$$

2. for a lossless two-port, applying (6.24):

   $$|\underline{s}_{11}|^2 + |\underline{s}_{21}|^2 = 1 \qquad 1 \qquad (6.45)$$

   $$|\underline{s}_{12}|^2 + |\underline{s}_{22}|^2 = 1 \qquad 1 \qquad (6.46)$$

   $$\underline{s}_{11}^* \underline{s}_{12} + \underline{s}_{21}^* \underline{s}_{22} = 0 \qquad 1 \qquad (6.47)$$

   the last expression obtained from (6.24) is merely the complex conjugate of (6.47);

3. if the two-port device is at the same time reciprocal and lossless, one finds when utilizing (6.44), (6.45), and (6.46) that

   $$|\underline{s}_{11}| = |\underline{s}_{22}| \qquad 1 \qquad (6.48)$$

   even though the two-port was not assumed to be symmetrical;

4. when a geometrical plane of symmetry can be placed between the two ports of the device, in such a way that the input and output ports cannot be distinguished, and when the reference planes are also symmetrically located, one has

   $$\underline{s}_{11} = \underline{s}_{22} \qquad 1 \qquad (6.49)$$

   The two port must also be reciprocal;

5. for a matched two-port device

   $$\underline{s}_{11} = \underline{s}_{22} = 0 \qquad 1 \qquad (6.50)$$

6.3.3 Matched Absorption Attenuator

A reciprocal device designed to reduce the power level is called an *attenuator.* The scattering matrix of a matched attenuator has the form

$$(\underline{s}) = \begin{pmatrix} 0 & \underline{s}_{12} \\ \underline{s}_{12} & 0 \end{pmatrix} \qquad 1 \qquad (6.51)$$

The attenuation level LA, expressed in dB (§ 5.6.9), is defined by the ratio of the input and output powers. When the output is ***connected to a matched termination*** ($\underline{a}_2 = 0$)

$$LA = 10 \log (P_1/P_2) = 10 \log (|\underline{a}_1|^2/|\underline{b}_2|^2) = -\ 20 \log |\underline{s}_{12}| \qquad \text{dB} \qquad (6.52)$$

The attenuator absorbs the power $P_1 - P_2$.

6.3.4 Matched Attenuators in Waveguide

In a ***matched*** attenuator, the power difference between input and output ***cannot be reflected***: the device must therefore contain absorbing material, most often a thin metal sheet deposited on a dielectric substrate. Several waveguide designs are sketched in Figure 6.11.

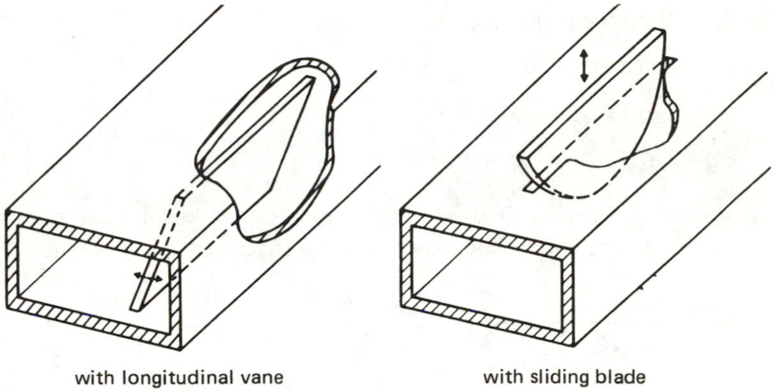
with longitudinal vane
with sliding blade

Fig. 6.11 Matched attenuators in waveguide.

6.3.5 Matched Attenuators in Coaxial Line

In a coaxial line attenuator, the energy loss may be produced by an absorbing layer deposited on the center conductor or by a center conductor made of a high resistivity metal, or also by a combination of lossy center conductor and resistive disk across the coaxial line (Fig. 6.12).

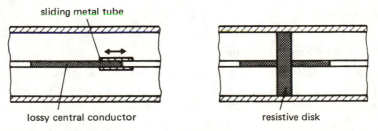

sliding metal tube
lossy central conductor
resistive disk

Fig. 6.12 Matched attenuators in coaxial line.

6.3.6 Particular Case: T Attenuator

When the lossy elements making up the attenuator of Figure 6.12 have dimensions much smaller than a wavelength, this attenuator may be represented by the equivalent schematic of Figure 6.13. The reference planes were chosen to have only resistive elements.

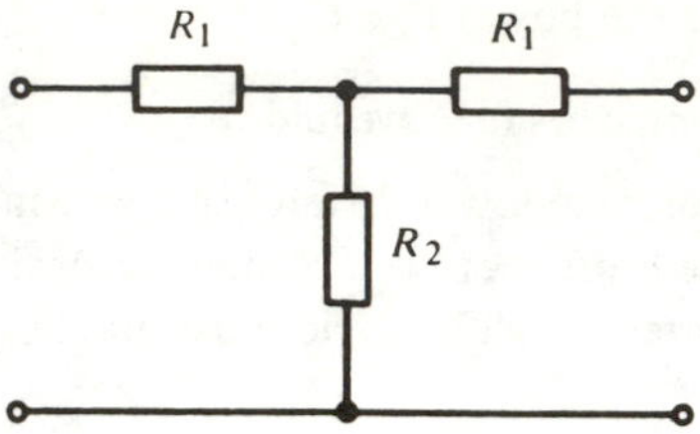

Fig. 6.13 Equivalent circuit of T attenuator.

The impedance matrix of the circuit has the form

$$(Z) = \begin{pmatrix} R_1 + R_2 & R_2 \\ R_2 & R_1 + R_2 \end{pmatrix} \qquad \Omega \qquad (6.53)$$

The scattering matrix is obtained in terms of (6.38), taking into account the fact that the input and output lines have the same characteristic impedance Z_c

$$(s) = \frac{1}{(R_1 + R_2 + Z_c)^2 - R_2^2} \begin{pmatrix} (R_1 + R_2)^2 - Z_c^2 - R_2^2 & 2R_2 Z_c \\ 2R_2 Z_c & (R_1 + R_2)^2 - Z_c^2 - R_2^2 \end{pmatrix} \qquad 1 \qquad (6.54)$$

For the attenuator to be matched, one must have $\underline{s}_{ii} = 0$, so that

$$R_2 = \frac{Z_c^2 - R_1^2}{2R_1} \qquad \Omega \qquad (6.55)$$

After some computations, one finds

$$s_{12} = \frac{Z_c - R_1}{Z_c + R_1} \qquad 1 \qquad (6.56)$$

6.3.7 Rotary Vane Attenuator

A precision variable attenuator is realized by means of a circular waveguide section propagating the spatially degenerate TE_{11} mode (§ 2.4.9). The absorbing element is a thin metal vane, which may rotate around the longitudinal axis (Fig. 6.14).

The signal is fed through a rectangular waveguide to port 1 (TE_{10} mode). A transition connects the rectangular waveguide to the circular guide (TE_{11} mode, § 2.4.9). Within the center section, the absorbing vane forms an angle θ with the plane normal to the electric field. The component $E \sin \theta$ of the field, which is located within the plane of the vane, is absorbed. The component $E \cos \theta$, normal to the vane, propagates without perturbation. In the output transition, the wave

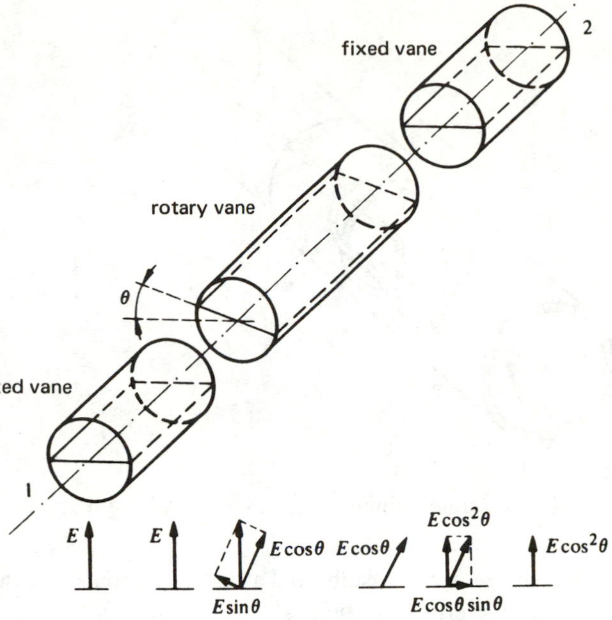

Fig. 6.14 Rotary vane attenuator in circular waveguide.

meets again an absorbing vane, and the component $E \sin \theta \cos \theta$, in the plane of the vane, is absorbed.

At the output at port 2, the field amplitude is $E \cos^2 \theta$, so that $|\underline{s}_{12}| = \cos^2 \theta$ and the attenuation level is

$$LA = -20 \log |\underline{s}_{21}| = -40 \log(\cos \theta) \qquad \text{dB} \qquad (6.57)$$

The attenuation is only a function of the angle of rotation, which is measured by mechanical means.

6.3.8 Reactive Mismatched Attenuator

Attenuation may also be produced by reflecting part of the incoming signal. An attenuator based on this approach is fabricated by connecting a section of waveguide below cutoff between two transmission lines (Fig. 6.15). The attenuation is directly related to the waveguidc length (evanescent mode, § 2.2.30) [120]. The reflection produced by this attenuator may, however, perturb the generator (§ 4.11.2); it is therefore recommended to insert an isolator in the line.

6.3.9 Applications of Attenuators

Attenuators may be of different types: fixed, adjustable, variable, with steps, precision, calibrated, etc. An electrically adjustable attenuator may be obtained

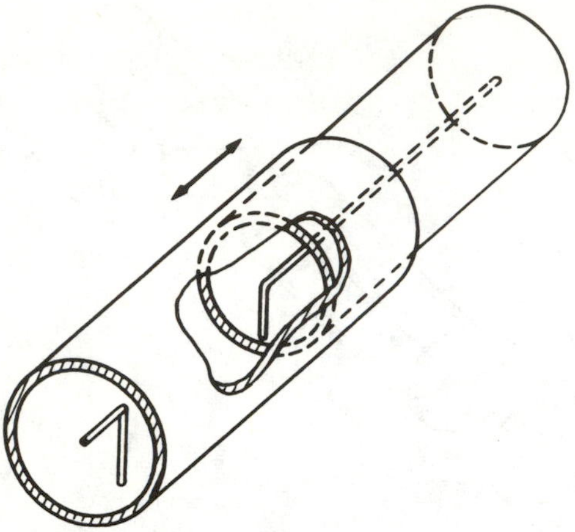

Fig. 6.15 Reactive cutoff attenuator.

using ferrites (§ 6.7.22) or semiconductors (§ 6.8.3). Their main applications are the following:

1. in attenuation measurements, a calibrated attenuator determines the difference between the levels of two signals,
2. a matched attenuator, most often of fixed value (pad), is used to reduce the reflection produced by a detector (difficult to match because it is a nonlinear device),
3. a matched attenuator permits the reduction of the level of a signal, to bring it within the measurement range of available instrumentation (§ 5.8.6).

6.3.10 Isolator

An *isolator* is a nonreciprocal attenuator, which lets a signal go through with a low attenuation in one direction ($|\underline{s}_{21}| \cong 1$), but presents an important attenuation in the opposite sense of propagation ($|\underline{s}_{12}| \cong 0$). Used to decouple successive stages of a system, it is realized with ferrites (§ 6.7.8 and § 6.7.15).

6.3.11 Phaseshifter

A reciprocal matched lossless two-port device is a *phaseshifter,* represented by the scattering matrix

$$(\underline{s}) = \begin{pmatrix} 0 & \exp(j\varphi) \\ \exp(j\varphi) & 0 \end{pmatrix} \qquad 1 \tag{6.58}$$

A phaseshifter is used to modify the phase of a signal, and thus is the counterpart of the attenuator, which modifies the amplitude. It is mostly utilized to determine the phaseshift produced by devices (§ 7.4.19). It may be realized in rectangular waveguide, loading it with a slab of dielectric material (§ 2.9.8). For precision measurements, a rotative phaseshifter is utilized (§ 6.3.12). Ferrite (§ 6.7.22) and semiconductor (§ 6.8.6) phaseshifters may be electrically controlled.

■ 6.3.12 Fox Rotative Phaseshifter, Quarter-Wave and Half-Wave Plates

A ***continuously variable phaseshift*** may be introduced into a system with an apparatus which is the analog of the rotative vane attenuator (§ 6.3.7). This is the *Fox rotative phaseshifter* [121]. The absorbing vanes, basic elements of the attenuator, are replaced by fixed *quarter-wave sections* and a rotative *half-wave section* in the middle of the device (Fig. 6.16).

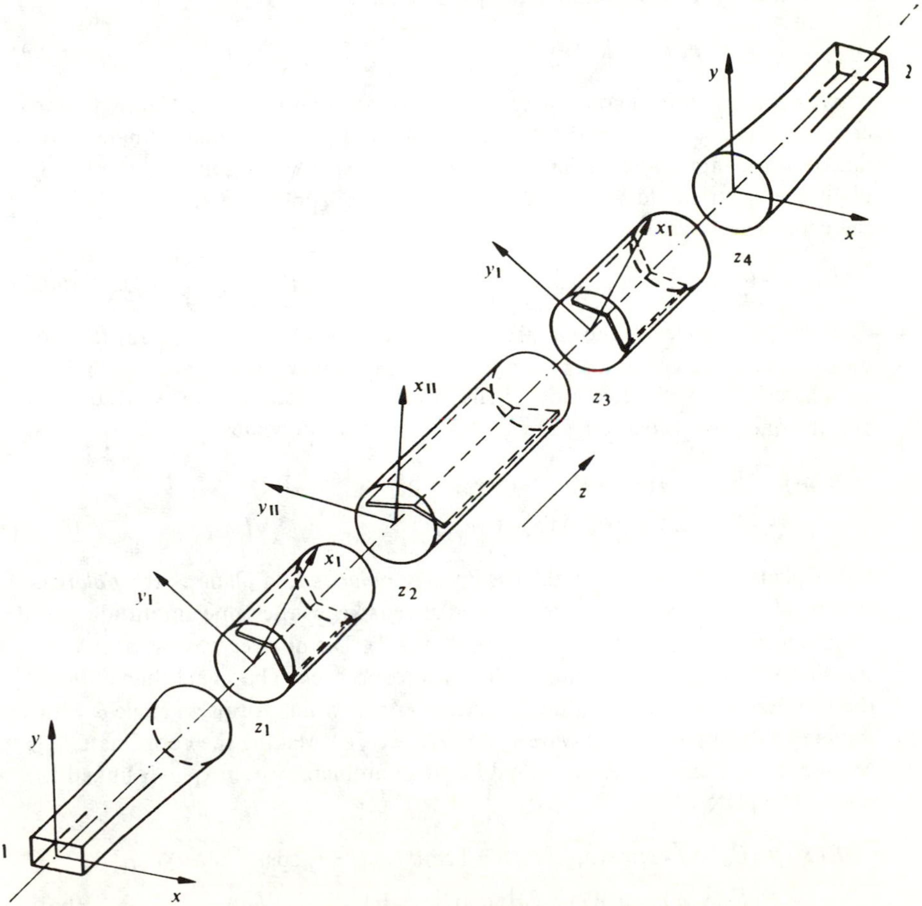

Fig. 6.16 Fox rotative phaseshifter.

These terms designate sections of waveguide through which two TE_{11} modes may propagate simultaneously, but with slightly different propagation velocities. The phaseshifts produced are thus different for the two modes. When the difference in phaseshift is $\pi/2$, the section is called a *quarter-wave plate.* A difference in phaseshift of π corresponds to a *half-wave plate.* In practice, such sections are realized by loading a circular waveguide with a dielectric plate. The latter modifies the propagation factors of the two modes differently, depending on whether their electric field is perpendicular or parallel to the plate. The two TE_{11} modes are then no longer degenerate.

The signal gets into port 1 in the rectangular waveguide and moves across the transition to the circular waveguide. At the center of the waveguide, the electric field of the input signal entering in z_1 is given by

$$\underline{E}(z_1) = \underline{E}_0 \cdot \boldsymbol{e}_y \qquad \text{V/m} \tag{6.59}$$

where $\boldsymbol{e}_y$ is the vertical coordinate of the rectangular waveguide. The first quarter-wave plate has its dielectric plate at 45° with respect to the coordinate system linked to the input port 1 (Fig. 6.17a). In the system $x_\mathrm{I}, y_\mathrm{I}$ linked to the fixed plate, the electric field at the center is obtained by applying a 45° rotation of the coordinate axes

$$\underline{E}(z_1) = (\underline{E}_0/\sqrt{2})\,(\boldsymbol{e}_{x\mathrm{I}} + \boldsymbol{e}_{y\mathrm{I}}) \qquad \text{V/m} \tag{6.60}$$

The two TE_{11} modes of the circular waveguide are excited, having electric fields directed, respectively, along x_I and y_I. The quarter-wave plate produces different phaseshifts for the two modes. The mode polarized along x_I is shifted by φ_1, and the one polarized along y_I by $\varphi_1 + \pi/2$, so that in plane z_2

$$\begin{aligned}\underline{E}(z_2) &= (\underline{E}_0/\sqrt{2})\left\{\boldsymbol{e}_{x\mathrm{I}} \exp(\mathrm{j}\varphi_1) + \boldsymbol{e}_{y\mathrm{I}} \exp[\mathrm{j}(\varphi_1 + \pi/2)]\right\}\\ &= (\underline{E}_0/\sqrt{2}) \exp(\mathrm{j}\varphi_1)\,(\boldsymbol{e}_{x\mathrm{I}} + \mathrm{j}\boldsymbol{e}_{y\mathrm{I}}) \qquad \text{V/m}\end{aligned} \tag{6.61}$$

In the plane z_1, the electric field was ***linearly polarized***, in plane z_2 the ***polarization is circular***. This means that the field always keeps the same amplitude, but its direction rotates as a function of time (§ 1.4.7). The quarter-wave section is a transition from linear to circular polarization and back. The signal then enters the rotative half-wave plate, the dielectric plate of which forms an angle θ with respect to the input quarter-wave plate (Fig. 6.17b). Making the coordinate change, one obtains the electric field in the coordinate system $x_\mathrm{II}, y_\mathrm{II}$ linked to the mobile plate.

$$\begin{aligned}\underline{E}(z_2) &= (\underline{E}_0/\sqrt{2}) \exp(\mathrm{j}\varphi_1)[(\cos\theta + \mathrm{j}\sin\theta)\,\boldsymbol{e}_{x\mathrm{II}} + (\mathrm{j}\cos\theta - \sin\theta)\,\boldsymbol{e}_{y\mathrm{II}}]\\ &= (\underline{E}_0/\sqrt{2}) \exp[\mathrm{j}(\varphi_1 + \theta)]\,(\boldsymbol{e}_{x\mathrm{II}} + \mathrm{j}\boldsymbol{e}_{y\mathrm{II}}) \qquad \text{V/m}\end{aligned} \tag{6.62}$$

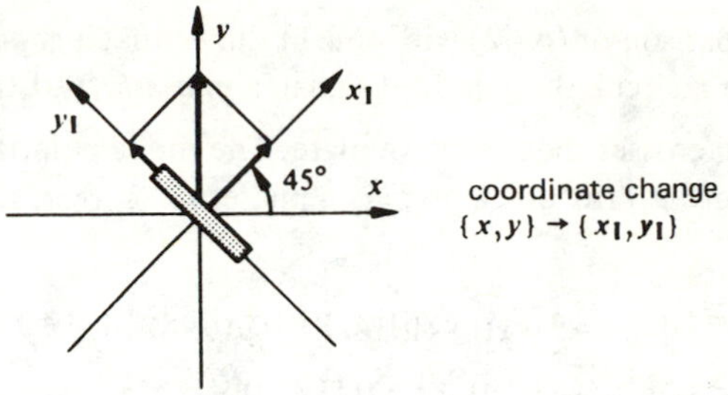

(a) transition from rectangular waveguide to fixed $\lambda/4$ section ($z = z_1$)

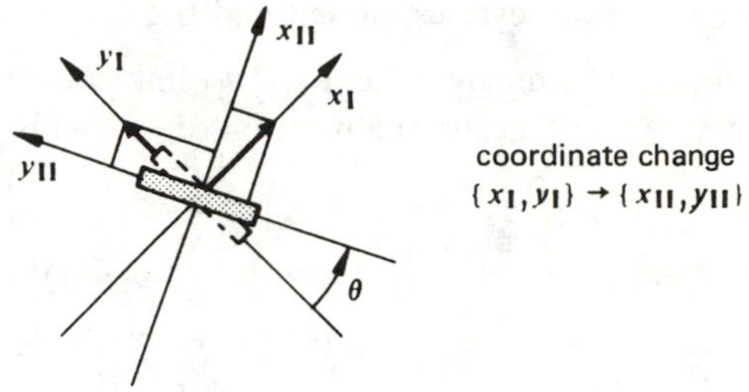

(b) transition from fixed $\lambda/4$ section to rotary $\lambda/2$ section ($z = z_2$)

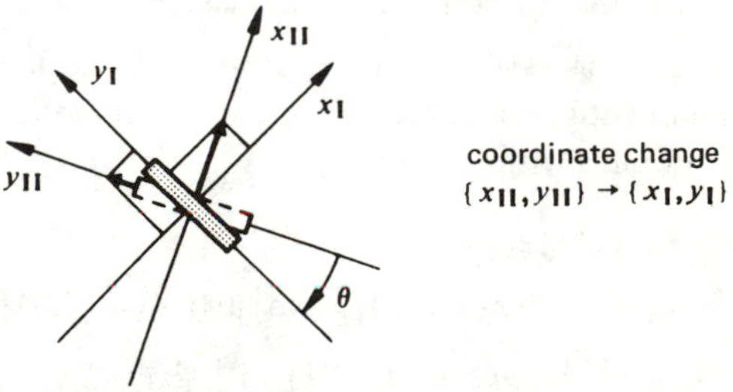

(c) transition from $\lambda/2$ rotary section to fixed $\lambda/4$ section ($z = z_3$)

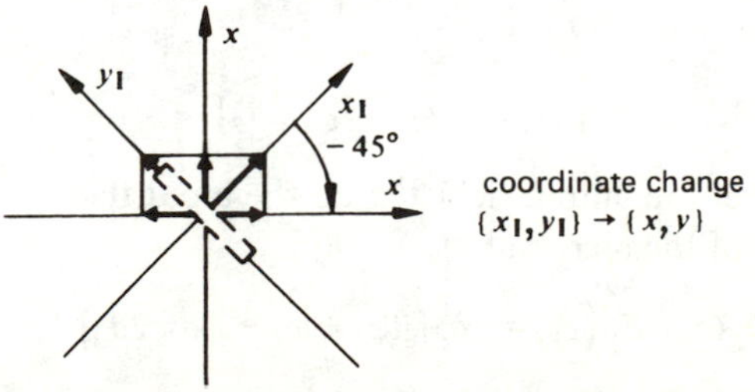

(d) transition from fixed $\lambda/4$ section to rectangular waveguide ($z = z_4$)

Fig. 6.17 Section changes in the Fox rotative phaseshifter.

The comparison of (6.62) with (6.61) shows that a ***geometrical rotation*** produces, for a circularly polarized signal, a ***phaseshift*** of the same value.

The signal crosses the half-wave plate. The mode polarized along x_{II} experiences a phaseshift φ_2, the one along y_{II} a phaseshift $\varphi_2 + \pi$, so that one obtains in plane z_3

$$\begin{aligned} \underline{E}(z_3) &= (\underline{E}_0/\sqrt{2})\left\{e_{xII}\exp(j\varphi_2) + je_{yII}\exp[j(\varphi_2+\pi)]\right\}\exp[j(\varphi_1+\theta)] \\ &= (\underline{E}_0/\sqrt{2})(e_{xII} - je_{yII})\exp[j(\varphi_1+\varphi_2+\theta)] \quad \text{V/m} \end{aligned} \tag{6.63}$$

The half-wave section reverses the sense of the circular polarization.

Moving from the coordinate system x_{II}, y_{II} linked to the half-wave plate, to the fixed system x_I, y_I by means of a rotation of $-\theta$ yields in plane z_3 (Fig. 6.17c)

$$\begin{aligned} \underline{E}(z_3) &= (\underline{E}_0/\sqrt{2})[(\cos\theta + j\sin\theta)e_{xI} \\ &\quad + (-j\cos\theta + \sin\theta)e_{yI}]\exp[j(\varphi_1+\varphi_2+\theta)] \\ &= (\underline{E}_0/\sqrt{2})(e_{xI} - je_{yI})\exp[j(\varphi_1+\varphi_2+2\theta)] \quad \text{V/m} \end{aligned} \tag{6.64}$$

Because the circularly polarized signal rotates in the direction opposite to z_2, the geometrical rotation of $-\theta$ produces a phaseshift $+\theta$ of the signal.

The output quarter-wave section restores the linear polarization of the signal. The x_I component gets a phaseshift φ_3, and that along y_I a phaseshift $\varphi_3 + \pi/2$, so that one obtains within plane z_4

$$\begin{aligned} \underline{E}(z_4) &= (\underline{E}_0/\sqrt{2})\left\{e_{xI}\exp(j\varphi_3) \right. \\ &\quad \left. - je_{yI}\exp[j(\varphi_3+\pi/2)]\right\}\exp[j(\varphi_1+\varphi_2+2\theta)] \\ &= (\underline{E}_0/\sqrt{2})(e_{xI} + e_{yI})\exp[j(\varphi_1+\varphi_2+\varphi_3+2\theta)] \quad \text{V/m} \end{aligned} \tag{6.65}$$

Finally, a rotation of $-45°$ brings back to the x,y coordinates of the rectangular waveguide (Fig. 6.17d)

$$\underline{E}(z_4) = \underline{E}_0 e_y \exp[j(\varphi_1+\varphi_2+\varphi_3+2\theta)] \quad \text{V/m} \tag{6.66}$$

The ratio of the output field at z_4 to the input field at z_1 provides the transfer factor $\underline{s}_{21}$ of the phaseshifter,

$$\underline{s}_{21} = \underline{E}_y(z_4)/\underline{E}_y(z_1) = \exp[j(\varphi_1+\varphi_2+\varphi_3+2\theta)] \quad \text{V/m} \tag{6.67}$$

The signal crossing the phaseshifter retains its original magnitude. The phaseshift produced has a constant component $\phi_1 + \phi_2 + \phi_3$, to which a variable phaseshift 2θ is added, which is directly proportional to the mechanical rotation θ of the half-wave plate. The measurement of this angle directly yields the electrical phaseshift, which may be varied in this manner between 0 and 2π.

6.3.13 Nonreciprocal Phaseshifter

Different phaseshifts along the two directions of propagation may be obtained by loading a line or a waveguide with gyrotropic material such as magnetized ferrite (Sec. 6.7): in this manner a *nonreciprocal phaseshifter* is realized.

6.3.14 Particular Case: Gyrator

When the difference between the two nonreciprocal phaseshifts is equal to π, the device is called a *gyrator* [122]. By choosing the reference planes at both ports so that $s_{21} = 1$, the following scattering matrix is obtained

$$(s) = \begin{pmatrix} 0 & -1 \\ 1 & 0 \end{pmatrix} \qquad\qquad 1 \qquad\qquad (6.68)$$

■ 6.3.15 Thin Symmetrical Obstacle, Iris

We consider an obstacle made of a thin metal blade or lossless *iris* (pec) having a thickness very much smaller than a wavelength. It is located transversally across the waveguide cross section. Figure 6.18 presents three particular geometries, considered in Sections 6.3.16, 6.3.17, and 6.3.18.

The obstacle is connected to sections of uniform waveguide extending on both sides to symmetrically located reference planes, at a distance d (Fig. 6.19). The

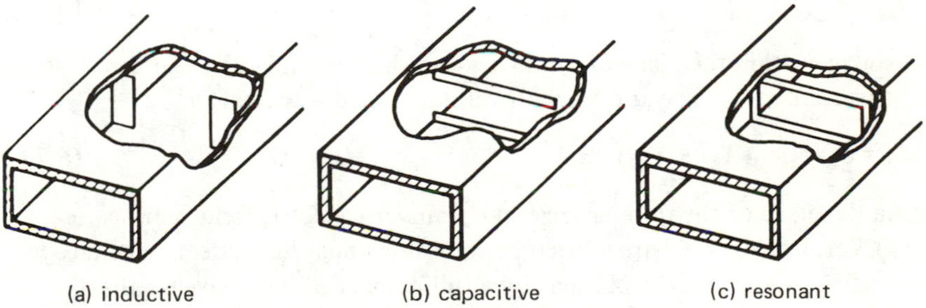

Fig. 6.18 Thin metal obstacles in rectangular waveguide.

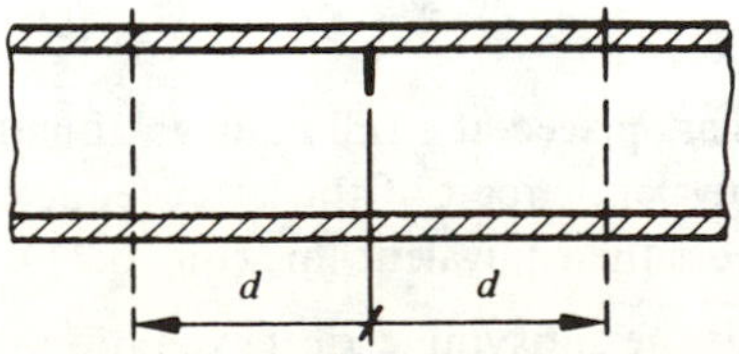

Fig. 6.19 Waveguide section containing a thin obstacle. Choice of the reference planes.

structure defined in this manner is isotropic (it does not contain any anisotropic material), symmetrical, and lossless (waveguide and obstacle in pec). Its impedance matrix thus has the form

$$(\underline{Z}) = \begin{pmatrix} \mathrm{j}X_{11} & \mathrm{j}X_{12} \\ \mathrm{j}X_{12} & \mathrm{j}X_{11} \end{pmatrix} \qquad \Omega \qquad (6.69)$$

Its equivalent T circuit is shown in Figure 6.20.

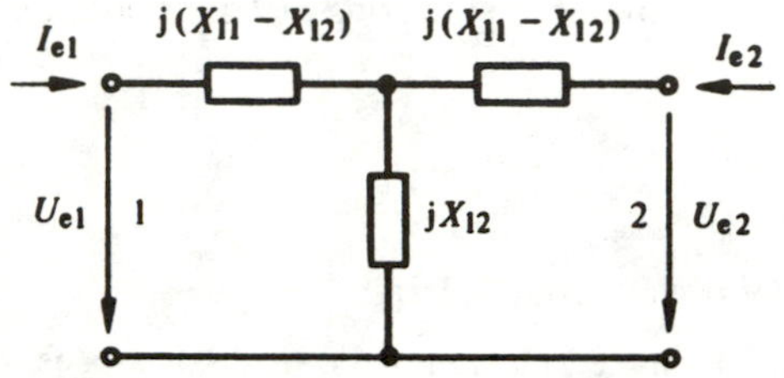

Fig. 6.20 Equivalent circuit of a lossless symmetrical obstacle in waveguide.

An ***antisymmetrical excitation*** is applied to the equivalent circuit; *i.e.*, at the two ports one applies

$$\underline{U}_{e1} = -\underline{U}_{e2} = \underline{U}_a \qquad \mathrm{V} \qquad (6.70)$$

$$\underline{I}_{e1} = -\underline{I}_{e2} = \underline{I}_a \qquad \mathrm{A} \qquad (6.71)$$

The voltage within the plane of symmetry vanishes, and no current crosses the shunt reactance X_{12}. The antisymmetrical impedance is then given by

$$\underline{Z}_a = \underline{U}_a/\underline{I}_a = \mathrm{j}(X_{11} - X_{12}) \qquad \Omega \qquad (6.72)$$

Within the plane of the thin obstacle, the transverse electric field vanishes (2.38). Zeros of the field are located periodically along the guide, at all transverse planes away from the obstacle by a full number of half wavelengths. Placing the reference planes at $d = n\lambda_g/2$, one finds $\underline{U}_a = 0$, which means that in (6.72) $\underline{Z}_a = 0$, and therefore,

$$X_{11} = X_{12} = X \qquad \Omega \qquad (6.73)$$

When the reference planes are placed at a full number of half-guide wavelengths from the obstacle, the equivalent circuit of the latter reduces to a single reactance shunt connected across the equivalent line (Fig. 6.21).

It is interesting to note that the antisymmetrical excitation is the single particular situation for which the dominant mode alone satisfies the boundary conditions, for any shape of the thin obstacle.

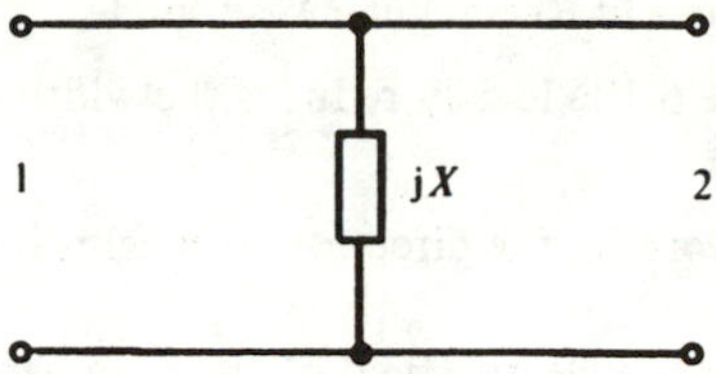

Fig. 6.21 Equivalent circuit of a thin lossless symmetrical iris in waveguide, for a particular choice of reference planes.

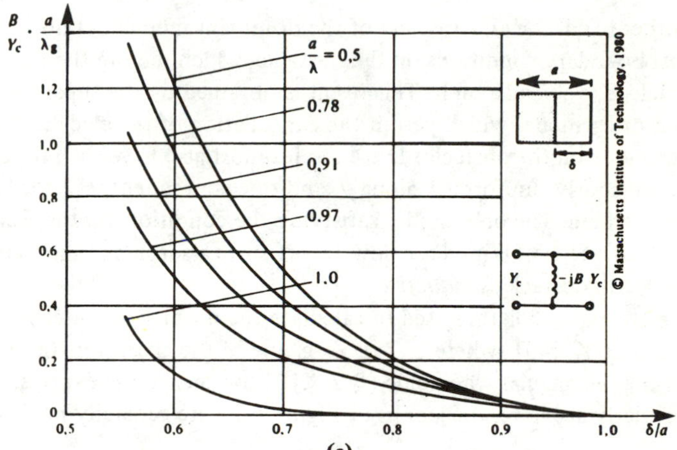

(a)

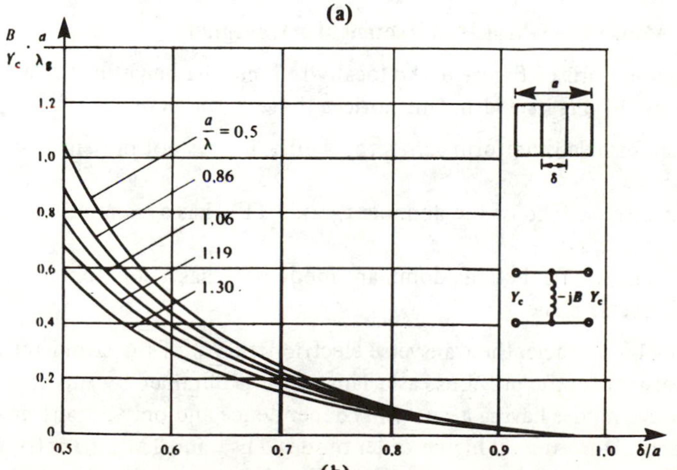

(b)

Fig. 6.22 Susceptance of an inductive iris in rectangular waveguide: (a) asymmetrical, (b) symmetrical.

6.3.16 Inductive Obstacle in Rectangular Waveguide

The vertical iris of Figure 6.18a locally reduces the width of the waveguide. One notices that

1. the obstacle is uniform in the direction of height, it does not present any variation along y;
2. the same is true for the electric field of the dominant mode;
3. the electric field of the dominant mode possesses only an $\underline{E}_y$ component. Only this component satisfies the uniformity along y and the boundary conditions at $y = 0$ and $y = b$ at the same time.

On the other hand, the electric field of the dominant mode does not satisfy the additional boundary conditions on the obstacle, which require the transverse electric field $\underline{E}_y$ to vanish on it. This result is obtained by the superposition of the higher order modes which permit the cancellation of the electric field of the dominant mode on the obstacle. These modes must also have the properties outlined previously: uniformity along y, and one component, $\underline{E}_y$, for the transverse electric field. The only modes satisfying the condition are the TE_{m0} modes (2.120). Since TE modes below cutoff store a surplus of magnetic energy (§ 2.2.30), this obstacle is *inductive.*
Approximate expressions are used to calculate the normalized susceptance of this obstacle B/Y_c [49], where $Y_c = 1/Z_c$ is one of the characteristic admittances of the rectangular waveguide (§ 2.3.16). Because these expressions are not particularly accurate, the graphs of Figure 6.22 are generally preferred.

6.3.17 Capacitive Obstacle in Rectangular Waveguide

The horizontal iris of Figure 6.18b locally reduces the height of the waveguide, without modifying its width. One notices that for this geometry:

1. the obstacle is uniform across the width, it does not present any variation along x,
2. the electric field of the dominant mode TE_{10} has a transverse dependence in $\sin(\pi x/a)$,
3. the electric field of the dominant mode only has a component $\underline{E}_y$ in the transverse plane.

On the metal obstacle, the transverse electric field $\underline{E}_y$ of the dominant mode must vanish, as in the previous case. This result is obtained by superposition of higher order modes having a $\sin(\pi x/a)$ dependence and only a transverse $\underline{E}_y$ component. However, all higher order modes TE_{1n} and TM_{1n} $(n \neq 0)$, which satisfy the first condition have non-zero $\underline{E}_x$ components. The second condition is satisfied by combining the TE_{1n} and the TM_{1n} modes so that their respective $\underline{E}_x$ components cancel out (same amplitudes, but opposite directions). Expressions (2.111) and (2.120) respectively give the factors π/a and $n\pi/b$ for the two

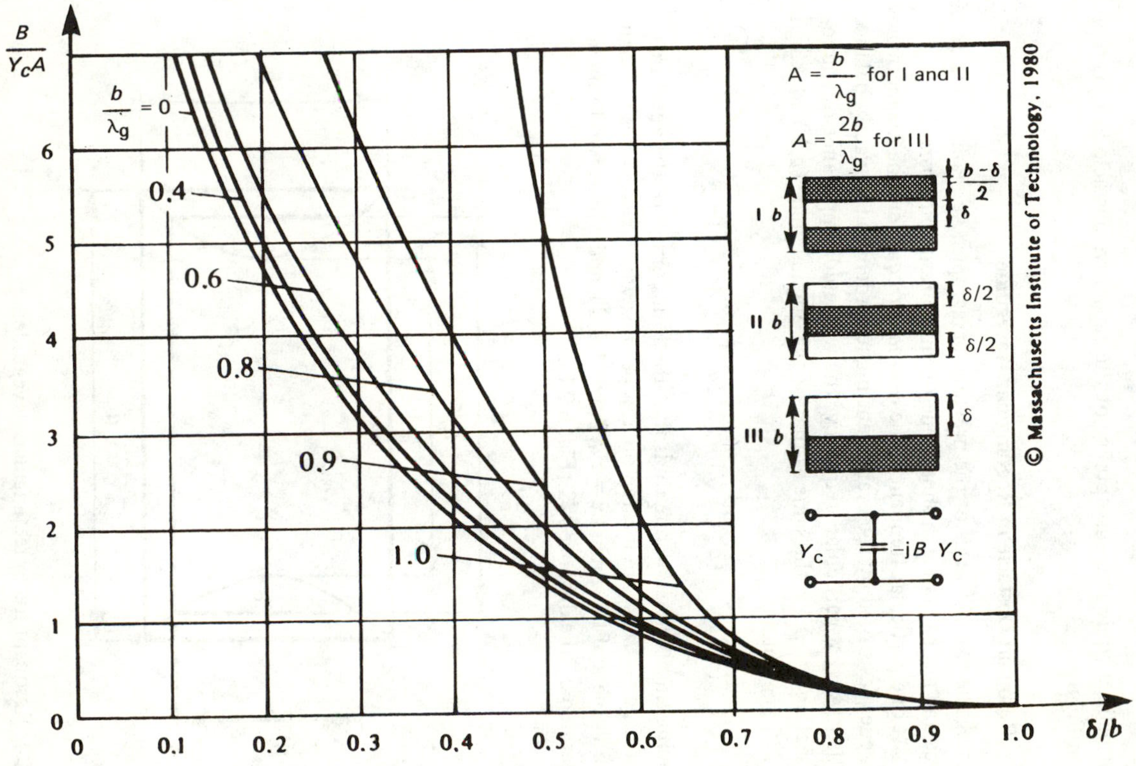

Fig. 6.23 Susceptance of a capacitive iris in rectangular waveguide.

modes. Their amplitudes must therefore be in a ratio na/b. The TM modes therefore predominate and the obstacle is *capacitive,* because these modes store an excedent of electrical energy.

The normalized equivalent susceptance of the capacitive obstacle B/Y_c is represented in Figure 6.23.

6.3.18 Resonant Obstacle in Rectangular Waveguide

Combining the two obstacles of the previous paragraphs, *i.e.,* connecting an inductance and a capacitance in shunt, one obtains a *resonant obstacle* (Fig. 6.18c). At one particular frequency, the ***resonant frequency***, the magnetic and electric energies stored by the evanescent higher order modes exactly balance each other. At that particular frequency, the waveguide impedance of the reduced section (a', b') is the same as that of the original guide (a, b). Making use of (2.131) one obtains

$$(a/b)\sqrt{1 - (\lambda/2a)^2} = (a'/b')\sqrt{1 - (\lambda/2a')^2} \qquad 1 \tag{6.74}$$

Developing this expression, one finds that the four corners of the aperture $(x = \pm a'/2, y = \pm b'/2)$ are located on a hyperbola which also passes through the four corners of the guide $(x = \pm a/2, y = \pm b/2)$, and whose two branches are separated by the distance $\lambda/2$ (Fig. 6.24).

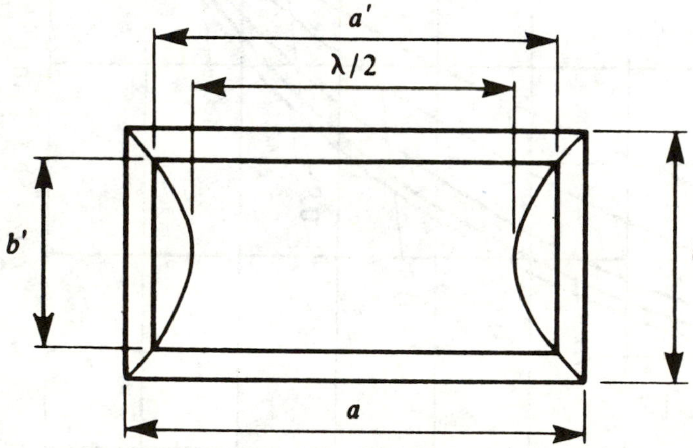

Fig. 6.24 Resonant iris in rectangular waveguide.

6.3.19 Application: Matching

Thin metal irises, whose equivalent circuit is a shunt-connected susceptance across the waveguide, are utilized to match terminations and devices, making use of the principle described in Section 7.2.12.

6.3.20 Tuning Screw

A screw inserted in one of the wide walls of the waveguide exhibits a series-connected *L-C* impedance shunting the line. For a small insertion of the screw, the capacitance predominates. The susceptance becomes infinite (short circuit) at resonance, obtained for an insertion of about $\lambda/4$ for the screw. For a larger insertion, the screw becomes inductive.

6.3.21 Slide-Screw Tuner

A device commonly used for matching in waveguides is made with a screw for which both the penetration depth and the position along the line may be adjusted, permitting matching by means of the principle described in paragraph 7.2.12 (Fig. 6.25).

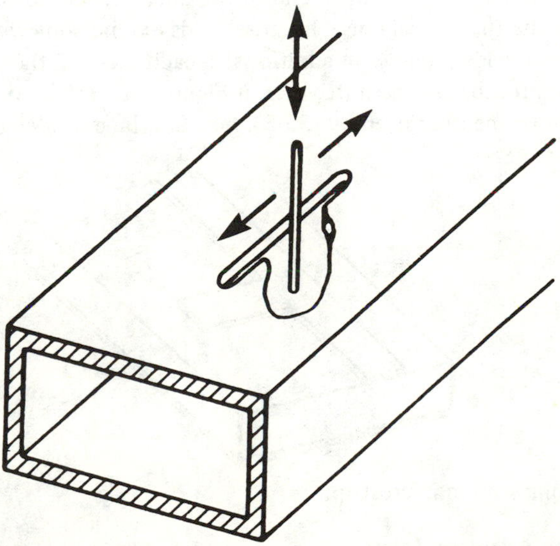

Fig. 6.25 Slide-screw tuner.

6.3.22 Reactive Obstacles in Circular Waveguides

An inductive iris is made of a circular opening centered on the waveguide axis. A circular disk centered on the axis is a capacitive obstacle (Fig. 6.26).

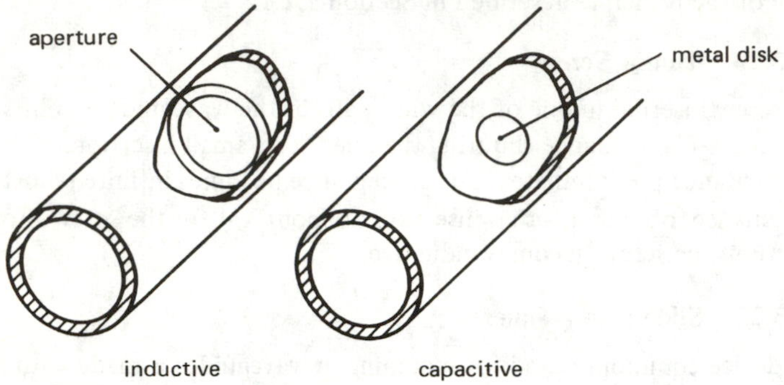

Fig. 6.26 Reactive irises in circular waveguide.

6.3.23 Shunt Susceptances in Microstrip: Tuning Stub

In microstrip, a shunt susceptance is realized with a section of line, called *stub,* as sketched in Figure 6.27. The open end of the microstrip is however ***not an open circuit*** in the theoretical sense, because fields extend somewhat beyond its extremity. These fields produce an additional capacitance C_0, the value of which is given for several substrate permittivities in Figure 3.15 [50]. An equivalent lengthening Δd of the line, given by (3.65), may also be considered.

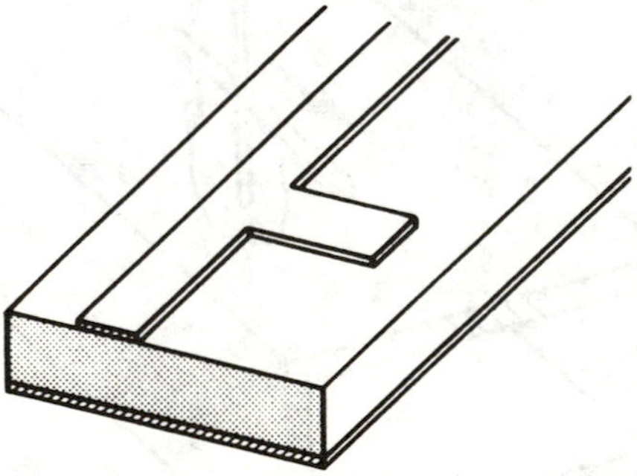

Fig. 6.27 Tuning stub in microstrip.

6.3.24 Twists, Adapters, Joints

Among the matched, reciprocal, and lossless two-ports should also be mentioned the many structures utilized to connect lines of different types and cross sections (for instance a rectangular waveguide to a coaxial line, or a microstrip to a circular waveguide, etc.), the sections used to modify the orientation (twists, bends) and the rotating joints, which connect a fixed generator to a rotating antenna (radar) [118].

6.3.25 Cavities and Filters

Two-ports with frequency-dependent properties and filters are realized with one or several resonant cavities connected for transmission (§ 3.5.15) or reaction (Fig. 5.2) [123].

6.4 THREE-PORT DEVICES

6.4.1 Scattering Matrix and Signal Flow Graph

The scattering matrix of a *three-port* device contains 9 terms

$$(\underline{s}) = \begin{pmatrix} \underline{s}_{11} & \underline{s}_{12} & \underline{s}_{13} \\ \underline{s}_{21} & \underline{s}_{22} & \underline{s}_{23} \\ \underline{s}_{31} & \underline{s}_{32} & \underline{s}_{33} \end{pmatrix} \quad 1 \qquad (6.75)$$

The corresponding signal flow graph is shown in Figure 6.28.

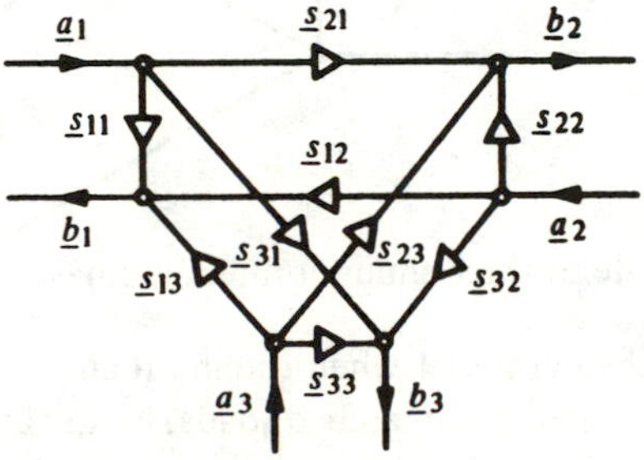

Fig. 6.28 Flow graph of a three-port.

6.4.2 Lossless Reciprocal Three Ports are Mismatched

Lossless reciprocal three-ports cannot be matched at the same time at their three ports, as can be demonstrated easily "ad absurdum". Letting all the $\underline{s}_{ii} = 0$ (matching, (6.26)), and $\underline{s}_{ij} = \underline{s}_{ji}$ (reciprocity, (6.18)), the relations for energy conservation (6.25) provide six independent expressions for a three-port:

$$|\underline{s}_{12}|^2 + |\underline{s}_{13}|^2 = 1 \quad 1 \qquad (6.76)$$

$$|\underline{s}_{12}|^2 + |\underline{s}_{23}|^2 = 1 \quad 1 \qquad (6.77)$$

$$|\underline{s}_{13}|^2 + |\underline{s}_{23}|^2 = 1 \quad 1 \qquad (6.78)$$

$$\underline{s}_{13}^{*}\underline{s}_{23} = 0 \quad 1 \qquad (6.79)$$

$$\underline{s}_{12}^{*}\underline{s}_{23} = 0 \quad 1 \qquad (6.80)$$

$$\underline{s}_{12}^{*}\underline{s}_{13} = 0 \quad 1 \qquad (6.81)$$

Let us assume that one term is different from zero, for instance $\underline{s}_{13} \neq 0$. From (6.79), one must then have $\underline{s}_{23} = 0$ and from (6.81) $\underline{s}_{12} = 0$, which are clearly inconsistent with (6.77). The same demonstration may be repeated, letting in turn $\underline{s}_{12}$ or $\underline{s}_{23}$ be non-zero. In all cases, the same inconsistency is obtained (Q.E.D.).

6.4.3 Alternate Demonstration

The same result is obtained when considering the equivalent circuit of Figure 6.29, which represents the junction of three shunt connected lines. The susceptance B accounts for the energy stored within the junction.

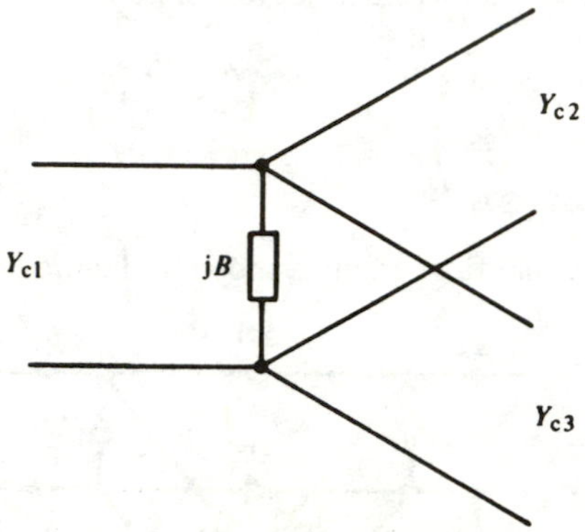

Fig. 6.29 Junction made of three shunt-connected lines.

The three-port is matched at port 1 when the input admittance, observed with the ports 2 and 3 terminated into matched loads, is equal to the admittance of line 1, so that

$$Y_{c1} = Y_{c2} + Y_{c3} + jB \qquad \text{S} \qquad (6.82)$$

For matching to be feasible, one should at least have $B = 0$. Proceeding in the same manner at ports 2 and 3

$$Y_{c2} = Y_{c3} + Y_{c1} \qquad \text{S} \qquad (6.83)$$

$$Y_{c3} = Y_{c1} + Y_{c2} \qquad \text{S} \qquad (6.84)$$

Summing the three expressions, one obtains

$$Y_{c1} + Y_{c2} + Y_{c3} = 2(Y_{c1} + Y_{c2} + Y_{c3}) \qquad \text{S} \qquad (6.85)$$

It is quite impossible to satisfy this equation with non-zero terms, and this confirms the conclusion of Section 6.4.2.

6.4.4 Circulator

A ***nonreciprocal*** three-port, on the other hand, may be adapted at its three ports. The energy conservation expressions (6.25) yield in this case

$$|\underline{s}_{21}|^2 + |\underline{s}_{31}|^2 = 1 \qquad 1 \qquad (6.86)$$

$$|\underline{s}_{12}|^2 + |\underline{s}_{32}|^2 = 1 \qquad 1 \qquad (6.87)$$

$$|\underline{s}_{13}|^2 + |\underline{s}_{23}|^2 = 1 \qquad 1 \qquad (6.88)$$

$$\underline{s}_{12}^{*}\underline{s}_{13} = 0 \qquad 1 \qquad (6.89)$$

$$\underline{s}_{21}^{*}\underline{s}_{23} = 0 \qquad 1 \qquad (6.90)$$

$$\underline{s}_{31}^{*}\underline{s}_{32} = 0 \qquad 1 \qquad (6.91)$$

Assuming that $\underline{s}_{13} \neq 0$, one obtains the following sequence

$$\underline{s}_{13} \neq 0 \underset{(6.89)}{\Rightarrow} \underline{s}_{12} = 0 \underset{(6.87)}{\Rightarrow} |\underline{s}_{32}| = 1 \underset{(6.91)}{\Rightarrow} \underline{s}_{31} = 0 \underset{(6.86)}{\Rightarrow} |\underline{s}_{21}| = 1 \underset{(6.90)}{\Rightarrow} \underline{s}_{23} = 0 \underset{(6.88)}{\Rightarrow} |\underline{s}_{13}| = 1 \qquad 1 \qquad (6.92)$$

The reference planes at the three ports are then placed in such a way that the three non-zero terms are positive real.

$$(s) = \begin{pmatrix} 0 & 0 & 1 \\ 1 & 0 & 0 \\ 0 & 1 & 0 \end{pmatrix} \qquad 1 \qquad (6.93)$$

The device obtained is an ideal *circulator* (Fig. 6.30). The principle of operation of the circulator is presented in Section 6.7.

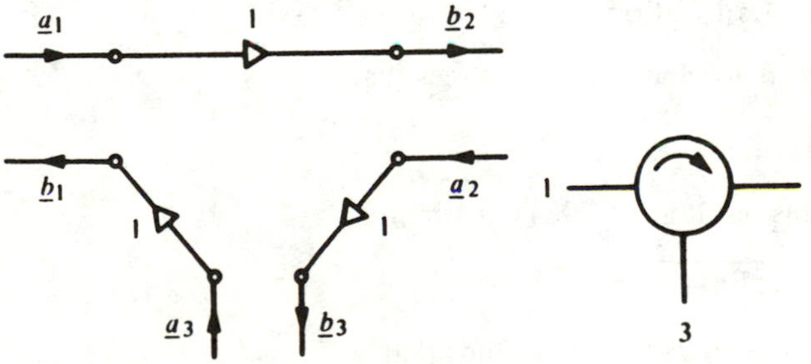

Fig. 6.30 Flow graph and symbol of a circulator.

6.4.5 Three-Port Matched at Two of its Ports

It is possible to match at the same time two of the three ports of a reciprocal lossless three-port. However, the solution obtained is decoupled: it consists of one matched two-port and one total reflection, or one-port (Fig. 6.31). The two devices are electrically separated from each other.

$$|\underline{s}_{12}| = |\underline{s}_{21}| = |\underline{s}_{33}| = 1 \qquad 1 \qquad (6.94)$$

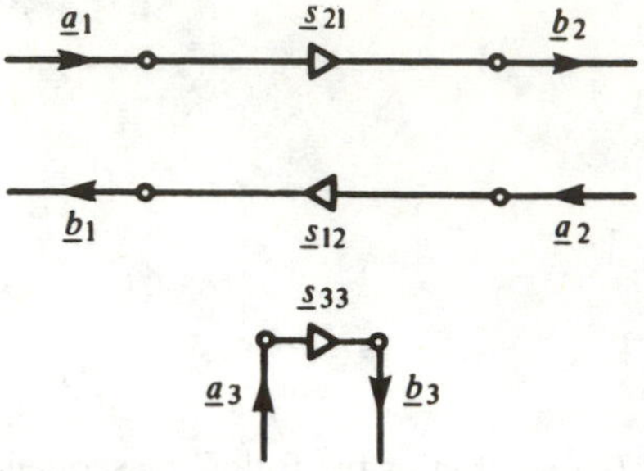

Fig. 6.31 Lossless three-port matched at two ports: it is formed of two physically independent devices.

6.4.6 Three-Port Almost Matched at Two of Its Ports

When the condition of Section 6.4.5 is slightly relaxed, so that a small mismatch is tolerated at ports 1 and 2, one obtains, choosing the reference planes which provide positive real terms on the diagonal,

$$s_{11} = s_{22} = \epsilon \ll 1 \qquad 1 \qquad (6.95)$$

The relation for energy conservation then yields

$$\epsilon^2 + |\underline{s}_{12}|^2 + |\underline{s}_{13}|^2 = 1 \qquad 1 \qquad (6.96)$$

$$\epsilon^2 + |\underline{s}_{12}|^2 + |\underline{s}_{23}|^2 = 1 \qquad 1 \qquad (6.97)$$

$$\epsilon(\underline{s}_{12} + \underline{s}_{12}^*) + \underline{s}_{13}\underline{s}_{23}^* = 0 \qquad 1 \qquad (6.98)$$

$$\epsilon\underline{s}_{13} + \underline{s}_{12}^*\underline{s}_{23} + \underline{s}_{13}^*\underline{s}_{33} = 0 \qquad 1 \qquad (6.99)$$

Comparing (6.96) and (6.97), one sees that

$$|\underline{s}_{13}| = |\underline{s}_{23}| \qquad 1 \qquad (6.100)$$

Utilizing this result and (6.98) one obtains

$$|\underline{s}_{13}| = \sqrt{2\epsilon \operatorname{Re}(\underline{s}_{12})} \qquad 1 \qquad (6.101)$$

With the aid of (6.96) one also finds that

$$|\underline{s}_{12}| = \sqrt{1 - 2\epsilon \operatorname{Re}(\underline{s}_{12}) - \epsilon^2} \cong 1 \qquad 1 \qquad (6.102)$$

Furthermore, (6.97) specifies the amplitude of $\underline{s}_{33}$

$$|\underline{s}_{33}| \cong |\underline{s}_{12}| \qquad 1 \qquad (6.103)$$

This three-port may be used to sample a small part of a signal, and does not produce large mismatches at ports 1 and 2. As for the third port, on the other hand, it is ***badly mismatched*** (6.103). An example of application is the slotted line (Sec. 7.2 and § 2.3.12).

6.4.7 T Junctions, Plane of the Junction

A junction is formed by joining two waveguides forming a straight angle, in the shape of a T. The two axes define the *plane of the junction.*

6.4.8 E-Plane T, Series T

In the *series T*, the small sides of the two waveguides are parallel to the plane of the junction (Fig. 6.32).

Since the electric field lines of the dominant TE_{10} mode in the three ports are parallel to this plane, the device is also known as an *E-plane T.* The junction possesses one plane of symmetry. Setting the reference planes symmetrically in ports 1 and 2, and considering the distribution of the electric field (Fig. 6.32), one finds that

$$\underline{S}_{11} = \underline{S}_{22} \qquad 1 \tag{6.104}$$

$$\underline{S}_{13} = -\underline{S}_{23} \qquad 1 \tag{6.105}$$

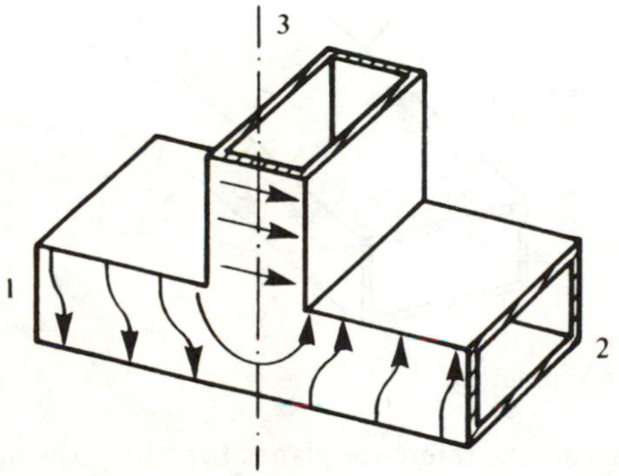

Fig. 6.32 E-plane T or series T.

6.4.9 H-Plane T, Shunt T

In the *shunt T*, the wide sides of the waveguides are parallel to the plane of the junction (Fig. 6.33).

Since the magnetic field lines of the dominant TE_{10} mode in the three ports are parallel to this plane, the device is called *H-plane T.* The symmetry of the junction yields, for a symmetrical selection of the reference planes

$$\underline{S}_{11} = \underline{S}_{22} \qquad 1 \tag{6.106}$$

$$\underline{S}_{13} = \underline{S}_{32} \qquad 1 \tag{6.107}$$

6.4.10 Shunt Y Junction

The *shunt Y junction,* formed by joining three waveguides, presents a threefold symmetry (Fig. 6.34).

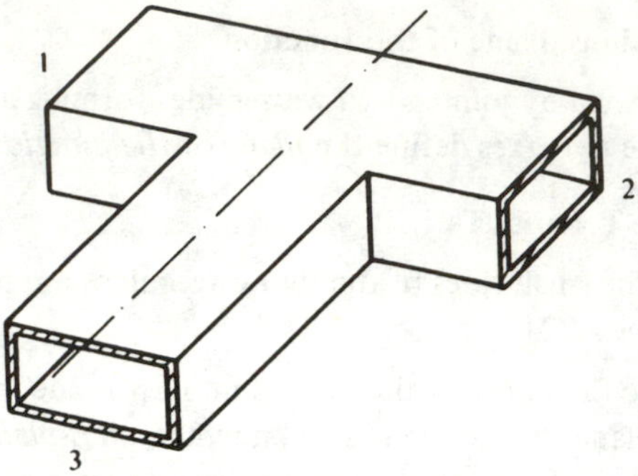

Fig. 6.33 H-plane T or shunt T.

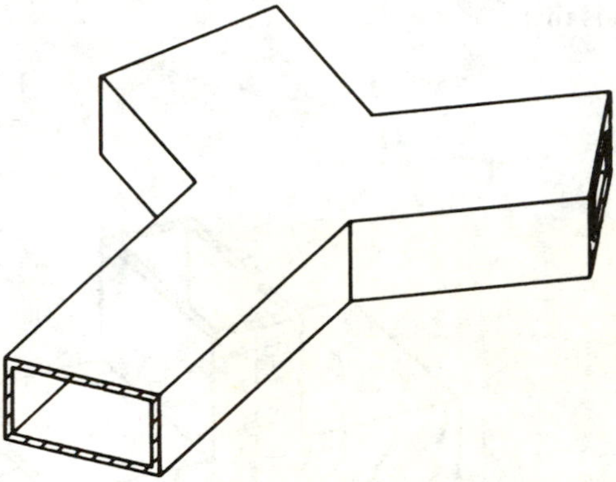

Fig. 6.34 Shunt Y junction.

In this case, choosing the reference planes for which the diagonal terms are real positive numbers

$$\underline{s}_{11} = \underline{s}_{22} = \underline{s}_{33} = A \qquad 1 \tag{6.108}$$

$$\underline{s}_{12} = \underline{s}_{13} = \underline{s}_{23} = B + \mathrm{j}C \qquad 1 \tag{6.109}$$

The energy conservation conditions then become

$$A^2 + 2B^2 + 2C^2 = 1 \qquad 1 \tag{6.110}$$

$$2AB + B^2 + C^2 = 0 \qquad 1 \tag{6.111}$$

These two relations define a closed curve in the coordinate system ABC (Fig. 6.35).

The largest possible reflection corresponds to three purely reactive terminations, completely uncoupled: $A = 1, B = C = 0$. The lowest possible reflection is obtained for $C = 0, A = 1/3$ and $B = -2/3$. The standing wave ratio is then

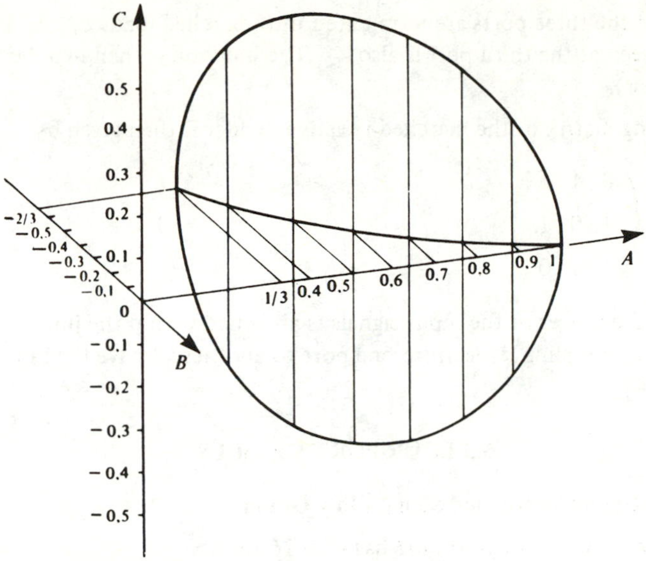

Fig. 6.35 Locus of solutions for a Shunt Y junction.

$$\text{VSWR} = \frac{1 + 1/3}{1 - 1/3} = 2 \tag{6.112}$$

6.4.11 Matched Resistive Divider

For certain particular applications, it is mandatory for a symmetrical threeport to be ***matched at each one of its three ports.*** One must then accept the dissipation of a part of the signal within the junction. A *matched resistive divider* is realized by placing resistors in the three ports (Fig. 6.36).

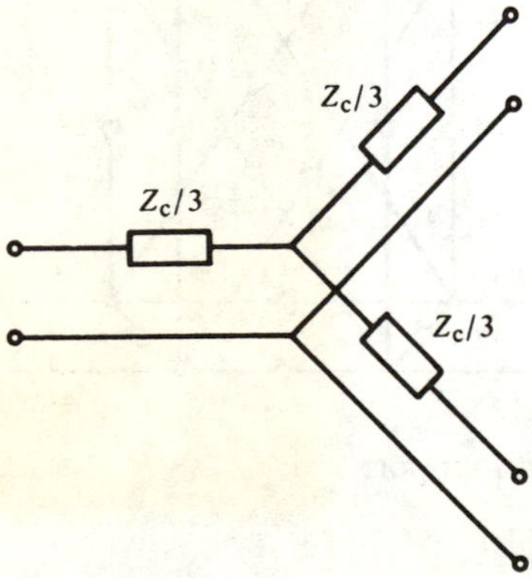

Fig. 6.36 Matched resistive divider.

When two of the three ports are terminated into matched loads Z_c, the input impedance seen at the third port is also Z_c. The junction is then matched (reflectionless).

The scattering matrix of the matched resistive divider is then given by

$$(\underline{s}) = \frac{1}{2}\begin{pmatrix} 0 & 1 & 1 \\ 1 & 0 & 1 \\ 1 & 1 & 0 \end{pmatrix} \qquad 1 \qquad (6.113)$$

One-half of the power of the input signals is absorbed within the junction. The attenuation or coupling levels from one port to another all have the same value of 6 dB [124].

6.5 FOUR-PORT DEVICES

6.5.1 Scattering Matrix and Signal Flow Graph

The scattering matrix of a *four-port* has $4^2 = 16$ terms

$$(\underline{s}) = \begin{pmatrix} \underline{s}_{11} & \underline{s}_{12} & \underline{s}_{13} & \underline{s}_{14} \\ \underline{s}_{21} & \underline{s}_{22} & \underline{s}_{23} & \underline{s}_{24} \\ \underline{s}_{31} & \underline{s}_{32} & \underline{s}_{33} & \underline{s}_{34} \\ \underline{s}_{41} & \underline{s}_{42} & \underline{s}_{43} & \underline{s}_{44} \end{pmatrix} \qquad 1 \qquad (6.114)$$

The signal flow graph of a four port is represented in Figure 6.37.

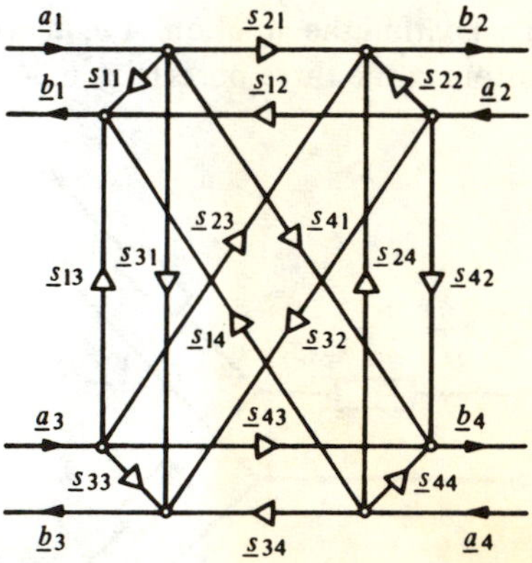

Fig. 6.37 Flow graph of a four-port.

■ 6.5.2 Directional Coupler

A reciprocal ($\underline{s}_{ij} = \underline{s}_{ji}$), matched ($\underline{s}_{ii} = 0$) and lossless four-port will be considered.

Among the 16 relations provided by the condition for energy conservation (6.25) let us consider the two equations

$$\underline{s}_{13}^{*}\underline{s}_{14} + \underline{s}_{23}^{*}\underline{s}_{24} = 0 \quad \text{(line 3} \times \text{column 4)} \qquad 1 \qquad (6.115)$$

$$\underline{s}_{13}^{*}\underline{s}_{23} + \underline{s}_{14}^{*}\underline{s}_{24} = 0 \quad \text{(line 1} \times \text{column 2)} \qquad 1 \qquad (6.116)$$

Multiplying them respectively by $\underline{s}_{14}^{*}$ and $\underline{s}_{23}^{*}$ and subtracting the resulting expressions, one gets

$$\underline{s}_{13}^{*}(|\underline{s}_{14}|^2 - |\underline{s}_{23}|^2) = 0 \qquad 1 \qquad (6.117)$$

There are two ways to satisfy this equation. First, one may let $\underline{s}_{13}$ equal zero. For two terms $\underline{s}_{14}$ and $\underline{s}_{23}$ different from zero, $\underline{s}_{24}$ must vanish in (6.115) and the device is a *directional coupler:*

$$(\underline{s}) = \begin{pmatrix} 0 & \underline{s}_{12} & 0 & \underline{s}_{14} \\ \underline{s}_{12} & 0 & \underline{s}_{23} & 0 \\ 0 & \underline{s}_{23} & 0 & \underline{s}_{34} \\ \underline{s}_{14} & 0 & \underline{s}_{34} & 0 \end{pmatrix} \qquad 1 \qquad (6.118)$$

A second possible way is to set

$$|\underline{s}_{14}| = |\underline{s}_{23}| \qquad 1 \qquad (6.119)$$

Reference planes may then be selected to have two purely imaginary terms

$$\underline{s}_{14} = \underline{s}_{23} = \mathrm{j}\beta \qquad (6.120)$$

One then considers two other expressions for energy conservation

$$|\underline{s}_{12}|^2 + |\underline{s}_{13}|^2 + |\underline{s}_{14}|^2 = 1 \quad \text{(line 1} \times \text{column 1)} \qquad 1 \qquad (6.121)$$

$$|\underline{s}_{13}|^2 + |\underline{s}_{23}|^2 + |\underline{s}_{34}|^2 = 1 \quad \text{(line 3} \times \text{column 3)} \qquad 1 \qquad (6.122)$$

With (6.119) and selecting reference planes to provide real terms, one gets

$$\underline{s}_{12} = \underline{s}_{34} = \alpha \qquad 1 \qquad (6.123)$$

Another conservation expression (line 1 $\times$ column 4) yields

$$\underline{s}_{12}^{*}\underline{s}_{24} + \underline{s}_{13}^{*}\underline{s}_{34} = 0 = \alpha(\underline{s}_{24} + \underline{s}_{13}^{*}) \qquad 1 \qquad (6.124)$$

On the other hand, by making use of (6.120), (6.115) becomes

$$\beta(\underline{s}_{13}^{*} - \underline{s}_{24}) = 0 \qquad 1 \qquad (6.125)$$

The system of equations (6.124), (6.125) admits two solutions. The first one,

$$\underline{s}_{13} = \underline{s}_{24} = 0 \qquad 1 \qquad (6.126)$$

corresponds to the directional coupler previously obtained (6.118). The second one

$$\alpha = \beta = 0 \qquad 1 \qquad (6.127)$$

is a decoupled four-port, made up of two two-ports without any connection between them.

One concludes that a lossless matched reciprocal four-port must always be a directional coupler, described by the scattering matrix of (6.118).

6.5.3 Comment

For a lossless reciprocal four-port to be matched at all its ports, every input must be coupled to only two other ports, the last one being isolated.

6.5.4 Properties of the Directional Coupler

The energy conservation condition (6.25) applied to the scattering matrix (6.118) yields 6 independent relations

$$|\underline{s}_{12}|^2 + |\underline{s}_{14}|^2 = 1 \qquad 1 \qquad (6.128)$$

$$|\underline{s}_{12}|^2 + |\underline{s}_{23}|^2 = 1 \qquad 1 \qquad (6.129)$$

$$|\underline{s}_{23}|^2 + |\underline{s}_{34}|^2 = 1 \qquad 1 \qquad (6.130)$$

$$|\underline{s}_{14}|^2 + |\underline{s}_{34}|^2 = 1 \qquad 1 \qquad (6.131)$$

$$\underline{s}_{12}\underline{s}_{23}^* + \underline{s}_{14}\underline{s}_{34}^* = 0 \qquad 1 \qquad (6.132)$$

$$\underline{s}_{12}\underline{s}_{14}^* + \underline{s}_{23}\underline{s}_{34}^* = 0 \qquad 1 \qquad (6.133)$$

It is readily apparent that

$$|\underline{s}_{12}| = |\underline{s}_{34}| = \alpha \qquad 1 \qquad (6.134)$$

$$|\underline{s}_{14}| = |\underline{s}_{23}| = \beta \qquad 1 \qquad (6.135)$$

with

$$\alpha^2 + \beta^2 = 1 \qquad 1 \qquad (6.136)$$

We then let

$$\underline{s}_{12} = \alpha \exp(j\varphi) \qquad 1 \qquad (6.137)$$

$$\underline{s}_{34} = \alpha \exp(j\eta) \qquad 1 \qquad (6.138)$$

$$\underline{s}_{14} = \beta \exp(j\psi) \qquad 1 \qquad (6.139)$$

$$\underline{s}_{23} = \beta \exp(j\theta) \qquad 1 \qquad (6.140)$$

Relations (6.132) and (6.133) require that arguments of the scattering matrix terms meet the following condition

$$(\varphi + \eta) - (\theta + \psi) = \pi \pm 2n\pi \qquad \text{rad} \qquad (6.141)$$

The amplitudes of the scattering matrix elements must satisfy (6.136), their arguments (6.141).

6.5.5 Selection of the Reference Planes

The distance between the reference planes at ports 1 and 2 may be set so that $\varphi = 0$, *i.e.*, $\underline{s}_{21}$ has a positive real value. Those at ports 3 and 4 may be set for $\eta = 0$. The scattering matrix then takes the form

$$(\underline{s}) = \begin{pmatrix} 0 & \alpha & 0 & \beta\exp(j\psi) \\ \alpha & 0 & \beta\exp(j\theta) & 0 \\ 0 & \beta\exp(j\theta) & 0 & \alpha \\ \beta\exp(j\psi) & 0 & \alpha & 0 \end{pmatrix} \qquad 1 \qquad (6.142)$$

with (6.136) and

$$\psi + \theta = \pi \pm 2n\pi \qquad \text{rad} \qquad (6.143)$$

Two of the degrees of freedom available in the choice of the reference planes have been used up. One may still specify ψ. Two particular situations present symmetry conditions.

6.5.6 Symmetrical Coupler

The terms having the amplitude β are taken with equal arguments

$$\psi = \theta = \pi/2 \qquad \text{rad} \qquad (6.144)$$

The scattering matrix then becomes

$$(\underline{s}) = \begin{pmatrix} 0 & \alpha & 0 & j\beta \\ \alpha & 0 & j\beta & 0 \\ 0 & j\beta & 0 & \alpha \\ j\beta & 0 & \alpha & 0 \end{pmatrix} \qquad 1 \qquad (6.145)$$

The corresponding signal flow graph is given in Figure 6.38.

6.5.7 Antisymmetrical Coupler

In this case, one lets

$$\psi = 0 \qquad \theta = \pi \qquad \text{rad} \qquad (6.146)$$

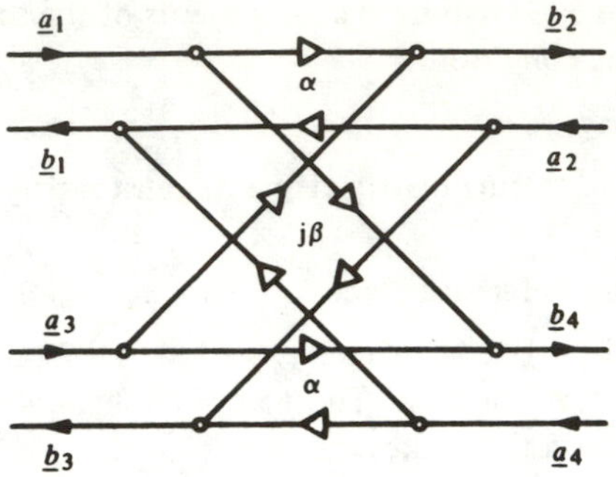

Fig. 6.38 Signal flow graph of a symmetrical coupler.

The scattering matrix then has the form

$$(\underline{s}) = \begin{pmatrix} 0 & \alpha & 0 & \beta \\ \alpha & 0 & -\beta & 0 \\ 0 & -\beta & 0 & \alpha \\ \beta & 0 & \alpha & 0 \end{pmatrix} \qquad 1 \tag{6.147}$$

The corresponding signal flow graph is sketched in Figure 6.39.

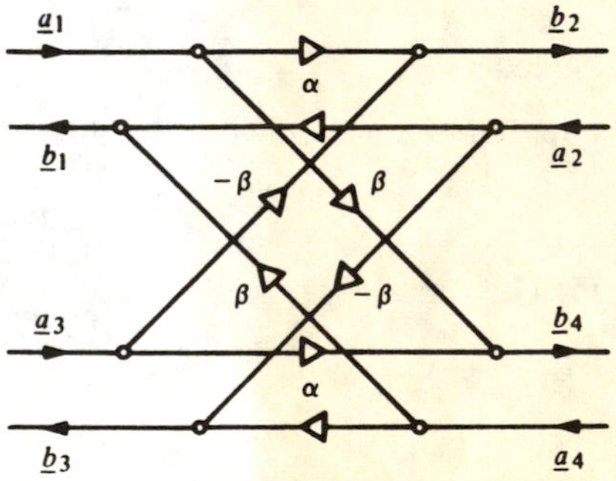

Fig. 6.39 Signal flow graph of an asymmetrical coupler.

6.5.8 Remark

The notion of a symmetrical coupler and an antisymmetrical coupler do not depend only upon the structure of the device itself, but also on the choice of the reference planes. Thus, it is always possible to transform a symmetrical coupler into an antisymmetrical one (and vice versa) by adding line sections modifying the location of the reference planes (§ 6.1.27) at the ports.

6.5.9 Convention

The numbering of the ports is done in such a way that

$$\alpha \geqslant \beta \qquad 1 \qquad (6.148)$$

6.5.10 Definition: Attenuation Level

In the same manner as for an attenuator (§ 6.3.3), the *attenuation level* is defined (in dB) from the ratio of output to input signals corresponding to the largest power transfer between two ports

$$LA = -\ 20 \log \alpha \qquad \text{dB} \qquad (6.149)$$

6.5.11 Definition: Coupling Level

Similarly, the *coupling level,* or more commonly the *coupling,* is defined from the ratio of signals from input to second coupled output

$$LC = -\ 20 \log \beta \qquad \text{dB} \qquad (6.150)$$

6.5.12 Design of Directional Couplers

The function of directional coupler represented by the scattering matrix (6.118) may be realized in practice by two different devices: couplers and junctions. The term ***coupler*** is generally associated with a device made of two transmission lines, most often of the same kind. The power transfer between the two lines takes place either at certain discrete locations (apertures, § 6.5.16) or in a distributed fashion all along the structure (§ 6.5.20). A ***junction*** is a point in space towards which several transmission lines converge (§ 6.5.24). These distinctions only relate to the structural design of the component. As for its operation, the two kinds of devices perform the same function.

6.5.13 Distinction: Practical Coupler

The performances of an ideal directional coupler (6.118) are only partially realizable in practice. On the one hand, any real device has some losses. The conditions for energy conservation (6.136) and (6.141) are therefore not entirely satisfied. In addition, it is not possible to perfectly match every port of the device, particularly over a broad frequency range. As a result, $\underline{s}_{ii} \neq 0$; in turn, a transfer of signal also takes place towards the fourth port, which was perfectly isolated in an ideal coupler. Thus, one has

$$|\underline{s}_{13}| \neq 0 \qquad 1 \qquad (6.151)$$

$$|\underline{s}_{24}| \neq 0 \qquad 1 \qquad (6.152)$$

6.5.14 Definition: Isolation

The transfer function for the residual signal between the two isolated ports of a coupler is called *isolation.* It is defined in a similar way as attenuation (§ 6.5.10) and coupling (§ 6.5.11).

$$\mathrm{L}I_{13} = -\ 20 \log |\underline{s}_{13}| \qquad \mathrm{dB} \qquad (6.153)$$

$$\mathrm{L}I_{24} = -\ 20 \log |\underline{s}_{24}| \qquad \mathrm{dB} \qquad (6.154)$$

There are two isolations, which may be different for a non-symmetrical coupler. For an ideal coupler, the isolation is infinite.

6.5.15 Definition: Directivity

The difference between isolation and coupling, which represents the amplitude ratio of signals coming out of the isolated port and out of the coupled port is called *directivity.*

$$\mathrm{L}D_{13} = \mathrm{L}I_{13} - \mathrm{L}C = -\ 20 \log |\underline{s}_{13}|/\beta \qquad \mathrm{dB} \qquad (6.155)$$

$$\mathrm{L}D_{24} = \mathrm{L}I_{24} - \mathrm{L}C = -\ 20 \log |\underline{s}_{24}|/\beta \qquad \mathrm{dB} \qquad (6.156)$$

The directivity determines the ***quality*** of a coupler. It defines the accuracy which may be obtained in reflectometry measurements (Sec. 7.3).

6.5.16 Localized Interaction Couplers in Waveguide

The transfer of energy between two waveguides may only take place at a discrete number of openings drilled into the common wall of the two waveguides (Fig. 6.40).

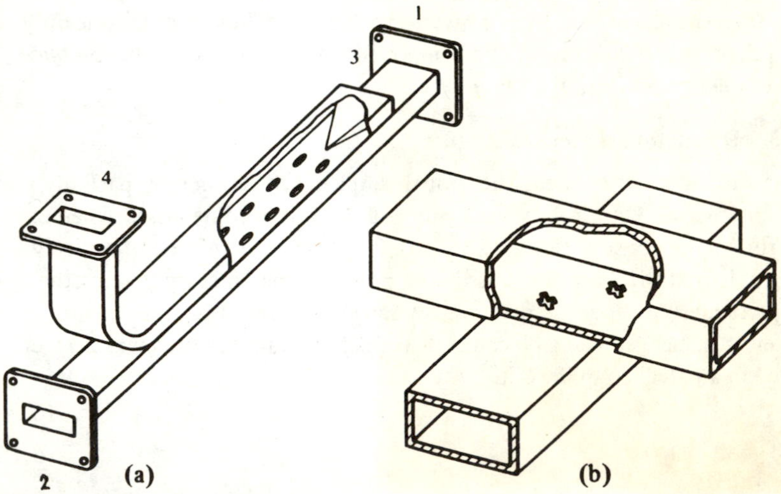

Fig. 6.40 Directional couplers in waveguide: (a) with array of openings, (b) cross-guide coupler.

The coupling is defined by the size and the positioning of the openings. A very large directivity may be obtained, up to 50-60 dB, by using a large number of holes, the structure then becoming rather long [125]. The manufacturing process requires a very good accuracy for this.

In practice, one of the four ports is generally terminated into a matched load, and the coupler becomes a ***lossy three-port***.

6.5.17 Localized Interaction Couplers in Microstrip: Branch Line Couplers

A localized interaction is obtained by inserting branch lines joining the two main microstrip lines (Fig. 6.41).

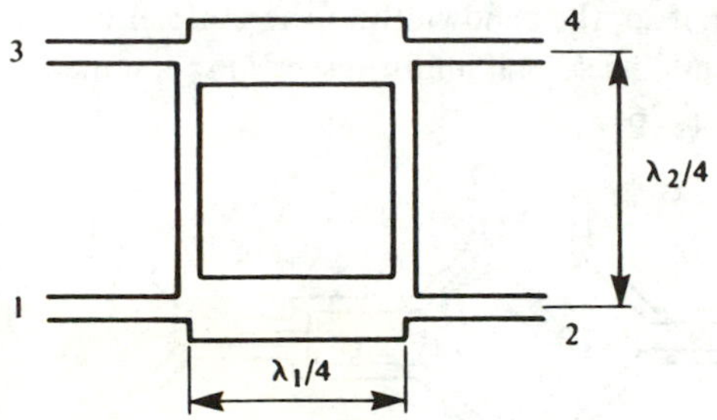

Fig. 6.41 Branch line coupler in microstrip.

Couplers of this type are designed considering the assembly of four three-port junctions joined by transmission lines. The coupling is adjusted by the relative widths of the different branches, which determine their characteristic impedance (§ 2.11.7). The operating frequency range is set by the lengths of the branches.

6.5.18 Particular Case: Hybrid Coupler

A particular choice of the line dimensions provides an equal division of the power to the two coupled ports, so that $\alpha = \beta$. The device is called a *hybrid coupler.* When the reference planes in the four ports (§ 6.5.5) are located symmetrically, the scattering matrix takes the form derived from (6.145) for an ideal coupler:

$$(\underset{\sim}{s}) = \frac{1}{\sqrt{2}} \begin{pmatrix} 0 & 1 & 0 & j \\ 1 & 0 & j & 0 \\ 0 & j & 0 & 1 \\ j & 0 & 1 & 0 \end{pmatrix} \qquad 1 \qquad (6.157)$$

Coupling and attenuation are equal here.

$$LA = LC = -\ 20 \log (1/\sqrt{2}) = 3.0103 \quad \text{dB} \qquad (6.158)$$

6.5.19 Definitions: Hybrid, 3 dB Coupler and Junction

The term *hybrid,* when referring to a coupler or junction, relates to an ***even power distribution*** between the two coupled ports. One also speaks of *3 dB couplers* or of *3 dB junctions.*

6.5.20 Distributed Interaction Couplers

Directional couplers made of open lines (stripline, § 2.6.3; microstrip, § 2.11.4) are realized by placing two lines close to each other. The electromagnetic fields of the first line induce a signal on the second one [68]. The interaction region may be uniform; if so, the bandwidth of the coupler is narrow. It may be widened by assembling several uniform sections, and making use of non-uniform lines (Fig. 6.42) [126].

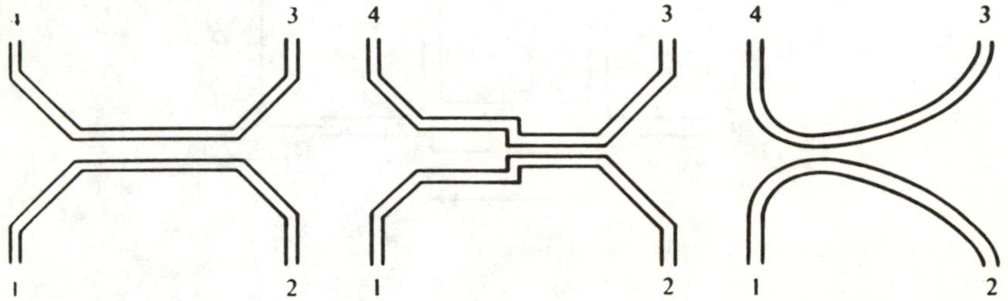

Fig. 6.42 Distributed interaction couplers.

6.5.21 Coupler in Homogeneous Line

When the structure of the coupler is homogeneous (only one medium of propagation), two linearly-independent TEM modes may propagate on the coupled line (§ 2.2.3). They are called ***even*** and ***odd mode***, depending on the respective polarity of the voltages on both lines (Fig. 6.43).

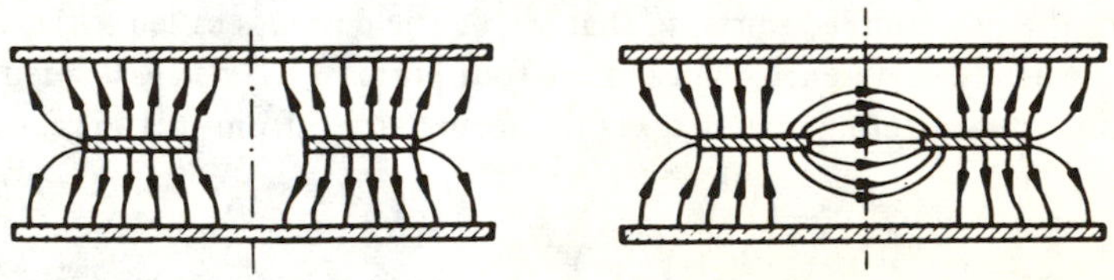

Fig. 6.43 Even and odd modes on a homogeneous coupled line.

The two modes have the same propagation factors, given by (2.26), and thus the same phase velocity. The study of the operation, carried out in terms of the superposition of the two modes, shows that the directivity of the coupled line is inherently very large. However, mismatches at the ends of the couplers may produce a signal at the isolated port, reducing the isolation.

6.5.22 Coupler in Inhomogeneous Line

In a coupled inhomogeneous line, for instance on a microstrip (§ 2.11.4), the velocities of propagation are different for even and for odd modes, because the fields of the two modes have different distributions. A significant part of the input signal then reaches the isolated port, and the directivity is small. In some situations, the signal coming out of this port actually becomes larger than that at the coupled port. This corresponds to a negative directivity, unless the numbering of ports is modified.

The addition of localized capacitances along the line allows the directivity of an inhomogeneous coupler to be improved [127] (Fig. 6.44).

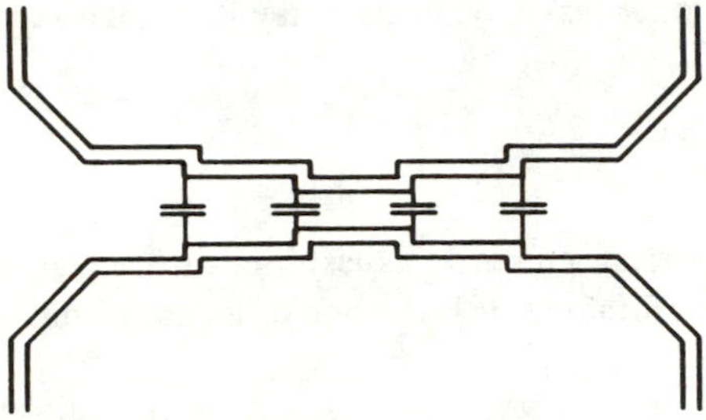

Fig. 6.44 Microstrip coupler with capacitors to improve the directivity.

6.5.23 Short Slot Hybrid Coupler, Riblet Coupler

In waveguide, the removal of part of the walls between the two guides (Fig. 6.45) yields a hybrid coupler, represented by the scattering matrix of (6.157). This is the *short slot hybrid* or *Riblet coupler.* In the section where the central wall has been removed, the TE_{10} mode is the even mode, and the TE_{20} mode is the odd mode. Coupling is obtained by combination of the two modes. The

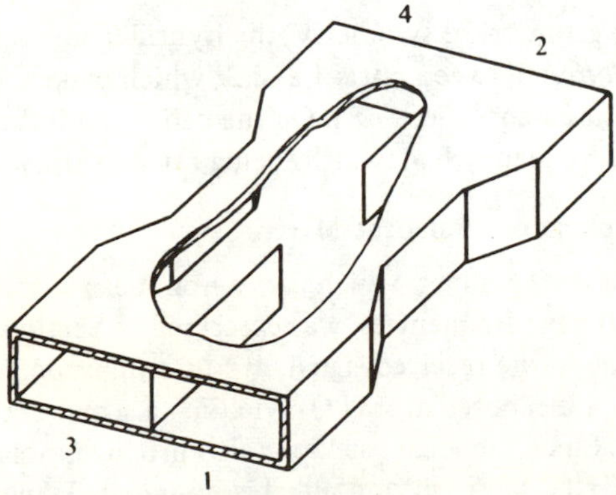

Fig. 6.45 Short-slot hybrid coupler or Riblet coupler.

width of the guide generally must be reduced to prevent the propagation of the higher order TE_{30} mode.

6.5.24 Hybrid Junctions

A junction is the ***meeting place*** of several lines or waveguides. It is called a *hybrid junction* when the input power is evenly distributed among the two coupled outputs (§ 6.5.19).

6.5.25 Hybrid T, Magic T

Several versions of hybrid T or of magic T junctions are shown in Figure 6.46.

The first version is basically the assembly of a series T (§ 6.4.8) and of a shunt T (§ 6.4.9), whose properties of symmetry may be recognized in the scattering matrix:

$$\underline{s}_{12} = \underline{s}_{14} \qquad 1 \tag{6.159}$$

$$\underline{s}_{23} = -\underline{s}_{34} \qquad 1 \tag{6.160}$$

By choosing symmetrical reference planes, the scattering matrix of an ideal matched hybrid T is obtained: it is the one of an antisymmetrical coupler (§ 6.5.7).

$$\underline{S} = \frac{1}{\sqrt{2}} \begin{pmatrix} 0 & 1 & 0 & 1 \\ 1 & 0 & -1 & 0 \\ 0 & -1 & 0 & 1 \\ 1 & 0 & 1 & 0 \end{pmatrix} \tag{6.161}$$

The different versions of hybrid Ts are obtained by folding the side arms (Fig. 6.46).

6.5.26 Comment

Due to their geometrical symmetry, the hybrid T junctions possess a ***very large intrinsic isolation*** between ports 1 and 3, which makes them particularly attractive for measurements (Sec. 7.3). On the other hand, they are difficult to match, the available frequency bandwidths being rather narrow.

6.5.27 Application: Balanced Mixer

The principle of the mixer, which uses a nonlinear semiconductor device to mix signals of different frequencies, was described in Section 5.3.14. In communications techniques, the received signal, of low amplitude, is mixed with the signal produced by a local oscillator (LO), yielding a signal at the intermediate frequency (IF). Since the local oscillator also produces noise at the intermediate frequency IF, the sensitivity obtained is bounded. Typically, powers at the nanowatt level may be detected.

The use of a hybrid T and of two paired mixers can greatly reduce the effect of the IF noise produced by the local oscillator.

The principle of operation is schematically presented in Figure 6.47. The signal, injected at port 3, appears at ports 2 and 4 in antiphase. The local oscillator,

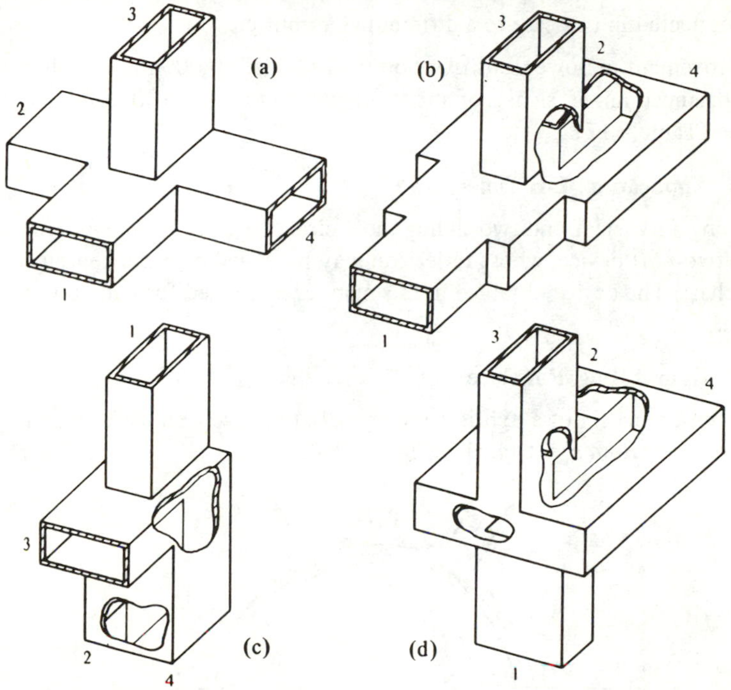

Fig. 6.46 Hybrid T couplers (a) standard, (b) with side arms bent in the H-plane, (c) with side arms bent in the E-plane, (d) Orthotee (Microwave Associates).

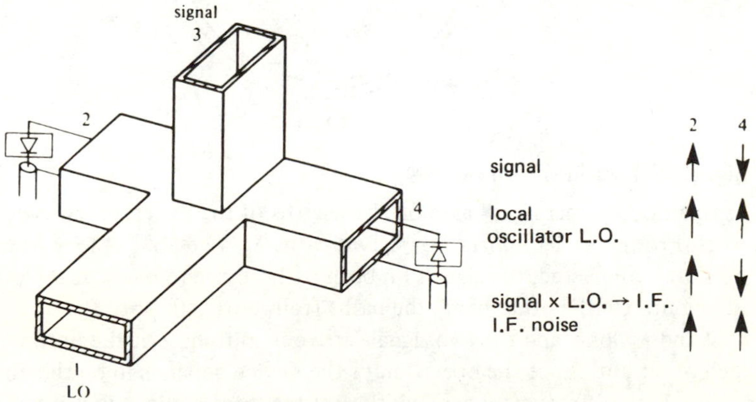

Fig. 6.47 Balanced mixer.

feeding port 1, provides signals in phase at ports 2 and 4. The mixer diodes, connected in opposition, yield IF signals in phase at the two ports, these signals then being added in an amplifier. As for the IF noise of the local oscillator, it appears in phase at ports 2 and 4, and is suppressed by addition after detection in the diodes connected in antiphase. Same polarity diodes may also be used, connected in this case to a differential amplifier.

An improvement in power sensitivity on the order of 10,000 may be obtained. With a balanced mixer, signals of about 0.1 picowatt, *i.e.*, −100 dBm, may be detected (Table 5.17).

6.5.28 Application: E-H Tuner

Combining a hybrid T and two sliding short circuits (§ 6.2.4), one obtains a lossless two-port device, whose reflection may be varied over the complete Smith chart. The device obtained in this manner is utilized for matching purposes.

6.5.29 Hybrid Ring, Rat Race

The scattering matrix of a hybrid coupler (6.161) is also obtained with the *hybrid ring* or *rat race*, sketched in Figure 6.48.

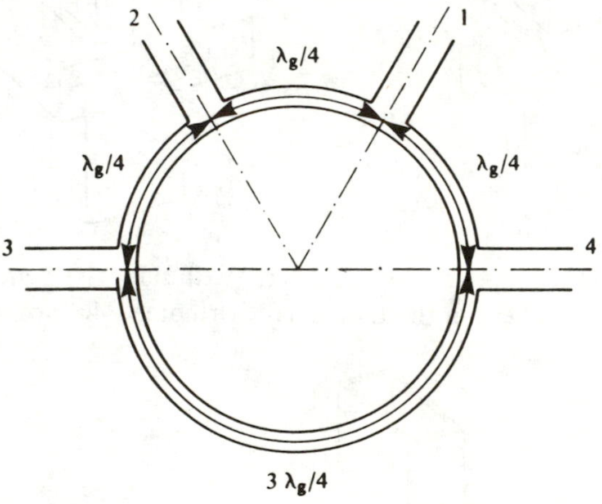

Fig. 6.48 Hybrid ring or rat race.

Its operation principle is based on the lengths of the different paths between ports. From port 1 to ports 2 or 4, two paths $\lambda_g/4$ and $5\lambda_g/4$ long, respectively, are shunt connected, the signals on both paths arrive in phase at the output, adding up. On the other hand, the paths from port 1 to port 3 are, respectively, $\lambda_g/2$ and λ_g long, and the two signals arrive in antiphase at the output, cancelling each other out. Since the operation of the device depends upon the dimensions, it has a limited frequency bandwidth. Rat races are realized in waveguide and in microstrip.

6.6 SIX-PORT DEVICES

6.6.1 General Remarks

Devices having any even number of ports larger than four may always be assembled by connecting four-port devices together. Thus, because reciprocal lossless four-ports may be matched (§ 6.5.2), it is always possible to realize a matched reciprocal $2n$ port. In practical applications, however, devices having more than four ports are seldom encountered, with the exception of *six-port devices,* now in use for the measurement of reflection (§ 7.3.13).

6.6.2 Purcell's Junction

A symmetrical six-port junction may be realized by assembling six waveguides (Fig. 6.49).

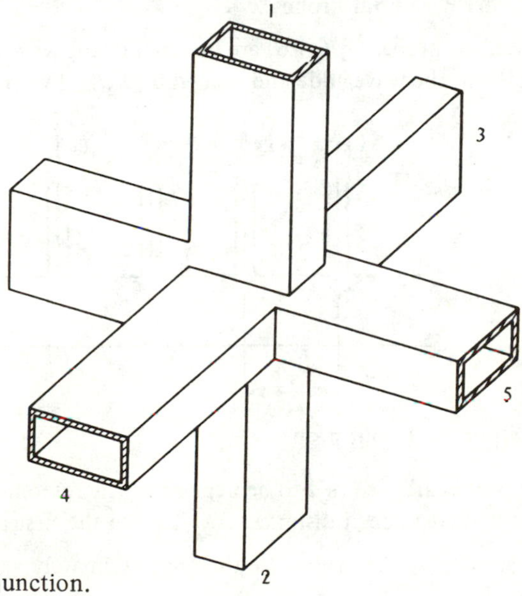

Fig. 6.49 Purcell junction.

Considering the geometrical symmetries, one finds the scattering matrix for a matched junction [116] to be

$$(\underline{s}) = \begin{pmatrix} 0 & \underline{\beta} & 0 & \underline{\delta} & 0 & \underline{\delta} \\ \underline{\beta} & 0 & \underline{\delta} & 0 & \underline{\delta} & 0 \\ 0 & \underline{\delta} & 0 & \underline{\beta} & 0 & \underline{\delta} \\ \underline{\delta} & 0 & \underline{\beta} & 0 & \underline{\delta} & 0 \\ 0 & \underline{\delta} & 0 & \underline{\delta} & 0 & \underline{\beta} \\ \underline{\delta} & 0 & \underline{\delta} & 0 & \underline{\beta} & 0 \end{pmatrix} \quad 1 \qquad (6.162)$$

with

$$1/3 \leqslant |\underline{\beta}| \leqslant 1 \qquad 1 \qquad (6.163)$$

$$0 \leqslant |\underline{\delta}| \leqslant 2/3 \qquad 1 \qquad (6.164)$$

Additionally, the energy conservation relations for a lossless device yield:

$$|\underline{\beta}|^2 + 2\,|\underline{\delta}|^2 = 1 \qquad 1 \qquad (6.165)$$

$$2\,\mathrm{Re}(\underline{\beta}^*\underline{\delta}) + |\underline{\delta}|^2 = 0 \qquad 1 \qquad (6.166)$$

Setting the reference planes to have real positive $\beta = A$ and letting $\underline{\delta} = B + \mathrm{j}C$, the expressions obtained are the same as those derived for a Y junction (6.110) and (6.111). The solutions are shown in Figure 6.35.

6.6.3 Six-Port With Four Probes

Four measurement probes (§ 6.4.6) are inserted into a waveguide. They are loosely coupled to the waveguide and spaced by $\lambda_g/8$ (Fig. 6.50). Interactions

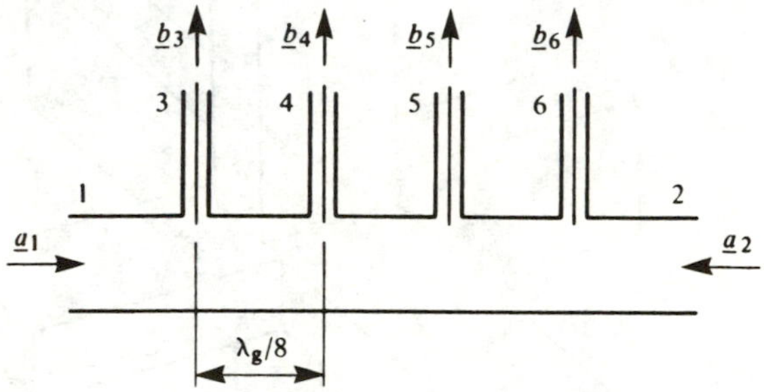

Fig. 6.50 Six-port with four probes.

between the probes are assumed to be negligible, the reference planes for $\underline{a}_1$ and $\underline{a}_2$ are set in the waveguide at distances $n\lambda_g/2$ from the first probe (port 3).

The signals sampled by the probes are then approximately

$$|\underline{b}_3| \cong K\,|\underline{a}_1 + \underline{a}_2| \qquad \mathrm{W}^{1/2} \qquad (6.167)$$

$$|\underline{b}_4| \cong K\,|\underline{a}_1 + \mathrm{j}\underline{a}_2| \qquad \mathrm{W}^{1/2} \qquad (6.168)$$

$$|\underline{b}_5| \cong K\,|\underline{a}_1 - \underline{a}_2| \qquad \mathrm{W}^{1/2} \qquad (6.169)$$

$$|\underline{b}_6| \cong K\,|\underline{a}_1 - \mathrm{j}\underline{a}_2| \qquad \mathrm{W}^{1/2} \qquad (6.170)$$

where $K \ll 1$ is the coupling factor, assumed to be the same for the four probes. The ratio $\underline{a}_2/\underline{a}_1$ is the reflection factor when the device to be measured is connected to the six-port at port 2. The amplitude and the phase of the reflection may both be determined from measurement of the ***signal amplitudes*** (only) at the four probes $\underline{b}_3, \underline{b}_4, \underline{b}_5$ and $\underline{b}_6$. This is the basic principle of the measurements using six-ports (§ 7.3.14). Expressions (6.167) to (6.170) are applicable only for the frequency at which the probe spacing is $\lambda_g/8$. This measurement technique is therefore restricted to the close vicinity of the design frequency.

6.6.4 Six-Port Design with Directional Couplers

The same result may be obtained by assembling symmetrical and antisymmetrical couplers (Fig. 6.51). To facilitate the understanding of the figure only one-half of the signal flow graph has been drawn, linking the inputs 1 and 2 to the outputs 3, 4, 5, and 6. The complete graph would be obtained by adding the same diagram with arrows pointing in the opposite direction. The numbering of the ports corresponds to expressions (6.167) to (6.170) with $K = 1/2$. This is, in essence, an eight-port device having two of its ports terminated into matched loads. The resulting assembly is a ***lossy device***.

The scattering matrix is obtained by inspection of the signal flow graph (Fig. 6.51), taking reciprocity into account

$$(\underline{s}) = \frac{1}{2}\begin{pmatrix} 0 & 0 & 1 & 1 & 1 & j \\ 0 & 0 & 1 & j & -1 & 1 \\ 1 & 1 & 0 & 0 & 0 & 0 \\ 1 & j & 0 & 0 & 0 & 0 \\ 1 & -1 & 0 & 0 & 0 & 0 \\ j & 1 & 0 & 0 & 0 & 0 \end{pmatrix} \qquad 1 \qquad (6.171)$$

The frequency limitations depend upon the couplers used in the assembly; they are less restrictive than those on the four probe device of Section 6.6.3.

6.7 NONRECIPROCAL FERRITE DEVICES

6.7.1 Introduction: Ferrites, YIG, Gyromagnetic Effect

The application of a magnetic field to certain ***compounds of metal oxides and rare earths***, known under the generic term of *ferrites* and *Yttrium Iron Garnets* (YIG) induces an anisotropy due to the *gyromagnetic effect* [128]. The phasor-vectors induction $\underline{\boldsymbol{B}}$ and magnetic field $\underline{\boldsymbol{H}}$ of an electromagnetic wave within a ferrite are not colinear, and their ratio is a complex permeability tensor $\underline{\bar{\bar{\mu}}}$ (§ 1.4.6).

$$\underline{\boldsymbol{B}} = \underline{\bar{\bar{\mu}}}\,\underline{\boldsymbol{H}} = \mu_0(\underline{\boldsymbol{H}} + \underline{\boldsymbol{M}}) \qquad \text{T} \qquad (6.172)$$

where $\underline{\boldsymbol{M}}$ is the magnetization.

The reciprocity theorem is ***not always satisfied***. Magnetized ferrites are utilized to manufacture ***nonreciprocal devices***, in particular isolators and circulators.

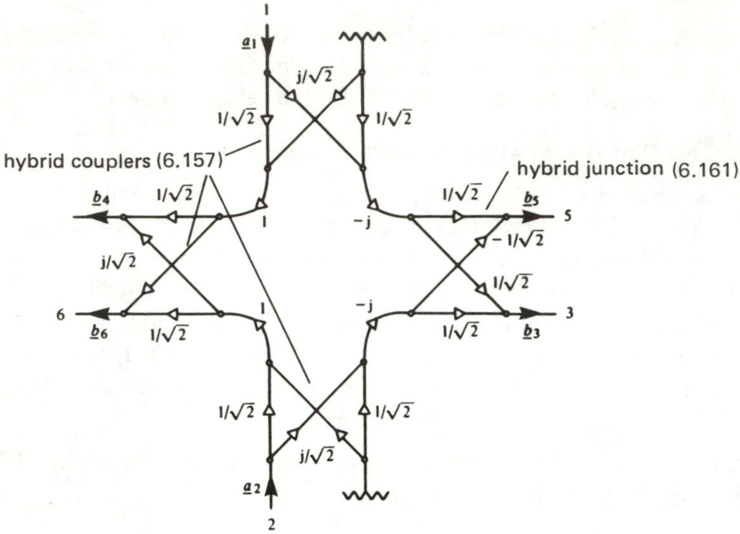

Fig. 6.51 Six-port design with hybrid couplers and junctions.

6.7.2 Gyromagnetic Effect for a Single Electron, Larmor Precession

The action of a magnetic field on the spin of an electron may be approximately described by a model of classical physics, even though the phenomenon is in the realm of quantum physics. The applied magnetic field $\boldsymbol{H}$, acting on the spin magnetic moment $\boldsymbol{m}$ of the electron, produces a torque which modifies the angular momentum of spin $\boldsymbol{p}$

$$\frac{\mathrm{d}\boldsymbol{p}}{\mathrm{d}t} = \mu_0 \boldsymbol{m} \times \boldsymbol{H} \qquad \text{VAs} \qquad (6.173)$$

In addition, $\boldsymbol{m}$ is proportional to $\boldsymbol{p}$. The quotient γ_g is called the *gyromagnetic ratio.* Its value is provided by quantum physics

$$\gamma_g = -gq/2m = 8.78\, g\, 10^{10} \qquad (\text{sT})^{-1} \qquad (6.174)$$

where g is the *Landé factor,* which takes a value of 2 for a single electron, and where q and m are, respectively, the charge and the mass of the electron.

The equation of movement for the magnetic moment $\boldsymbol{m}$ is obtained with the two expressions (6.173) and (6.174), taking into account the fact that $\boldsymbol{m}$ and $\boldsymbol{p}$ have opposite directions

$$\frac{\mathrm{d}\boldsymbol{m}}{\mathrm{d}t} = -\gamma_g \mu_0 \boldsymbol{m} \times \boldsymbol{H} \qquad \text{Am}^2/\text{s} \qquad (6.175)$$

The time variation is in a direction perpendicular to both the vector $\boldsymbol{m}$ and to the applied magnetic field $\boldsymbol{H}$. When the latter is time-invariant, the vector $\boldsymbol{m}$ rotates around the direction of the magnetic field, just like a gyroscope (Fig. 6.52). This

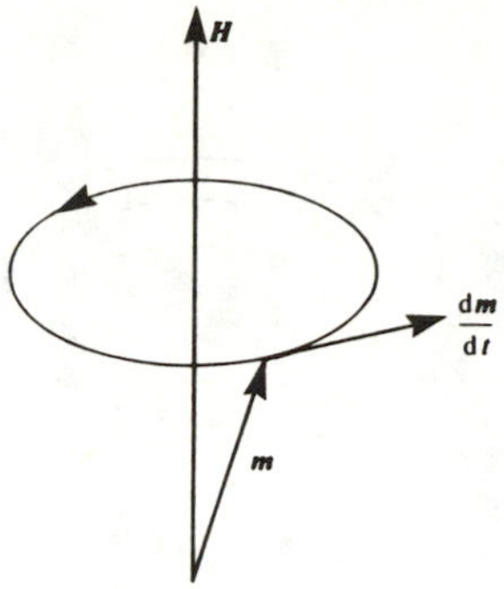

Fig. 6.52 Larmor precession for a single electron.

phenomenon, well known in physics, is called *Larmor precession.*

In most solids, electrons tend to group in pairs having opposite spins, the overall interaction with a magnetic field becoming negligible, or vanishing altogether. Only in the materials known as ferromagnetic (Fe, Co, Ni, and certain rare earths) does the presence of unpaired electrons provide a significant interaction [128]. Furthermore, if the electromagnetic field of a wave is to interact with unpaired electrons, this wave should be able to penetrate within the material, which is only possible when the conductivity σ is small. Both these conditions are met in ferrites, compounds of ferromagnetic metal oxides, and in YIG.

6.7.3 Gyromagnetic Effect in an Infinite Homogeneous Ferrite

The unpaired electrons in a ferrite interact with the magnetic field in much the same manner as a single electron does. Taking an average of all the particles, and taking into account additional interactions between electrons and crystal lattice, one obtains the relation [128]

$$\frac{\mathrm{d}\boldsymbol{M}}{\mathrm{d}t} = -\gamma_g \mu_0 \boldsymbol{M} \times \boldsymbol{H} + \frac{\alpha}{|\boldsymbol{M}|} \boldsymbol{M} \times \frac{\mathrm{d}\boldsymbol{M}}{\mathrm{d}t} \qquad \text{A/ms} \qquad (6.176)$$

where $\boldsymbol{M}$ is the magnetization (volume density of magnetic moments $\boldsymbol{m}$) and where α is a dimensionless factor accounting for spin-spin and spin-lattice interactions; it is generally dependent upon the magnetic field (amplitude and frequency). Its value for commonly utilized materials is at most only a few hundredths.

The last term in (6.176) introduces a restoring force, which tends to line up the magnetization $\boldsymbol{M}$ with the magnetic field $\boldsymbol{H}$. The vector $\boldsymbol{M}$, whose length remains constant, describes a spiral on a portion of a sphere of radius $|\boldsymbol{M}|$ (Fig. 6.53).

The Landé factor g, which appears within the gyromagnetic ratio γ_g, is no longer equal to 2 for a ferrite. For most materials, it takes values between 1.9 and 2.1, while values between 1.5 and 2.7 appear in nickel ferrites. When the magnetic

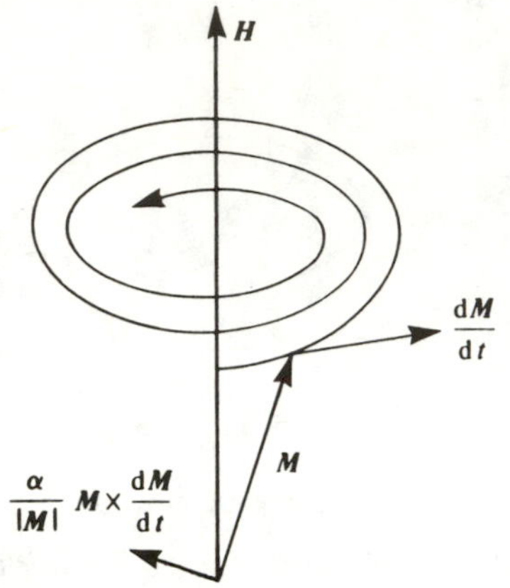

Fig. 6.53 Gyromagnetic effect in a lossy ferrite.

field $\boldsymbol{H}$ varies in time, the same is true for the magnetization $\boldsymbol{M}$, the two vectors being connected through the magnetic properties of the ferrite. As a result, (6.176) is a nonlinear expression, as it contains a product of time-dependent quantities.

■ 6.7.4 Study of Small Signals, Larmor and Magnetization Angular Frequencies

In practice, most applications of ferrites consider small time variations of $\boldsymbol{H}$ and $\boldsymbol{M}$, so that second order terms may be neglected. Equations may then be linearized and small signals studied, as in Section 4.4.3.

The magnetic field $\boldsymbol{H}$ possesses a direct component $\boldsymbol{H}_0$, along which one places the y-axis, and a sinusoidal alternating component, represented in complex notation as

$$\boldsymbol{H} = \boldsymbol{H}_0 + \mathrm{Re}[\sqrt{2}\,\underline{\boldsymbol{H}}\exp(\mathrm{j}\omega t)] \qquad \text{A/m} \qquad (6.177)$$

The direct component of the magnetization is also directed along y (see Fig. 6.53). The magnetization is then

$$\boldsymbol{M} = \boldsymbol{M}_0 + \mathrm{Re}[\sqrt{2}\,\underline{\boldsymbol{M}}\exp(\mathrm{j}\omega t)] \qquad \text{A/m} \qquad (6.178)$$

The two expressions are introduced into (6.176) and the terms in $\exp(\mathrm{j}\omega t)$ identified, yielding

$$\mathrm{j}\omega\underline{\boldsymbol{M}} - (\omega_{\mathrm{L}} + \mathrm{j}\omega\alpha)\boldsymbol{e}_y \times \underline{\boldsymbol{M}} = -\omega_M \boldsymbol{e}_y \times \underline{\boldsymbol{H}} \qquad \text{A/ms} \qquad (6.179)$$

where the *Larmor angular frequency* has been defined as

$$\omega_{\mathrm{L}} = \mu_0 \gamma_{\mathrm{g}} H_0 \qquad \text{rad/s} \qquad (6.180)$$

and the *magnetization angular frequency* by

$$\omega_M = \mu_0 \gamma_g M_0 \qquad \text{rad/s} \qquad (6.181)$$

A ***saturated ferrite*** is considered here, in which the magnetic properties are homogeneous. Relation (6.179) may then be written in matrix form, developing $\underline{\boldsymbol{H}}$ and $\underline{\boldsymbol{M}}$ in the rectangular coordinate system

$$\begin{pmatrix} j\omega & 0 & -\omega_L - j\omega\alpha \\ 0 & j\omega & 0 \\ \omega_L + j\omega\alpha & 0 & j\omega \end{pmatrix} \begin{pmatrix} \underline{M}_x \\ \underline{M}_y \\ \underline{M}_z \end{pmatrix} = \begin{pmatrix} 0 & 0 & -\omega_M \\ 0 & 0 & 0 \\ +\omega_M & 0, & 0 \end{pmatrix} \begin{pmatrix} \underline{H}_x \\ \underline{H}_y \\ \underline{H}_z \end{pmatrix} \qquad \text{A/ms} \qquad (6.182)$$

It may be seen in (6.182) that $\underline{M}_y = 0$.

6.7.5 Permeability Tensor, Linewidth

The matrix in the left-hand term of (6.182) is inverted and the relation $\underline{\boldsymbol{M}}(\underline{\boldsymbol{H}})$ written in explicit form. The value obtained for $\underline{\boldsymbol{M}}$ is introduced into (6.172), yielding the *permeability tensor* $\underline{\bar{\bar{\mu}}}$

$$\underline{\bar{\bar{\mu}}} = \mu_0 \begin{pmatrix} \underline{\mu}_r & 0 & -j\underline{K} \\ 0 & 1 & 0 \\ j\underline{K} & 0 & \underline{\mu}_r \end{pmatrix} \qquad \text{Vs/Am} \qquad \mathbf{(6.183)}$$

with

$$\underline{\mu}_r = 1 + (\omega_L + j\omega\alpha)\omega_M / [(\omega_L + j\omega\alpha)^2 - \omega^2] \qquad 1 \qquad (6.184)$$

$$\underline{K} = \omega\, \omega_M / [(\omega_L + j\omega\alpha)^2 - \omega^2] \qquad 1 \qquad (6.185)$$

In the absence of losses ($\alpha = 0$), these two terms are real. The matrix $\underline{\bar{\bar{\mu}}}$ is then ***hermitian*** [53].

Figure 6.54 shows the real and imaginary parts of $\underline{\mu}_r$ and $\underline{K}$ for $\alpha = 0.059$. The imaginary parts μ_r'' and K'', which determine absorption, both pass through a maximum at $\omega = \omega_L$ (gyromagnetic resonance) and decrease rapidly as one moves away from the resonance.

The *resonance linewidth* ΔH is defined by the difference between the magnetizing fields for the two points at which absorption is one-half the value at the maximum. The damping factor α is related to the linewidth by (6.184) and (6.185)

$$\alpha \cong \gamma_g \mu_0 \Delta H / 2\omega_L \qquad 1 \qquad (6.186)$$

6.7.6 Note: Actual Ferrites

In fact, a ferrite is a polycrystalline assembly of magnetic domains, for which the homogeneity assumed in Section 6.7.3 is only an approximation. The presence of material inhomogeneity widens the resonance line but, fortunately, does not increase as much the losses. The situation gets still more involved when the ferrite is not saturated, for instance in devices operating at the remanence (§ 6.7.12). In this case, magnetic domains are not all oriented in the same direction, and statistical methods are required to determine the properties of the ferrite [129].

The assumed homogeneity is encountered in practice only within single crystals of YIG having a good surface finish (§ 6.7.23).

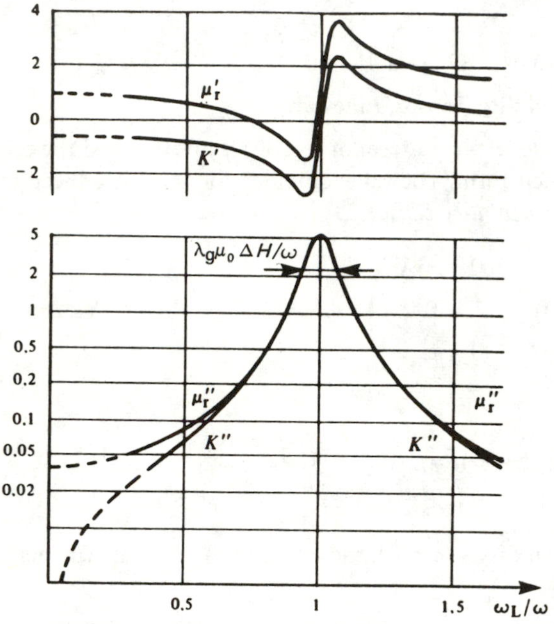

Fig. 6.54 Field-dependence of the terms of the permeability tensor of a ferrite.

6.7.7 Circular Polarization

Since the gyromagnetic resonance possesses a rotating behavior (Fig. 6.53), the application of a rotating magnetic field produces particular phenomena within the ferrite. A magnetic field circularly polarized in a plane perpendicular to the direct magnetic field $\boldsymbol{H}_0$ (directed along $\boldsymbol{e}_y$) is represented by the vector-phasor (§ 1.4.7)

$$\underline{\boldsymbol{H}}_{\pm} = \underline{H}_{\mathrm{c}}(\boldsymbol{e}_x \mp \mathrm{j}\boldsymbol{e}_y) \qquad \mathrm{A/m} \qquad (6.187)$$

where $\underline{H}_{\mathrm{c}}$ is a constant and where signs $\pm$ correspond to the two directions of rotation. Utilizing (6.183) one obtains the induction

$$\underline{B}_{\pm} = \underline{\mu}_{\pm}\underline{H}_{\pm} = \mu_0(\underline{\mu}_r \pm \underline{K})\underline{H}_{\pm} \qquad \text{T} \qquad (6.188)$$

These are two scalar relations: the phasor-vectors $\underline{H}_{\pm}$ are proportional to the eigenvectors of the tensor $\underline{\overline{\overline{\mu}}}$, the corresponding eigenvalues being

$$\underline{\mu}_{\pm} = \mu_0[1 + \omega_M/(\omega_L \mp \omega + j\omega\alpha)] \qquad \text{Vs/Am} \qquad \mathbf{(6.189)}$$

Figure 6.55 shows the values taken, respectively, by the real and imaginary parts as a function of ω_L/ω. The gyromagnetic resonance only appears for the positive circular polarization. If a direction of propagation could be associated with a sense of rotation, a nonreciprocal effect would be obtained. It would be a unidirectional attenuation close to the resonance, a differential phaseshift for a biasing magnetic field further away from the resonance.

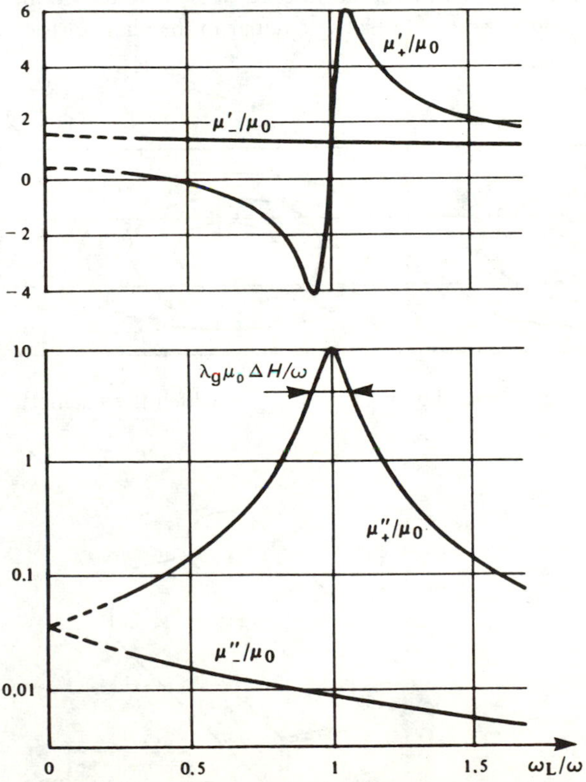

Fig. 6.55 Properties of a ferrite in a rotating magnetic field.

6.7.8 Rectangular Waveguide Isolator

The $\underline{H}_x$ and $\underline{H}_z$ components of the dominant TE_{10} mode in rectangular waveguide are in phase-quadrature (§ 2.3.11). Furthermore, their amplitudes are equal in the two planes $x = x_{cp}$ and $x = a - x_{cp}$, where x_{cp} is determined from (2.126) and (2.127)

$$x_{cp} = (a/\pi) \arctan(\pi/a\beta_{10}) \quad \text{m} \tag{6.190}$$

In these two planes, the magnetic field of the microwave signal is ***circularly polarized*** (cp) (Fig. 6.56). The direction of rotation is related to the direction of propagation of the wave (forward or reverse). The condition outlined in paragraph 6.7.7 is then satisfied. A thin ferrite blade is placed within one of these planes and submitted to a direct magnetic field $H_0 = \omega/(\gamma_g \mu_0)$ directed along $\boldsymbol{e}_y$ (Fig. 6.57). The positive circularly polarized wave exhibits a significant attenuation (Fig. 6.55): this is the blocked or isolated direction. The other wave, whose polarization is negative circular in the plane of the ferrite is only

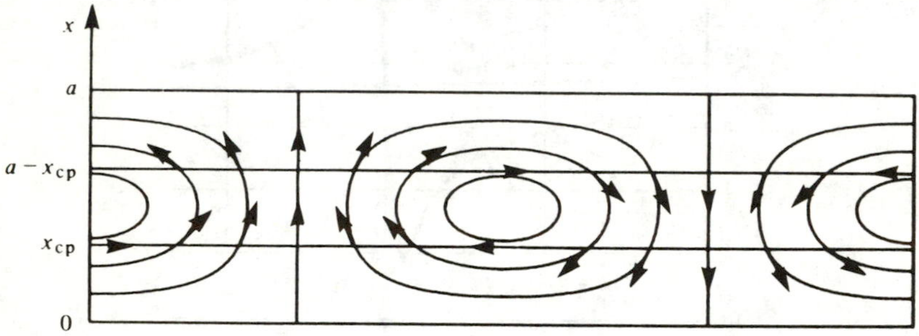

Fig. 6.56 Location of the planes $x = x_{cp}$, in which the magnetic field is circularly polarized.

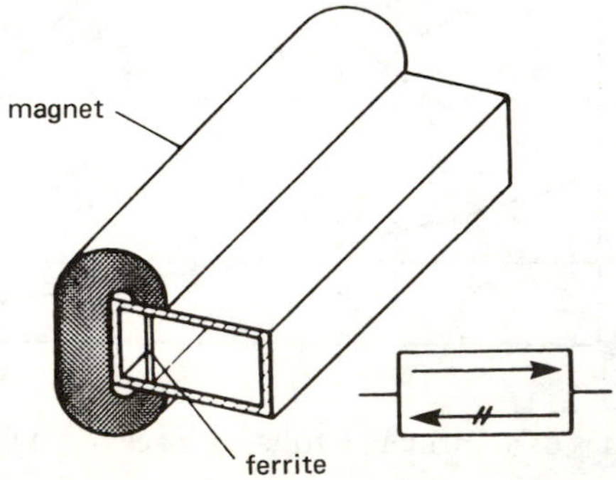

Fig. 6.57 Isolator in rectangular waveguide.

slightly attenuated (forward direction). The device obtained is an *isolator* (§ 6.3.10). The direct magnetic field H_0 is provided by a permanent magnet.

As in the coupler, an ***attenuation*** (§ 6.5.10, low-loss direction) and an ***isolation*** (§ 6.5.14, high-loss direction) are defined.

The simple structure of Figure 6.57 presents several drawbacks:

1. The ferrite blade perturbs the field within the waveguide, so that the field components are only approximately those of the empty waveguide. The isolation is smaller, the attenuation larger than those derived by means of the perturbation method (Sec. 2.7).
2. The frequency range is very narrow, as both the absorption (Fig. 6.54) and the position of the plane x_{cp} of circular polarization (6.178) are frequency-dependent.
3. The magnetic properties of both the ferrite and of the magnet vary with temperature.
4. The removal of the heat dissipated in the ferrite is difficult, due to the low thermal conductance of ferrite materials.

A more evolved isolator structure, shown in Figure 6.58, utilizes a dielectric slab, which sets the plane of circular polarization within the ferrite at all frequencies [130].

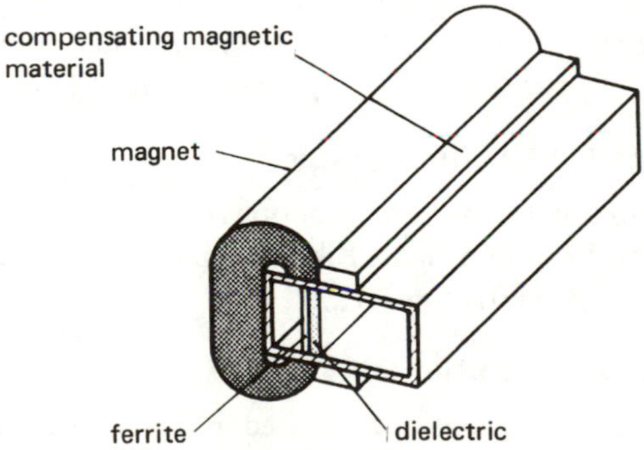

Fig. 6.58 Broadband isolator in rectangular waveguide.

A biasing magnetic field H_0 dependent upon the position along z is obtained with a non-uniform air gap along the structure. The variation due to temperature is partially compensated for by the addition of magnetic material having a low Curie point shunting the field.

Another isolator design has ferrite slabs on the broad walls of the waveguide (H-plane), placed in an inhomogeneous biasing field H_0 (Fig. 6.59). The ferrite is

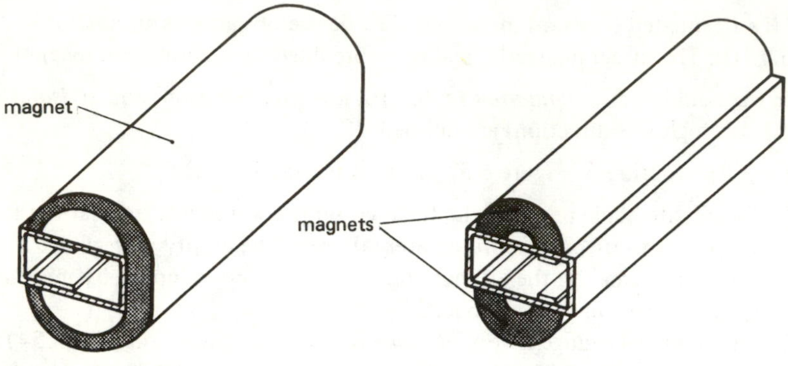

Fig. 6.59 H-plane ferrite isolators in rectangular waveguide.

then in good thermal contact with the metal wall, so that the evacuation of the heat dissipated is more efficient, and the device can operate at higher power levels. This structure, however, requires a larger and consequently heavier magnet.

Isolators were designed to cover the usable frequency band of a waveguide (§ 2.2.35) with more than 40 dB of isolation and a 0.5 to 1 dB attenuation.

6.7.9 Nonreciprocal Phaseshifters

Away from the gyromagnetic resonance, the attenuation becomes small for both directions of propagation. The two phaseshifts, linked to the real part of the permittivity μ'_+ and μ'_- (Fig. 6.55) depend upon the direction of propagation: a *nonreciprocal phaseshifter* is obtained (§ 6.3.13) and, when the phaseshift difference $\Delta\varphi = |\varphi_+ - \varphi_-|$ is equal to 180°, a ***passive gyrator*** (§ 6.3.14).

6.7.10 Nonreciprocal Phaseshift Circulator

Combining one gyrator and two hybrid junctions (§ 6.5.25) or two half-gyrators ($\Delta\varphi = 90°$), one hybrid junction, and a Riblet coupler (§ 6.5.23) one obtains a four-port circulator (Fig. 6.60) (§ 6.4.4).

6.7.11 Circulator Used as Isolator

A circulator having its ports 3 and 4 terminated into matched loads (§ 6.2.2) is an isolator. Since the energy is then dissipated in the loads (but not in the ferrite) this isolator is capable of handling high power levels.

6.7.12 Latching Phaseshifters

A ferrite toroid (often of rectangular cross section) with a center conductor within a waveguide is a ***bistable*** device, the *latching phaseshifter* (Fig. 6.61) [131].

A current pulse crossing the center conductor latches the ferrite in one of its two stable remanence states. The switching time is on the order of a few microseconds. This phaseshifter is nonreciprocal.

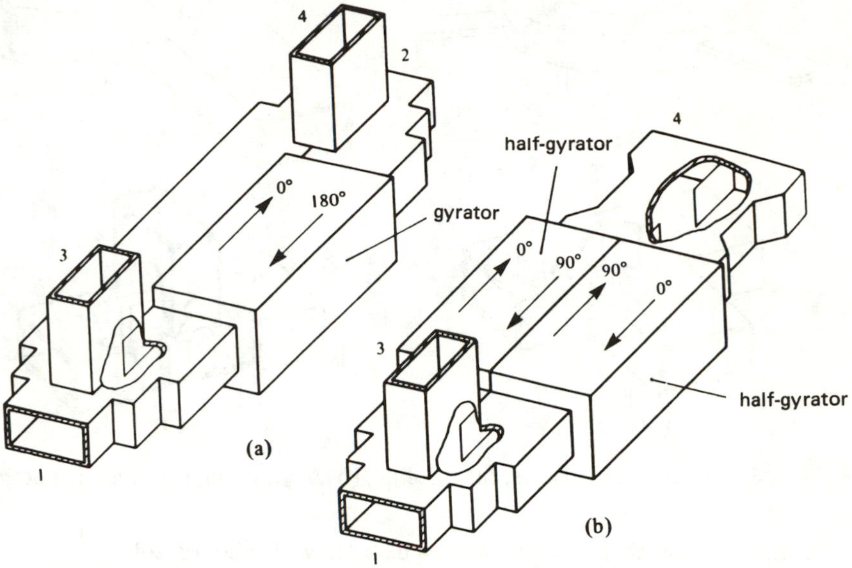

Fig. 6.60 Nonreciprocal phaseshift circulators.

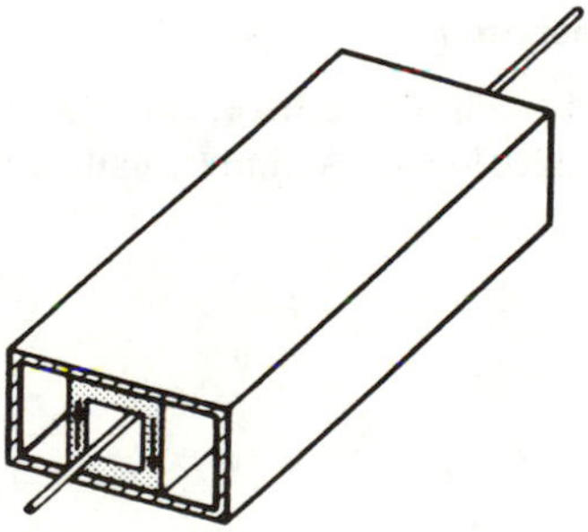

Fig. 6.61 Latching phaseshifter.

6.7.13 Junction Circulators

A piece of ferrite (cylinder, disk, triangle, slab) is introduced in a three-port junction, in either waveguide, stripline, or microstrip, and is magnetized perpendicularly to the plane of the junction. The magnetization of the ferrite produces a rotation of the field structure, which is adjusted to decouple one of the three ports [132, 133] (§ 6.7.14). In this manner, one obtains a junction circulator, of small size (§ 6.4.4). Several structures are shown in Figure 6.62.

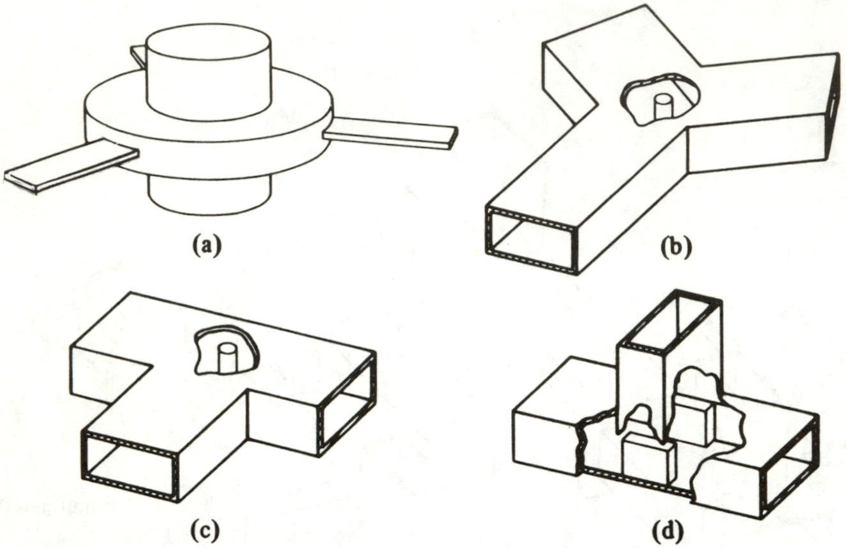

Fig. 6.62 Junction circulators: (a) stripline, (b) Y junction, (c) shunt T junction, (d) series T junction.

Waveguide circulators cover the waveguide band with a 20 dB isolation and about 0.3 dB of attenuation. In stripline, an octave bandwidth with a 17 dB isolation and about 0.4-0.5 dB of attenuation is typical.

6.7.14 Principle of Operation

A ***simplified model*** will be considered here, formed of a cylindrical cavity coupled to three waveguides located symmetrically with respect to the cavity (Fig. 6.63).

The waveguide at port 1 propagates the dominant TE_{10} mode and excites mag-

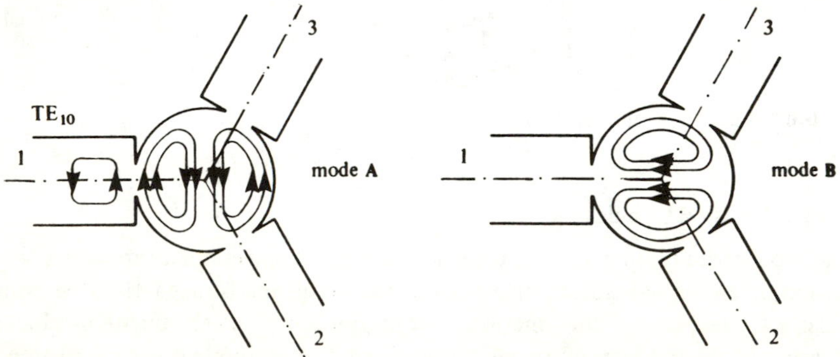

Fig. 6.63 Circular cylindrical cavity resonating in the TM_{110} mode, coupled to three waveguides.

netically the modes of resonance of the cavity through an aperture (§ 3.6.4). The resonant TM_{110} mode of the cavity, which is spatially degenerate, is considered for this application (Table 2.18). The magnetic field lines for the two degenerate modes, designated as A and B, are sketched in Figure 6.63.
The dominant waveguide mode couples to the A mode of the cavity. The B mode is not excited, the inductive coupling term (3.108) vanishing because of the geometrical asymmetry. Since any linear combination of two degenerate resonant modes is still a resonant mode (§ 3.3.7) the A mode (linearly polarized) may be developed over two circularly polarized modes, symbolically designated as $1/2(A + jB)$ and $1/2(A - jB)$ (§ 1.4.7).

The introduction of a cylinder of magnetized ferrite at the center of the cavity lifts the degeneracy of the modes. For the mode having a positive (negative) circular polarization, the ferrite exhibits a scalar permeability $\underline{\mu}_+$ ($\underline{\mu}_-$). Two different resonant frequencies f_+ (f_-) are then obtained for the two modes (3.83). If the frequency f of the signal is located between the two resonances, the input impedance has, respectively, a capacitive (inductive) component for the mode of lower (upper) resonant frequency. The two modes are thus shifted in phase, respectively ahead and behind in time. The vector sum of the two circularly polarized modes is a linearly polarized mode, which is rotated with respect to the original A mode of the unloaded cavity [133]. The size and characteristics of the ferrite are selected in such a way that, together with a suitable biasing magnetic field, they provide a 30° rotation (Fig. 6.64).

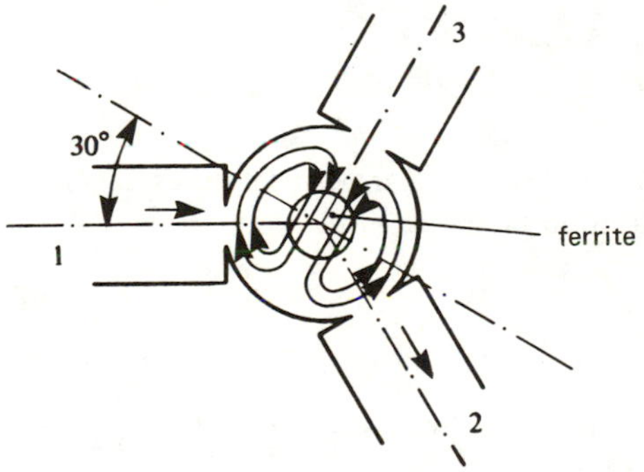

Fig. 6.64 Magnetic field in the cavity of Fig. 6.63 loaded with a magnetized ferrite.

The field structure, rotated by the introduction of the magnetized ferrite, no longer couples to port 3, the coupling term (3.108) vanishing because of the asymmetry of the field. The signal entering port 1 is entirely coupled to guide 2. Since the junction is symmetrical, the same process is repeated with 120° rotations. A signal applied to port 2 comes out of port 3, port 1 then being isolated. Similarly, a signal applied to port 3 comes out of port 1. The loaded junction is a circulator (§ 6.4.4).

The model considered here is greatly oversimplified, its purpose being to describe in a way as simple as possible the basic principle of operation. Coupling was assumed to be loose, and one admits that the ferrite does not perturb the field structure. Quite on the contrary in an actual circulator, the waveguide is tightly coupled to the cavity, which is a basic requirement for small attenuation and broad band. The complete study of an actual device becomes quite complex. Nevertheless, the requirements of symmetry and of field rotation described here remain valid.

6.7.15 Thin Isolator

Terminating one of the three ports of a junction circulator with a matched load, one obtains an isolator, in the same manner as in section 6.7.11. Pushing the load right into the junction, an extremely compact design is obtained at the cost of some performance (about 1 cm long), for low power applications (Fig. 6.65). This kind of device is mostly realized in waveguide.

□ 6.7.16 Faraday Rotation

When a wave propagates in a ferrite medium, assumed to be infinite in the z-direction, along the biasing magnetic field H_0 (longitudinal magnetization), a rotation of the polarization plane of the wave is observed, called *Faraday rotation.* A linearly polarized plane wave may be separated into two circularly polarized waves (§ 1.4.7).

$$\underline{H}(z=0) = e_x H_1 = \frac{H_1}{2}\left[(e_x - \mathrm{j}e_y) + (e_x + \mathrm{j}e_y)\right] \qquad \text{A/m} \qquad (6.191)$$

The ferrite presents different scalar permeabilities $\underline{\mu}_+$ and $\underline{\mu}_-$ for these two waves. Thus, their propagation factors are different

$$\beta_+ = \omega\sqrt{\epsilon\mu_+} \qquad \text{m}^{-1} \qquad (6.192)$$

$$\beta_- = \omega\sqrt{\epsilon\mu_-} \qquad \text{m}^{-1} \qquad (6.193)$$

where the effect of losses has been neglected. The magnetic field $\underline{H}$ of a forward wave is then given by

$$\underline{H}(z) = \frac{H_1}{2}\left[(e_x - \mathrm{j}e_y)\exp(-\mathrm{j}\beta_+ z) + (e_x + \mathrm{j}e_y)\exp(-\mathrm{j}\beta_- z)\right]$$

$$= H_1 \exp(-\mathrm{j}\beta_m z)\left[e_x \cos\Delta\beta z - e_y \sin\Delta\beta z\right] \qquad \text{A/m} \qquad (6.194)$$

with

$$\beta_m = (\beta_+ + \beta_-)/2 \qquad \text{m}^{-1} \qquad (6.195)$$

$$\Delta\beta = (\beta_+ - \beta_-)/2 \qquad \text{m}^{-1} \qquad (6.196)$$

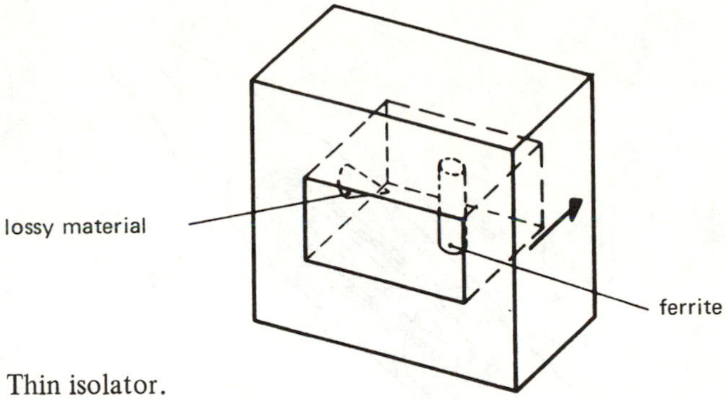

Fig. 6.65 Thin isolator.

This is a linearly polarized wave, having a phaseshift per unit length β_m and a plane of polarization rotating as a function of z. Within the plane $z = l$, the rotation angle θ with respect to the original direction (in $z = 0$) is (Fig. 6.66)

$$\theta = \Delta\beta l = \arctan (H_y/H_x) \qquad \text{rad} \tag{6.197}$$

The field of the backward wave is obtained by changing the sign of the propagation factor in (6.194). The same angle θ is obtained in the plane $z = -l$ (Fig. 6.66). The sense of rotation of the plane of polarization is independent of the direction of propagation: ***rotation is nonreciprocal***.

A similar rotation takes place for the dominant TE_{11} mode in a circular waveguide loaded with a cylinder of longitudinally magnetized ferrite.

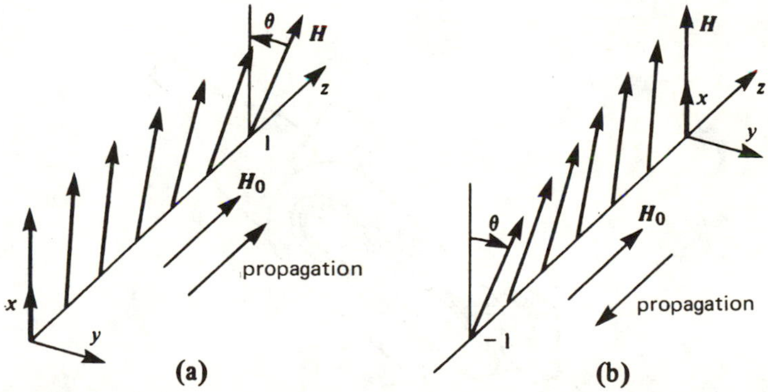

Fig. 6.66 Faraday rotation in a magnetized ferrite: (a) forward wave, (b) backward wave.

□ 6.7.17 Faraday Gyrator

The angle of Faraday rotation is adjusted so that $\theta = 90°$, and transitions from rectangular to circular waveguide are connected at both ends. The device obtained is a *Faraday gyrator* [122] (Fig. 6.67). The plane of the input waveguide may be restored by adding a twist.

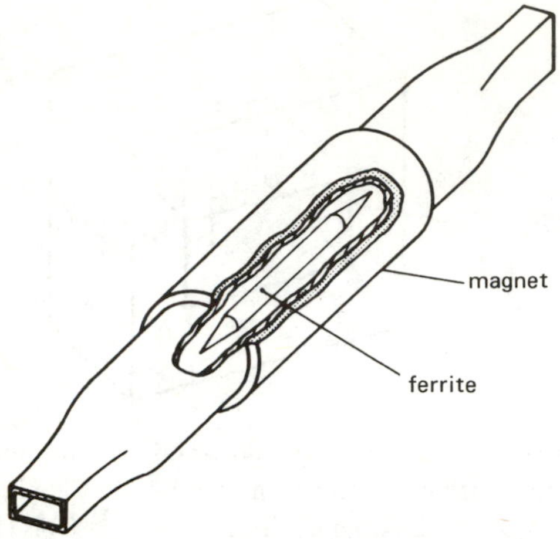

Fig. 6.67 Faraday rotation gyrator in circular waveguide.

□ 6.7.18 Faraday Circulator

The rotation angle is set at 45° and the guide is connected to two orthogonal mode couplers, which allow the two degenerate TE_{11} modes to couple selectively with two rectangular waveguides. The device obtained in this manner is a four-port *Faraday circulator* (Fig. 6.68). Due to the particular geometrical location and orientation of the ports, this device is rather awkward to use.

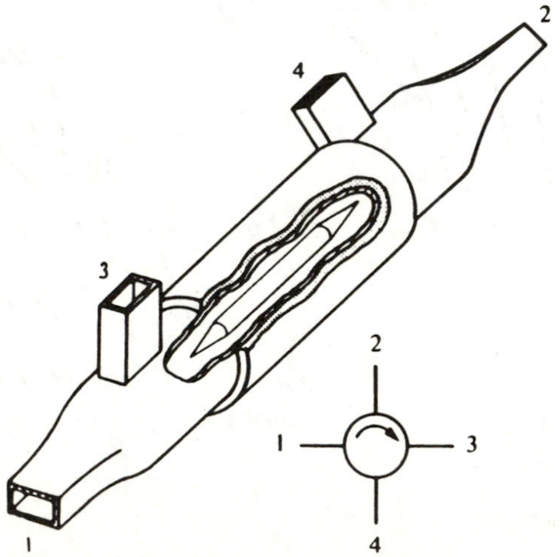

Fig. 6.68 Faraday rotation circulator.

□ 6.7.19 Faraday Isolator

Ports 3 and 4 of the circulator are terminated into matched loads (as in § 6.7.11). Loads are made with resistive vanes, located within the device itself, which absorb the signal when the electric field is in the plane of the vane (§ 6.3.7) (Fig. 6.69). The device realized in this manner is a *Faraday isolator.*

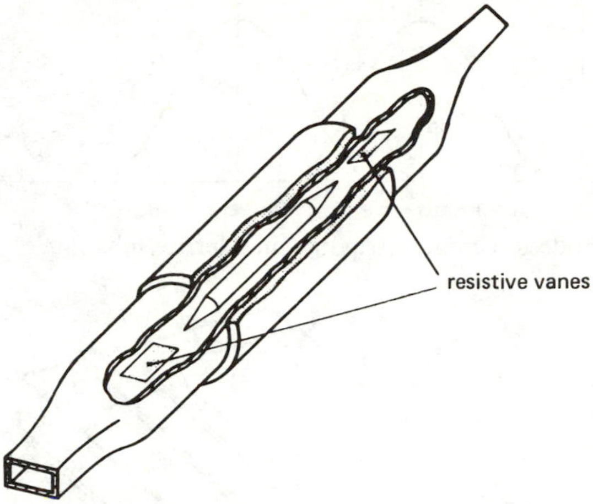

Fig. 6.69 Faraday rotation isolator.

6.7.20 Drawbacks

Since the angle of rotation θ varies with frequency (6.197), the frequency band of the Faraday rotators is limited. The dissipated power is difficult to remove and thermal effects cannot be effectively compensated for. As a result, these devices, which were the first nonreciprocal passive devices ever designed, are seldom utilized nowadays, except at millimeter waves.

6.7.21 Peripheral Mode Devices

In a microstrip line deposited on a magnetized ferrite substrate, the gyromagnetic effect produces a lateral displacement of the fields. For the forward wave, they concentrate under one edge of the upper conductor; those of the backward wave concentrate under the opposite edge. These modes are called peripheral or *edge modes* (Fig. 6.70).

Widening the upper conductor leads to a spatial separation of the two waves. It is then possible to produce different effects on the two of them. A nonreciprocal phaseshifter is made with edges of different lengths. Placing a lossy material close to one edge attenuates one of the waves, producing an isolator (Fig. 6.71).

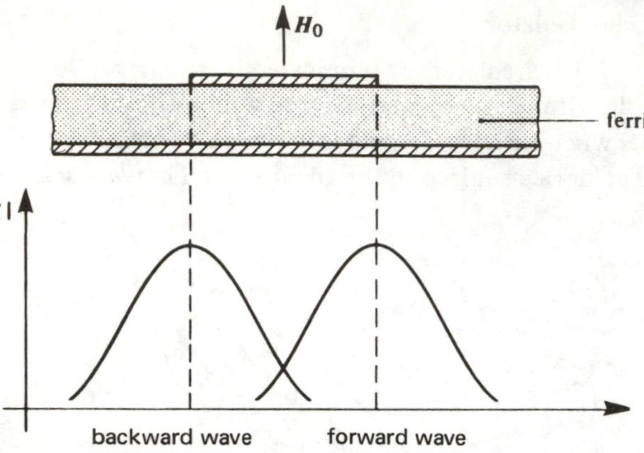

Fig. 6.70 Edge modes on a microstrip line on a ferrite substrate.

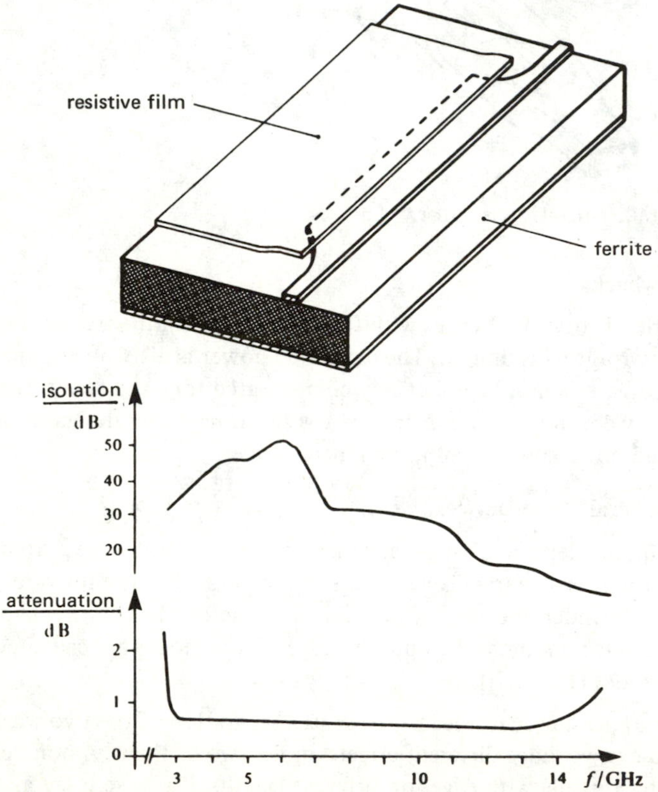

Fig. 6.71 Edge mode isolator in microstrip.

The frequency band is extremely wide. Since the ferrite substrate possesses non-negligible losses, the forward attenuation remains rather high [134].

Circulators having an arbitrary number of ports may be designed with peripheral modes (Fig. 6.72) [135].

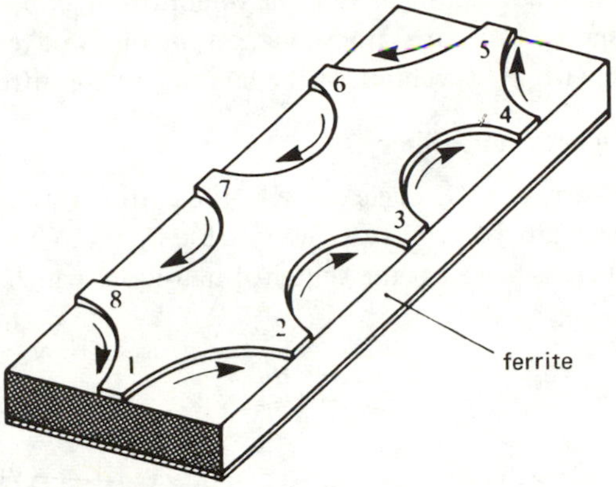

Fig. 6.72 Principle of an eight-port circulator with edge modes.

6.7.22 Ferrite Switches and Modulators

When the biasing magnetic field varies in time ($\boldsymbol{H}_0(t)$), a ferrite may be used to switch or to modulate a signal. For instance, the inversion of the biasing field ($\boldsymbol{H}_0$ is replaced by $-\boldsymbol{H}_0$) in a circulator produces the reversal of the sense of rotation, and thus of the numbering sequence of the ports. An input may thus be switched to two outputs, by changing the current fed to an electromagnet. The switch obtained remains a nonreciprocal device. A variable field isolator similarly becomes an electronically adjustable attenuator, a switch or an amplitude modulator. A nonreciprocal phaseshifter behaves as a phase modulator, and so on.

In principle, any one of the structures presented in Section 6.7 permits the design of control or modulation devices. However, some additional requirements should be considered for selection:

1. The energy stored within the magnetic circuit. The driving circuit must supply the energy required to modify the parameters (attenuation or phaseshift) in a modulator, or to switch between two states in a bistable switch. It is thus recommended to select a structure operating with a small biasing magnetic field.
2. Eddy current losses. During the switching operation, eddy currents are induced in the waveguide walls or in the conductors of the line (stripline, microstrip). In the switching circuit, these conductors appear as a trans-

former having its output windings short-circuited This effect may be reduced by making use of a small magnetic field and by reducing as much as possible the thickness of the conductors. For instance, metallized ceramic or thermoplastic parts may be used.

Ferrite switches and modulators permit one to control high power microwave signals. Their response times are rather long, on the order of tens to hundreds of microseconds, with the exception of the latching phaseshifter (§ 6.7.12).

6.7.23 Gyromagnetic YIG Filter

In Yttrium-Iron Garnet (YIG) single crystals, the gyromagnetic resonance has a very narrow line (Fig. 6.54), which allows tunable filters to be realized. A polished YIG sphere is placed at the center of half-loops which are orthogonally located in space (Fig. 6.73).

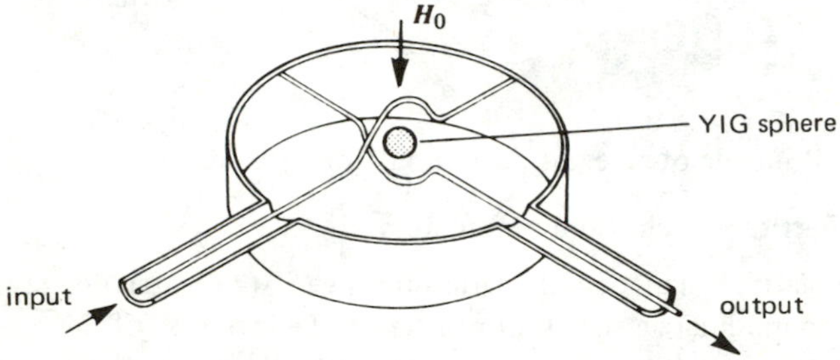

Fig. 6.73 YIG gyromagnetic filter.

Outside of the resonance, the two lines are not coupled to each other, because they are placed at right angles. At resonance, the off-diagonal term $\underline{K}$ of the permeability tensor $\underline{\bar{\bar{\mu}}}$ produces coupling between the orthogonal loops. Once the device is matched, it is a very narrow band gyrator. The resonance frequency is a function of the magnetic field $\boldsymbol{H}_0$ produced by an electromagnet [136].

The resulting device is an electronically tunable *gyromagnetic filter.* By acting on the current biasing the electromagnet, the center frequency of the bandpass filter is adjusted. Typically, such filters cover the band from 0.5 to 40 GHz. They are used in particular for swept frequency generators, frequency counters (§ 5.3.5), and spectrum analyzers (Sec. 5.4).

6.8 SOLID STATE CONTROL DEVICES

6.8.1 Introduction

Microwave signals of low power levels may be electrically controlled, switched and modulated by solid state junction devices: switches, attenuators, phaseshifters, tunable filters, and limiters. These devices are characterized by:

1. a low drive power, on the order of 100 mW at the most;
2. a fast response time, on the order of 100 nanoseconds, or less;
3. a small bulk and low weight;
4. a long lifetime and a high reliability (as long as they are properly taken care of).

Signals having up to 1 kW of average power and 100 kW of pulse power (§ 5.8.3) may be controlled within the frequency bands extending from 30 MHz up to 30 GHz [137].

The semiconductor components either present a capacitance which varies with the bias voltage (varactor, § 4.9.3) or a current-dependent resistance (PIN diode, § 6.8.2).

6.8.2 PIN Diode

The PIN junction, made of a layer of ***intrinsic semiconductor*** sandwiched between two highly doped regions, respectively p and n, presents a variable resistance for microwave signals (Fig. 6.74).

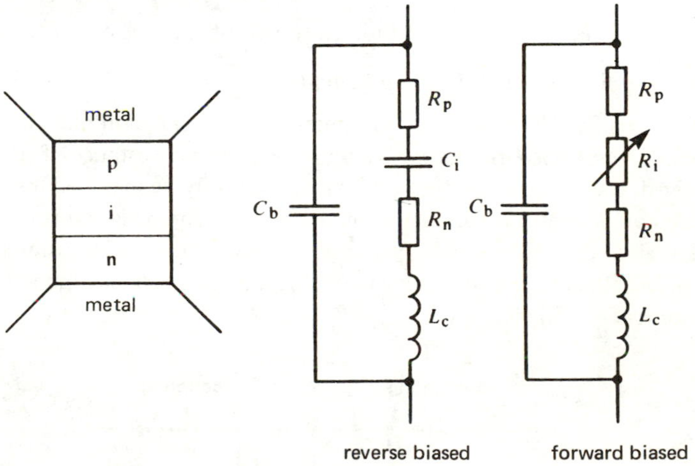

Fig. 6.74 PIN diode: geometry and equivalent circuit (reverse and forward biased).

When inversely biased, the PIN diode is practically a capacitor: the region i is the dielectric, regions p and n the electrodes. The equivalent circuit of the semiconductor is a RC series circuit (C_i: region i; R_p, R_n: regions p and n). Parasitic elements must also be taken into account: the series inductance L_c of the conductors, and the shunt capacitance C_b of the encapsulation.

When biased in the forward direction, the intrinsic region becomes conductive, since carriers are injected into it from both sides. The equivalent circuit is then formed from a variable resistor R_i for the intrinsic region, the other elements remaining all the same. The resistor R_i of a PIN diode can be made to vary roughly from zero to infinity as a function of the bias current (Fig. 6.75).

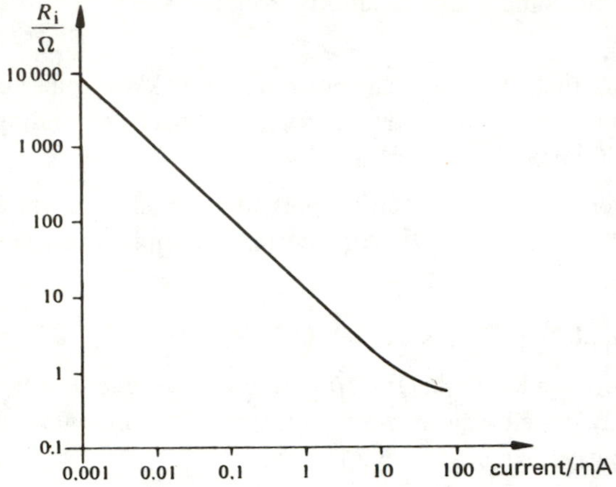

Fig. 6.75 Resistance R_i of a PIN diode in terms of the bias current.

6.8.3 Application: PIN Diode Attenuator

Connecting a PIN diode across a waveguide or a transmission line yields an ***electrically controlled attenuator***. However, since the resistance of the diode varies with the bias voltage, this attenuator may only be matched by using other variable components. The diode mismatch may be compensated for by cascading several diodes of the same type, separated by sections of transmission line (Fig. 6.76). PIN diode attenuators are utilized in particular within automatic level control (ALC) circuits.

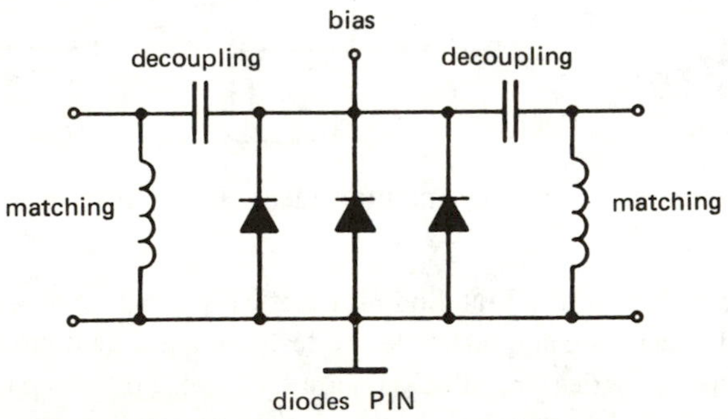

Fig. 6.76 PIN diode attenuator.

6.8.4 Application: PIN Diode Switch

The two extreme values taken by the resistance of the PIN diode correspond approximately to the open line and the short circuit. PIN diodes, connected either in series or in shunt on the line may be used to design a switch. Since in both states the diode only absorbs a small part of the microwave signal, this switch can control rather large power levels. More elaborate switches are fabricated by combining series and shunt mounted diodes on the transmission lines (Fig. 6.77).

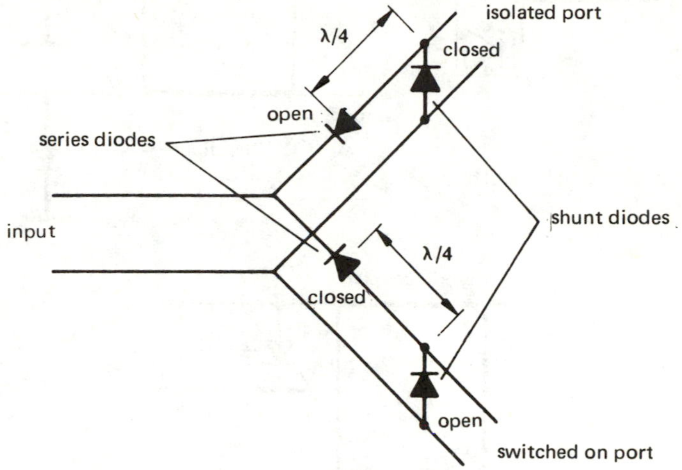

Fig. 6.77 Line switch.

In any practical design, the parasitic elements (encapsulation) must be taken into account. They may be matched, but they still degrade and limit the performance of the device. The best results are obtained by using unencapsulated diodes (chips) directly implanted on a microstrip substrate.

6.8.5 Application: PIN Diode Phaseshifter

Several kinds of bistable logic phaseshifters are realized by combining PIN diode switches with sections of line and junctions (Fig. 6.78). Phaseshifters of this kind are mainly utilized to electronically drive the phase array antennas of surveillance radars (§ 8.1.9).

6.8.6 Application: Varactor Phaseshifter

A continuous variation of the phase (analog phaseshifter) may be obtained by connecting two varactors at the ports of a hybrid coupler (Fig. 6.79). The phases of the reflected signals are adjusted by biasing the varactors. Due to the particular properties of the coupler (6.157), the reflected signals add in phase at the output and cancel out at the input, when the reflections of the two diodes are identical.

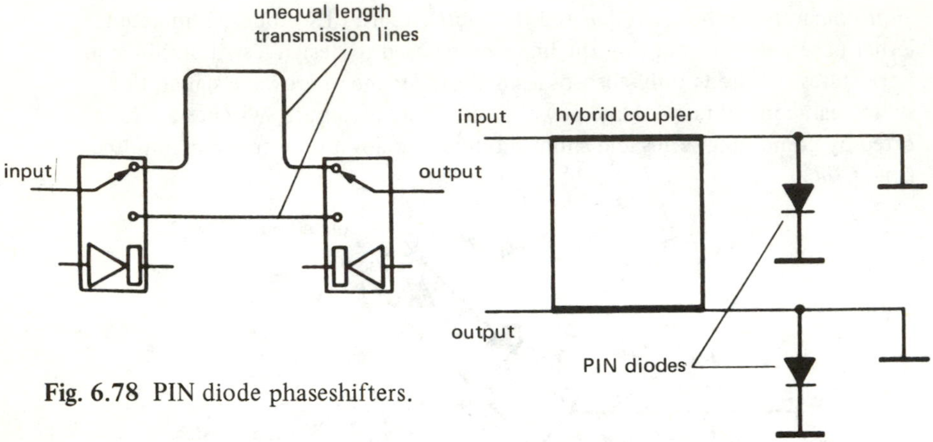

Fig. 6.78 PIN diode phaseshifters.

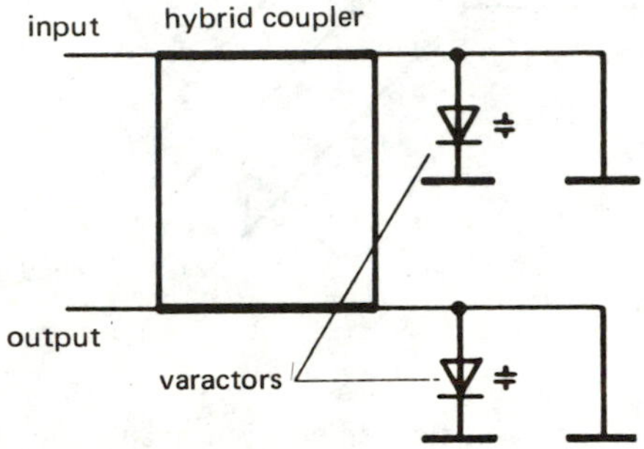

Fig. 6.79 Varactor analog phaseshifter.

6.8.7 Application: Tunable Filter

With a varactor connected to the end of a resonator the resonance frequency of the resonator can be changed. Sections 3.3.15 and 3.3.16 permit one to determine the frequency in terms of the capacitance(s).

6.8.8 Varactor Limiter

The capacitance of a varactor (§ 4.9.3) is not only a function of the applied direct voltage, but also of the amplitude of the microwave signal. This nonlinear property is utilized to realize passive limiters. An unbiased varactor, connected in shunt across the waveguide, presents a small capacitance at low signal levels. At large powers, the voltage across the varactor junction increases. Then the average value of its capacitance also increases (4.56). The resulting mismatch produces a reflection of the signal, in such a way that the transmitted signal remains at an almost constant level, called plateau, independently of the level of the input signal. This plateau is in turn limited, and the output signal rises again as the input power exceeds a threshold. The device realized in this manner is a *varactor limiter.* Beyond the threshold, the limiter exhibits a significant attenuation (Fig. 6.80). Varactor limiters are used to protect detectors and input stages of receivers, particularly for pulse radars.

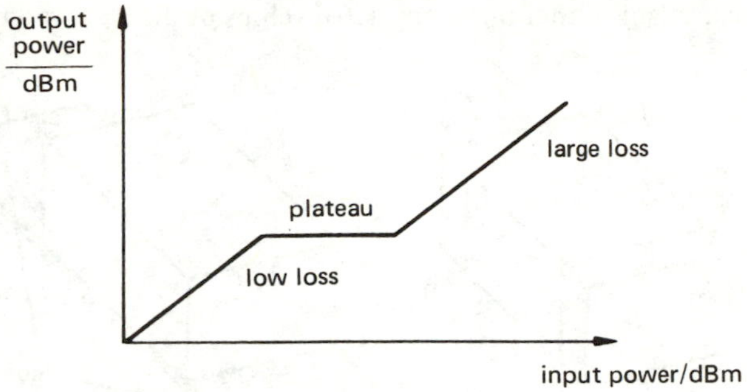

Fig. 6.80 Limiter characteristic (ideal).

6.9 INSERTION OF ELECTRONIC COMPONENTS

6.9.1 Introduction

Several signal processing functions (modulation, phaseshifting, amplification) require, in addition to sections of transmission lines, couplers, and junctions, several electronic components, which must be inserted within the circuit. Two main kinds of elements may be encountered: lumped elements (§ 6.9.2) and concentrated components (§ 6.9.5).

6.9.2 Lumped Elements

This name is used to designate components available on the market, ready to be inserted within the system:

1. Capacitors, most often of ceramic, of very small size. Capacitors are available with values ranging from a fraction of picofarad up to several nanofarads. Connections are made over metallized surfaces or with thin strips (Fig. 6.81).
2. Resistors, similarly of very small sizes, have values ranging from a few ohms up to several kiloohms. Connections are made in the same way as for capacitors.
3. Attenuators and terminations are manufactured with resistors (Fig. 6.82).
4. Inductances are made, either as small molded coils, or with a metal strip deposited over a dielectric substrate.
5. Miniature junction circulators or isolators (§ 6.7.13) are small disks, a few

centimeters in diameter, associated with a permanent magnet and thin metal strips connecting them to the circuit. A circular opening must generally be cut in the substrate to position them (Fig. 6.83).

6. Diodes of all kinds, Schottky (§ 7.1.5), PIN (§ 6.8.2), Gunn (§ 4.6.2), IMPATT (§ 4.7.4), TRAPATT (§ 4.7.5), and varactor (§ 4.9.3). These diodes are available either unencapsulated (chips or dice), or mounted into

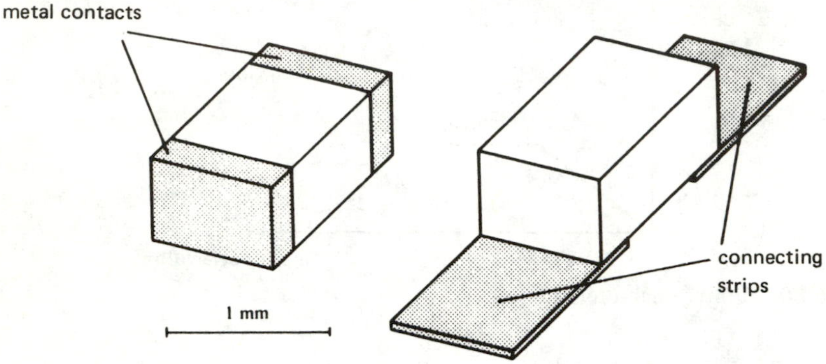

Fig. 6.81 Miniature capacitors for planar lines.

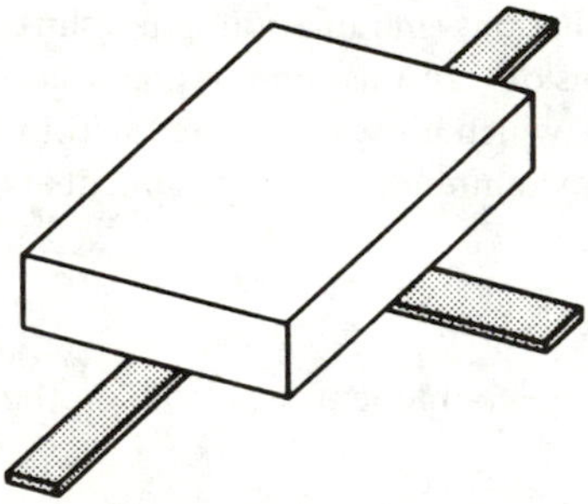

Fig. 6.82 Miniature attenuator for planar lines.

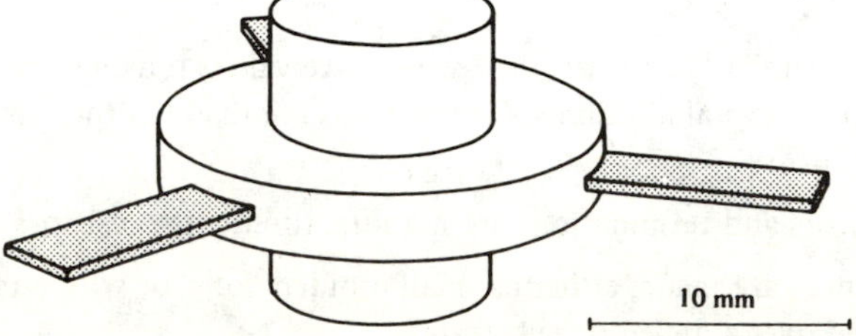

Fig. 6.83 Miniature circulator for microstrip.

enclosures of various types [138]. Particular encapsulations found are Leadless Inverted Devices (LID), beam leads, and cylindrical with or without setting screw. Several case types are represented in Figure 6.84.

7. Transistors, bipolar or MESFET. They also are available unencapsulated (chips or dice), in coaxial cases, or in specifically designed enclosures to mount within planar circuits (Fig. 6.85).

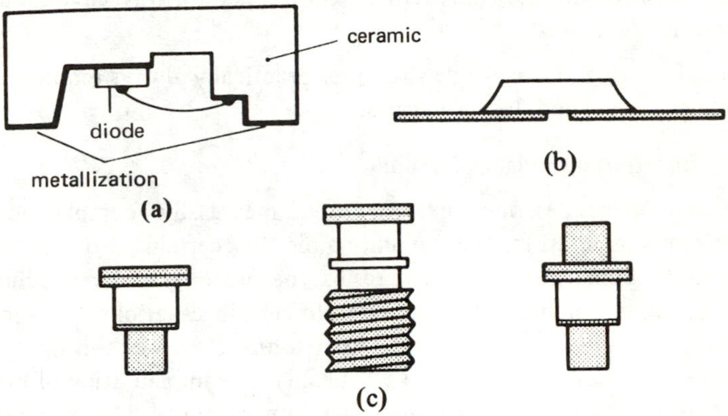

Fig. 6.84 Examples of diode encapsulations: (a) LID, (b) beam-lead, (c) various cylindrical cases.

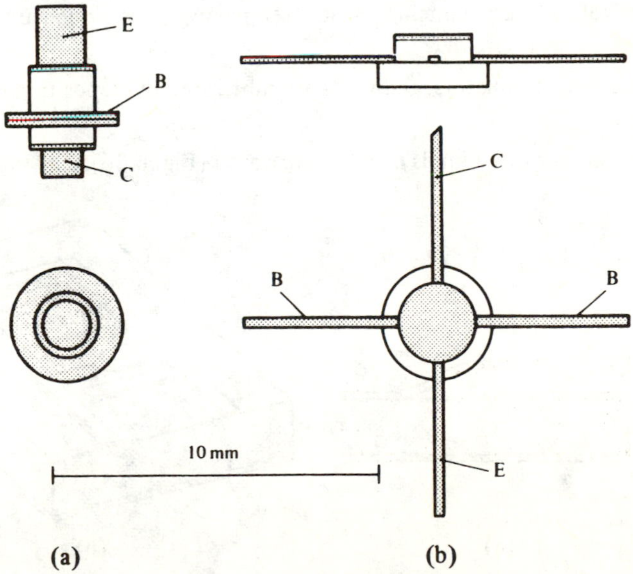

Fig. 6.85 Microwave transistors (a) coaxial encapsulation, (b) case for planar lines.

6.9.3 Waveguide Mounts

Almost without exception, the only lumped elements utilized within waveguides are diodes, generally encapsulated into a cylindrical case. Waveguide sections must be specifically made to insert the diodes. They contain a structure for mounting and biasing, as well as matching elements. In certain situations, reduced section waveguides or ridged waveguides are required The other functions are fulfilled with devices specifically developed for waveguides, outlined in the previous sections.

The manufacturing of waveguide structures practically always requires the machining of metal and dielectric parts.

6.9.4 Insertion into Planar Circuits

All the components described in Section 6.9.2 may easily be implanted into planar circuits (Sec. 2.11). Their mounting requires certain particular techniques for assembly: gluing with conductive resins, thermocompression welding, ultrasonic welding, or bonding with gold wires. In certain situations, holes must be drilled into the substrate to permit the insertion of elements; this may present difficulties with ceramic substrates (§ 2.11.23). The implantation of lumped elements is much easier in planar lines than in waveguides. This is one reason for the rapid growth of planar lines in recent years.

6.9.5 Concentrated Components

This name refers to components, most often passive ones, which are fabricated directly on a planar circuit:

1. Resistors, attenuators, and loads are fabricated by deposition of resistive layers.
2. Capacitors of two kinds may be fabricated (Fig. 6.86). The sandwich

(a)

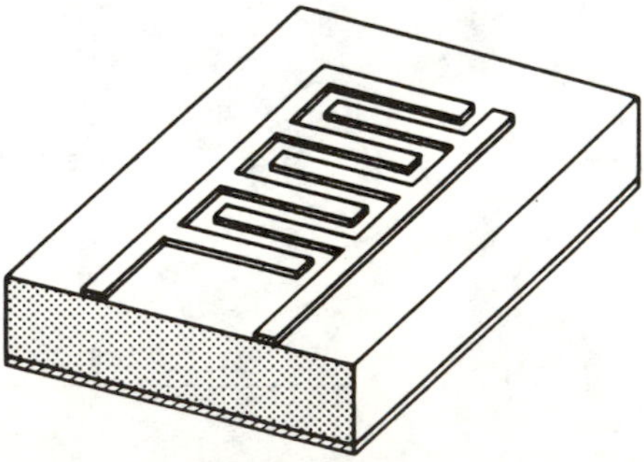

(b)

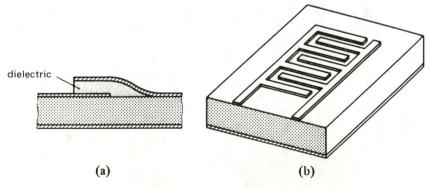

Fig. 6.86 Concentrated capacitors: (a) "sandwich" structure, (b) interdigital structure.

structure is obtained by deposition of a dielectric layer, then metallizing its upper surface. Capacitance values between a fraction of a picofarad and several hundred picofarads are obtained. Smaller capacitance values are provided by interdigital structures made using photoetching techniques (§ 2.11.22) [38].

3. Inductances shaped as spirals and loops are obtained by photoetching (Fig. 6.87). Typical values range from 1 to 10 nanohenries [139].

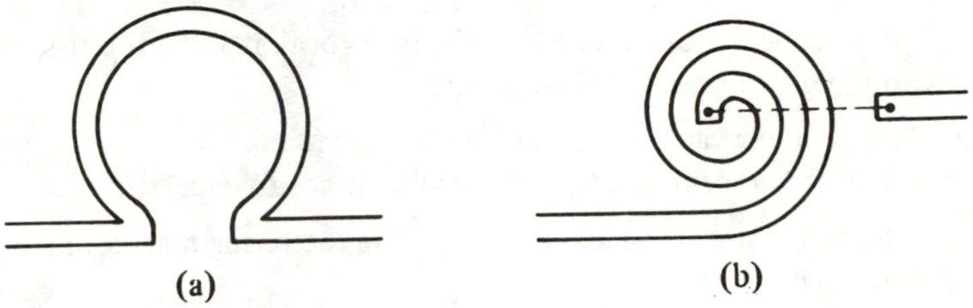

Fig. 6.87 Printed inductances: (a) loop, (b) spiral.

6.10 PROBLEMS

6.10.1 Two two-port devices represented by the following scattering matrices are cascaded:

$$\begin{bmatrix} 0.1 & 0.8 \\ 0.8 & 0.1 \end{bmatrix} \quad \begin{bmatrix} 0.4 & 0.6 \\ 0.6 & 0.4 \end{bmatrix}$$

Find the scattering matrix of the resulting device. Determine its properties: symmetry, reciprocity, losses, match.

6.10.2 Utilizing the energy conservation relation for the scattering matrix (6.24), determine the equivalent condition for the impedance matrix.

6.10.3 How does a shift of the reference planes affect the impedance matrix of a device?

6.10.4 A negative resistance amplifier provides a 13 dB gain. It is connected with a circulator having a 20 dB isolation and a 0.5 dB attenuation. Determine the scattering matrix of the assembly.

6.10.5 One wishes to realize a reflection varying both in amplitude and phase, by using a variable attenuator and a sliding short circuit. Determine the reflection factor $\underline{\rho}$ in terms of the attenuation and of the location of the short-circuit.

6.10.6 Determine which resistors are required to fabricate a T attenuator, knowing that it must meet the following requirements:

1. matching to a line of characteristic impedance 60 Ω;
2. attenuation of 27 dB.

6.10.7 A generator produces a 100 W signal, but cannot operate properly when the reflected signal has a power larger than 0.2 W. For a load which may become purely reactive, what device should one use to protect the generator? Give its minimal characteristics. The power reaching the load should be as large as possible.

6.10.8 Determine in terms of frequency the reactance produced by a microstrip stub (Fig. 6.27). Its length is 5 mm, its width 2 mm. The substrate has a relative permittivity $\epsilon_r = 9.5$, and a thickness h = 0.6 mm.

6.10.9 Is it possible to construct a lossless, reciprocal, matched four-port device by assembling two identical three-ports?

6.10.10 What coupling can one obtain from port 1 to port 3 of a three-port almost matched at two of its ports, if the VSWR is not to exceed 1.15?

6.10.11 Determine the equivalent circuit and the scattering matrix of a 5-port matched resistive divider.

6.10.12 An ideal directional coupler is provided at its fourth port with a load having a 1.05 VSWR. What is the directivity of the resulting assembly?

6.10.13 An ideal hybrid coupler has its ports 2 and 4 connected to identical sliding short-circuits. Determine the scattering matrix of the assembly, as a function of the position of the short-circuits.

6.10.14 An ideal hybrid tee is terminated by a short-circuit within the reference plane at port 2 and with a sliding short-circuit, having a reflection $\underline{\rho} = \exp(j\varphi)$ at port 4. Determine the transmission and the reflection of the resulting assembly. Indicate its possible uses.

6.10.15 One wishes to fabricate a power divider by assembling two hybrid couplers and a phaseshifter. Determine which assembly of the devices yields the desired behavior. Determine the ratio of output to input power in terms of the phaseshift. All the component devices used are matched.

6.10.16 In order to obtain a high power signal, two amplifiers are connected in parallel with hybrid junctions, assumed to be ideal ones. Determine the output powers as a function of the input power for the following situations:

1. two identical amplifiers having a 20 dB gain,
2. two amplifiers having respectively 20 dB and 26 dB gains, but the same phaseshifts,
3. two amplifiers having the same 20 dB gains, but phaseshifts of respectively 67 and 77 degrees.

In all cases, the amplifiers are matched.

6.10.17 Determine the frequency response of the Pound discriminator shown in Figure 5.3. All devices are assumed to be ideal.

6.10.18 Demonstrate the principle of operation of the circulators shown in Figure 6.60.

6.10.19 A junction circulator has a 17 dB isolation between ports 1 and 3. Which device could be used to increase this isolation, and where should one place it?

CHAPTER 7

DEVICE MEASUREMENTS

7.1 PRINCIPLES OF COMPARATIVE MEASUREMENTS

7.1.1 Definitions

A device is characterized by the terms of its scattering matrix (Sec. 6.1). These are always complex ratios of signals having different amplitudes and phases, but always the same frequency. Comparative measurements are of two distinct types:

1. reflection measurements, determination of the $\underline{s}_{ii}$ terms located on the diagonal of the scattering matrix. The quotient of incoming and outgoing signals at one port of the device must be determined; this is the ratio of backward to forward waves on the transmission line connected to port i. The other ports of the device must be terminated on matched loads. The measurements are made with a slotted line (Sec. 7.2) or a reflectometer (Sec. 7.3);
2. transmission measurements, determination of the off-diagonal terms $\underline{s}_{ij}$ $(i \neq j)$ of the scattering matrix. Here, the ratio of two signals at different locations within the same circuit must be measured (input and output). When only the amplitude ratio is considered, the parameter measured is the attenuation (or the gain). The measurement of the phaseshift usually requires an interferometry bridge.

Chapter 7 is completed by sections dealing with particular techniques for the measurements of resonant cavities (Sec. 7.5) and of the noise factor (Sec. 7.6).

7.1.2 Definition: Comparative Measurements

Comparative measurements deal with amplitude and phase differences of two ***signals which exist inside of the same circuit***. None of the two signals is known accurately. In contrast, signal measurements specifically look for the characteristics of a single signal: frequency, power, etc. (Ch. 5).

7.1.3 Example

It would of course be possible to measure separately the two signals and then take their ratio. As an example, the amplitudes of the two signals, determined by power measurements, yield the following values, taking into consideration the main sources of error:

$$P_1 = 10 \pm 1.2 \quad \text{mW} \tag{7.1}$$

$$P_2 = 9 \pm 1.1 \quad \text{mW} \tag{7.2}$$

The difference and the quotient are, respectively,

$$P_1 - P_2 = 1 \pm 2.3 \quad \text{mW} \tag{7.3}$$

$$P_2/P_1 = 0.9 \begin{matrix} +0.248 \\ -0.195 \end{matrix} \quad 1 \tag{7.4}$$

The resulting error is obviously not acceptable. In the present case, which is only slightly exaggerated, one does not even know which one of the two signals is the larger one! Such a situation is encountered whenever differences among large numbers are considered.

7.1.4 Conclusion

The accuracy of the power measurements could be increased, but this is not a practical alternative: it would increase cost, measurement time, etc. Consequently, comparative measurements are seldom carried out by taking two signal measurements.

Specific techniques were developed for comparative measurements. Most often, the signals are combined to determine their sum, difference, or still other linear combinations of the two. For instance, in a comparison bridge, one looks for a zero, adjusting respectively the attenuation and the phaseshift in one channel until two signals of equal amplitude and opposite polarity just cancel one another. The required attenuation and phaseshift are then measured.

7.1.5 Crystal Detector

The basic device used for most comparative measurements is the crystal detector. It is basically a solid-state rectifier, designed specifically for operation at microwaves: the transit time of the carriers must be short, and the parasitic capacitance small. Metal semiconductor junctions and also p-n junctions in heavily doped semiconductors (back diodes) are commonly used [140].

The area of the junction has to be small, ***to reduce the parasitic capacitance***. Point-contact diodes permit particularly small sizes, at the cost of a great fragility (Fig. 7.1). Planar processes also permit the realization of small size junctions, which are more rugged, in Schottky diodes (Fig. 7.2) [140].

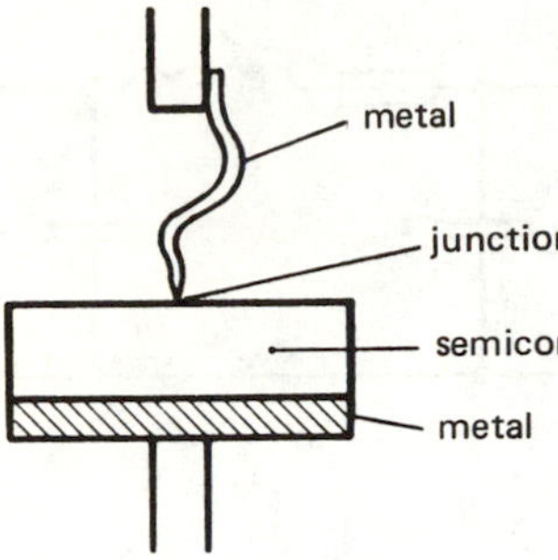

Fig. 7.1 Point-contact diode (greatly enlarged, without case).

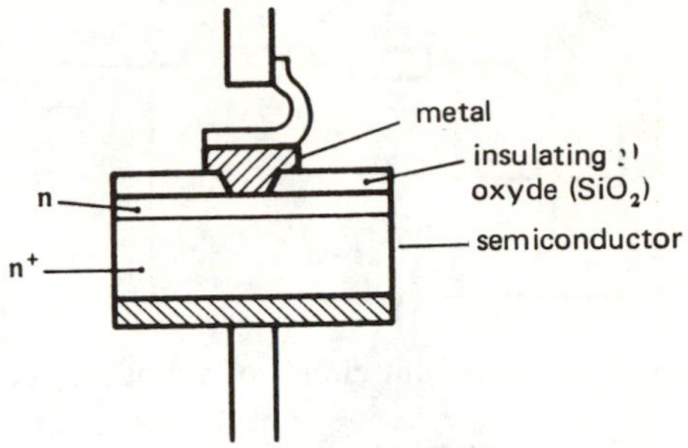

Fig. 7.2 Schottky diode (greatly enlarged, without case).

A large number of diode types is available for different frequency ranges, power levels, and applications. The same kind of diode is used in mixers (§ 5.3.14).

7.1.6 Principle of Operation: Square-Law Detector

In the junction, the current to voltage relationship at small signal levels is of the form:

$$I = I_s[\exp(\alpha U) - 1] = I_s\left[\alpha U + \frac{(\alpha U)^2}{2!} + \frac{(\alpha U)^3}{3!} + \ldots\right] \quad \text{A} \qquad (5.7)$$

A sine wave voltage $U = U_1 \sin \omega t$ is applied to this junction, and the resulting development of (5.7) yields a dc component produced by the even terms of the series:

$$I_c = I_s\left[\frac{(\alpha U_1)^2}{4} + \frac{(\alpha U_1)^4}{64} + \frac{(\alpha U_1)^6}{2304} + \ldots\right] \quad \text{A} \qquad (7.5)$$

Only the term in U_1^2 has a significant amplitude. The current flowing through the junction also has time-varying components at radial frequencies $\omega, 2\omega, \ldots n\omega, \ldots$

The diode is modelized by an equivalent circuit, in which the junction is a current source (5.7), to which are added the parasitic capacitances of the encapsulation C_b and of the junction C_c, the series inductance of the connecting wires L_c and the series resistance R_s of the semiconductor substrate (Fig. 7.3).

This model is a low-pass filter, which dampens the high frequency signals: only the dc component appears at the output.

The current is proportional to the square of the microwave voltage at the diode's terminals, and thus to the microwave power. This device is, therefore, called a *square law detector.* The measuring instruments to which it is connected are calibrated in terms of this characteristic.

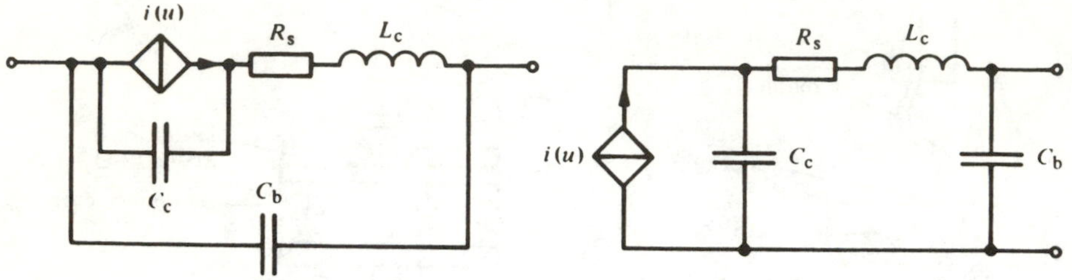

Fig. 7.3 Equivalent circuit of a diode detector.

7.1.7 Restriction

When the level of the input signal increases, the relation (5.7) is no longer applicable. The characteristic curve then tends to saturate, and the current becomes proportional to the voltage (a linear region). The transition from the square-law region to the linear region depends upon the type of device, its polarization and the input resistance of the measuring apparatus connected to the diode. Before any measurement is made, it is recommended to check whether the detector actually operates within the square-law region of its characteristic. If this is not the case, either the signal level must be reduced with an attenuator (§ 6.3.3 to 6.3.9), or the instrument must be calibrated to use the detector as a non-square-law element [141]. In the latter situation, the measurement is no longer simply comparative, as the calibration depends upon the signal level.

7.1.8 Square Wave Modulation

When carrying out measurements at a single frequency, the microwave signal is most often modulated by a square wave (Fig. 7.4). The modulation frequency generally used is 1 kHz. In this manner, the sensitivity of the measurement is increased: the detected signal envelope passes through a narrow-band filter which cuts down the noise level (§ 7.6.1). The problem of drift in dc amplifiers is also avoided [142].

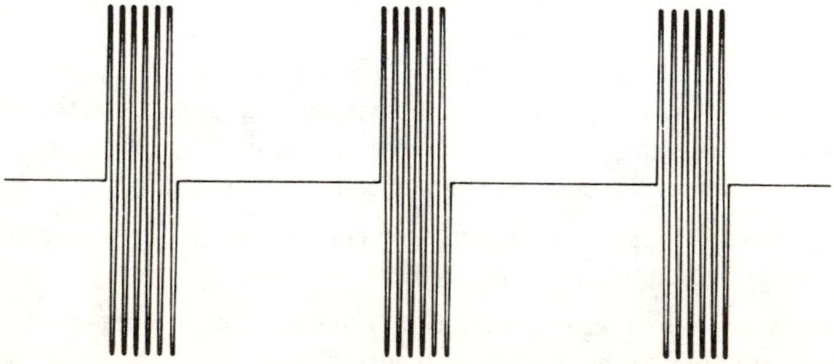

Fig. 7.4 Square wave modulated microwave signal.

7.2 REFLECTION MEASUREMENTS: SLOTTED LINE

7.2.1 Voltage Along a Line

A transmission line is connected to the port i of the device measured in reflection. Every other access is terminated into a matched load ($\underline{a}_j = 0, j \neq i$, Fig. 7.5).

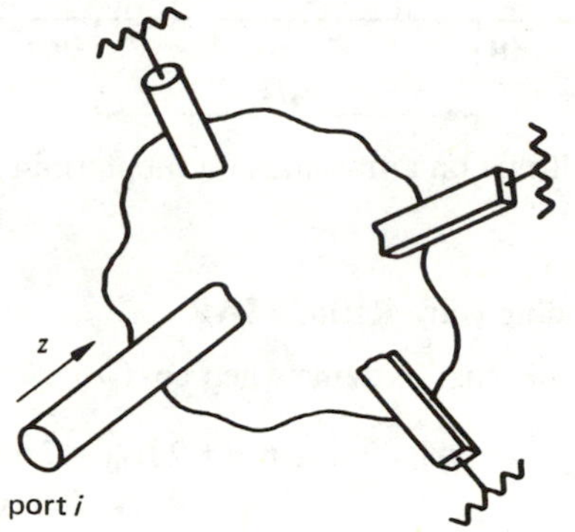

Fig. 7.5 Device connection for the measurement of $\underline{s}_{ii}$.

The voltage along the line i, represented by the complex normalized signals (§ 6.1.7), takes the value (6.3)

$$\underline{U}_i = \sqrt{Z_{ci}}\ (\underline{a}_i + \underline{b}_i) = \sqrt{Z_{ci}}\ \underline{a}_i\,(1 + \underline{s}_{ii}) \qquad \text{V} \tag{7.6}$$

in the reference plane.

It is here assumed that the device is passive and that therefore $|\underline{s}_{ii}| \leqslant 1$. Moving along the line, one has (§ 6.1.26):

$$\underline{U}_i(z) = \sqrt{Z_{ci}}\ \underline{a}_i(z)[1 + \underline{s}_{ii}\exp(2\mathrm{j}\beta_i z)] \qquad \text{V} \tag{7.7}$$

The voltage amplitude is given by:

$$|\underline{U}_i(z)| = \sqrt{Z_{ci}}\ |\underline{a}_i(z)|\sqrt{[1 + |\underline{s}_{ii}|\cos(\varphi + 2\beta_i z)]^2 + |\underline{s}_{ii}|^2\sin^2(\varphi + 2\beta_i z)}$$
$$= \sqrt{Z_{ci}}\ |\underline{a}_i(z)|\sqrt{1 + |\underline{s}_{ii}|^2 + 2|\underline{s}_{ii}|\cos(\varphi + 2\beta_i z)} \qquad \text{V} \tag{7.8}$$

where

$$\underline{s}_{ii} = |\underline{s}_{ii}|\exp(\mathrm{j}\varphi) \qquad 1 \tag{7.9}$$

The voltage shape is shown in Figure 7.6.

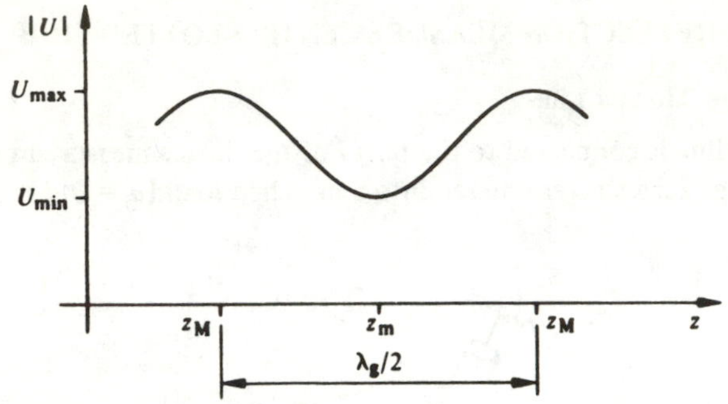

Fig. 7.6 Voltage amplitude on a transmission line terminated into a mismatched load.

7.2.2 Voltage Standing Wave Ratio, VSWR

Along the line, the voltage has extrema when $\cos(\varphi + 2\beta z) = \pm 1$ (§ 9.5.1)

$$U_{max} = \sqrt{Z_{ci}}\,|\underline{a}_i|\,(1 + |\underline{s}_{ii}|) \qquad \text{when } \varphi + 2\beta z_M = 2n\pi \qquad \text{V} \qquad (7.10)$$

$$U_{min} = \sqrt{Z_{ci}}\,|\underline{a}_i|\,(1 - |\underline{s}_{ii}|) \qquad \text{when } \varphi + 2\beta z_m = (2n+1)\pi \qquad \text{V} \qquad (7.11)$$

The ratio of the two extreme voltages is called the *voltage standing wave ratio* or *VSWR*, and designated as s:

$$s = \frac{1 + |\underline{s}_{ii}|}{1 - |\underline{s}_{ii}|} \qquad 1 \leqslant s \leqslant \infty \qquad 1 \qquad \textbf{(7.12)}$$

The magnitude of the diagonal term $\underline{s}_{ii}$ of the scattering matrix is obtained in terms of the voltage standing wave ratio:

$$|\underline{s}_{ii}| = \frac{s - 1}{s + 1} \qquad 1 \qquad (7.13)$$

In certain countries (like the U.K.), the voltage standing wave ratio is defined as the ratio U_{min}/U_{max} [216].

7.2.3 Comment

In order to determine $|\underline{s}_{ii}|$ by means of a VSWR measurement, one can either vary z, β, or both terms. In the first situation, a traveling probe determines the local voltage along the line (§ 7.2.4 to 7.2.15). The second approach utilizes one or two fixed probes. The frequency is then varied (by a sweeper), resulting in a variation of β (§ 7.2.16).

7.2.4 Waveguide Slotted Line

In a rectangular waveguide propagating the dominant TE_{10} mode, the surface current at the center of the broad walls is directed lengthwise (§ 2.3.12). A slot can be machined along the center, without cutting across any line of current, and therefore without significantly perturbing the electromagnetic field structure within the guide (Fig. 7.7).

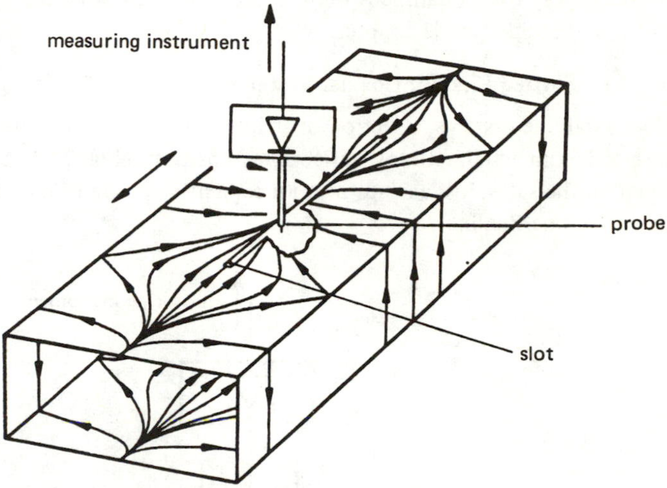

Fig. 7.7 Slotted line in waveguide.

A capacitive probe (§ 3.6.3) introduced into the slot samples a signal proportional to the transverse electric field, and therefore to $\underline{U}_i(z)$. The probe is connected to a crystal detector (§ 7.1.5) and to a measuring instrument. In this manner, the standing wave ratio VSWR and the location of the extrema z_m and z_M are determined by sliding the probe. The phase φ (§ 7.2.9) and the guide wavelength λ_g (§ 5.2.9) can be then calculated [109], [143].

7.2.5 Slot Mode

The change in conducting boundaries, caused by the cutting of a slot, allows another mode to propagate, the slot mode (Fig. 7.8).

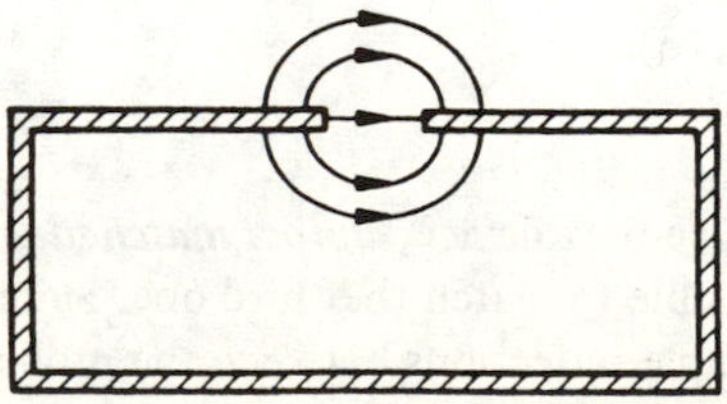

Fig. 7.8 Slot mode.

This is the dominant mode in slotline (§ 2.11.18). In theory, this mode is orthogonal to the TE_{10} mode; it may nevertheless be excited at discontinuities, in particular at the extremities of the slot. Since this mode has a propagation velocity different from that of the TE_{10} mode, its presence may give rise to beats, any measurement then becoming hopeless. The excitation of this mode must be prevented. This is done by placing lossy material around the probe tip, in such a way that it dampens the slot mode without affecting the TE_{10} mode in the waveguide (mode suppressor).

7.2.6 Slotted Line for Coaxial Systems

In coaxial line systems, a section of stripline with circular center conductor is inserted, connected with low-reflection transitions (Fig. 7.9). As the stripline is open on the sides, it is easy to insert a capacitive probe to measure the voltage pattern along the line.

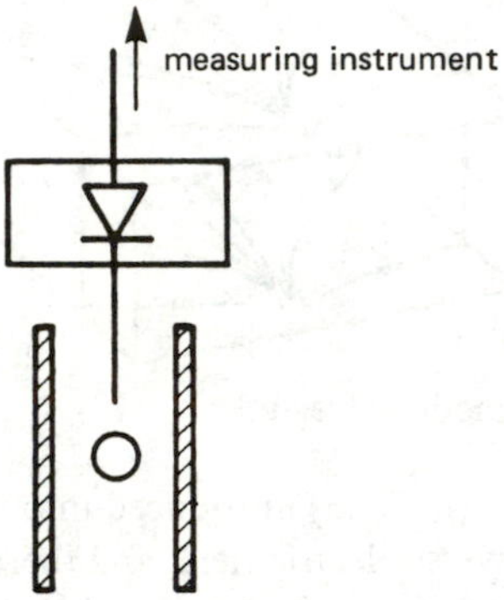

Fig. 7.9 Cross section of a slotted line in stripline, for measurements in coaxial sytems.

7.2.7 Necessary Precaution

The fields within the slotted line must not be perturbed, as that would lead to measurement errors; the probe must therefore penetrate inside the guide as little as possible. On the other hand, the detected signal must remain measurable. When very small signal amplitudes are encountered, a particular approach may be utilized (§ 7.2.13).

7.2.8 Comment

The slotted line is a three-port device, ***almost matched*** at two of its ports (§ 6.4.6). It is not possible to match the third one. Suitable precautions must be taken to avoid multiple reflections between the probe and the detector connected to it.

7.2.9 Measurement of the Phase

The phase φ of the reflection $\underline{s}_{ii}$ is defined at a reference plane, ***which must first be specified***. To measure this phase, the position z_m of a voltage minimum (9.33) is determined in the slotted line connected to the device measured. In theory, the distance between this minimum and the reference plane can be measured, and then φ calculated with (7.11). However, this distance is often quite difficult to measure accurately. It is determined by an ***additional measurement***, placing a short-circuit in the reference plane. The position z_c of a voltage minimum in the line is determined (Fig. 7.10).

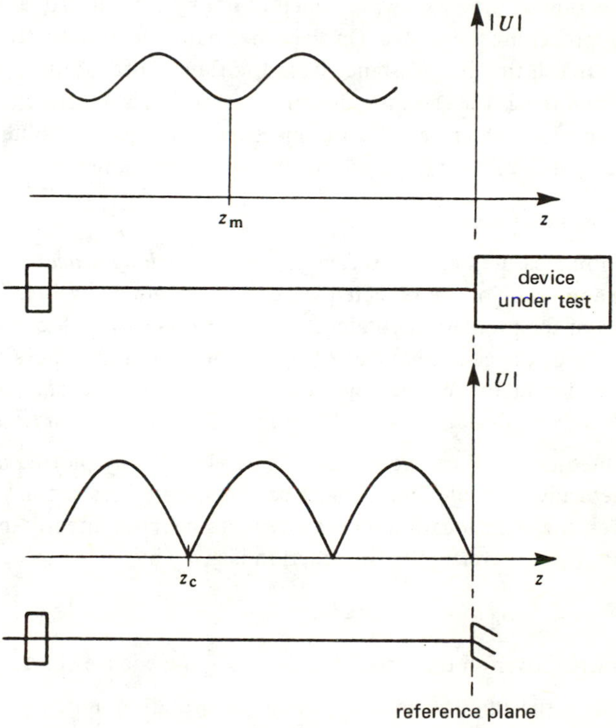

Fig. 7.10 Measurement of the phase of the reflection factor.

The minimum in z_c is known to correspond to $\varphi = \pi$, and thus, using (7.11):

$$\pi + 2\beta z_c = (2n + 1)\pi \qquad \text{rad} \qquad (7.14)$$

and consequently,

$$\beta z_c = n\pi \qquad \text{rad} \qquad (7.15)$$

The value of φ is then obtained from (7.11).

$$\varphi = 2\beta(z_c - z_m) + (2n + 1)\pi = 4\pi(z_c - z_m)/\lambda_g + \pi \pm 2n\pi \quad \text{rad} \qquad (7.16)$$

Due to the periodicity, the phase φ is only defined to n periods ($\pm 2n\pi$). In practice, the value of n is selected to yield the principal value for the phase $-\pi \leqslant \varphi \leqslant \pi$.

7.2.10 Comment: Direction of z

The direction of the z-axis was defined in Figure 6.1: the value of z increases when approaching the device. On the other hand, on a measurement bench, the scale on the slotted line (distance from the flange) often points away from the load. When this is the case, the sign of z within (7.16) must be modified. It is quite important, when carrying out measurements, to determine the geometrical layout of the line with respect to the device measured.

■ 7.2.11 Smith Chart

Reflection measurements are often part of a ***matching process***, in view of reducing the amplitude of the reflected wave to an acceptable value. The graphical solution of the problem is obtained in the *Smith chart* (§ 9.5.3 to § 9.5.7). It is a bijection of the complex plane of the normalized impedances on the plane of the reflection factor. With it, the value of the input impedance or of the input admittance of a device can be determined when $\underline{s}_{ii}$ is known (Fig. 7.11).

The values measured on the slotted line can be directly plotted on the chart. Commercially available Smith charts have associated scales for VSWR and phase. However, the angle location on the chart can be more directly indicated as a fraction of the wavelength. This fraction is given by:

$$\varphi_{\lambda Z} = (z_c - z_m)/\lambda_g \pm n/2 = \varphi/4\pi - 1/4 \quad 1 \qquad (7.17)$$

to be carried over on the internal scale of the impedance chart.

When using the admittance chart, a half-turn rotation must be added (§ 9.5.5), in which case:

$$\varphi_{\lambda Y} = \varphi_{\lambda Z} + 1/4 = (z_c - z_m)/\lambda_g + 1/4 \pm n/2 \quad 1 \qquad (7.18)$$

7.2.12 Transmission Line Matching

The main approaches available to match a load to a transmission line are described within section 9.6 and in several specialized books [87, 144]. In microwaves, shunt-connected reactive elements are used most often for matching (Fig. 7.12). The susceptances are realized using thin irises in waveguides (§ 6.3.15 to 6.3.20) or stubs in microstrip (§ 6.3.23).

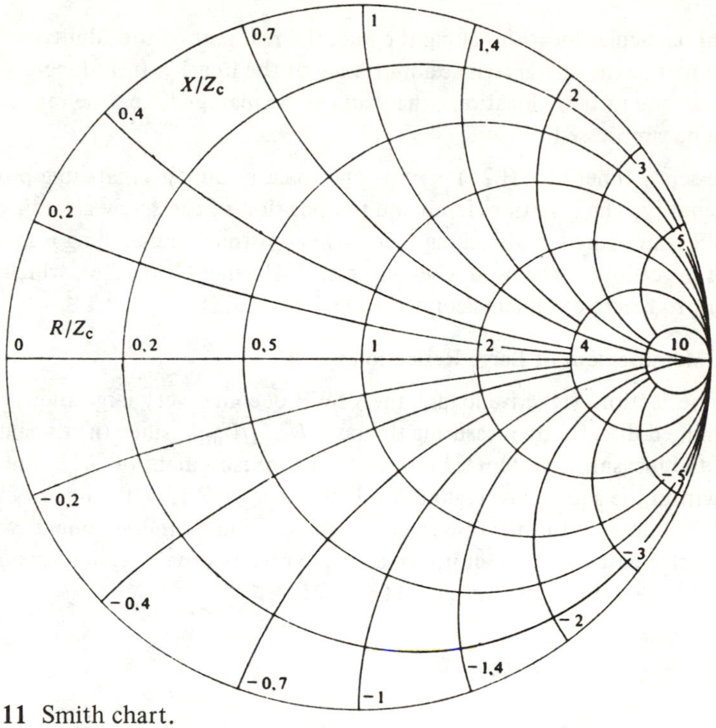

Fig. 7.11 Smith chart.

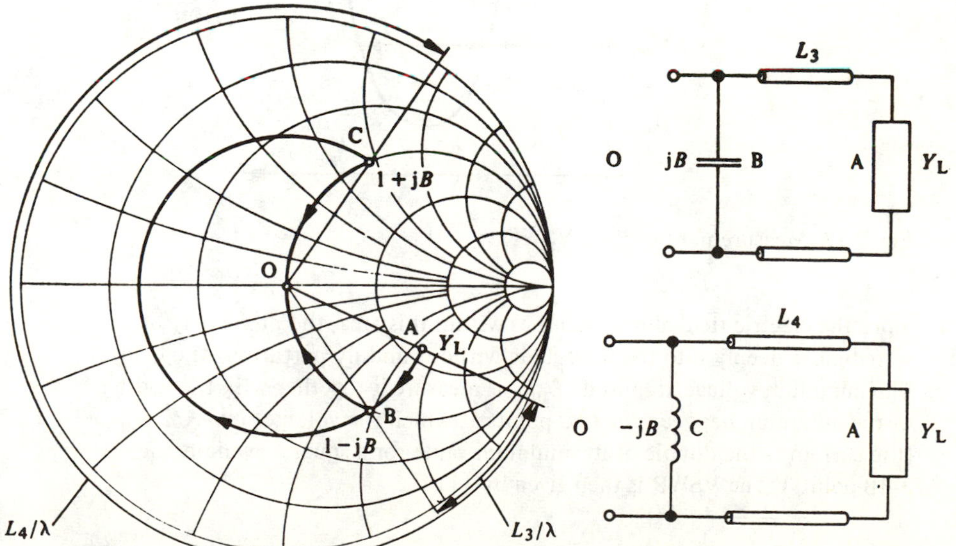

Fig. 7.12 Single frequency matching with a shunt susceptance.

At several particular locations along the line, the real part of the admittance $\underline{Y}_t(z)$ is equal to the characteristic admittance of the line Y_c. Introducing a susceptance at one of these locations, the load can be matched, i.e., the reflected wave can be suppressed.

The slide-screw tuner (§ 6.3.21) permits one to carry out this matching process experimentally. The insertion depth and the position of the screw are adjusted, the VSWR is measured, and making successive adjustments, matching is made. A similar procedure is followed when using an E-H tuner (§ 6.5.28), which is equivalent to reactive elements connected in T (§ 9.5.7).

7.2.13 Measurement of Large Reflections

For large reflections (reactive loads), the VSWR becomes very large and cannot be determined directly by measuring the ratio U_{max}/U_{min}, since the dynamic range of the measuring system is limited. As the measurement of U_{max} must remain within the square-law region of the detector (§ 7.1.7), the one of U_{min} may well fall down into the noise level, where no accurate measurement can be made. An alternate way was set up to take care of this situation: only the close vicinity of the minimum is considered (Fig. 7.13).

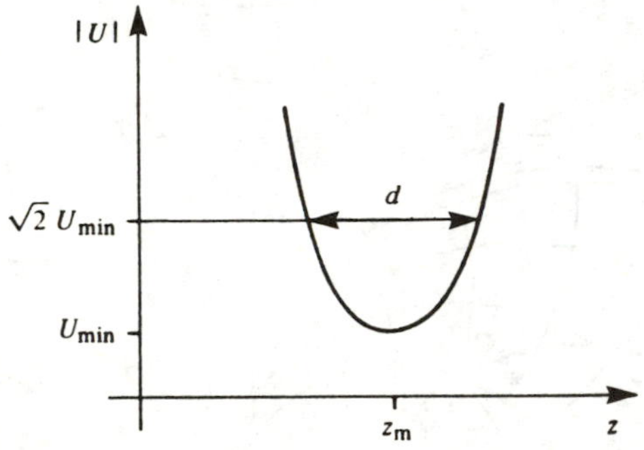

Fig. 7.13 Measurement of large VSWR.

Since the electric field almost vanishes within this area, the probe may be inserted more deeply into the waveguide without unduly perturbing the fields. The minimum voltage amplitude U_{min} is measured, together with the width d of the minimum between the two points at which the voltage $|\underline{U}| = \sqrt{2}\, U_{min}$ (the current is the double of its minimum value for a square-law detector, 3-dB points). The VSWR is then given by [111]:

$$s = \sqrt{1 + \frac{1}{\sin^2(\pi d/\lambda_g)}} \cong \frac{\lambda_g}{\pi d} \qquad 1 \qquad (7.19)$$

7.2.14 Source of Error: Residual VSWR

The slight mechanical irregularities and the discontinuities at the ends of the measurement region (slotted line or stripline) produce a residual VSWR which is less than 1.01 for a high quality line. The resulting error on $|\underline{s}_{ii}|$ is then smaller than ± 0.005.

7.2.15 Sequence of Measurements

The measurement of the reflection with a slotted line is carried out in the sequence:

1. tuning of the generator to the measurement frequency;
2. sliding (by hand) of the probe to the location of maximum voltage;
3. adjustment (by hand) of the measuring instrument to full scale reading;
4. sliding (by hand) of the probe to the location of minimum voltage;
5. reading of the VSWR on the instrument. Microwave instruments are usually graduated directly in VSWR;
6. determination of the position of a minimum measurement on the scale of the slotted line;
7. measurement of the position of a minimum with the short-circuit in the reference plane;
8. calculations;
9. the whole sequence is repeated for each frequency.

This measurement requires a significant amount of handling and does not allow for a continuous monitoring of the reflection. In particular, it is poorly suited to the experimental matching procedure, either using a slide-screw tuner (§ 6.3.21) or an E-H tuner (§ 6.5.28).

7.2.16 Response from a Fixed Probe

A detector connected to a fixed probe, located at a distance L from the reference plane, yields a current (7.8):

$$I_c \sim 1 + |\underline{s}_{ii}|^2 + 2|\underline{s}_{ii}| \cos(\varphi + 2\beta L) \qquad \text{A} \qquad (7.20)$$

A frequency sweep produces on the one hand a variation of $|\underline{s}_{ii}|$ and of φ, which is slow for non-resonant devices, and on the other hand a variation of β. When the line is long ($L \gg \lambda$), the last term of (7.20) varies quite rapidly as a function of the signal frequency. This term can be separated by a high-pass filter, and its trace observed on an oscilloscope. The amplitude $|\underline{s}_{ii}|$ corresponds directly to the envelope of the curve. Phase information is more difficult to extract [145].

The use of beats between incident and reflected waves of frequency-varying signals is the basic principle of the chirp radar (§ 8.1.10).

Two fixed probes, staggered along the line, can also be used to eliminate the slow varying terms of 7.20 [146]. The resulting signal is then further amplitude-modulated, an effect which is difficult to compensate.

As it can provide a very good accuracy, particularly at small values of $|\underline{s}_{ii}|$, this measurement technique looks most attractive. Unfortunately, it requires a rather long line (typically several meters), which makes its use awkward, and decreases its practical interest.

7.2.17 Swept Frequency Slotted Line

The frequency of the signal and the position of the probe in the slotted line may be varied at the same time, while recording the detected signal on an oscilloscope or on a plotter (Fig. 7.14).

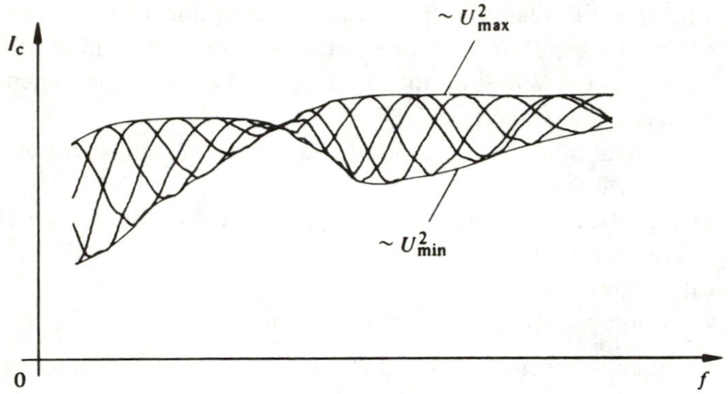

Fig. 7.14 Swept frequency response of slotted line.

A set of curves is obtained in this manner: one trace for each sweep in frequency. The envelopes yield, respectively, U_{max} and U_{min} as a function of frequency. The VSWR is given by their ratio. Displaying these curves on a logarithmic scale further simplifies the calculations: the VSWR is then directly related to the spacing between the two envelopes [147].

7.3 REFLECTOMETRY

7.3.1 Fundamentals

Reflectometric methods utilize one or two directional couplers (Sec. 6.5) to separate the reflected signal from the incident one. The comparison of the two yields the reflection factor. This approach provides a faster way to measure reflection, which is, however, affected by errors resulting from the finite directivity of the couplers. Such errors can be compensated for or corrected.

■ 7.3.2 Single Coupler Reflectometer

A directional coupler, having one port terminated into a matched load, is connected to port i of the device to be measured (Fig. 7.15). All other ports of the device are terminated into matched loads ($\underline{a}_j = 0, j \neq i$). The measuring instrument connected at port 3 of the coupler must also be matched.

Only the channels contributing to the reflectometer measurements are shown in the signal flow graph of Figure 7.15. The output signal at port 3 of the coupler is:

$$\underline{b}_3 = \underline{a}_1(\underline{s}_{21}\underline{s}_{ii}\underline{s}_{32} + \underline{s}_{31}) \qquad \mathrm{W}^{1/2} \qquad (7.21)$$

where $\underline{s}_{ii}$ is the reflection factor being measured, and $\underline{s}_{21}$, $\underline{s}_{32}$ and $\underline{s}_{31}$ are the respective transfer functions of the coupler (assumed to be matched).

Replacing the device measured by a short-circuit in the reference plane, the output signal at port 3 becomes:

$$\underline{b}_{3c} = \underline{a}_1(-\underline{s}_{21}\underline{s}_{32} + \underline{s}_{31}) \qquad \mathrm{W}^{1/2} \qquad (7.22)$$

The input signal $\underline{a}_1$ is eliminated, and taking the ratio of the two expressions:

$$\frac{\underline{b}_3}{\underline{b}_{3c}} = \frac{\underline{s}_{21}\underline{s}_{ii}\underline{s}_{32} + \underline{s}_{31}}{-\underline{s}_{21}\underline{s}_{32} + \underline{s}_{31}} \qquad 1 \qquad (7.23)$$

For an ideal coupler (infinite directivity), $\underline{s}_{31} = 0$, so that the ratio yields directly:

$$\frac{\underline{b}_3}{\underline{b}_{3c}} = -\underline{s}_{ii} \qquad 1 \qquad (7.24)$$

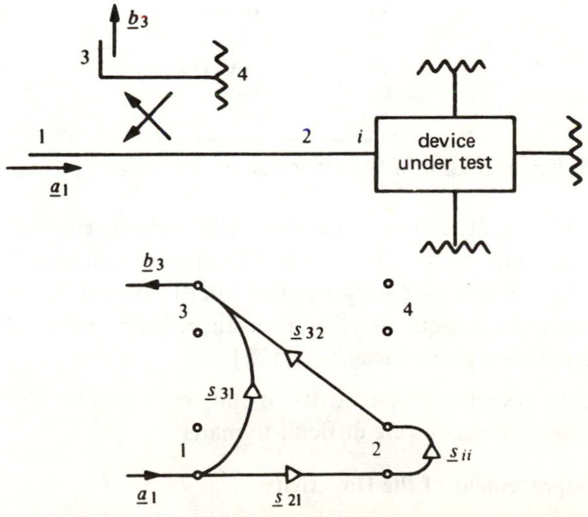

Fig. 7.15 Single coupler reflectometer.

For a real coupler, the term $\underline{s}_{31}$ is small, but not small enough to be neglected in the numerator of (7.23), in particular when $\underline{s}_{ii}$ itself is small. In the denominator, since $|\underline{s}_{31}| \ll |\underline{s}_{21}\underline{s}_{32}|$ for good quality couplers, this term can be neglected to first order, resulting in:

$$\frac{\underline{b}_3}{\underline{b}_{3c}} \cong -\underline{s}_{ii} - \frac{\underline{s}_{31}}{\underline{s}_{21}\underline{s}_{32}} \qquad 1 \qquad (7.25)$$

However, the measurements of $\underline{b}_3$ and of $\underline{b}_{3c}$ must be carried out in sequence. Only the amplitudes of these terms are given by the measurements, but not their arguments. The amplitude of $\underline{s}_{ii}$ is then known to be located between two limits:

$$\left|\frac{\underline{b}_3}{\underline{b}_{3c}}\right| - \left|\frac{\underline{s}_{31}}{\underline{s}_{21}\underline{s}_{32}}\right| \leqslant |\underline{s}_{ii}| \leqslant \left|\frac{\underline{b}_3}{\underline{b}_{3c}}\right| + \left|\frac{\underline{s}_{31}}{\underline{s}_{21}\underline{s}_{32}}\right| \qquad 1 \qquad (7.26)$$

7.3.3 Range of the Uncertainty

For a small coupling (large number of LC in dB), $|\underline{s}_{21}| \cong 1$. The remaining error term $|\underline{s}_{31}/\underline{s}_{32}|$ appears in the definition of the directivity (§ 6.5.15). The limits of the error range are thus directly related to the directivity of the coupler (Table 7.16).

Directivity LD/dB	$\lvert\underline{s}_{31}/\underline{s}_{32}\rvert$
10	0.3162
20	0.1
30	0.03162
40	0.01
50	0.003162
60	0.001
∞	0

Table 7.16 Upper bound of the error due to directivity.

In waveguides, the directivity of commercially available couplers with arrays of holes is on the order of 40 dB. The resulting error is then superior by a factor of two to the one obtained using a slotted line. To obtain the same accuracy, a 46 dB directivity is required. High precision couplers, having directivities of 50-60 dB, provide a greater accuracy [125].

In coaxial line systems, the directivity of couplers is smaller (20-30 dB), in particular because connectors are difficult to match.

7.3.4 Improvement of the Directivity

When carrying out measurements at a single frequency, the directivity can be improved by placing a tuner between the coupler and the device to be measured (Fig. 7.15). This tuner produces a fixed reflection, which can be used to cancel out $\underline{s}_{31}$. The tuner is adjusted using a sliding load.

7.3.5 Correction of the Error

The error produced by the finite directivity of the coupler can be evaluated and corrected, introducing a section of precision line between the coupler and the device [148]. The reflected signal can then be separated from the spurious signal by considering the beats of the two signals over frequency (Fig. 7.17).

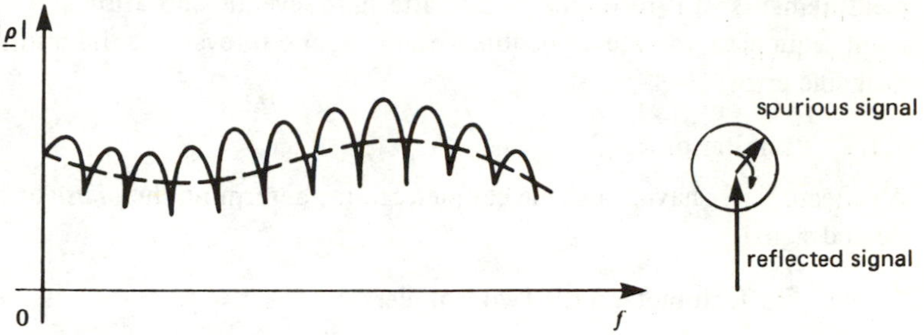

Fig. 7.17 Principle of correction of the error due to directivity.

7.3.6 Interest of the Method

Despite the error introduced by the coupler, reflectometry presents a great interest in practice, in particular when devices are to be matched. It is most often used with swept frequency techniques. The reflection is displayed on an oscilloscope screen or on a plotter, as a function of frequency (Fig. 7.18). A fixed short-circuit (§ 6.2.3) and a precision variable attenuator (§ 6.3.7) are used to calibrate the measurement setup.

The limits of the error, given in Table 7.16, can be directly reported on the diagram. One can then determine immediately whether a device meets a maximum

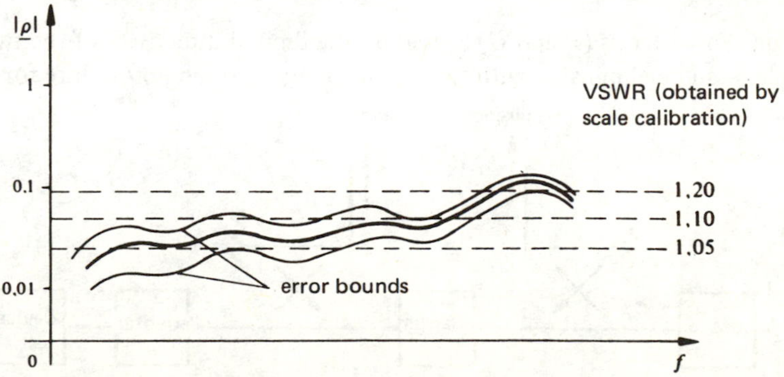

Fig. 7.18 Reflection versus frequency.

reflection specification or not, and detect possible anomalies (resonances and so on) which would elude point by point measurements. It is also possible to determine possible trouble areas (maxima of the reflection); more precise measurements are then carried out at those particular frequencies.

7.3.7 Condition: Stable Generator

To carry out accurate reflectometry measurements with a single coupler, a highly stable signal source, both in amplitude and in frequency, is required: the signal must not fluctuate during the time period separating the calibration from the measurements. It is recommended to alternate several calibration and measurement sequences, to detect possible variations, and to evaluate the resulting probable error.

7.3.8 Limitation

A reflectometer having a single coupler cannot determine the ***phase*** of the reflected signal.

7.3.9 Reflectometer with Two Couplers

The complete reflectometer assembly, formed of two couplers and associated tuning elements, is schematically represented in Figure 7.19.

The analysis of the reflectometer yields the following expression [149]:

$$\frac{\underline{b}_s}{\underline{b}_r} = \frac{\underline{A}\,\underline{s}_{ii} + \underline{B}}{\underline{C}\,\underline{s}_{ii} + \underline{D}} \qquad 1 \qquad (7.27)$$

$$\underline{A} = \underline{s}_{21}\underline{s}_{32} - \underline{s}_{31}\underline{s}_{22} \qquad 1 \qquad (7.28)$$

$$\underline{B} = \underline{s}_{31} \qquad 1 \qquad (7.29)$$

$$\underline{C} = \underline{s}_{21}\underline{s}_{42} - \underline{s}_{41}\underline{s}_{22} \qquad 1 \qquad (7.30)$$

$$\underline{D} = \underline{s}_{41} \qquad 1 \qquad (7.31)$$

The unwanted terms ($\underline{B}$ and $\underline{C}$) caused by the limited directivity of the two couplers can be eliminated with the use of tuners. A tuning procedure for single

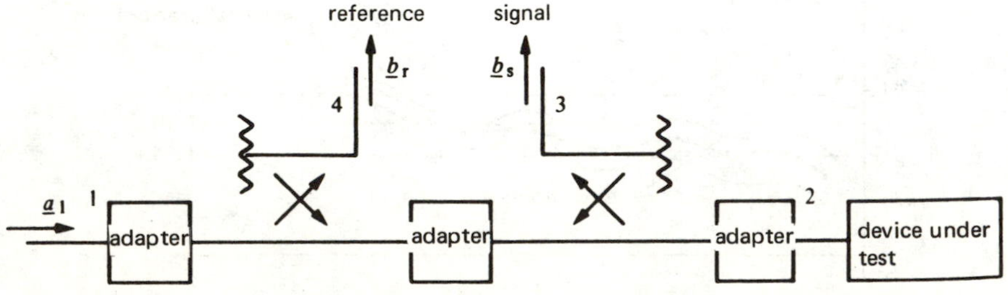

Fig. 7.19 Two-coupler reflectometer.

frequency operation was developed at the National Bureau of Standards (NBS, Boulder, Colorado, USA), leading to very high accuracies within the 2.6 to 18 GHz range.

$$\pm (0.00013 + 0.0032\ |\underline{s}_{ii}|) \qquad 1 \tag{7.32}$$

The reflectometer brings about two signals, $\underline{b}_s$ and $\underline{b}_r$, which are to be compared, analyzed, or used to control the generator. Several approaches are available:

1. Make use of the signal $\underline{b}_r$ to stabilize the generator (feedback). The measurement is then made with a single coupler (§ 7.3.2). No information is obtained on the phase.
2. Take the ratio of the two amplitudes in a ratio meter. The phase of the reflection is not measured in this approach either.
3. Complete the reflectometer with a microwave interferometry bridge (§ 7.3.10).
4. Frequency shift (§ 5.3.4) the two signals and complete their study (ratio, phaseshift) at lower frequencies by electronic means (§ 7.3.11).
5. Combine the two signals in a six-port (Sec. 6.6) and determine ratio and phaseshift in terms of the four detected amplitudes (§ 7.3.13 to 7.3.15).

7.3.10 Interference Bridge

Ports 3 and 4 of the reflectometer are connected to a variable phaseshifter (§ 6.3.12) and a variable attenuator (§ 6.3.7), respectively. The output signals are then combined in a hybrid tee (Fig. 7.20). When the two signals reaching the hybrid tee are identical (in amplitude and phase), no signal comes out of the difference arm. Knowing the attenuation and phaseshift required for this result, the value of the complex ratio $\underline{b}_s/\underline{b}_r$ is obtained, and then (7.27) yields $\underline{s}_{ii}$.

The attenuation of the phaseshifter and the phaseshift of the attenuator cannot be neglected. The two instruments must be calibrated accordingly.

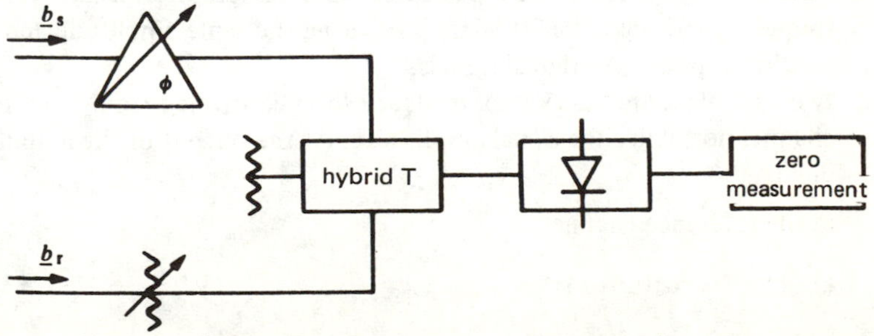

Fig. 7.20 Phase bridge complementing the reflectometer of Figure 7.19.

7.3.11 Network Analyzer

In a *network analyzer*, the microwave signals in both channels are replaced by lower frequency signals. The frequency shift, in which the amplitude and phase-shift relationships are carefully maintained between the signals, is carried out by mixing (§ 5.3.4) or by sampling (§ 5.9.3). ***Fixed frequency*** signals are obtained; they are analyzed by analog or digital electronic methods [143, 150]. The schematic diagram of a network analyzer is shown in Figure 7.21.

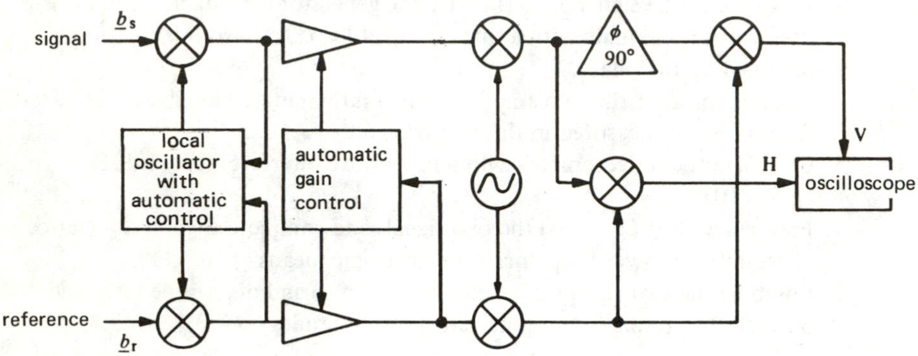

Fig. 7.21 Network analyzer.

Let us assume that the reflectometer was calibrated and tuned to yield a quotient $\underline{b}_s/\underline{b}_r$ precisely equal to the reflection factor:

$$\frac{\underline{b}_s}{\underline{b}_r} = \underline{\rho} = |\underline{\rho}| \exp(j\varphi) \qquad 1 \qquad (7.33)$$

The sequence of the operations is the following:

1. two sampling units carry out the first frequency shift. The variable microwave frequency signals are replaced by signals at an intermediate fixed frequency (for instance 20 MHz) presenting the same amplitude and phase relationships as the original signals;
2. two variable amplifiers with a feedback loop controlled from the reference channel normalize the signal amplitudes. At the output of the amplifiers, one has:

   in the reference channel,

   $$U_r(t) = U_0 \cos(\omega_{IF} t) \qquad \text{V} \qquad (7.34)$$

   and in the signal channel,

   $$U_s(t) = U_0 |\underline{\rho}| \cos(\omega_{IF} t + \varphi) \qquad \text{V} \qquad (7.35)$$

where ω_{IF} is the fixed intermediate angular frequency;

3. a second frequency shift is accomplished with mixers and a local oscillator, bringing the signal down to the low frequency (around 300 kHz). The amplitude and phaseshift relations are again maintained (7.34 and 7.35);
4. the signal and the reference are applied to a mixer, yielding a difference signal $|\underline{\rho}| \cos \varphi$;
5. the signal is phaseshifted by 90 degrees and mixed with the reference, yielding a signal $|\underline{\rho}| \sin \varphi$;
6. applying the two signals to a cathode-ray tube, the representation in the complex plane of $\underline{\rho}$ appears on the screen (Smith chart, § 7.2.11);
7. the ratio of the two signals yields $\tan \varphi$;
8. the two low frequency signals can be rectified, yielding $|\underline{\rho}|$.

When the ratio $\underline{b}_s/\underline{b}_r$ is not exactly equal to $\underline{\rho}$, the amplitude and phase differences can be compensated (change of reference plane) with attenuators and phaseshifters in the low frequency section of the instrument.

Last but not least, the analyzer can be coupled to a computer, programmed to correct the errors introduced by the couplers. A complex calibration procedure, using sliding short-circuits and loads, allows the determination of $\underline{A}$, $\underline{B}$ $\underline{C}$, and $\underline{D}$ in (7.27) as functions of frequency, and the storage of them. Once the measurement of $\underline{b}_s/\underline{b}_r$ is made, the exact value of $\underline{s}_{ii}$ is extracted from (7.27).

7.3.12 Comment

Network analyzers are extremely practical instruments, and fairly easy to use. They quickly and continuously provide the parameters to be measured. They are, however, rather expensive instruments, due to the highly sophisticated and sensitive apparatus they are composed of. Typically, they are specialized laboratory instruments.

7.3.13 Six-Port Measurements

A more recently developed approach linearly combines the microwave signals in the two channels (reference and signal). The combined signals obtained are rectified with crystal detectors (§ 7.1.5). From four amplitude measurements, the relationships linking the original microwave signals (amplitude and phase) can be determined.

7.3.14 Use of an Ideal Six-Port

An ideal six-port is shown in Figure 6.51. The signals $\underline{b}_s$ and $\underline{b}_r$ coming out of the reflectometer of Figure 7.19 are fed to ports 1 and 2, respectively, of the six-port. The four output signals at ports 3 to 6 are then proportional to the sum, the difference, and the quadrature combinations of the input signals. Detecting these four signals, the dc components of the currents in the diodes are:

$$I_3 \sim |\underline{b}_s + \underline{b}_r|^2 = |\underline{b}_s|^2 + |\underline{b}_r|^2 + 2|\underline{b}_s\underline{b}_r| \cos\theta \quad \text{W} \tag{7.36}$$

$$I_4 \sim |\underline{b}_s + \mathrm{j}\underline{b}_r|^2 = |\underline{b}_s|^2 + |\underline{b}_r|^2 - 2|\underline{b}_s\underline{b}_r| \sin\theta \quad \text{W} \tag{7.37}$$

$$I_5 \sim |\underline{b}_s - \underline{b}_r|^2 = |\underline{b}_s|^2 + |\underline{b}_r|^2 - 2|\underline{b}_s\underline{b}_r| \cos\theta \quad \text{W} \tag{7.38}$$

$$I_6 \sim |\underline{b}_s - \mathrm{j}\underline{b}_r|^2 = |\underline{b}_s|^2 + |\underline{b}_r|^2 + 2|\underline{b}_s\underline{b}_r| \sin\theta \quad \text{W} \tag{7.39}$$

where θ is the phase difference between $\underline{b}_s$ and $\underline{b}_r$.

The desired relations are then calculated. From (7.36) to (7.39) one obtains:

$$\theta = \arctan\left(\frac{I_6 - I_4}{I_3 - I_5}\right) \quad \text{rad} \tag{7.40}$$

$$\left|\frac{\underline{b}_s}{\underline{b}_r}\right| + \left|\frac{\underline{b}_r}{\underline{b}_s}\right| = \frac{2(I_3 + I_5)}{\sqrt{(I_6 - I_4)^2 + (I_3 - I_5)^2}} \quad 1 \tag{7.41}$$

The value of $|\underline{b}_s/\underline{b}_r|$ is obtained from (7.41). However, this equation has two possible solutions. To suppress the ambiguity, one has to know which one of the two terms is the larger one. One can see that, while the microwave network is fairly simple, made up of passive elements and detectors only, the more difficult aspects of the measurements are shifted towards the mathematical processing of the acquired data. While the basic principle has been known for a long time, the actual use of six-port techniques only became of interest with the advent of microprocessors.

The measurement can be simplified somewhat by adding couplers and detectors to determine directly the amplitude of $|\underline{b}_r|$ and/or that of $|\underline{b}_s|$, simplifying the calculation of $|\underline{b}_s/\underline{b}_r|$ and, in particular, eliminating the ambiguity in (7.41). The resulting elements are then 7-ports or 8-ports respectively. A certain redundancy is obtained, the equations outnumbering the unknowns. This allows one to cross-check the accuracy of the results obtained and to detect possible malfunctions (such as defective detectors).

7.3.15 Use of a Real Six-Port

The practical implementation becomes more complex when dealing with actual components, as opposed to the ideal ones of the previous paragraph: the phase-shifts between channels are no longer exactly $n \cdot \pi/2$ ($n = 0, \ldots 3$), the power split is not the same in different channels, the omnipresent mismatches produce spurious feedback loops, the responses of the detectors are different. The practical result is that (7.35) takes a more general form:

$$I_3 = K_3|\underline{s}_{31}\underline{b}_s + \underline{s}_{32}\underline{b}_r|^2 = K_3[|\underline{s}_{31}\underline{b}_s|^2 + |\underline{s}_{32}\underline{b}_r|^2 + 2|\underline{s}_{31}\underline{s}_{32}\underline{b}_s\underline{b}_r| \cos(\theta + \varphi_{31} - \varphi_{32})] \quad \text{A} \tag{7.42}$$

where K_3 is the conversion factor of the detector connected to port 3, and $\underline{s}_{31}$ and $\underline{s}_{32}$ are the transfer functions from ports 1 and 2 to port 3, with arguments φ_{31} and φ_{32}. All these terms can be determined by applying known signals to ports 1 and 2 (the calibration of the 6-port).

Similar relations are obtained for the currents in the three other outputs I_4, I_5, and I_6. It is not possible, in this general situation, to derive explicit expressions for θ and for $|\underline{b}_s/\underline{b}_r|$. Four transcendental equations are to be solved, and one tries, using an optimization process, to find the values of θ, $|\underline{b}_r|$ and $|\underline{b}_s|$ which satisfy all of them at the same time [151]. When computations are made quickly enough, this approach provides an almost real-time measurement (a swept frequency). Sufficient memory storage must be available to store the values of $\underline{A}$, $\underline{B}$, $\underline{C}$, and $\underline{D}$ of (7.27), together with the four conversion factors K_i of the detectors and the eight transfer functions of the six-port, at all the frequencies used in the measurements.

7.3.16 Time-Domain Reflectometer, TDR

The echo technique, currently used in pulsed radars (§ 8.1.6) produces a short pulse of microwave signal and analyzes the reflection coming back. The very same principle is utilized in the *time domain reflectometer* or *TDR*. This apparatus detects discontinuities along a transmission line system, localizes them, and characterizes them. In contrast to the previously described techniques, which make use of a continuous wave signal (CW), the TDR is based on steps or pulses of signal.

7.3.17 Time-Domain Reflectometer in a Two-Conductor Line

Within systems made up entirely of two-conductor lines in which propagation takes place over a very broad frequency range without lower limit, a unit voltage step is emitted on the line and the line voltage is monitored on an oscilloscope (Fig. 7.22).

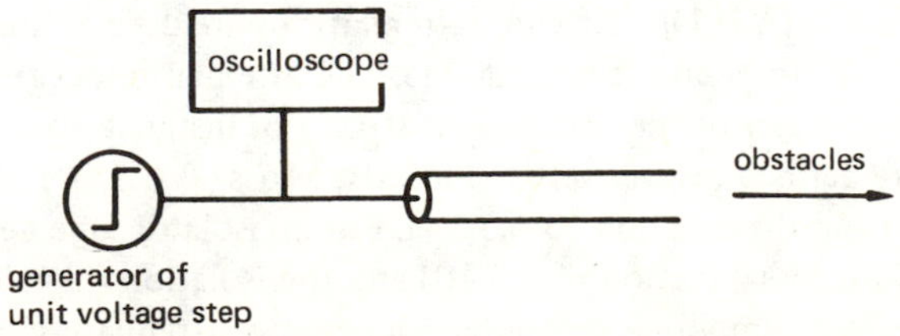

Fig. 7.22 Time-domain reflectometer.

Discontinuities give rise to echoes, which appear on the screen with a delay (position along the horizontal axis of the screen), proportional to the back and forth transmission time, and thus to the distance d to the disturbance, as per relation:

$$t = 2d/c = 2d\sqrt{\epsilon_r}/c_0 \qquad \text{s} \qquad (7.43)$$

where c_0 is the velocity of light ($\cong 3 \cdot 10^8$ m/s = 300 m/μs = 300 mm/ns = 0.3 mm/ps) and ϵ_r the relative permittivity of the dielectric material filling the line. On an air line ($\epsilon_r \cong 1$), the scale factor is thus 6.66 ps/mm. When a system is formed of several cables having different dielectrics, each section has its own scale factor. The accuracy of the measurement of the position is related to the risetime of the unit step.

The nature of the discontinuity is related to the shape of the voltage observed on the scope. Response curves for various loads are shown in Figure 7.23.

The reflection factor can also be determined as a function of the frequency, taking the ratio of the Fourier transforms of the emitted and received signals [152].

This method cannot be used when one of the conductors is broken off or missing, for instance, when connecting to a waveguide system.

7.3.18 Time-Domain Reflectometer in Waveguide

A similar method is available to study limited bandwidth systems (waveguides, etc.). It utilizes a pulse-modulated microwave signal (Fig. 5.22).

7.4 ATTENUATION AND PHASESHIFT

7.4.1 Definition: Attenuation, Insertion Loss

For any component having two or more ports, the *attenuation* or *insertion loss* between ports i and j is defined as the amplitude ratio of input signal at j to output signal at i. The other ports are all terminated into matched loads. The ***attenuation level*** in dB is defined as:

$$LA_{ji} = -20 \log |\underline{s}_{ji}| \qquad \text{dB} \qquad \textbf{(7.44)}$$

In an N-port, $N(N-1)/2$ different attenuations are defined for a reciprocal device, $N(N-1)$ for a non-reciprocal one. The attenuations correspond to the non-diagonal terms of the scattering matrix. For instance, in a directional coupler are defined respectively the insertion loss (§ 6.5.10), the coupling (§ 6.5.11) and the isolation (§ 6.5.14). For an isolator, one defines the insertion loss (low loss direction, § 6.3.10) and the isolation (high loss direction, § 6.3.10). In an amplifier, one obtains a negative attenuation, called gain or amplification (Ch. 4).

7.4.2 Measurement

The insertion loss is determined by taking the ratio of the power levels with and without the device inserted into the network. The generator and load are adapted

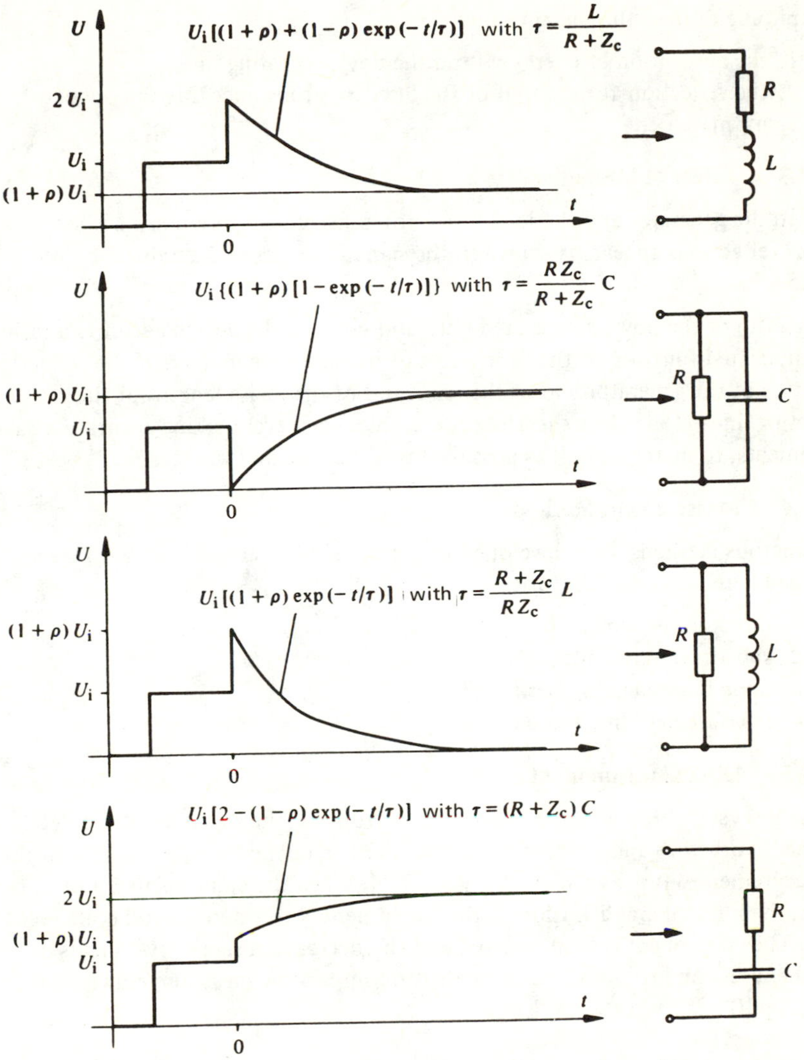

Fig. 7.23 Scope responses for complex load impedances. In each case $\rho = (R - Z_c)/(R + Z_c)$.

to the transmission lines, all other ports of the device are terminated into matched loads (Fig. 7.24).

The attenuation is then obtained by:

$$LA = -\ 10 \log(P_2/P_1) \qquad \text{dB} \qquad (7.45)$$

It must be noted that the device measured does not have to be matched. The amplitude drop of the signal results from:

1. the absorption of energy within the device (heating up);
2. the reflection at the input of the device (for instance due to connector mismatch).

7.4.3 Effect of Mismatches

When the generator and the load (measuring instrument) are not matched, multiple reflections appear, as shown in the signal flow-graph diagram of Figure 7.25.

The ratio of the powers measured with and without the device becomes a rather complicated function of the four terms of its scattering matrix, of the reflection factors of the generator and of the load, and of the phase lengths of the connecting lines [153]. It is therefore preferable, whenever making accurate measurements, to match as well as possible both the load and the source.

7.4.4 Measurement Methods

Numerous methods were developed to measure the insertion loss; they are arranged here into four categories:

1. direct measurement,
2. substitution method,
3. impedance measurement, and
4. measurement in a resonant cavity.

7.4.5 Direct Measurement

With a crystal detector and a measuring instrument, the signal levels are determined, both with and without the component to be measured inserted into the measurement setup, as shown in Figure 7.24. When the component is not in the line, the internal amplification of the instrument is adjusted for full scale reading. The component is then inserted and the decrease in signal level yields directly the insertion loss. Most instruments for comparative measurements are graduated in dB.

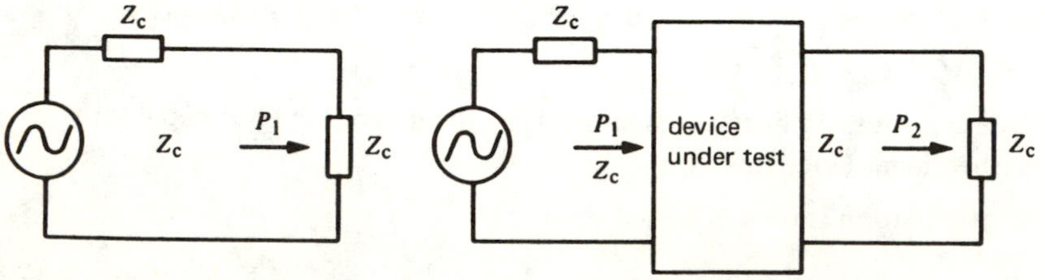

Fig. 7.24 Definition of powers in the measurement of attenuation.

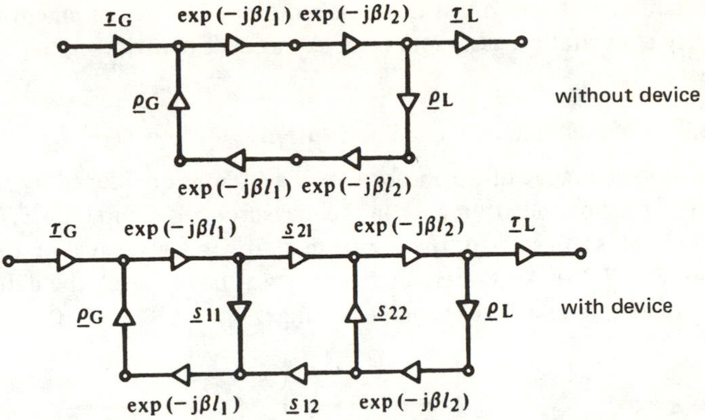

Fig. 7.25 Signal flow graphs of the setup for the measurement of attenuation.

7.4.6 Sources of Errors

The direct measurement becomes inaccurate when the following conditions are not satisfied:

1. The detector must be matched to the transmission line. To do that, one often places an attenuator (pad) or an isolator in front of the detector. The first element reduces the sensitivity of the measurement, the second one doesn't, but may be bulky and heavy.
2. The generator must be matched to the line. This is done by inserting an isolator between the source and the measurement system.
3. The detector must operate within the square-law region of its characteristic (§ 7.1.7). This limits the dynamic range of the measurement.
4. The connections must be reproduced in a manner as identical as possible in the different steps of the procedure.
5. The signal must remain steady during the whole measurement. As this takes a certain length of time, amplitude and frequency fluctuations may perturb the measurements. It is therefore highly recommended that the measurement sequence be repeated two or three times, in order to detect possible instabilities and evaluate the resulting error.

7.4.7 Limitations Due to the Dynamical Range

The accuracy of the measurement is directly related to the square-law characteristic of the detector. The direct method is therefore only suitable for small changes in the signal level (up to about 10 dB). The use of bolometers (§ 5.6.6), which have a broader dynamical range than crystal detectors, permits one to widen somewhat the measurement scale. A diode can also be utilized outside of its square-law region: an additional calibration is however necessary, and a constant amplification instrument is required. The measurement is no longer purely comparative, as a more involved procedure involving the power level must be used.

7.4.8 Measurements Using Substitution

The dynamic range of the measurement system is broadened by the introduction of a calibrated attenuator in the measurement setup (§ 6.3.7). The attenuation level is adjusted for the crystal detector to always receive the same signal level (Fig. 7.26). As it operates at a constant power level, the detector no longer has to operate only in its square-law range.

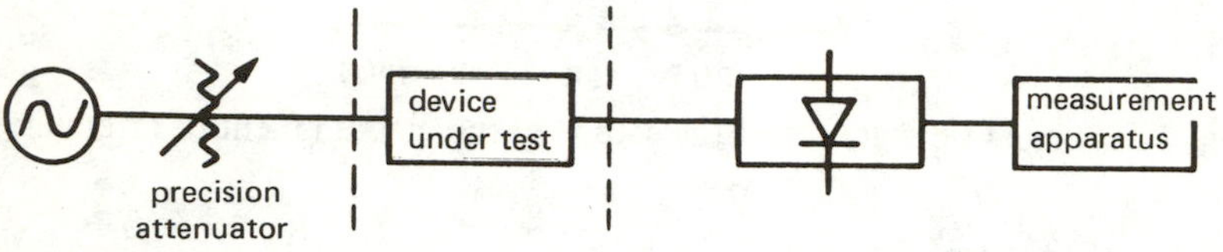

Fig. 7.26 Substitution measurement setup.

The variable attenuator can be inserted at various locations along the measurement setup:

1. In the microwave part. This situation, represented in Figure 7.26, is the one most commonly encountered in waveguide systems. A rotary vane attenuator is utilized, yielding a 1-2% accuracy over the complete frequency band of the waveguide. The measurement range then becomes on the order of 50-60 dB.
2. At the intermediate frequency. In coaxial line systems, no broad band calibrated attenuators are available. A frequency shift is then made (§ 5.3.4), yielding a signal at a fixed intermediate frequency. A cutoff reactive attenuator is then utilized (§ 6.3.8), also providing a dynamic range around 50-60 dB.
3. In the audio modulation. It is also possible to introduce a substitution after the detector, acting on the audio square wave envelope (§ 7.1.8) with an electronic attenuator. In this situation, however, the signal level at the diode varies, and the power range remains limited due to the diode characteristic.

7.4.9 Measurements Using Couplers

The substitution methods of Section 7.4.8 make use of a sequence of measurements, sensitive to variations in time of the signal. It is possible to avoid this time-dependence by making use of couplers (Fig. 7.27). The remainder of the measurements follows the procedure used in reflectometry (§ 7.3.9 to 7.3.15). To some extent, the second coupler compensates the frequency dependence of

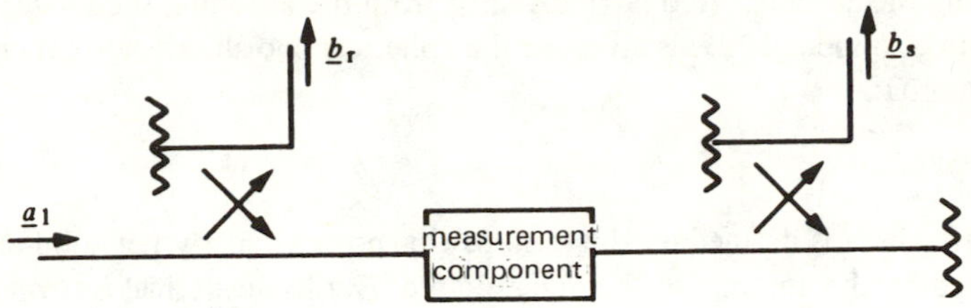

Fig. 7.27 Transmission measurement with two couplers.

the first one. It may also be left out and the signal $\underline{b}_s$ detected directly at the output of the device.

The ratio $\underline{b}_s/\underline{b}_r$ is proportional here to the transfer function of the device measured. The multiplying factor is determined by taking a measurement without the device. Afterwards, the values of $\underline{b}_s$ and $\underline{b}_r$ are always measured at the same time: any variation of $\underline{a}_1$ simultaneously affects both output signals. Their quotient remains unchanged.

The finite directivity of the couplers does not lead here to significant errors, except when the device to be measured is badly mismatched.

The measurement with couplers, associated with an interferometry bridge, a network analyzer, or a six-port setup determines both the attenuation and the phaseshift of the device.

7.4.10 Example: Antenna Measurements

Antennas provide the transition between guided propagation (transmission line or waveguide) and the radiated wave. At microwaves, antennas are most often directional: they emit and, reciprocally, receive a signal in a particular direction, related to their geometry (§ 1.2.5). By concentrating a signal in a narrow beam pointed towards the receiver, the total emitted power required for information transfer is greatly reduced. An antenna located outside of the beam does not pick up the signal, reducing cross talk and unauthorized listening. In addition, a directional antenna only picks up noise coming from the general direction of the transmitter, increasing the signal to noise ratio (Sec. 7.6). High directivity is a must for radars (Sec. 8.1) and communications (Sec. 8.2).

Antenna measurements are typically attenuation measurements: the ratio of received power to transmitted power is to be determined, first when the antennas are pointed towards one another (maximum power transfer, maximum gain, § 7.4.11), then as a function of pointing direction (radiation diagram, § 7.4.12).

7.4.11 Reminder: Power Gain

A power P_f (watt) is fed to an antenna. This power is distributed over the space surrounding the antenna. At a large distance from the antenna, the power radiated within a solid angle $d\Omega$ is given, in the spherical coordinate system centered on the antenna:

$$P_\Omega(\theta,\varphi)\,d\Omega \qquad \text{W} \qquad (7.46)$$

The power gain G is defined by the ratio of this power density per solid angle P_Ω radiated by the antenna to the one radiated by a hypothetical ***isotropic radiator***, which would evenly distribute the power across all directions:

$$G(\theta,\varphi) = \frac{P_\Omega(\theta,\varphi)}{P_f/4\pi} \qquad 1 \qquad (7.47)$$

The concept of gain expresses the fact that the antenna operates in a non-isotropic way, i.e., it concentrates the radiated power over certain directions. The maximum gain is defined as:

$$G_M = \max[G(\theta,\varphi)] \qquad 1 \qquad (7.48)$$

7.4.12 Reminder: Radiation Pattern

The following function is plotted in polar coordinates:

$$r(\theta,\varphi) = G(\theta,\varphi)/G_M \qquad 0 \leqslant r \leqslant 1 \qquad 1 \qquad (7.49)$$

It is the ratio of the power radiated in one direction to its largest value (Fig. 7.28).

7.4.13 Principle of the Measurement

To measure an antenna, one must have at least two antennas, or an antenna and a reflector of known scattering cross section. The distance between the two an-

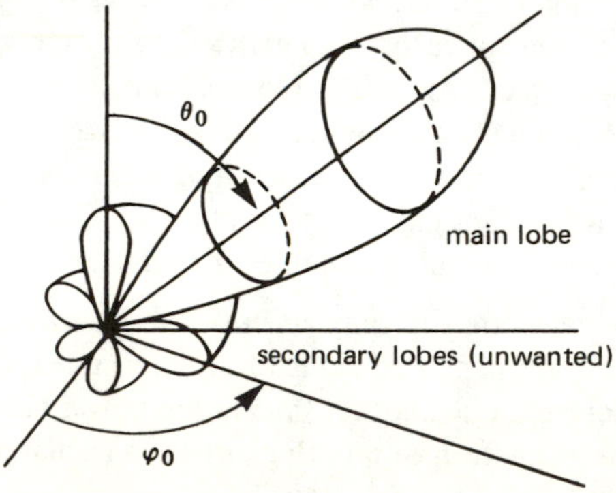

Fig. 7.28 Radiation diagram for a directive antenna.

tennas must be large enough to permit extrapolation of the results obtained: the second antenna must be placed within the far field of the first one, in which region only the radiation terms remain significant ($1/r$ decay of the fields). In the general case of an antenna having as largest dimension d, this condition is approximately satisfied when the distance between antennas is:

$$L \geqslant 2d^2/\lambda \qquad \text{m} \qquad (7.50)$$

As an example, for a parabolic dish antenna having a 10 m diameter operating at 10 GHz (λ = 3 cm), measurements should be carried out at a distance

$$L \geqslant 2(10)^2/0.03 = 6.66\ldots \times 10^3 \qquad \text{m} \qquad (7.51)$$

The measurement setup is schematically represented in Figure 7.29.

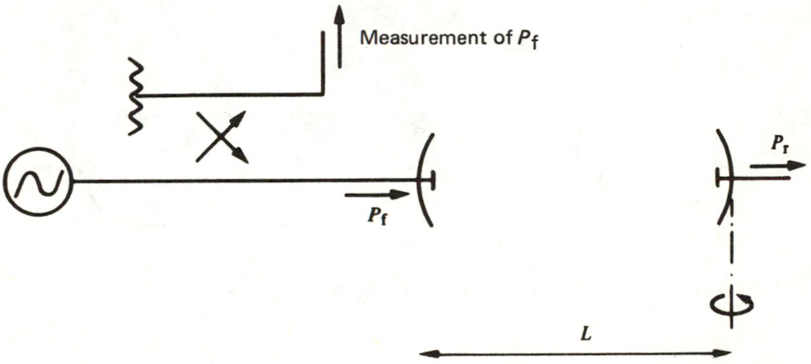

Fig. 7.29 Setup for the measurement of antennas. The receiving antenna is placed on a rotating mount.

The ratio of the power fed to the transmitting antenna by the one picked up by the receiving antenna is determined. The product of the two antenna gains is then given by the expression [214]

$$G_1 G_2 = \frac{P_r}{P_f}\left(\frac{4\pi L}{\lambda}\right)^2 \qquad 1 \qquad (7.52)$$

To determine the gain G_2 of the antenna being measured, one has to:

1. know G_1, utilizing a previously calibrated antenna;
2. specify $G_1 = G_2$, making use of two identical antennas;
3. make three measurements with three different antennas. In this way, the three gain products $G_{ij} = G_i G_j$, $i,j = 1, 2, 3$, $i \neq j$ are measured. The gains are then given by:

$$G_1 = \sqrt{\frac{G_{12}G_{13}}{G_{23}}}, G_2 = \sqrt{\frac{G_{12}G_{23}}{G_{13}}}, G_3 = \sqrt{\frac{G_{13}G_{23}}{G_{12}}} \quad 1 \qquad (7.53)$$

The radiation pattern is obtained by rotation of the antenna being measured: it is placed on a turntable (Fig. 7.29) or some other mount, orientable in both ϕ and θ directions. In the latter situation, the power radiated by the antenna in all directions can be measured. The total emitted power is then determined by integration and used to evaluate the efficiency of the antenna and its directivity [214].

7.4.14 Reflection Measurement

The attenuation and the phaseshift of a ***matched reciprocal device*** can be determined from reflection measurements (Sections 7.2 and 7.3). A fixed short circuit is connected to the output of the device, measurements are carried out at its input (Fig. 7.30).

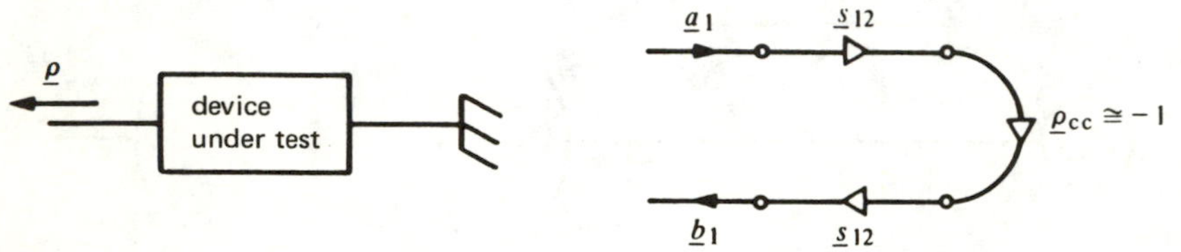

Fig. 7.30 Measurement of attenuation through reflection.

$$\underline{\rho} = \frac{\underline{b}_1}{\underline{a}_1} = \underline{s}_{12}^2 \underline{\rho}_{cc} \cong -\underline{s}_{12}^2 \quad 1 \qquad (7.54)$$

The attenuation and the phaseshift are then given respectively by:

$$LA = -10 \log |\underline{\rho}| + 10 \log |\underline{\rho}_{cc}| \cong -10 \log |\underline{\rho}| \quad \text{dB} \qquad (7.55)$$

$$\varphi = 1/2 \arg(\underline{\rho}) - 1/2 \arg(\underline{\rho}_{cc}) \pm n\pi \cong 1/2 \arg(\underline{\rho}) + \pi/2 \pm n\pi \quad \text{rad} \qquad (7.56)$$

Only reciprocal and matched devices may be measured in this manner.

7.4.15 Generalization: Method of Deschamps

The characteristics of reciprocal but mismatched devices can be determined in a similar manner, making use of a sliding short circuit (§ 6.2.4). The phase ψ of the signal injected at the second port is then shifted (Fig. 7.31). One obtains then, assuming that the short-circuit is lossless:

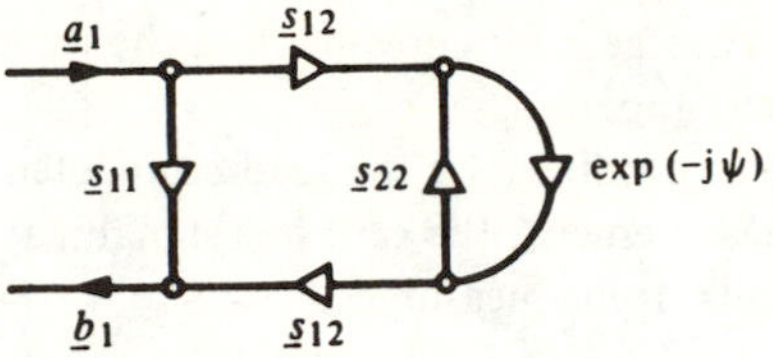

Fig. 7.31 Two-port terminated by a sliding short-circuit.

$$\rho = \frac{b_1}{a_1} = s_{11} + \frac{s_{12}^2 \exp(-j\psi)}{1 - s_{22}\exp(-j\psi)} \qquad 1 \qquad (7.57)$$

This expression actually defines a conformal mapping. When ψ varies, the point corresponding to the input in the complex ρ plane (Smith chart) moves over a circle. The values of s_{11}, s_{12} and s_{22} can be determined from the circle parameters, either by using a graphical technique devised by Deschamps [155, 111], or a computer program [156].

7.4.16 Resonant Cavity Measurement

When very small attenuations are to be measured, for instance those produced by sections of waveguide, a resonant cavity is formed, placing the device to be measured between two short-circuits (Fig. 7.32).

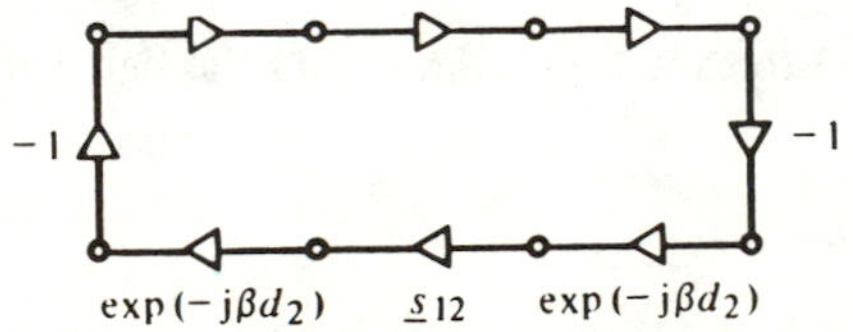

Fig. 7.32 Resonant cavity containing the device under test.

Resonances are observed when the total phaseshift of the chain is equal to $2n\pi$:

$$2\beta d_1 + 2\beta d_2 + 2\arg(s_{12}) = 2n\pi \qquad \text{rad} \qquad (7.58)$$

The resonance frequencies are measured; the phaseshift is determined from them. The attenuation is obtained from the measured value of the quality factor (Sec. 7.5). Like the measurement of reflection (§ 7.4.14), this approach is of interest only for matched reciprocal devices. It is limited to certain particular frequencies, those at which the cavity resonates.

■ 7.4.17 Particular Case: Directivity of a Coupler

The isolation and the directivity of high grade couplers (§ 6.5.14 and 6.5.15) are difficult to determine from direct measurements of the attenuation, due to the fact that:

1. they are very large, often greater than the dynamic range of the available measurement apparatus;
2. the reflection produced by the load connected to the third port may disturb the measurement: the spurious signal may well be as large or larger than the signal to be measured.

Nevertheless, accurate directivity measurements can be made using standard laboratory equipment. The coupled output is connected to an adjustable mismatch: a slide-screw tuner (§ 6.3.21) followed by a matched load (Fig. 7.33).

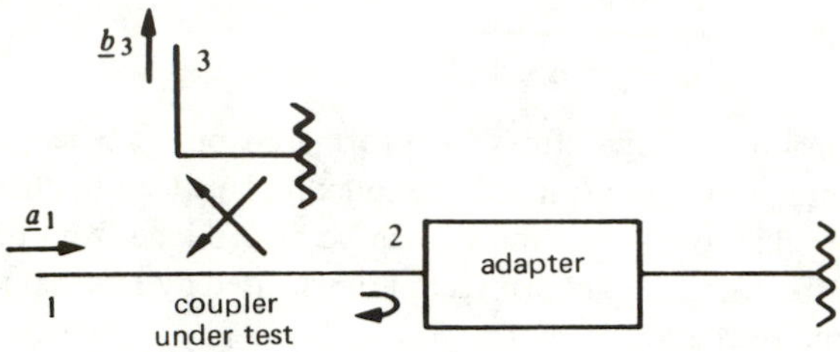

Fig. 7.33 Setup for the measurement of directivity.

The output signal at port 3 is given by (7.21):

$$\underline{b}_3 = \underline{a}_1 \left[\underline{s}_{21}\underline{\rho}\underline{s}_{32} + \underline{s}_{31}\right] \qquad \text{W}^{1/2} \qquad (7.59)$$

The tuner is adjusted to cancel out completely the signal at port 3. This happens when:

$$\underline{\rho} = -\underline{s}_{31}/(\underline{s}_{21}\underline{s}_{32}) \qquad 1 \qquad (7.60)$$

The tuner and the matched load are then removed, taking great care not to modify the settings, and $\underline{\rho}$ is determined by a reflection measurement (slotted line, Sec. 7.2). The terms $\underline{s}_{21}$ and $\underline{s}_{32}$ are obtained with regular attenuation measurements; the value of $\underline{s}_{31}$ is then determined from (7.60).

7.4.18 Comments

In most practical situations, one wishes either to reduce or to increase the attenuation, for the following reasons:

1. In the ***low-loss direction***, the attenuation reduces the signal reaching the user. The power that the generator must produce is consequently increased. This means increased cost. For a system to be economically viable, it must operate at the lowest possible power input.
2. In a high power system, losses produce significant heating of the system. The heat produced must be evacuated (by cooling fans, or running water). This complicates both the design and the operation of the system. Therefore, one wishes to reduce the losses to their lowest possible level.

3. In a low power system (receiver), the attenuation degrades the signal to noise ratio (§ 7.6.8), and thus reduces the sensitivity.
4. In the ***high-loss direction*** (isolation), the residual signal going through the component introduces an error, bringing forward cross talk; it can even damage power-sensitive elements. In this case, one wishes to increase the losses as much as possible.

This means that in many practical situations one does not have to know the exact value of the attenuation, but only whether it is lower than (low-loss direction) or greater than (high-loss direction) a specified limit. For instance, in an isolator, one may wish to have an insertion loss smaller than 0.5 dB and an isolation greater than 30 dB over the operating frequency band (template, Fig. 7.34).

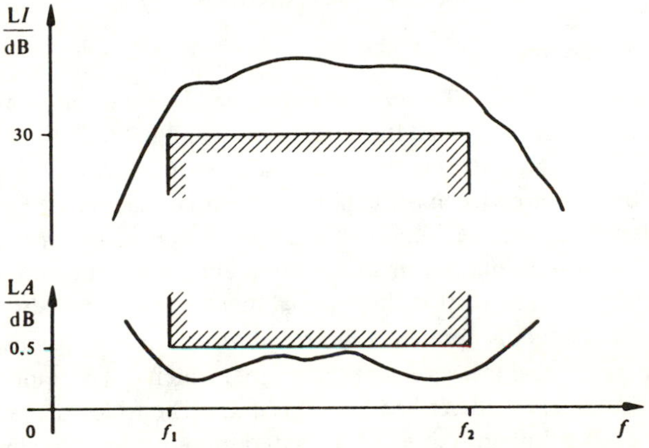

Fig. 7.34 Isolator gauge.

The measurements must indicate whether the requirements are satisfied: the curves must not penetrate into the hatched zones. Accurate measurements are only required when the device attenuation is close to the limit.

For a directional coupler, a minimum insertion loss and a maximum isolation are required, while the coupling must be as close as possible to the specified value. The true attenuation must be determined accurately for all components taking part in measurements (attenuators, couplers, etc.).

7.4.19 Comment: Measurement of the Phaseshift

While the measurement of attenuation is very common at microwaves, the phaseshift is seldom determined. Only in certain particular situations does one need to know the phaseshift.

1. In communications, the phaseshift must vary linearly with frequency. Otherwise, the signal undergoes phase distortion.
2. The power-handling capability of a system may be increased by connecting identical elements in parallel with hybrid junctions and couplers. For the signals to actually add up at the output, the phaseshifts and the attenuations (or gains) of the two channels must be as identical as possible.
3. Surveillance radars utilize phased-array antennas (§ 8.1.9) made up of an assembly of fixed radiating elements, fed with signals having variable phases. The emitted beam can be pointed by purely electronic control, without need for any mechanical movement of the antenna. The phase of the signal is adjusted by phaseshifters, either using PIN diodes (§ 6.8.5), varactors (§ 6.8.6), or ferrites (§ 6.7.12).

7.5 CAVITY MEASUREMENTS

7.5.1 Introduction

For every resonant mode, a cavity is characterized by its resonant frequency f_0, its loaded quality factor Q_c and its factor(s) of coupling β_c (Ch. 3). The determination of these parameters requires comparative measurements: the reflection factor is to be measured when a one-port cavity is connected at the end of a line; the transmission factor is required for reactively mounted or for two-port cavities (Fig. 5.2). In addition, one or several frequencies must be determined and in certain methods, a relaxation time: signal measurements are also used (Ch. 5). Cavity measurements thus present a ***hybrid nature***, requiring particular techniques. The most usual measurement technique, which only requires standard laboratory equipment, is described first (§ 7.5.2). The following sections present the general outlines of more sophisticated methods, in some instances for automated measurements.

7.5.2 Measurement Principle

The test setup is shown in Figure 7.35, respectively for a reflection measurement (with directional coupler) and one in transmission.

The microwave signal is provided by a swept frequency generator (sweeper), the frequency band being adjusted to bracket the resonance of the cavity. A voltage proportional to the signal frequency is applied to the horizontal scale of the oscilloscope, so that the detected signal can be observed on the screen as a function of frequency (§ 5.9.2). Part of the signal is sampled and used to automatically control the level (§ 6.8.3). An isolator (§ 6.7.15) prevents the reflections from reaching the generator: outside of the resonance, the cavity is a reactive load. A wavemeter is used to determine the frequency (§ 5.2.1).

The signal reflected or transmitted by the cavity is detected and displayed on

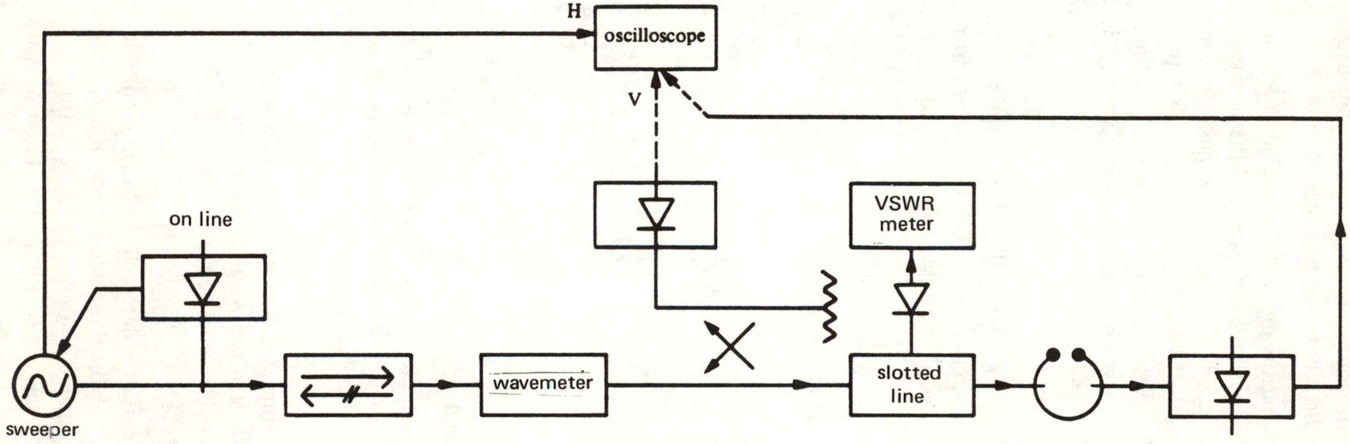

Fig. 7.35 Block diagram of a cavity test setup.

the screen of an oscilloscope (Fig. 7.36). With a square-law detector, the vertical scale (linear) corresponds directly to the power level.

The wavemeter produces a narrow dip on the trace observed on the screen. This dip is first made to coincide with the extremum of the resonance curve, and the resonant frequency f_0 is determined from the wavemeter reading. The dip is then placed successively on the two half-power points, on either side of the resonance. The line width $\Delta f_{1/2}$ is then evaluated. The scale on the scope can also be calibrated in terms of frequency. The ***loaded quality factor*** Q_c is then given by

$$Q_c = f_0/2\Delta f_{1/2} \quad 1 \qquad \textbf{(3.101)}$$

Frequency markers of the sweeper can similarly be utilized, when available.

For a reflecting cavity, the coupling factor β_c is determined from Sections 3.5.13 (power ratio) and 3.5.14 (slotted line extrema). The unloaded quality factor Q_0 is then determined from (3.97). For a loosely coupled cavity $Q_0 \cong Q_c$.

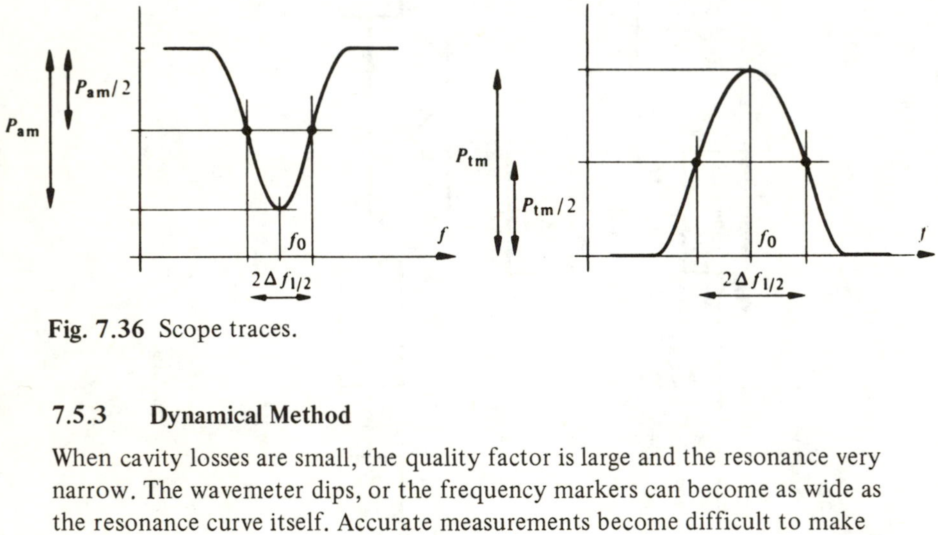

Fig. 7.36 Scope traces.

7.5.3 Dynamical Method

When cavity losses are small, the quality factor is large and the resonance very narrow. The wavemeter dips, or the frequency markers can become as wide as the resonance curve itself. Accurate measurements become difficult to make under such conditions.

The measurement can be improved by using an auxiliary signal source of adjustable frequency. This signal is mixed with the swept frequency signal of Figure 7.35, then fed to an adjustable selective narrow band filter (Fig. 7.37), for instance a calibrated receiver.

The signal at the filter's output is detected, and then used to locally intensify the scope trace. This can only be done when trace intensification, sometimes called *z*-modulation, is available on the scope used. In this manner, two bright dots appear on the trace displayed on the screen. These dots are positioned by

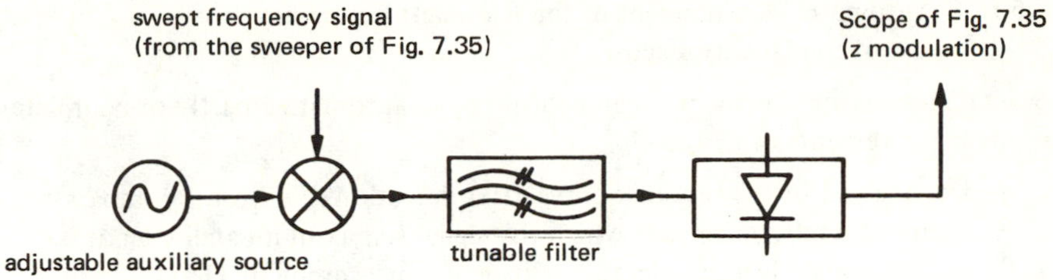

Fig. 7.37 Dynamic method: additional components.

varying the frequency of the auxiliary source, their spacing by tuning the selective filter. When the two dots coincide with the half-power points (Fig. 7.36), the frequency of the auxiliary source is f_0 and the selective filter is adjusted at $\Delta f_{1/2}$. The frequency f_0 is measured with a frequency counter [113].

7.5.4 Comment

The methods presented so far require several delicate manipulations: oscillators must be tuned, half-power points on an oscilloscope trace must be located, and the frequency corresponding to the spacing must be determined. These operations may lead to rather large inaccuracies, particularly for low-loss cavities. When a large number of cavity measurements are foreseen, as happens when determining materials properties (Sec. 8.5), one would like to have a test setup in which measurements are easier to make.

7.5.5 Automatic Measurement of the Resonant Frequency

The resonant frequency of a cavity can be determined in a simple manner by placing the cavity within the feedback loop of an oscillator (§ 4.1.2), with a broadband amplifier [157] (Fig. 7.38).

The phaseshift of the loop is adjusted for maximum output power. The oscillator then produces a signal at the frequency f_0. The signal amplitude increases with the quality factor Q_c. The power dependence is related to the characteristics of the amplifier.

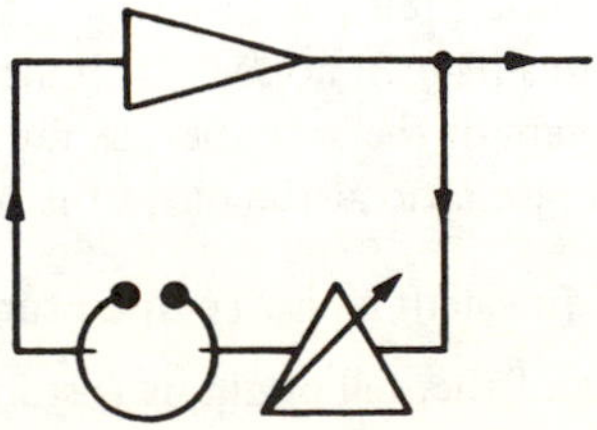

Fig. 7.38 Active cavity measurement.

7.5.6 Automatic Measurement of the Resonant Frequency and of the Quality Factor

Several measurement principles can be utilized to automatize partly or completely the measurements of f_0 and Q_c.

1. The analysis of the response curve after detection. Electronic comparator circuits detect the times at which the signal is maximum and when it is half this value, leading to a fully automatic measurement [158].
2. The measurement of phaseshift close to the resonance. Signals modulated in amplitude or in phase are employed [143].
3. The measurement of the transient response of the cavity when a pulsed microwave signal is applied (§ 5.8.2) [109] (Fig. 3.2).

7.6 MEASUREMENT OF THE NOISE FIGURE

7.6.1 Noise

It is not possible to indefinitely keep amplifying very low amplitude signals, due to the existence of random perturbations known as *electromagnetic noise* [111]. For a signal to be detectable, its power level must be superior (by a factor of 5 to 10 at least) to the average noise level. Under certain particular conditions, signals can be extracted from a noise actually having a larger amplitude: this can only be done when the signal itself possesses certain particular code characteristics, for instance a large number of repetitions (redundancy).

The electromagnetic noise results from two sources:

1. the external noise, picked up by the antenna, which comes from cosmic, atmospheric, industrial sources;
2. the internal noise which is produced ***within the receiver***. Particularly worth mentioning is the thermal noise in resistive elements and the shot noise in transistor p-n junctions.

7.6.2 Comment

Any signal amplification produces ***some degradation***. While flowing through an amplifier, a signal sees its amplitude multiplied by the power gain G of the device. The input noise is also multiplied by the factor G but in addition, the amplifier also contributes its own noise, which may be quite significant. Consequently, the apparent gain of the amplifier, as far as noise is concerned, is larger than G. The signal-to-noise ratio at the output is smaller than it was at the input.

7.6.3 Definition: Equivalent Noise Temperature

Since the noise is often of thermal origin, it is studied using the black body concept [159]. This element completely absorbs the incident radiation and reemits it entirely. The spectral power density (W/Hz) of the black body emission is directly proportional to its absolute temperature T(K), the multiplying factor being Boltzmann's constant k_B:

$$k_B = 1.3804 \ldots \times 10^{-23} \qquad \text{J/K} \qquad (7.61)$$

The average noise power N emitted by a black body over the frequency band B is thus given by:

$$N = k_B T B \qquad \text{W} \qquad \mathbf{(7.62)}$$

This expression is only valid for $k_B T \gg hf$, where $h = 6.62 \ldots \times 10^{-34}$ Js in Planck's constant (the limit is due to quantum effects).

By analogy with the black body, *equivalent noise temperatures* are respectively defined for the antenna T_a and for the receiver T_r (noise carried back to the input). The equivalent schematic is shown in Figure 7.39. ***The equivalent noise temperature of an element is generally different from its physical temperature.***

At the entrance of the receiver, there is a signal S and a noise level $k_B T_a B$, where B is the bandwidth of the receiver, assuming that all stages have the same bandwidth. At the output, the signal power is GS, and the noise is $Gk_B(T_a + T_r)B$.

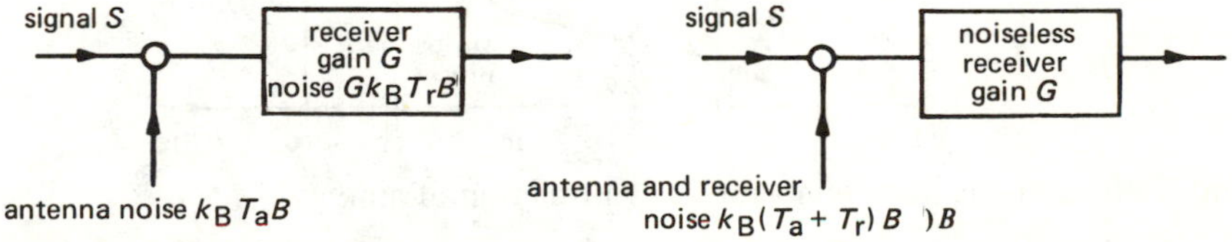

Fig. 7.39 Equivalent circuits of a receiver connected to an antenna.

7.6.4 Antenna Noise Temperature

The equivalent noise temperature of the antenna, which defines the average noise power picked up by this antenna, depends both upon frequency and orientation of the antenna (Fig. 7.40). In particular, an antenna should not be pointed towards the sun. An antenna pointed skyward (90°) has a noise temperature of only a few Kelvins, the minimum is situated near 3 GHz.

7.6.5 Receiver Noise Temperature

The equivalent temperature of several receivers is shown in Figure 7.41. Vacuum tube or transistor amplifiers appear towards the upper part of the graph (noisy). Considerably smaller noise levels are obtained using parametric amplifiers and MASERS (Sec. 4.10), devices which are particularly critical to operate.

7.6.6 Definition: Noise Figure

The *noise figure* F is defined as the quotient of the signal to noise ratios at the

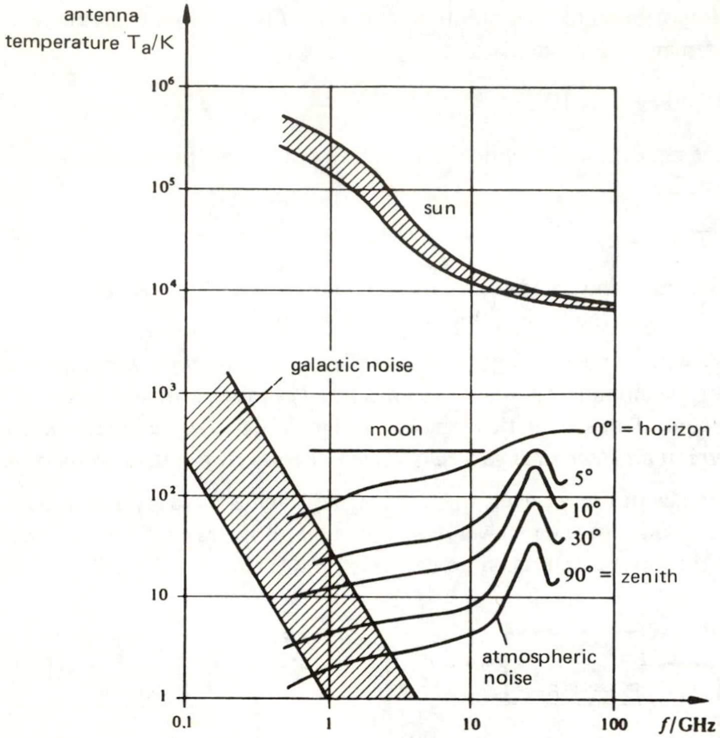

Fig. 7.40 Antenna noise temperature as a function of aim.

input and at the output of the device considered (amplifier, receiver, etc.) ***when the input is connected to a noise source at*** $T_0 = 290$ K.

$$F = \frac{(S/N)_{\text{input}}}{(S/N)_{\text{output}}} = \frac{S}{GS} \quad \frac{Gk_B(T_0 + T_r)B}{k_B T_0 B} = 1 + \frac{T_r}{T_0} \quad 1 \qquad \textbf{(7.63)}$$

The noise figure F indicates by how much the signal to noise ratio is reduced. A low value corresponds to a high quality, and thus to a large sensitivity.

The average equivalent noise power referred to the input, when connected to an antenna having an equivalent noise temperature T_a, is given by (Fig. 7.39):

$$N = k_B(T_a + T_r)B = k_B T_0 F B + k_B(T_a - T_0)B \qquad \text{W} \qquad (7.64)$$

The noise figure is often expressed in decibels:

$$LF = 10 \log F \qquad \text{dB} \qquad (7.65)$$

■ 7.6.7 Cascaded Components

Several two-ports, having power gains G_i, equivalent noise temperatures T_i, and

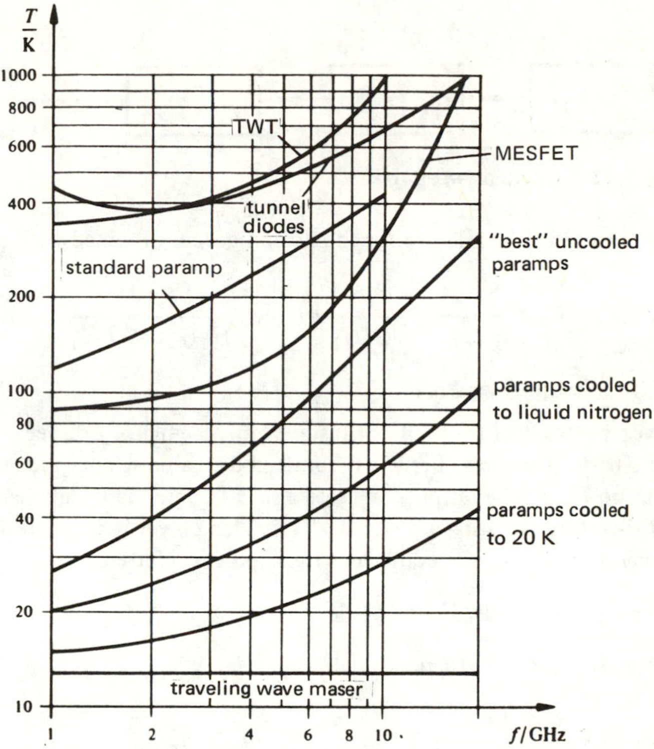

Fig. 7.41 Noise temperature of receivers.

noise figures F_i (the latter two quantities related through (7.63)) are connected in cascade (Fig. 7.42). The two-ports are all matched and all have the same frequency bandwidth; alternately, the output two-port has the narrowest bandwidth.

The equivalent noise temperature of the whole assembly is obtained by simply adding the equivalent temperatures of all the elements referred to the input of the chain, i.e., divided by the gains of all the preceding elements:

$$T_{1\ldots n} = T_1 + \frac{T_2}{G_1} + \frac{T_3}{G_1 G_2} + \ldots + \frac{T_n}{G_1 G_2 \ldots G_{n-1}} \quad \text{K} \tag{7.66}$$

When the gain of the first two-port in the chain is large, the resulting noise temperature is roughly that of this first component. The noise of all two-ports following it is divided by the gain of the first element. When the first two-port is lossy ($G_1 < 1$), the noise produced by the following components is emphasized. To avoid this, the receiver must be connected as closely as possible to the antenna.

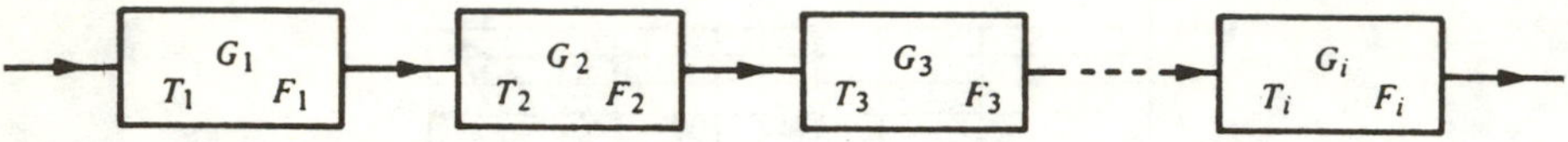

Fig. 7.42 Cascaded two-ports.

The corresponding relation for the noise figure is obtained with (7.63):

$$F_{1\ldots n} = F_1 + \frac{F_2 - 1}{G_1} + \frac{F_3 - 1}{G_1 G_2} + \ldots + \frac{F_n - 1}{G_1 G_2 \ldots G_{n-1}} \qquad 1 \qquad (7.67)$$

7.6.8 Passive Two-Port

A passive two-port ($G < 1$), for instance an attenuator or a section of cable, with all its parts at temperature T_1, is connected to a noise source at the same temperature T_1. At the output, one sees a set of components, all at temperature T_1, and therefore radiating a power $k_B T_1 B$. This power is equal to the two-port gain G, multiplying the equivalent noise power referred to the input (7.64):

$$k_B T_1 B = G[k_B T_0 F B + k_B(T_1 - T_0)B] \qquad \text{W} \qquad (7.68)$$

The noise figure F is then defined as

$$F = 1 + \frac{T_1}{T_0}\,\frac{1 - G}{G} = 1 + (L - 1)(T_1/T_0) \qquad 1 \qquad (7.69)$$

where $L = 1/G$ is the power loss factor.

An antenna should, therefore, always be connected to the receiver with a transmission line having losses L as small as possible, or the line should be cooled.

7.6.9 Measurement of the Noise Figure

The device to be characterized is connected successively to two noise sources having accurately known noise temperatures T_1 and T_2. The output powers measured in the two situations are (7.64):

$$N_{si} = F G k_B T_0 B + G k_B (T_i - T_0) B \qquad i = 1, 2 \qquad \text{W} \qquad (7.70)$$

The value of F is drawn from these two relations, yielding:

$$F = \frac{1}{Y - 1}\left[\frac{T_2}{T_0} - 1 - Y\left(\frac{T_1}{T_0} - 1\right)\right] \qquad 1 \qquad (7.71)$$

where $Y = N_{s2}/N_{s1}$. This measurement involves the ratio of two power levels: it is a comparative measurement.

7.6.10 Noise Sources

To obtain good accuracy, very different values are required for T_1 and T_2. The sources must also be matched to the transmission lines.

7.6.11 Hot and Cold Sources

For high precision measurements, for instance, primary references in a standards laboratory, a resistor is placed within a thermostatic enclosure (calorimeter, § 5.6.2). The physical temperature of the resistor is its noise temperature. Liquid nitrogen ($T_1 = 77.3$ K) is used for the cold source, boiling water ($T_2 = 373.1$ K) for the hot source. Such sources are quite bulky and used mostly to calibrate secondary references.

7.6.12 Discharge Tube Source

A gas discharge tube crosses the waveguide, which is terminated further on by a matched load (Fig. 7.43).

When the tube is off, the noise temperature T_1 is the ambient temperature. When the tube is lit, its noise temperature T_2 is roughly the same as the electrons' within the discharge (Table 7.44).

The same principle is used to realize coaxial noise sources.

7.6.13 Other Noise Sources

In a saturated vacuum diode, the noise temperature is proportional to the anode current, which can be adjusted by modifying the filament bias.

Noise sources make use of the shot noise produced by a current crossing a semiconductor p-n junction.

A signal generator can also be used to measure the noise figure. In this case, it is necessary to know the frequency band B of the device to be measured.

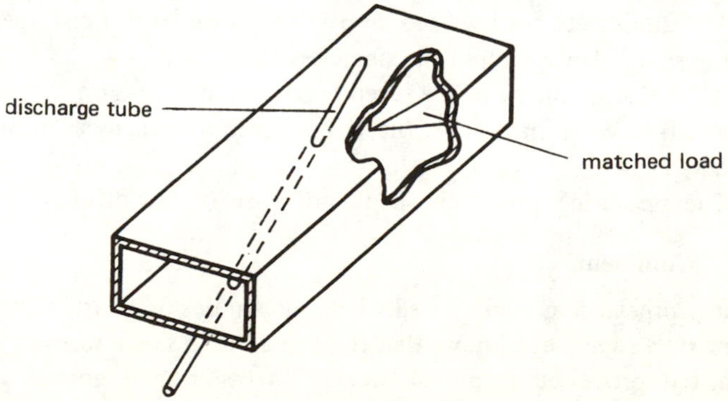

Fig. 7.43 Discharge noise source in waveguide.

Gaz	T_2 (K)
Hg	11 000
A	15 000
H	32 000
Xe	9 000
Ne	25 000
He	29 000
N_2	11 500

Table 7.44 Electron temperature in a gas discharge [160].

7.6.14 Sources of Errors and Inaccuracies

The measurement of noise can be affected by several sources of errors:

1. The inaccuracy of the temperature measurement of the noise sources. These must be calibrated in terms of primary references, hot and cold sources (§ 7.6.11).
2. Source mismatch. The noise figure is defined assuming that all devices are matched. When this requirement is not met, errors difficult to evaluate appear, depending on the kind of noise source used. This effect is not too important in the measurement of usual amplifiers ($LF \cong$ 8-10 dB). It becomes quite significant, however, in the measurement of very low noise receivers.
3. Parasitic elements in the sources. The equivalent circuit of a noise source is not simply a signal source and a resistor, but also contains reactive elements: interelectrode capacitances and conductor inductances. The noise temperature is then frequency-dependent.
4. Theory considers only linear components. But noise is a random process, in which the instantaneous power can become very much larger than its average value over short periods of time. The presence of nonlinearities in the system gives rise to errors.
5. The diode detector used to compare the noise levels must operate within the square-law region of its characteristic.
6. All the components of the receiver are assumed to have the same bandwidth B. When this is not true, a more complete analysis must be executed [111].
7. The measuring instruments can modify the bandwidth B.

7.6.15 Comment

Particular precautions are required when one studies or measures noise. When working on a signal, one knows that this signal has ***a single source***, that it is then transmitted, processed, amplified, etc. On the other hand, noise is produced ***by every part of the system:*** there are multiple sources, distributed throughout the circuit, which must all be considered.

7.6.16 Comparison

At 10 MHz, the equivalent noise temperature of a transistor is roughly 10 K. The atmospheric noise at that frequency causes an antenna temperature of about 10^7 K. The outside noise is clearly predominant.

At 5 GHz, on the contrary, the noise temperature of a skyward pointed antenna is only 8 K. A TWT amplifier then produces a noise temperature of 300 K. In this situation, the internal noise of the amplifier is the more important one.

7.7 PROBLEMS

7.7.1 A slotted line measurement yields VSWR = 1.8. Two adjacent minima are located at $z_1 = 8.48$ cm and $z_2 = 10.48$ cm. Placing a short circuit in the reference plane, a minimum is obtained for $z_c = 9.98$ cm. Determine the guide wavelength and the reflection factor $\underline{s}_{ii}$ (amplitude and phase).

7.7.2 For which VSWR values can the simplified expression of (7.19) be used with less than 1% error? Determine the corresponding values of d/λ_g.

7.7.3 A load is formed from an iris followed by a matched termination. At 10 GHz the normalized iris susceptance is $BZ_c = -2$. How can this load be matched with a second iris? At what distance from the first one should it be located?

7.7.4 A reflectometer measurement yields $|\underline{b}_3|/|\underline{b}_{3c}| = 0.2$. Knowing that the coupler used has a 46 dB directivity, determine the modulus of the reflection factor and the error limits.

7.7.5 A coupler with a 34 dB directivity is utilized to measure the VSWR of a component. Which is the largest value that one can measure to be sure that the component's VSWR does not exceed 1.25?

7.7.6 The four detectors of an ideal six-port (§ 7.3.14) measure, respectively, the four following currents:

$$I_3 = 25\ \mu\text{A} \qquad I_4 = 15\ \mu\text{A} \qquad I_5 = 50\ \mu\text{A} \qquad I_6 = 5\ \mu\text{A}$$

Knowing, in addition, that $|\underline{b}_r| > |\underline{b}_s|$, determine the ratio of these terms, both in amplitude and in phase.

7.7.7 Can the Purcell's six-port junction (§ 6.6.2) be used to make measurements of the reflection factor? At which ports should the signal and the reference be connected?

7.7.8 A two-port device absorbs 20% of the incident power and has a VSWR of 4. Knowing that it is reciprocal and symmetrical, determine the components of its scattering matrix and its insertion loss.

7.7.9 The introduction of a thin lossless iris within a waveguide produces a 10 dB decrease in the output signal. Determine the VSWR produced by this iris.

Determine its normalized susceptance (within the plane of the iris).

7.7.10 A reciprocal and matched two-port having a 1.5 dB attenuation is measured with a detector whose VSWR is 2. The generator has itself a reflection factor of 0.3. What is the range of values that can be obtained when measuring the attenuation of this device?

7.7.11 The gains of three antennas are being measured at 6 GHz. The distance between two antennas is set at 5 meters, the emitted power is 10 mW for all measurements. As for the received power, it is

1. with antennas 1 and 2: 4.43 μW
2. with antennas 1 and 3: 5.70 μW
3. with antennas 2 and 3: 9.97 μW

Determine the gain of each antenna.

7.7.12 The attenuation A, in dB, of a matched element is measured by terminating it into a short circuit. A VSWR of 5 is measured at the input. Determine the value of A.

7.7.13 The attenuation of a symmetrical reciprocal component is measured by connecting it to a short-circuit. The input VSWR measured is 2.5 A second measurement, with the element terminated into a matched load, yields a VSWR of 1.5. What can be said regarding the attenuation A?

7.7.14 The directivity of a coupler is determined with the measurement technique outlined in Section 7.4.17. After the adjustment, the tuner has a VSWR of 1.21. The insertion loss of the coupler is 1 dB. What is its directivity?

7.7.15 Two cables, having insertion losses of 1 dB and of 2 dB, respectively, are available to connect an antenna to a receiver. These cables must be cascaded, one outside of the building, where the ambient temperature is −5 degrees Celsius, the other one inside of the receiving station where the temperature is 20 degrees. Determine the order of connection of the cables that yields the more favorable noise figure.

7.7.16 A discharge tube has an equivalent noise temperature of 10,000 K. It is used to measure the noise figure of a receiver. The powers measured, respectively, with the tube on and off (ambient temperature of 290 K), are in a 12 to 1 ratio. Determine the noise figure F of the receiver and its equivalent noise temperature.

CHAPTER 8

APPLICATIONS

8.1 RADAR

8.1.1 Description

Radar stands for *RAdio Detection And Ranging,* based upon the echo from a target located across the trajectory of an electromagnetic wave. In most instances, the transmitter and the receiver are connected to a common antenna: this situation will be considered here. The time required for the wave to go from the emitter to the target and back to the receiver is measured, and yields the distance R. The frequency variation of the signal is related to the radial velocity of the target with respect to the source (Doppler shift, § 8.1.13). The direction of the target is determined by pointing a narrow beam in the direction from which the reflected signal is the largest. The radar system is schematically represented in Figure 8.1, in which the relevant terms are introduced. Radar systems are very well covered in the technical literature [161-163].

8.1.2 The Radar Equation

The emitter feeds the antenna with a power P_f; the antenna radiates it into space, concentrating it along one or several privileged directions, where the antenna gain G is the largest (§ 7.4.11). The power density decreases as $1/4\pi r^2$, where r is the distance from the antenna. Part of the signal reaches the target, which reflects it partially towards the antenna. The target is represented by its effective scattering cross section σ (§ 8.1.4). The density of the reflected power decreases in turn inversely as the square of the distance. The antenna picks up a part of the reflected signal; its receiving properties are represented by the effective reception area A_e, which is itself related to the antenna gain by the relation $A_e = G\lambda^2/4\pi$, where λ is the wavelength.

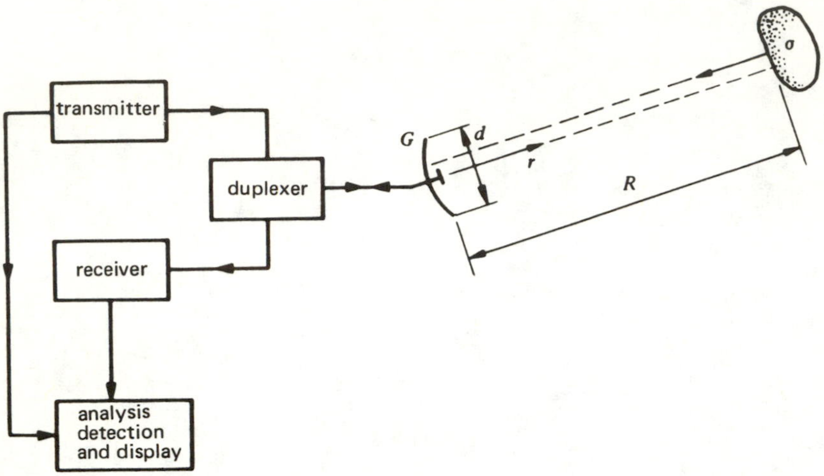

Fig. 8.1 Block diagram of a radar.

The ratio of the received power P_r to the power P_f supplied by the source is given by the *radar equation:*

$$\frac{P_r}{P_f} = G \times \frac{1}{4\pi R^2} \times \sigma \times \frac{1}{4\pi R^2} \times \frac{G\lambda^2}{4\pi} = \frac{G^2\lambda^2\sigma}{(4\pi)^3 R^4} \qquad 1 \qquad \textbf{(8.1)}$$

The received power is fed to the receiver, also connected to the antenna. The attenuation produced by atmospheric losses (§ 8.2.11) is neglected here. The target is supposed to be located beyond the near field of the antenna, as per (7.50):

$$R > 2d^2/\lambda \qquad \text{m} \qquad (8.2)$$

with d the *overall* antenna dimension.

8.1.3 Signal Analysis

In the receiver, the amplified signal is fed to a data processing unit, which compares the received signal with the transmitted one. However, the presence of noise complicates the detection process: the signal-to-noise ratio (§ 7.6.3) must be large enough to permit a correct operation. As noise is a random process, its amplitude can take large values, well above its average value, over short periods of time. A noise spike might well be interpreted as an echo by the system. To avoid false alarms, the evolution with respect to time of the received signal is also carefully monitored. In this manner, true echoes can be distinguished from spurious noise spikes [164].

8.1.4 Effective Scattering Cross Section σ

A target is characterized by its *effective scattering cross section* σ, defined by the ratio of the power reflected back towards the radar to the incident power density. The reflection produced by the target may, in principle, be calculated by solving Maxwell's equations and applying the boundary conditions on the scatterer's surface. The treatment of the equations obtained in this manner is, however, most complicated; the problem could only be solved for some simple geometries [161]. One such geometry is a perfectly reflecting metal (pec) sphere of radius a. The ratio of the effective scattering cross section σ to the geometrical cross section πa^2 is sketched in Figure 8.2 in terms of the normalized size a/λ. When the radius a is very small with respect to the wavelength λ, the effective scattering cross section is much smaller than the geometrical one. One obtains roughly:

$$\sigma/\pi a^2 \cong 10^5 (a/\lambda)^4 \qquad \text{for } a/\lambda < 1/2\pi \qquad 1 \qquad (8.3)$$

The apparent size of the target, as seen by the radar, decreases quite fast when the target is smaller than the wavelength. At larger values of a/λ, resonances are observed, the two cross sections becoming equal at the upper limit (optical region). This behavior remains valid for targets having different shapes. The limiting values, given in Table 8.3 for several geometries, can be used to first order for elements having a large size in terms of wavelengths.

All the values in Table 8.3 are for metallic scatterers. For dielectric targets, the reflection and, consequently, the effective scattering cross section will be smaller. This can be accounted for by introducing the reflection factor for a uniform plane wave at normal incidence [214].

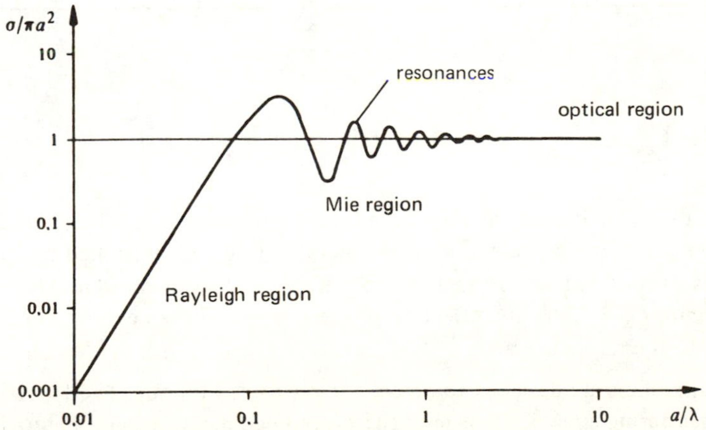

Fig. 8.2 Effective cross section of a metal sphere.

Table 8.3 Approximation for the effective scattering cross section of a metallic target much larger than the wavelength λ.

target		Effective scattering cross section σ (limited optics)
Sphere	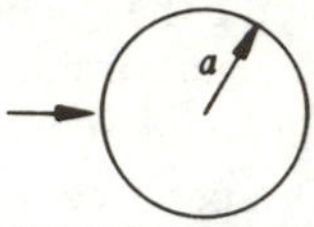	πa^2
Cone (axial incidence)	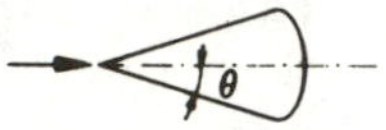	$\frac{\lambda^2 \, \mathrm{tg}^4 \theta}{4\pi}$
Disk		$\pi a^2 \cot^2 \theta J_1^2 \left(\frac{4\pi a}{\lambda} \sin \theta \right)$
Flat surface of large size (normal incidence)	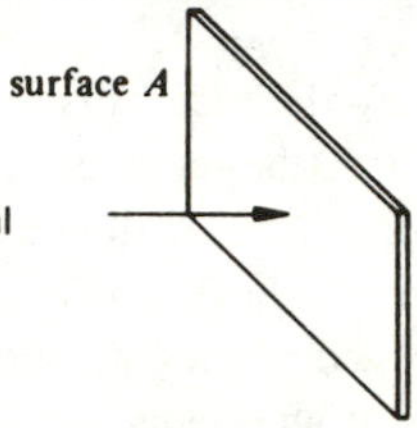	$\frac{4\pi A^2}{\lambda^2}$
circular cylinder	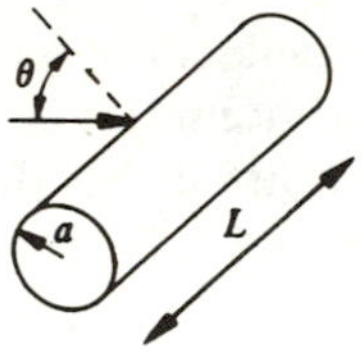	$\frac{a\lambda}{2\pi} \frac{\cos\theta \sin^2\left(\frac{2\pi L}{\lambda^2} \sin\theta\right)}{\sin^2\theta}$

$$\underline{\rho} = \frac{\sqrt{\underline{\epsilon}_r} - 1}{\sqrt{\underline{\epsilon}_r} + 1} \qquad 1 \qquad (8.4)$$

With a lossy dielectric, the reflection factor $\underline{\rho}$ is complex. To first order, the effective scattering cross section of the metallic target is multiplied by $|\underline{\rho}|^2$. This yields a rough estimate for a dielectric target. The latter ***could however be a resonator*** (§ 3.3.18): its reflection properties would then exhibit large variations close to the resonances.

Certain radars are designed to detect the presence of humans (§ 8.1.16). Effective scattering cross sections have therefore been measured and Figure 8.4 presents values obtained at several frequencies [165]. These vary widely.

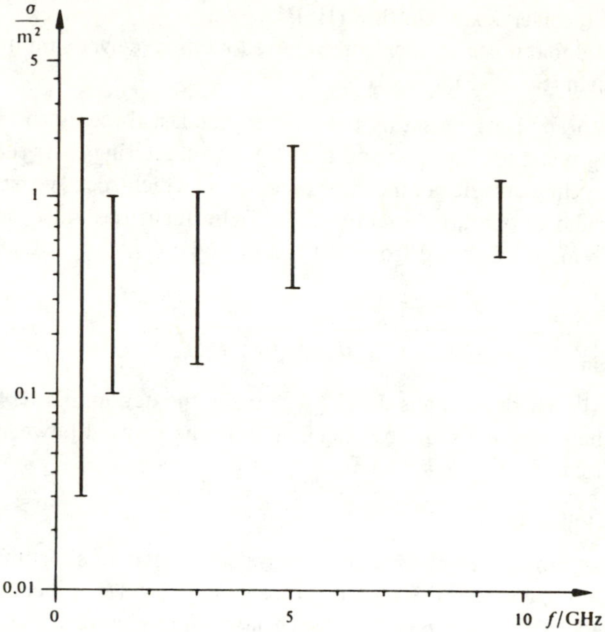

Fig. 8.4 Effective scattering cross section of a man, in terms of frequency.

□ 8.1.5 Particular Case: Short-Circuited Antenna

When the target is an antenna having its output short-circuited, the effective scattering cross section is given by:

$$\sigma = GA_e = \frac{G^2\lambda^2}{4\pi} = \frac{4\pi A_e^2}{\lambda^2} \qquad \text{m}^2 \tag{8.5}$$

It is possible to measure with a radar the gain and the radiation pattern of an antenna, without the need to feed it.

8.1.6 Remark

The foregoing shows the main problem encountered in radar design: the selection of the parameters. One must indeed take into account:

1. the power P_f provided by the generator (W),
2. the antenna gain G (1),
3. the frequency f (GHz) or the wavelength λ (m),
4. the range R_{max} (m), which is the farthest distance for detection,
5. the effective scattering cross section of the target σ (m^2),
6. the lowest signal-to-noise ratio (P_r/N) required at the input of the receiver, quantity defined by the data processing unit,

7. the receiver's bandwidth B (Hz),
8. the global noise temperature referred to the receiver's input $(T_a + T_r)$ in Kelvin (Sec. 7.6).

The first three terms are set by the emitter, the last three by the receiving and processing system. The range and the effective scattering cross section define the target, they are the geometrical parameters which specify the whole system. For the radar to operate correctly, these eight quantities must meet the following requirement, obtained from (8.1) and (7.64):

$$\left.\frac{P_r}{N}\right|_{min} \leqslant \frac{\sigma P_f G^2}{f^2 R_{max}^4 (T_a + T_r) B} \times \frac{c_0^2}{(4\pi)^3 k_B} \quad 1 \tag{8.6}$$

Particularly worth noting is the R_{max}^4 term in the denominator. If one wishes to double the range of a radar by changing only the emitted power, the latter must be increased by a factor $2^4 = 16$.

8.1.7 Pulse Radar

The radars most often utilized to measure distances emit short microwave pulses of duration τ, with a pulse repetition frequency $f_r = 1/\tau_r$ (Sec. 5.8). In this manner, a high pulse power is obtained, with only a moderate average power: ***the ratio of average to pulse power*** is the product τf_r. Typical values are $\tau = 1$ μs and $f_r = 1$ kHz, for which the power ratio is 1/1000.

The transmitted wave covers the distance R between the transmitter and the target, then returns towards the radar with a time delay t_{ar} (Fig. 8.5). As it is

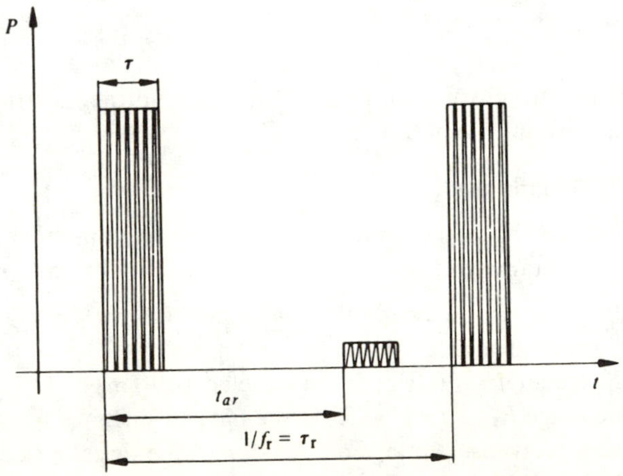

Fig. 8.5 Principle of pulse radar. The scale factor is 150 meters per microsecond.

an electromagnetic wave propagating at the velocity of light c_0, the distance R is obtained with:

$$R = c_0 t_{ar}/2 \qquad \text{m} \qquad (8.7)$$

8.1.8 TR and ATR Cell Duplexer

The power emitted by a pulse radar is larger, by several orders of magnitude, than the power that the receiver input can tolerate. It would only take a very small portion of the emitted signal to be reflected into the system, for instance, by a mismatched antenna, for the sensitive input stages of the receiver to be irreparably damaged. A certain amount of protection could be introduced by placing a varactor limiter in front of the receiver (§ 6.8.8). This protection is, however, not sufficient for high power levels. Very fast acting switches, called *TR* (Transmit-Receive) and *ATR* (anti-TR) *cells* are actually required for adequate protection.

The central part of the cells is a ***voltage breaker***, made up of two electrodes of conical shape placed on the two broad waveguide walls, within a waveguide section filled with rarefied gases (Fig. 8.6). The electric field concentrates between the electrodes. When the signal exceeds the disruptive voltage, a discharge takes place. The cell is completed by matching elements. A TR cell lets small amplitude signals go through, but reflects almost entirely the high amplitude ones. The opposite occurs in ATR cells. Combining the two devices, a duplexer is realized (Fig. 8.7).

The transmitted pulse triggers discharges in both cells: the signal then travels directly from the emitter to the antenna, the receiver being completely isolated. At the end of the pulse, both arcs go off: the emitter is then isolated and the

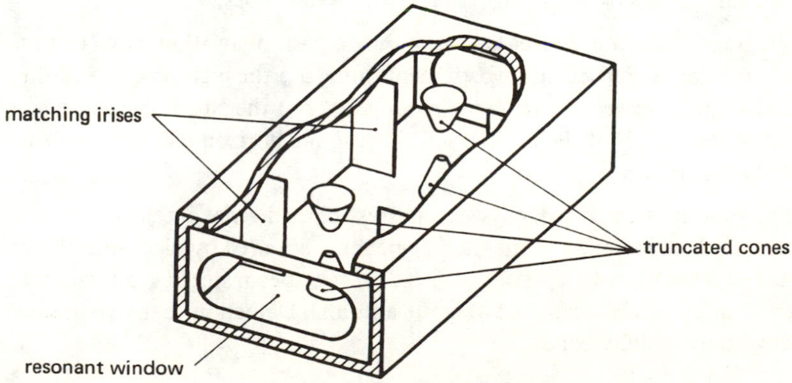

Fig. 8.6 TR cell.

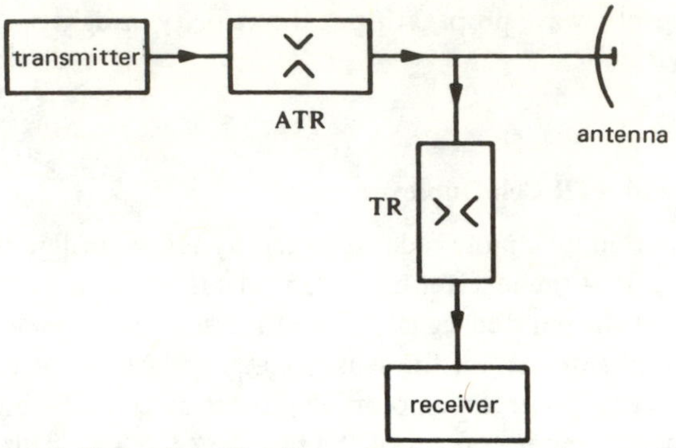

Fig. 8.7 Diplexer with TR and ATR cells.

receiver gets back the return signal picked up by the antenna. The distances between the two cells are carefully adjusted to avoid reflections back to the emitter in the emitting step, or back to the antenna in the receiving one.

For an effective protection of the receiver, a very fast triggering of the cells is required (around 10 ns). The extinction must also be quick, to allow the returning echo signal to be picked up. Recovery times of the order of 50 μs are typical, corresponding to a shortest measurable distance of 7.5 km for a high power pulse radar with a TR-ATR duplexer.

8.1.9 Application: Surveillance Radar

Radars commonly employed for air and sea surveillance possess a rotating antenna (Fig. 8.8). The received signal is displayed, most often in polar coordinates, on the screen of a cathode ray tube (Fig. 8.9). One can visualize on it, like on a geography map, the position of obstacles and vehicles. Remanence screens permit moving targets to be distinguished from fixed echoes.

Surveillance radars are utilized on a worldwide basis in aviation, navy and territory surveillance. Particularly worth mentioning are the high power, and thus very long range,radars sweeping from the arctic ranges of the North American continent to the islands of the Pacific, scrutinizing the horizon searching for some hypothetical missile.

The more sophisticated radar systems possess phased array antennas, formed of numerous fixed radiating elements. The beam is oriented by electronically controlled phase shifters (§ 6.7.12 and § 6.8.5). The beam steering is then much faster than by mechanically rotating the antenna. Detection then becomes on the order of only a millisecond.

8.1.10 Frequency- Swept Radar, Chirp Radar

When operating over shorter distances, a continuous wave microwave signal is

Fig. 8.8 Antenna of airport radar.

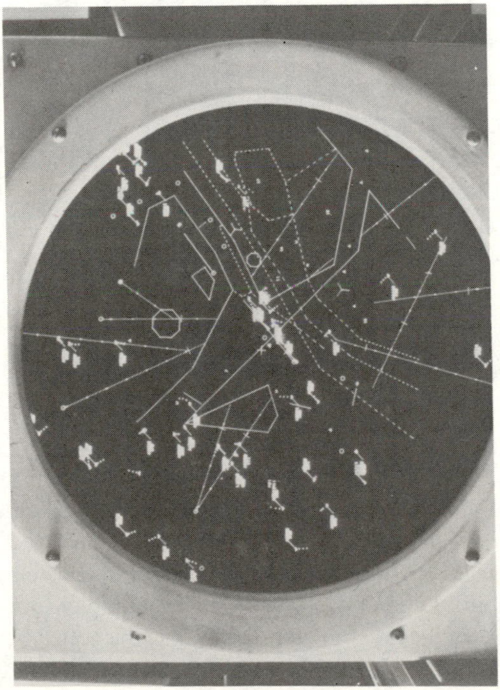

Fig. 8.9 Oscilloscope display of airport radar.

emitted, its frequency shifting linearly with time. The basic principle is illustrated in Figure 8.10: a signal of frequency f_1, emitted at time t_1, comes back after reflection on a target, at time $t_1 + t_{ar}$. At that time, the emitter signal has another frequency f_2. Mixing the two signals (§ 5.3.14) yields a signal at the frequency difference $\Delta f = f_2 - f_1$, which is proportional to the time t_{ar} and thus to the distance covered. When the slope of the $f(t)$ line is m, the distance R is obtained in terms of the frequency difference by the relation:

$$R = \frac{c_0 t_{ar}}{2} = \frac{c_0 \Delta f}{2m} \qquad \text{m} \qquad (8.8)$$

The frequency sweep is repeated with a period τ_r.

For an unambiguous measurement of distance, one must have $t_{ar} < \tau_r/2$, which limits the range R_{max} to $c_0 \tau_r/4$. Adjusting the slope m in terms of the distance to be measured, intermediate frequencies Δf in the audible range are obtained. It is then possible to evaluate the distance by the pitch of the output sound. For this reason, radars based on this principle are known as *chirp radars.*

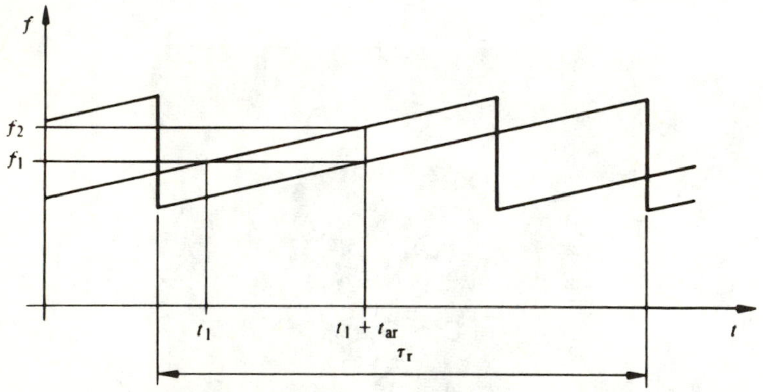

Fig. 8.10 Principle of the chirp radar.

8.1.11 Application: Altimeter

Classical altimeters are, in fact, barometers. Knowing the local air pressure and the pressure at sea level, the altitude can be determined. On board a plane, however, this may not be sufficient: planes are known to have crashed while flying at the correct altitude, but over the wrong topography (due to an unnoticed lateral drift). The radar altimeter determines the actual height over the ground below, it provides an information which is different from the one provided by the classical instruments.

Lunar modules of the Apollo project used radar altimeters to soft land on the moon. As the moon has no atmosphere similar to earth's, barometers would be quite useless.

8.1.12 Application: Measurement of Levels

The use of a radar is of particular interest in hostile environments, filled with dust, where neither mechanical methods (touch) nor optical ones (laser) work satisfactorily. Devices based on the chirp radar principle provide an accuracy of the order of one centimeter; they are utilized to measure levels, among other things: molten metal in blast furnaces, corrosive liquids, and ore in mines [166].

■ 8.1.13 Doppler Radar: Measurement of Speed

When a radar illuminates an object in radial motion, the frequency of the reflected signal is shifted, with respect to that of the emitted signal, by an amount proportional to the ***radial velocity of the object***: this is called *Doppler shift* (Fig. 8.11).

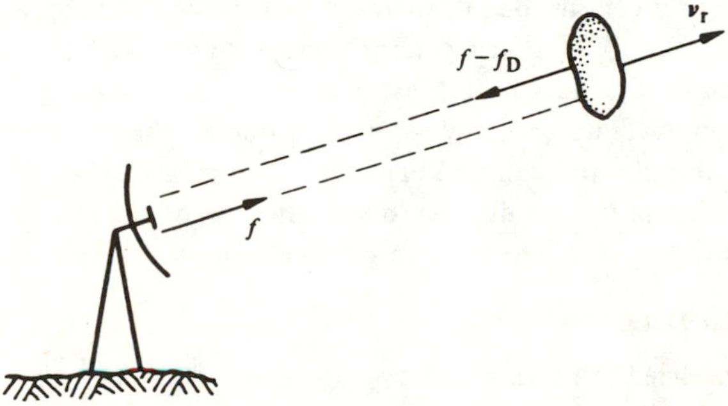

Fig. 8.11 Principle of the Doppler effect.

The transmitted signal, of constant frequency, has a time-dependence of the form sin (ωt); the reflected one has sin $(\omega t - 2\beta R)$ when reaching the receiver, with $\beta = \omega / c_0$. For a steadily moving body, the distance has a linear dependence of time $R = R_0 + v_r t$. The received signal then has the time-dependence:

$$\sin[(\omega - 2\beta v_r)t - 2\beta R_0] \qquad 1 \qquad (8.9)$$

Mixing the received signal with part of the transmitted one (§ 5.3.14), a signal at the intermediate frequency, or *Doppler frequency* $f_D = \beta v_r / \pi$ is obtained. The ***radial velocity*** is then obtained by measuring this frequency:

$$v_r = c_0 f_D / 2f \qquad \text{m/s} \qquad (8.10)$$

The data processing unit must here separate echoes produced by moving objects from those of targets at rest or in slow motion (clouds). This discrimination is carried out by a high-pass filter, since targets at rest yield a dc signal. The lowest speed that can be measured is then limited by the filter cutoff.

For accurate speed measurements, the frequency of the generator must remain quite stable. Also, the radar antenna must remain steady during the measurement. Should the antenna vibrate, the processing system would consider this vibration frequency as a Doppler shift: the radar then sees all fixed objects as being in motion, which may lead to rather amazing results.

8.1.14 Comment: Sources of Error

Doppler radars are routinely used to measure the speed of moving vehicles for speed limit enforcement by police. The accuracy of the measurement is now generally acknowledged, even by drivers directly involved. The principle of the measurement itself yields an exact relationship (8.10), with all factors measurable to high accuracy (frequency measurement, Ch. 5). Still, the measured signal should actually be the one reflected by the vehicle. The case where trees produced spurious reflections, due to their leaves rustling in the breeze, was mentioned; so was the case of a farm tractor apparently clocked at 150 km/hr! The echo measured by the radar in this last case was due to the large fan in front of the engine. Some US designed devices were found to phase-lock on some harmonic of the Doppler frequency. Very great care must be taken, first in the design and manufacture of the devices (discrimination of the return signal), and then when installing them (to avoid any possible source of spurious reflections).

8.1.15 Radar Detectors

With the generalized utilization of Doppler radars to detect speed violators, radar detectors appeared on the market. They are small devices made up of an antenna, a diode detector (§ 7.1.5) and an electronic and acoustical apparatus which emits a whistle when a microwave signal is detected. The protection provided by such devices is open to question. Radars have narrow beams, so that the device may well have ended its measurement by the time the detector emits its warning – except when important sidelobes and reflections are present around the radar. In certain countries, radar detectors are strictly forbidden.

8.1.16 Burglar Detection

A simple Doppler radar is made up of a Gunn diode oscillator (Sec. 4.6) connected to an antenna. A second diode mixes the transmitted wave and the one reflected by any moving target, providing a signal at the ***Doppler frequency***. The Gunn diode itself may even double as a mixer: the sensitivity is then smaller. The large series manufacture of such devices led to considerable cost reductions. They are commonly used nowadays in burglar alarms, reacting to any motion. Similar devices are part of door-opening systems.

8.1.17 Medical Application

A radar monitoring small movements in a non-invasive manner can be a great help in medicine, in particular to check continuously respiratory and heart rates [167]. As an example, a setup devised to study the small movements of a foot is shown in Figure 8.12. The detected signal displays the arterial blood pulses.

8.1.18 Radar Astronomy

A very large power pulse radar was used to measure planetary distances and rotation speeds. In this way, it was established that the planet Mercury does not keep the same side facing the sun, as was believed earlier. Echoes produced by Ganymede, the largest moon of Jupiter, about 10^9 kilometers away were detected [168]. As the distance appears in the denominator of the radar equation (8.1) at the fourth power, the return signal is exceedingly small, of the order of 10^{-15} W down to 10^{-18} W (femto- to attowatts) (§ 8.9.2).

8.1.19 Countermeasures

The development of jamming techniques in military applications closely parallels that of detection systems. A simple approach is to drop chaff, such as aluminum foil, near the plane or missile one wishes to camouflage. The radar then picks up, in the place of a single trace, a large number of reflections, among which it may be rather difficult to locate the target. More complex methods pick up the signal transmitted by the radar, amplify it, process it in a suitable manner, and return it to the radar. It is possible, in this manner, to provide wrong information for distance and trajectory. To avoid being confused, the most sophisticated radars emit variable frequency pulses of very short duration (pulse compression), which are difficult to detect fast enough and, above all, to modify in an adequate manner.

The field of countermeasures is a most active one, yielding some interesting applications in the civilian area. Due to the top-secret nature of the field, it is very difficult to find information about it [169].

8.2 COMMUNICATIONS

8.2.1 Introduction

Microwaves are used extensively in communications, most often for point to point transmissions over free space, from one emitter to one receiver. When both stations are located at ground level, the system is called a ***microwave link*** (§ 8.2.12). ***Satellite communications*** are realized with microwaves (§ 8.2.13), as are links with space probes and expeditions (§ 8.2.14).

Circular waveguides (§ 2.7.6) are also utilized for communications. Since the low-loss mode is not the dominant one, large difficulties had to be overcome.

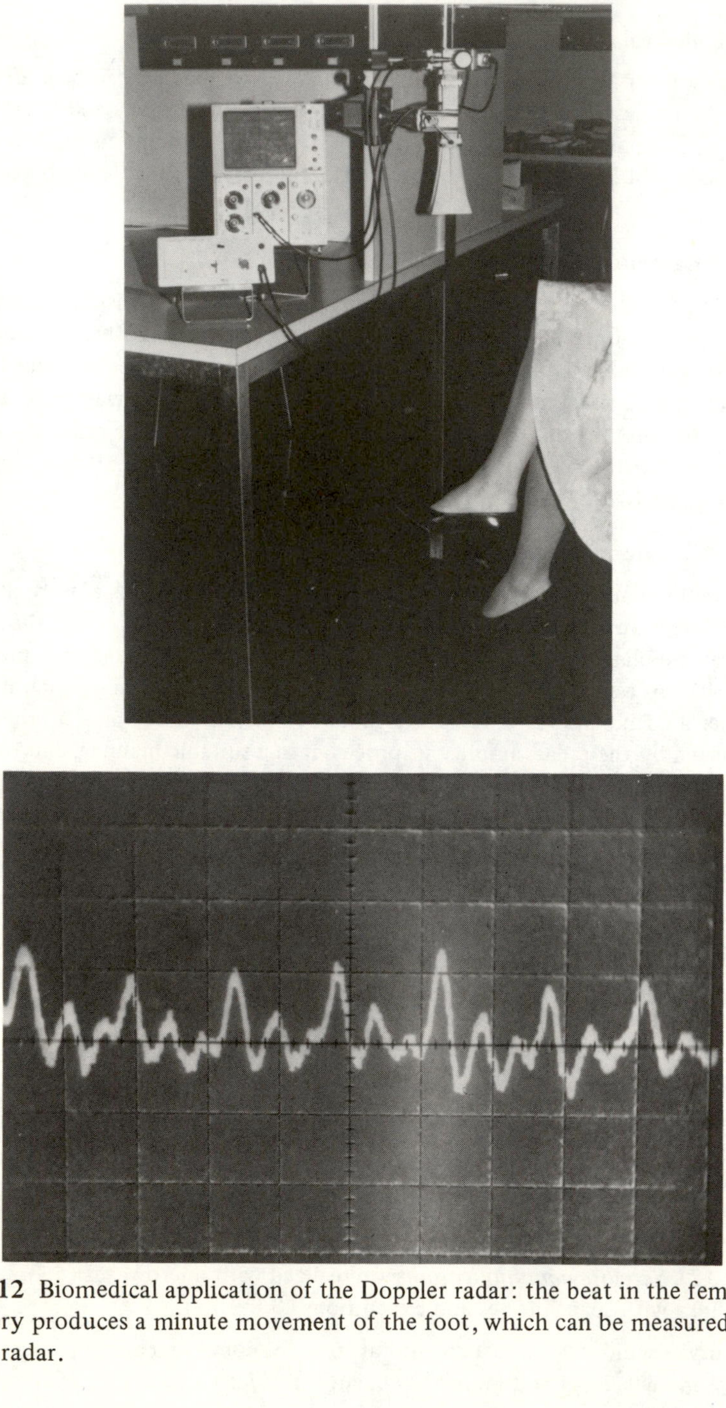

Fig. 8.12 Biomedical application of the Doppler radar: the beat in the femoral artery produces a minute movement of the foot, which can be measured with the radar.

Only a few prototype sections became operational before the method became technically obsolete with the advent of fiber optics (Sec. 2.10 and § 8.2.15).

8.2.2 Power Ratio

In transmission systems with two antennas and a single path (Fig. 8.13), the received power P_r is given by:

$$P_r = P_f G_1 G_2 (\lambda / 4\pi L)^2 \qquad \text{W} \qquad (8.11)$$

where P_f is the power fed to the emitting antenna, G_1 and G_2 are the power gains (§ 7.4.11) of the two antennas and L is the distance between them. Atmospheric losses are neglected (§ 8.2.11). Knowing the noise power picked up, the one added by the receiver (Sec. 7.6), and the signal-to-noise ratio required for a transmission of acceptable quality, the minimum value for the generated power P_f can be determined in terms of the parameters of all other components of the system. Their actual selection results from an evaluation of the costs.

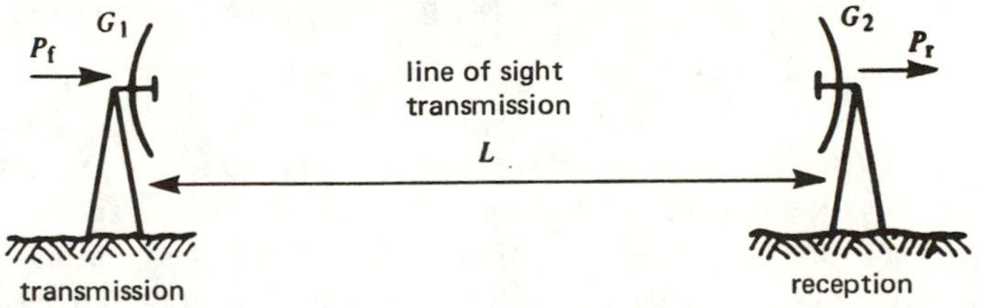

Fig. 8.13 Transmission link.

■ 8.2.3 Atmospheric Propagation, Standard Atmosphere

While the electromagnetic properties of air are very close to those of a vacuum ($\epsilon_r \cong 1, \mu_r = 1$), they are not rigorously identical. When considering a long trajectory within the atmosphere, the variations due to the pressure p, the temperature T, and the moisture level ν have to be considered. The relative permittivity of air at microwaves is given by an experimental expression:

$$\epsilon_r = n^2 \cong \left[1 + \left(\frac{79p}{T} - \frac{11\nu}{T} + \frac{3.8 \cdot 10^5 \nu}{T^2}\right) 10^{-6}\right]^2 \qquad 1 \qquad (8.12)$$

where p is the barometric pressure in millibar, T the temperature in Kelvin, ν the water vapor pressure in millibar and n the refractive index.

Waves in the visible spectrum are not affected by moisture: their refractive index is given by (8.12), letting $\nu = 0$.

A large number of measurements led to the establishment of an average profile for ϵ_r as a function of height h, called the *standard atmosphere* (Fig. 8.14) [170].

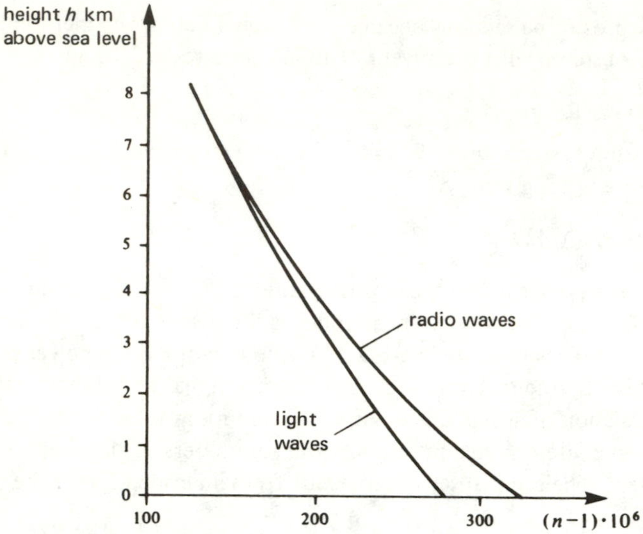

Fig. 8.14 Standard atmosphere.

The study of atmospheric propagation is a problem in spherical geometry: the refractive index n is a function of height h, and thus of the distance $r = h + R_T$ to the center of the earth ($R_T \cong 6370$ km is the earth radius). Let us consider three homogeneous layers, having respectively indices n_0, n_1 and n_2 (Fig. 8.15). On the very large spheres of radius r_0 and r_1 separating two media, Snell's law specifies that:

$$n_0 \sin \theta_0 = n_1 \sin \theta_1' \qquad 1 \qquad (8.13)$$

$$n_1 \sin \theta_1'' = n_2 \sin \theta_2 \qquad 1 \qquad (8.14)$$

Besides that, geometrical considerations on Figure 8.15 show that:

$$r_1 \sin \theta_1'' = r_0 \sin \theta_1' \qquad \text{m} \qquad (8.15)$$

Combining the two relations yields

$$n_0 r_0 \sin \theta_0 = n_1 r_0 \sin \theta_1' = n_1 r_1 \sin \theta_1'' = n_2 r_1 \sin \theta_2 \qquad \text{m} \qquad (8.16)$$

The result is generalized for continuous variation, yielding

$$nr \sin \theta = n_0 r_0 \sin \theta_0 \qquad \text{m} \qquad (8.17)$$

Contrary to general belief, electromagnetic rays do not travel in a straight line within the atmosphere, but follow slightly curved trajectories. For microwave links, the incidence angle θ is most often close to $\pi/2$, so that one generally uses

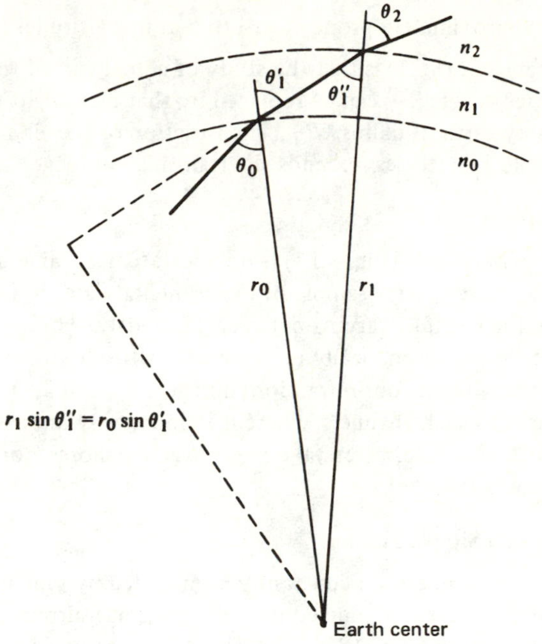

Fig. 8.15 Propagation in spherically stratified media.

the grazing angle $\alpha = \pi/2 - \theta$, in terms of which (8.17) becomes:

$$nr \cos \alpha = n_0 r_0 \cos \alpha_0 \qquad \text{m} \qquad (8.18)$$

Taking sea level as the reference, where $r_0 = R_T$ (earth radius) and developing approximately the functions:

$$n(R_T + h) \cos \alpha = n_0 R_T \cos \alpha_0 \qquad \text{m} \qquad (8.19)$$

$$n\left(1 + \frac{h}{R_T}\right)\left(1 - \frac{\alpha^2}{2}\right) \cong n_0 \left(1 - \frac{\alpha_0^2}{2}\right) \qquad 1 \qquad (8.20)$$

neglecting the second-order terms, this becomes:

$$\frac{1}{2}(\alpha^2 - \alpha_0^2) \cong (n - n_0) + \frac{h}{R_T} = \frac{h}{kR_T} \qquad 1 \qquad (8.21)$$

where

$$k = \frac{1}{1 + (n - n_0) R_T/h} \cong 4/3 \qquad 1 \qquad (8.22)$$

noting that $n - n_0$ is approximately proportional to h at low altitudes (Fig.8.14). The practical significance of (8.21) is that the study of actual curved trajectories over the earth, of actual radius R_T, can be replaced by that of straight trajectories over an imaginary earth of radius kR_T. The variation of the air's permittivity, while quite small, nevertheless extends the transmission range.

8.2.4 Real Atmosphere

The standard atmospheric profile (Fig. 8.14) is only a statistical value, representing an average taken over a large sample of experimental data. In fact, the term k depends upon the latitude, varying between 1.2 and 1.5 [170]. When the atmosphere fluctuates, n may even locally increase with altitude: waves are then trapped within a channel (anomalous refraction, mirages, Section 2.9). Shadow regions are located next to such channels, where it is apparently impossible to emit or to receive. Such anomalies often take place over sea shores (for instance in the notorious Bermuda triangle).

▪ 8.2.5 Propagation in a Simple Plasma

The high layers of the atmosphere are constantly bombarded by ionizing solar radiation, so that spherical layers of ionized particles (plasma) surround the earth: they are called the ***ionosphere*** (§ 8.2.6). The effect produced on communications is determined by considering a plane wave propagating in a *simple plasma.* The ionized particles forming this plasma have no average drift; there is no applied static magnetic field. Linearized relations for small signals are considered (§ 4.4.3).

The plasma is made up of a homogeneous density n_0, of electrons of mass m, and of positive monovalent ions of mass M. The current density (phasor-vector) is related to the speed of the particles $\underline{\boldsymbol{v}}_e$ and $\underline{\boldsymbol{v}}_p$ for electrons and for positive ions, respectively.

$$\underline{\boldsymbol{J}} = n_0 q(\underline{\boldsymbol{v}}_e - \underline{\boldsymbol{v}}_p) \qquad \text{A/m}^2 \qquad (8.23)$$

These speeds are proportional to the electric field $\underline{\boldsymbol{E}}$ (4.23):

$$\underline{\boldsymbol{v}}_e = q\underline{\boldsymbol{E}}/\mathrm{j}\omega m \qquad \text{m/s} \qquad (8.24)$$

$$\underline{\boldsymbol{v}}_p = -q\underline{\boldsymbol{E}}/\mathrm{j}\omega M \qquad \text{m/s} \qquad (8.25)$$

Combining these expressions, Ohm's law for the plasma is obtained:

$$\underline{\boldsymbol{J}} = \frac{n_0 q^2}{\mathrm{j}\omega m}\left(1 + \frac{m}{M}\right)\underline{\boldsymbol{E}} = \sigma\underline{\boldsymbol{E}} \qquad \text{A/m}^2 \qquad (8.26)$$

Since $M \cong 1830\, Z\, m$ (Z = atomic number), the contribution of ions is quite negligible. This result is in turn introduced into Maxwell's equations, yielding the propagation equation:

$$\nabla \times \nabla \times \underline{E} = -\,\beta \times \beta \times \underline{E} = \mu_0 \epsilon_0 (\omega^2 - \omega_p^2)\,\underline{E} \qquad \text{Vs/m}^3 \qquad (8.27)$$

with

$$\omega_p = \left(\frac{n_0 q^2}{m\epsilon_0}\right)^{1/2} = 56.4\sqrt{n_0\,[\text{m}^{-3}]} \qquad \text{rad/s} \qquad (4.32)$$

Propagation takes place:

1. For longitudinal waves ($\beta \times \underline{E} = 0$) when $\omega = \omega_p$. These are plasma oscillations, without energy transfer, as $v_g = 0$.
2. For transverse waves ($\beta \cdot \underline{E} = 0$) for $\omega > \omega_p$. The propagation behavior of β is identical to that of a propagating mode in a waveguide with a cut-off angular frequency ω_p (§ 2.2.31).

The presence of a plasma prevents the propagation of signals at frequencies lower than $\omega_p/2\pi$. When space vehicles come back into the atmosphere, a dense plasma sheath surrounds them, stopping all communications during one of the most critical phases of the whole event.

8.2.6 Ionosphere

While the ionosphere is by no means a simple plasma, as it fluctuates and is placed within the earth's magnetic field, results obtained in the previous paragraph remain approximately valid. The ionosphere is made up of layers D, E, F_1 and F_2 (Fig. 8.16). The upper layer F_2 varies considerably, in density and height, depending further on seasons and solar cycles [171].

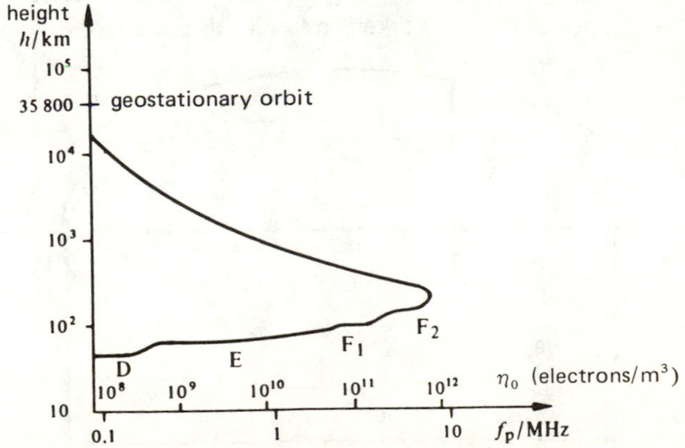

Fig. 8.16 Average electron density in the ionosphere and corresponding plasma frequency versus height.

The cutoff frequency is located, in the average, around 8 MHz: microwave signals, at much higher frequencies, cross the ionosphere without reflection and almost without dispersion. Their amplitude must, however, remain small enough for small signal theory to apply.

8.2.7 Ground Effect, Fresnel Zones

The electromagnetic wave impinging upon the soil is partially reflected [214]. The receiver may thus receive an indirect wave, the ground reflection, which is added to the direct wave. The two waves cover different distances: they have then different phases when reaching the receiver. When the difference in path lengths is less than $\lambda/2$, a signal of perpendicular (horizontal) polarization is reduced, and if the angle of incidence is large, the parallel (vertical) polarization is enhanced; the opposite takes place for differences in path lengths between $\lambda/2$ and λ (and so on). The regions in space where these reflections take place are respectively called first, second, and so on, *Fresnel zones.* The first Fresnel zone is the locus of points in space for which all indirect paths differ by half a wavelength at most from the direct path length. It is located within an ellipsoid of revolution, with the two antennas located at the focal points (Fig. 8.17). The short half-axis is $1/(2\sqrt{\lambda L})$.

At a location between the two antennas, the radius h_0 of the ellipsoid's cross section is given by:

$$h_0 = \sqrt{\frac{\lambda L_1 L_2}{L}} \quad \text{m} \qquad (8.28)$$

It is found in practice that only signals reflected within the first Fresnel zone have a large enough signal amplitude to produce significant interference. As much as possible, precautions are taken to keep this zone free of any obstacles.

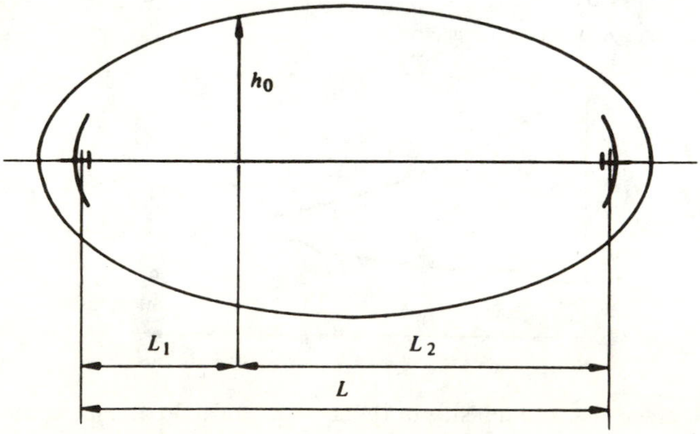

Fig. 8.17 First Fresnel zone.

8.2.8 Fading and Scintillation

In contrast to what is often thought, line of sight links are not stable, but may present significant variations in time, mostly due to interference.

The received signal is produced by the interference of the direct path and the reflected path. Depending upon the relative phases of the two signals, the field may be enhanced or reduced; maxima and minima are, respectively,

$$E_0(1 + |\underline{\rho}|) \qquad \text{and} \qquad E_0(1 - |\underline{\rho}|) \qquad \text{V/m} \tag{8.29}$$

with E_0 = amplitude of the direct field

$|\underline{\rho}|$ = modulus of the ground reflection factor.

The atmosphere is inhomogeneous and fluctuating in time (§ 8.2.4), so that the relative phases of the two signals and their amplitudes keep changing. Long term fluctuations are called *fading.* Rapid variations are known as *scintillation.*

Even when there is no ground reflection, such phenomena can still take place; the direct path may be divided into several different paths by atmospheric inhomogeneities. The most common source of fading remains, however, the interference between the direct path and the reflections. It can be avoided to a large extent by keeping the first Fresnel zone free from any obstacles (§ 8.2.7).

The effect of changes in path length increases with frequency. For this reason, fading is more frequent and pronounced at high frequencies.

8.2.9 Propagation over the Horizon, Diffraction Zone

Propagation over the horizon takes place when the straight line from emitter to receiver intersects the terrestrial globe. When two stations are somewhat beyond the real horizon, the curvature of the rays produced by the atmospheric inhomogeneity leads to the definition of a ***radio horizon*** different from the actual one (effect of atmospheric diffraction, § 8.2.3). A transmission can thus take place at a distance superior to the optical bearing, but beyond this radio horizon, the signal decreases quite fast within the *diffraction zone* (Fig. 8.18). As a matter of fact, ***diffraction*** of the signal by the terrestrial globe can take place, and thus the signal is transmitted beyond the radio horizon.

In the same way as it does in optics, a sharp edge, for instance a mountain ridge, produces diffraction. Signals transmitted in this manner are strongly attenuated, but may nevertheless be used, as they are steady and independent from the atmospheric conditions. A transmission using diffraction over a ridge requires a large emitted power. An obstacle such as a mountain can be taken advantage of when setting up a long range link.

8.2.10 Propagation Trans-Horizon, Troposcatter

In Figure 8.18, one sees that, as soon as the radio horizon line is crossed, the

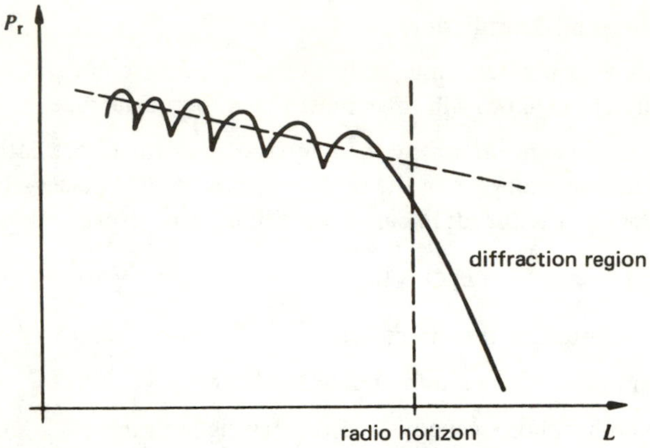

Fig. 8.18 Diffraction region.

signal decreases quite rapidly. However, when the receiver keeps moving further away from the transmitter, the decrease becomes slower after a certain distance, not as would be expected from diffraction theory (Fig. 8.19). The field is then small, but remains usable with large gain antennas and a powerful transmitter.

This propagation can be explained by the diffusion of the wave in a part of the troposphere, in direct line of sight of both transmitter and receiver. This diffusion appears to be produced by atmospheric turbulence, which gives rise to fluctuating local inhomogeneities. The received field is the result of many small contributions, produced by diffusion over a very wide region of the atmosphere (Fig. 8.20).

8.2.11 Effects of Weather

At frequencies above 10 GHz, the atmosphere produces absorption and diffrac-

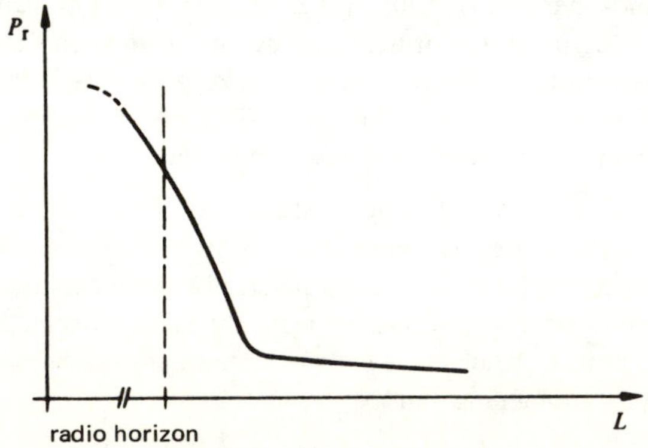

Fig. 8.19 Trans-horizon propagation.

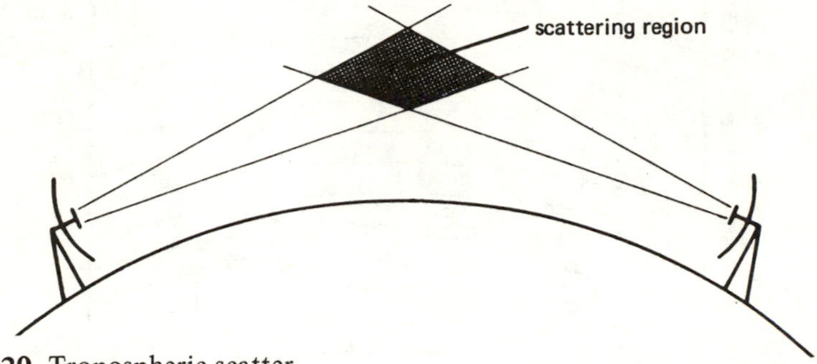

Fig. 8.20 Tropospheric scatter.

tion of the signal, caused by oxygen, water vapor, rain, clouds and snow. Numerous studies have been carried out, or are in progress. These phenomena are quite complex and measured data present a wide spread of values [172, 173]. The practical results are an additional attenuation of the signal, a rotation of its polarization plane, as well as noise (scintillation). As communications bands become increasingly saturated, it becomes imperative to also operate at frequencies above 10 GHz. Statistical studies of meteorology are made to predict the reliability of a transmission [174].

8.2.12 Microwave Links

Two questions must be answered when installing a line of sight microwave link:

1. Choose a path undisturbed by obstacles, taking into account the atmospheric diffraction (§ 8.2.3) and the ground effect (§ 8.2.7). The topographic profile is drawn on a height-position chart (Fig. 8.21). On the same diagram, the first Fresnel zone is drawn, and the locations and heights of the antennas can then be determined [175].
2. The power ratio (§ 8.2.2) is used to select the most suitable components for the distance considered.

A microwave link has telephone and television channels. It is easier to install than a standard telephone line, particularly, over rough terrain: antennas can be brought into position on mountain tops by helicopter, the repeater stations are powered by batteries, which may even be charged with solar cells.

When installing a microwave link, care must be taken to avoid interference between consecutive sections. This is done by assigning different carrier frequencies for adjacent sections.

8.2.13 Communications Satellites

The utilization of stationary satellites for communications was first advocated in 1945 by A.C. Clarke [176]. The first satellite, Sputnik I, was placed in orbit

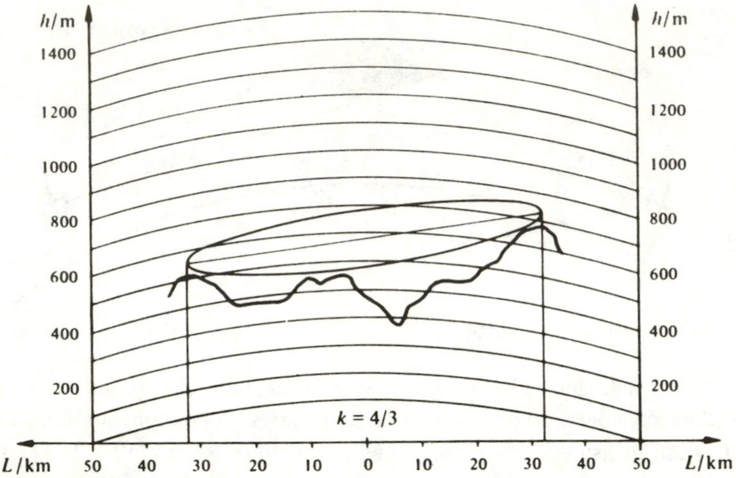

Fig. 8.21 Example of utilization of a height-position chart to set up a microwave link above an imaginary Earth of radius kR_T with $k = 4/3$.

on October 4 1957. In 1962 the first communications satellite, Telstar I was launched; it was a fly-past satellite, on a low-earth orbit. The first stationary satellite, Early Bird, dates to 1965 and was followed by several generations of Intelsat satellites.

At a height of 35,800 kilometers, a satellite orbits the earth in 24 hours: it remains steady over a fixed point on the equator. This is what is called the stationary orbit (Fig. 8.22).

The use of a stationary orbit presents very significant advantages for communications:

1. The antenna of the ground station points towards a fixed location in the sky, only small corrections being required to compensate for atmospheric diffraction and orbital variations.
2. As the distance to the satellite remains constant, there is no Doppler shift (§ 8.1.13). As a result, the frequency does not change. For a fly-past satellite, on the contrary, the radial motion of the satellite with respect to the antenna produces a frequency shift, which must be taken into consideration.
3. As the satellite is located very far out, it is way above the upper part of the atmosphere, and therefore practically not slowed down by it. It could remain in orbit for more than a million years. Its operation is, however, limited in time by the fuel required for stabilization.

On the other hand, the stationary orbit also presents drawbacks:

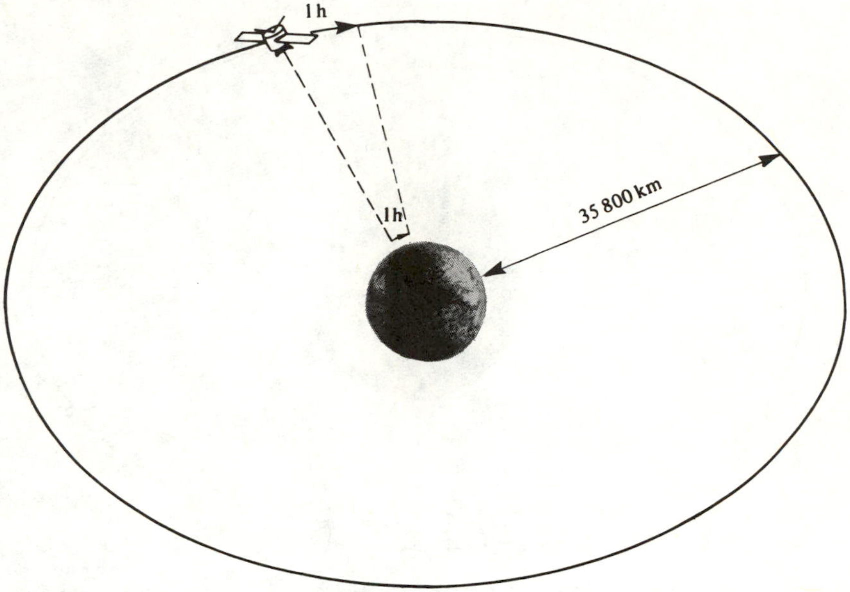

Fig. 8.22 Geostationary satellite orbit.

1. The time needed for a one-way transmission from the ground to the satellite is greater or equal to 35,800 km/c_0 = 119 ms. A back-and-forth telephone transmission (four ways) thus takes more than 476 ms. This is a rather long delay, which might become awkward.
2. The large distance to cover reduces considerably the signal level (§ 8.2.2) The receiving station must utilize a large size antenna, on the order of a 30 meter diameter, and a high sensitivity receiver (parametric amplifier § 7.6.5 and Sec. 4.10).
3. Since the antenna installed on the satellite has a limited size, the region illuminated by the beam is quite large (a continent). The signal from the emitter is picked-up, then shifted and reemitted on a different carrier frequency by the satellite. Part of the returning signal may well be picked up by the emitting station, producing an unpleasant echo. Echo-suppressors are then required.

Communication satellites play a very significant role in information transfer, for television, telephone, and computer data. In addition to international systems, numerous domestic communications satellites are by now in orbit: Molnya, Anik, Palapa and many others (Fig. 8.23). In 1980, 93 stationary locations are taken [177]. The utilization of satellites for the direct broadcasting of television is being considered. The 12 GHz frequency range may be utilized for this purpose.

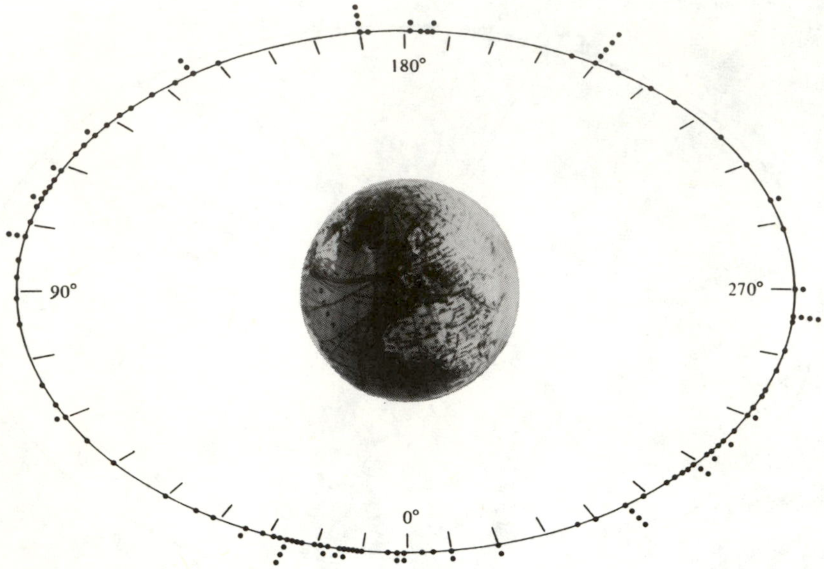

Fig. 8.23 Positions of the geostationary orbit above the equator taken up by communications satellites.

8.2.14 Space Communications

At much further distances, communications with space expeditions and probes are always carried by microwaves. The distance from earth to the moon being 384,000 kilometers, the signal transmitted to Apollo's astronauts or to Lunokhod's drives suffered a 1.28 second delay. Images from Jupiter and its moons were transmitted by the space probes Voyager 1 and 2 (Fig. 8.24) [178]. The average distance from Jupiter to the sun is 778 millions of kilometers, or 43 light-minutes. The signal received from such a probe is exceedingly small.

8.2.15 Applications of Fiber Optics

Who has not seen, at the window of some shop, those light-fountains formed of multicolored fibers gently swinging in air currents? This ornamental application should not overshadow the very great potential of optical fibers in numerous areas, due to their very many advantages:

1. small bulk,
2. insensitivity to radioelectrical perturbations,
3. lightness,
4. very small attenuations (< 5 dB/km),
5. negligible radiation (very small cross-talk),

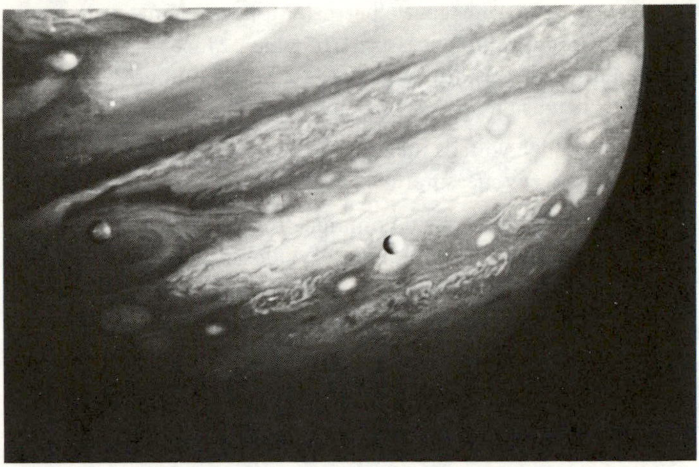

Fig. 8.24 Picture of Jupiter and of two of its satellites, Io (on the left) and Europa (on the right), transmitted by the Voyager I space probe.

6. basic material (silica) available in large quantities,
7. small cost.

Their use will thus be favored whenever one of these advantages is significant.

In the communications area, by far the most promising, numerous experimental and some commercial links have been installed. Optical cables are utilized for heavy-traffic telephone connections over short distances, favored by the small cross sections (particularly appreciated in towns), the insensitivity to perturbations and the negligibly small crosstalk [179]. The progressive setting up of optical fibers in telephone systems is linked to the development of reliable and inexpensive signal sources (light emitting diodes and lasers) and the optical treatment of information. The latter is still carried out by conversion into an electrical signal, electronic processing, and conversion to light. Fiber optic transatlantic cables are being considered: they would be lighter, and would have better performances at a lower cost than conventional cables.

Several cable distribution networks have chosen optical fibers for capacity reasons. The subscribers to a revolutionary Japanese system (HI OVIS) have available, in addition to multiple radio and television programs, a computer terminal to carry out calculations or to access data banks (weather forecast, telephone directory, newspapers, etc.) and even make reservations [180].

As for the military, they are quite pleased with the near impossibility of spying on or jamming communications, as well as on the reduced weight on board aircraft.

In computer systems, electromagnetic perturbations introduce a limitation on the velocity of traffic, to avoid excessive error rates. Since optical cables are not affected by such disturbances, the data flow between computers and interfaces can be increased when using optical fibers.

Last but not least, in high voltage measurements, optical fibers eliminate all insulation problems between measurement point and instrumentation.

Among applications outside of the communications field, one of the first successes of fiber optics was the development of endoscopy in medicine [181]. The same procedure was extended to mechanical engineering, in particular to the car industry, where fibers permit control of certain hidden parts [182].

8.3 MICROWAVE HEATING

8.3.1 Historical Background

In 1945, a manufacturer of radar magnetrons found out, supposedly by chance, that microwaves can produce heating [183]. This effect was rapidly utilized in the design of ovens, heating food and all kinds of materials. For a long time, such ovens remained typical cafeteria and restaurant equipment. Since the beginning of the seventies, microwave ovens increasingly appeared in family kitchens in several countries. In the United States alone, more than two million microwave ovens are sold every year; the total number of microwave ovens installed exceeds 10 million. More microwave ovens are sold in the U.S. than all other types of stoves combined.

8.3.2 Operating Principle

A magnetron (Sec. 4.2), operating most often in the 2.45 GHz band (Industrial, Scientific and Medical band: I.S.M.) generates microwave power between a few hundred watts and a few kilowatts, depending upon the application. It is connected by means of a waveguide to a resonant cavity (oven), which contains the material to be heated or dried: food, wood, paper, plastics, chemicals, textiles, building materials, etc. A ***mode stirrer***, which is a kind of fan with metallic vanes, distributes the microwave energy among the different resonant modes of the cavity, ensuring a homogeneous heating. Figure 8.25 schematically shows a microwave oven.

8.3.3 Characterization of Different Heating Methods

When using ***hot air***, the heat is ***generated outside of the object*** to be heated (by flame, resistive heater) and conveyed by conduction or convection (hot air). The ***surface*** of the object is heated first, and heat then flows towards the inside by conduction only: a temperature gradient from surface to center is required, so that the inside of the object always remains colder than the surface. The heating

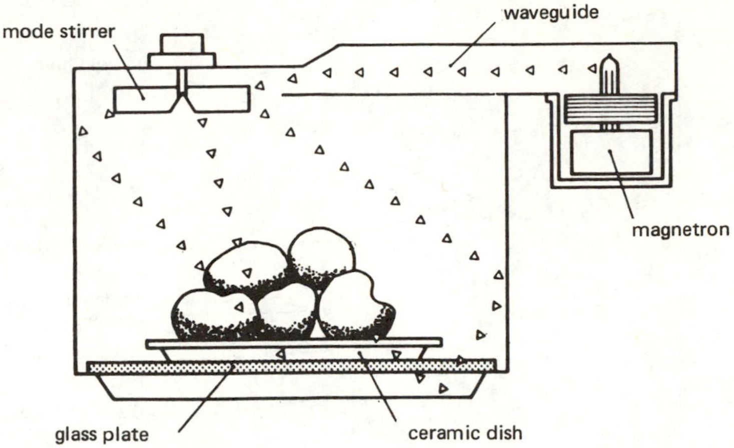

Fig. 8.25 Microwave oven.

process may become long if one wishes to bring the inside up to a cooking temperature without charring the surface.

Infrared heating, which also produces heat on the surface is generated by electromagnetic radiation in the submillimeter range. Within the object, the heating process is basically the same as in the previous situation. The only difference is that the surrounding air is not quite as hot, since it is heated indirectly by the surface of the object.

Microwave radiation penetrates more deeply within the object heated (as long as the material is not metallic). Within the material itself, the electromagnetic energy is transformed into heat by means of several complex conversion mechanisms such as dipole rotation and stretching of large molecules, interface polarization and ionic conduction. Heat is thus generated ***in a distributed manner inside of the material to be treated***, allowing for a more uniform and faster heating. The surface, which is in contact with the cold surrounding air, remains cooler than in previous techniques (in the case of vegetables, vitamins are less affected by cooking). Little heat is lost to the environment. The surface itself may still be heated, if desired, by insulating it from the surrounding air (wrapping, lossy gravy, etc.).

Roughly speaking, while traditional techniques heat a surface, microwaves heat the whole volume of the object treated [184].

8.3.4 Thermal Conversion Mechanism at Microwaves

The average power converted into heat for a unit volume is given by:

$$P_{abs} = 2\pi \epsilon_0 f |\underline{E}|^2 \epsilon_r'' \qquad \text{W/m}^3 \qquad (8.30)$$

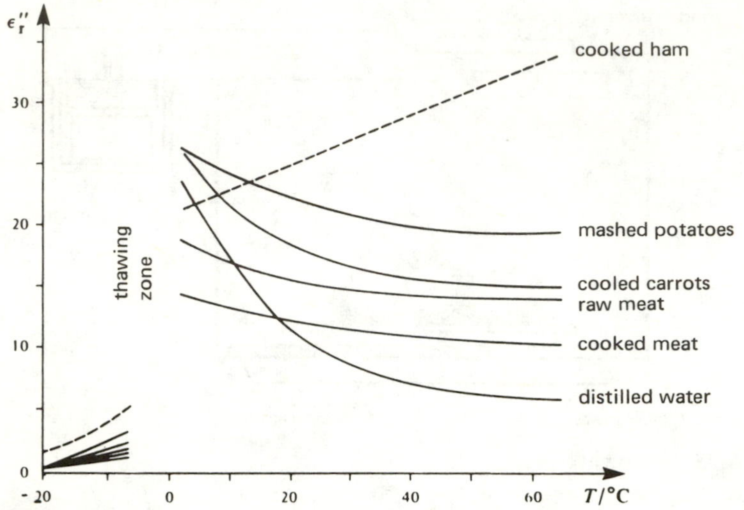

Fig. 8.26 Lossy component of the relative permittivity ϵ_r'' at 2.45 GHz for several foods, versus temperature.

Measured values of ϵ_r'' for several foodstuffs are given in Figure 8.26 [185].

In most instances, ϵ_r'' decreases with increasing temperature: the cooking process is then self-limiting. Only certain fatty foods follow a different dependence: when heating such foods, precautions must be taken to avoid thermal runaway. Dielectric losses are quite small in frozen foods.

The electromagnetic properties of materials vary considerably during the heating process (cooking, drying, etc.). Water is evaporated out of the material, which then becomes less lossy. The heating process practically stops when all the water has been removed. The load seen by the generator may vary considerably during the process: generators must be able to withstand large mismatches.

8.3.5 Applicators

The device used to transfer energy from the microwave source to the load to be treated is called the *applicator.* These devices must be selected with great care, according to the utilization considered. They may be classified into four categories:

1. The resonant cavity (Fig. 8.25) is utilized for the cooking of food and for other industrial applications. In the latter, the material is most often allowed to continuously flow through the cavity, and end sections must be designed to prevent the microwave energy from leaking away towards the outside. In chemical applicators, arcing in a reactant gas takes place without any physical contact, pollution or wear of the material (there is no need to introduce metal electrodes) [186].

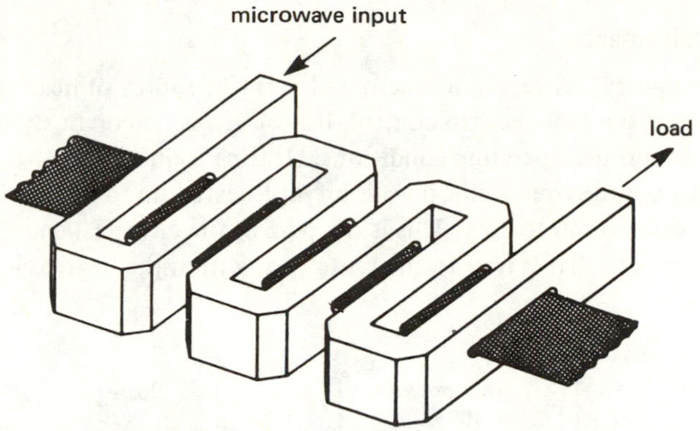

Fig. 8.27 Traveling wave applicator.

2. Traveling wave applicators (Fig. 8.27). Sheet materials or textile threads cross a slotted rectangular waveguide (§ 2.3.12) [187].
3. Slow wave applicator (Fig. 8.28). The material moves close to an open transmission line, interacting in a continuous fashion with the fringing electromagnetic field of the line. This field is quite inhomogeneous; however, as the objects move continuously, the average heat may be kept approximately constant [188].
4. Free space applicator. This last kind of applicator is, for all practical purposes, an antenna. It is designed to irradiate rather bulky elements, which cannot be fitted within an enclosure. Particular precautions must be taken to avoid personal exposure of the operators. Applicators of this kind were shown to reduce concrete and hard stone to powder, the exposure to microwaves producing very large thermal gradients [186]. At lower power levels, similar applicators are used to locally heat patients undergoing hyperthermia treatments (§ 8.3.8).

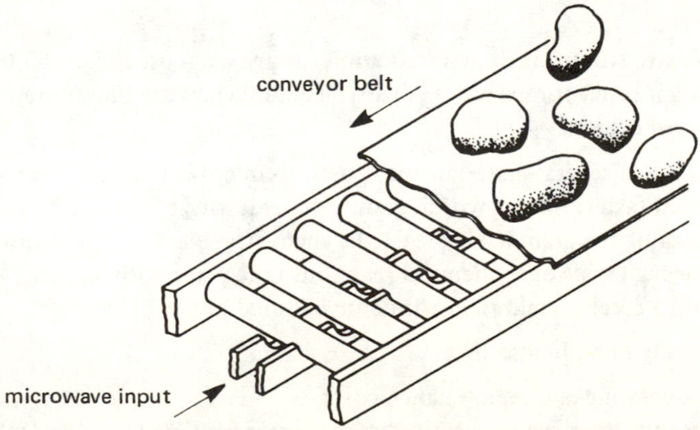

Fig. 8.28 Slow wave applicator.

8.3.6 Advantages

Microwave energy is a very convenient and versatile source of heat, quite flexible, reacting instantaneously to control. It is clean, as no combustion products appear under proper operating conditions. Heating is quite effective, since the surrounding air, the oven or local walls are not heated up. Average efficiencies are on the order of 40 to 50%. This is the part of the electrical energy drawn from the network that is transformed into heat within the heated object (Fig. 8.29).

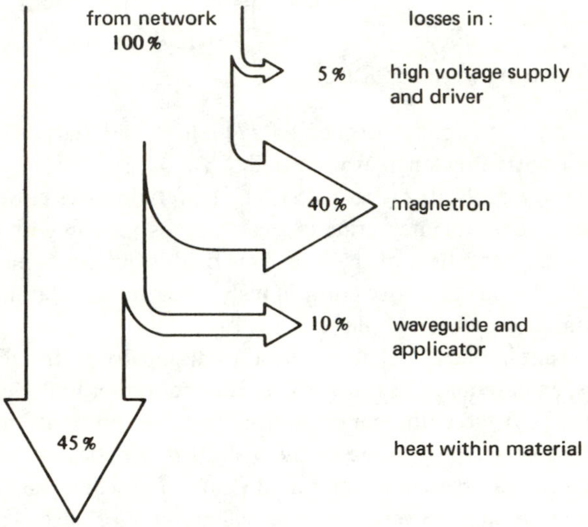

Fig. 8.29 Energy balance of microwave heating.

A comprehensive study has shown that a microwave oven used from 60 to 70% less energy than conventional stoves [189]. These savings are quite interesting during energy shortages.

When a mixture of several materials is exposed to microwaves, the more lossy ones are heated faster: insects within grain can be destroyed by microwaves, without damaging the grain itself [190]. In chemistry, electrons are heated up faster than heavy ions: some chemical reactions take place with less energy expenditure and a greater yield than with conventional heating [186].

On the plus side, one can also list:

1. lower operating and maintenance costs,
2. savings in storage space resulting from shorter heat cycle,

3. reduction of in-process shrinkage and loss, and
4. very fast response.

In certain particular applications, the finished product has a better quality. Alternatively, a high grade product can be obtained using lower cost raw materials.

8.3.7 Drawbacks

The advantages mentioned are to some extent compromised by several disadvantages, which should never be overlooked:

1. high capital costs: magnetrons and other electron tubes are more expensive than resistors;
2. inability to grill surfaces (the basic requirement for a rare steak!);
3. difficulty in drying low-loss solvents, which may remain after water has been removed;
4. more complex technology devices, difficult to repair by unskilled personnel;
5. the need to shield very carefully all equipment, in particular the applicators, to avoid any hazard to their operators. Ovens are equipped with safety latches, which cut off the power as soon as the door is opened. Doors are further equipped with chokes (Fig. 8.30), which reduce very significantly any leakage towards the outside.

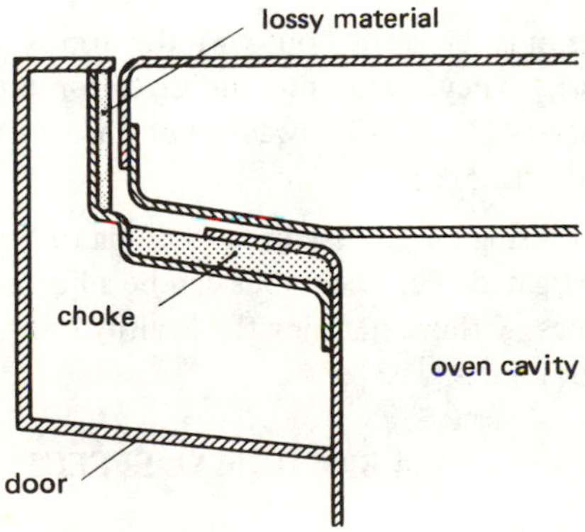

Fig. 8.30 Detail of an oven door joint.

8.3.8 Medical Application: Hyperthermia

The application of heat is a therapeutic procedure commonly used in medicine. A local temperature rise speeds up the metabolic processes, producing a dilatation of the blood vessels. An increase in blood flow results, together with a better irrigation of the tissues, with a faster removal of wastes and heat. Tissues warmed up in this manner receive more nutrients and antibodies, and the healing process is thus speeded up. At the same time, the pain is reduced and sedative effects are noted. Classical ways to apply heat (by hot packs, hot baths, paraffin baths and infrared) reach the surface only. For thermal treatments in depth, for instance to heat stiff joints or muscles, very high frequency *hyperthermia* (13, 27, or 40 MHz), ultrasonics, and microwaves are utilized.

Microwave hyperthermia was introduced in 1947. A generator at the ISM frequency of 2.45 GHz feeds an adjustable power up to about 200 W to an applicator (antenna) which radiates it to the patient. The applicator is placed a few centimeters away from the region treated. Radiation levels for the treatment are from 100 mW/cm^2 up to several watts per square centimeter. The duration, which varies from 15 to 30 minutes, may be repeated several times per day or per week, depending upon the particular ailment.

Microwave hyperthermia is used, among other things, for:

1. orthopedics: arthroses, arthritis, bruises, sciatica, articular rheumatism;
2. internal medicine: asthma, bronchitis, infarctus, pleuresia, urology;
3. dermatology: chilblains, boils, carbuncles, sores;
4. oto-rhyno-laryngology: inflammations, otitis, abscesses, laryngitis;
5. dental care;
6. ophthalmology.

These treatments must be carried out with the utmost care, as an excessive dose may produce burns. They are counter-indicated for patients suffering from certain vascular illnesses [191]. Microwave hyperthermia was successfully utilized to facilitate childbirth [192].

The possibility of using microwave hyperthermia to treat cancer tumors is presently being investigated. This technique can be adjusted to selectively heat up the diseased tissues, without harming the healthy ones. Some encouraging results have been obtained [193].

8.4 BIOLOGICAL EFFECTS

8.4.1 Thermal Effects

The exposure to an excessive level of radiation can produce hazards. The microwave radiation is non-ionizing (§ 1.2.8), its main effect being of a *thermal nature,* commonly used in medical applications (§ 8.3.8).

The body absorbs radiation and automatically adapts to the resulting temperature increase, excess heat being removed by the blood flow. However, should the radiation become too intense, the thermal balance no longer could be restored by the body processes, and burns would then occur. As microwaves tend to heat deeply into the body, one might fear deep burns would occur while the surface temperature remained acceptable. There exists a certain radiation threshold, beyond which irreversible changes (cumulative) do occur.

A considerable number of studies were carried out to determine this threshold. No permanent effect was observed for power levels lower than 100 mW/cm^2 [194]. The most sensitive organ, the eye, was found to possess a threshold around 150 mW/cm^2 for the development of cataracts after a 1½ hour continuous exposure [195]. The male genital organs are sensitive to heat: microwave radiation may thus produce a temporary sterility. The sun provides us with the level of 100 mW/cm^2 in the infrared range, at noon on a sunny summer day.

Introducing a safety factor of 10, the value of 10 mW/cm^2 was adopted, first in the U.S., then in most Western European countries, as the upper limit tolerable for a microwave irradiation of indefinite length. This value should never be exceeded in the vicinity of radar or communications stations, nor near microwave ovens or applicators. A somewhat larger level is tolerated for irradiations shorter than 6 minutes (Fig. 8.31).

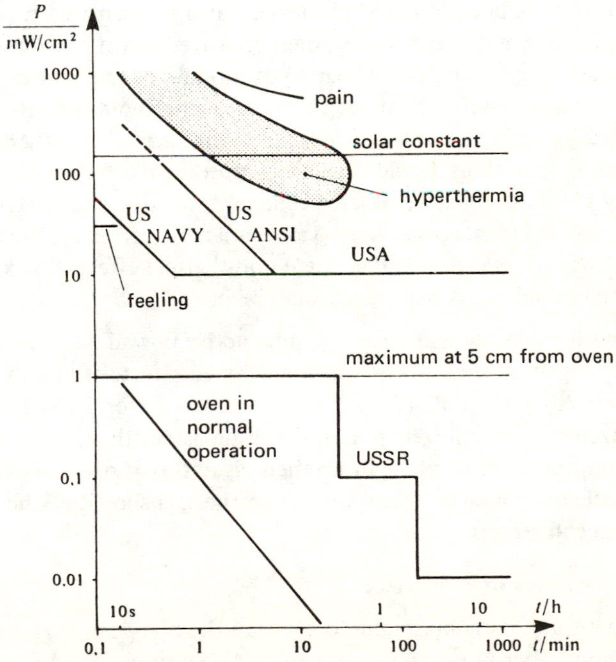

Fig. 8.31 Radiation levels and limiting values.

8.4.2 Comment

The limit of 10 mW/cm^2 is very much below the density levels utilized in hyperthermia (Fig. 8.31). This difference may be accounted for from the surrounding conditions of application. Medical treatments are of a local nature, they are carried out under continuous monitoring. To obtain any kind of therapeutical effect, the lower threshold of 100 mW/cm^2 must be exceeded [191]. On the opposite side, the 10 mW/cm^2 is the upper limit to which an uninformed bystander may be exposed.

8.4.3 Non-Thermal Effects

In the Soviet Union and in Eastern Europe, the upper limit for microwave exposure was set at 10 $\mu W/cm^2$ for long term exposures: this value is one thousand times smaller than the one prescribed in the U.S. and elsewhere. A somewhat higher level is tolerated for short term exposures (and, supposedly, in the vicinity of a certain embassy. . .). This very low limit is justified by the existence of non-thermal effects, produced by other interactions of the microwave radiation with the organism. Some researchers actually feel that hazards might occur even at very low signal levels. Health problems attributed to microwaves include nervousness, hormonal imbalance, malformations, anomalous brain activity, etc. Certain symptoms, such as chromosomal aberrations and direct excitation of neural cells have been observed in the laboratory on isolated tissue samples and organs. These could not, however, be reproduced as yet on complete organisms. Other phenomena, such as the hearing of modulated microwave signals, actually appear at power levels well above the thermal threshold. Effects noticed and published by some researchers could not be duplicated by others under supposedly identical conditions. Couldn't some symptoms attributed to microwaves be actually produced by some other mechanism, for instance X-rays emitted by a poorly shielded magnetron, or by ozone emanating from a defective electrical supply? While it is probable that some non-thermal effects do exist, it has not been established as yet that they would be hazardous.

The hunt for non-thermal effects is quite active nowadays. However, one might wonder why if, indeed, hazardous non-thermal effects of microwaves should exist, were they not noticed on radar operators of World War II? Precautions taken against radiation were practically nonexistent then. It was even mentioned that some operators would light up their cigarettes at operating radar antennas! Also, patients treated by microwave hyperthermia should be the first ones to experience ill effects.

8.4.4 Wearers of Pacemakers

The presence of metallic conductors within the body can trigger other complex interactions. Microwave radiation, penetrating deeply within the body, may induce a current in cardiac pacemakers, creating a hazard to their wearers. Experiments were carried out with signals at several frequencies and operating regimes. The most significant effects were noted for radar signals around 9 GHz. At the industrial 2.45 GHz frequency, only small changes of rhythm have been observed up to the maximum level of the experiment, set at 25 mW/cm^2.

8.4.5 Remark: Oven Safety

It may seem surprising that the device whose safety is most often questioned should be the domestic microwave oven! The great majority of ovens presently built are carefully shielded, their leakage being much lower than the 1 mW/cm^2 limit at 5 cm from the door prescribed in several countries at the time of purchase. On the other hand, one seldom hears of hyperthermia treatments, where patients are exposed (for their own good) to radiation levels 100 to 10,000 times larger than those received in the vicinity of a microwave oven (Fig. 8.31). Microwave ovens have been around for more than 30 years; some 10 million of them are routinely operated in the U.S. Still, no major accident has been reported to date, except maybe some minor burns when touching hot dishes. Which other consumer appliance could offer such a high safety record? An American physicist, James Van Allen, summarized the situation by saying: "in my judgment, [the oven's] it's hazard is the same as the likelihood of getting a skin tan from moonlight."

8.5 MEASUREMENTS OF MATERIALS

8.5.1 Objectives

The knowledge of a material's properties at microwave frequencies is required for several reasons:

1. To design and manufacture microwave equipment for radar and communications,
2. To treat materials by microwave irradiation (cooking and heating, Fig. 8.26). It is then necessary to know the properties of materials at all the different steps in the processing, in order to design applicators providing an overall optimum power transfer (§ 8.3 5);
3. To monitor a physical parameter, such as moisture, during a treatment;
4. For research in physics and chemistry, by determining resonance lines (absorption spectra) of materials;
5. In medicine, it is possible to diagnose certain illnesses, which produce a change in the fluid content of tissues.

8.5.2 Classification

Two different approaches are encountered for the measurement of materials:

1. destructive methods. In all classical techniques, a sample of the material is removed, then placed within a cavity or a waveguide section [111];

2. non-destructive methods. In this case, one or several microwave probes are placed on the surface of the material. The properties are derived from a measurement of the reflection or of the transmission.

■ 8.5.3 Long Sample in a Waveguide

When the material presents a ***high absorption*** at microwaves, the signal rapidly decays within the material, so that the effect of the back end becomes negligible. A sample of the material fills the waveguide cross section (Fig. 8.32).

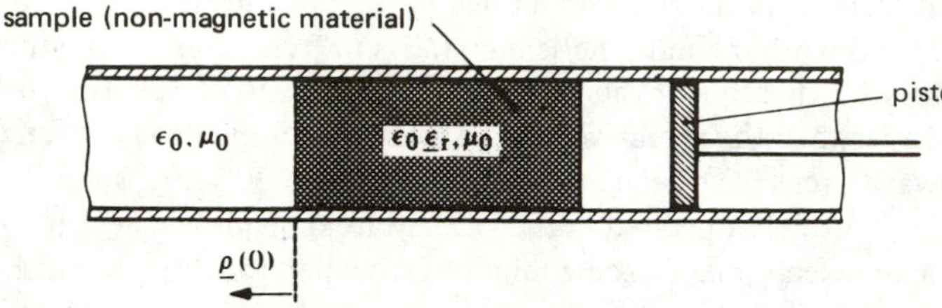

Fig. 8.32 Long sample in waveguide.

When a motion of the piston at the back end does not change the signal reflected into the waveguide, this clearly means that the back end effect may be neglected. Wave impedances for the empty (v) and loaded (c) guides, for the dominant TE_{10} mode in rectangular waveguide are respectively given by (2.97):

$$Z_{ev} = \frac{\omega\mu}{\beta_v} = \frac{\omega\mu_0}{\sqrt{\omega^2 \epsilon_0 \mu_0 - (\pi/a)^2}} \qquad \Omega \qquad (8.31)$$

$$\underline{Z}_{ec} = j\,\frac{\omega\mu}{\underline{\gamma}_c} = \frac{\omega\mu_0}{\sqrt{\omega^2 \epsilon_0 \underline{\epsilon}_r \mu_0 - (\pi/a)^2}} \qquad \Omega \qquad (8.32)$$

Taking the ratio of the two quantities:

$$\frac{Z_{ev}}{\underline{Z}_{ec}} = \sqrt{\frac{\omega^2 \epsilon_0 \underline{\epsilon}_r \mu_0 - (\pi/a)^2}{\omega^2 \epsilon_0 \mu_0 - (\pi/a)^2}} \qquad 1 \qquad (8.33)$$

from which one obtains, after a few manipulations:

$$\underline{\epsilon}_r = \left(\frac{Z_{ev}}{\underline{Z}_{ec}}\right)^2 \left[1 - \left(\frac{\lambda}{2a}\right)^2\right] + \left(\frac{\lambda}{2a}\right)^2 \qquad 1 \qquad (8.34)$$

The impedance ratio is directly related to the reflection factor $\underline{\rho}(0)$ within the interface (9.18), determined by the measurement of the reflection (Sec. 7.2 and 7.3):

$$\frac{\underline{Z}_{ev}}{\underline{Z}_{ec}} = \frac{1 - \underline{\rho}(0)}{1 + \underline{\rho}(0)} \qquad 1 \qquad (8.35)$$

When studying liquids or powders, the method may be extended to low loss materials, inserting a matched termination into the loaded section [197].

■ 8.5.4 Short Circuited Sample

When ***absorption within the material is low***, the effect of the back end cannot be neglected. In this case, a sample of length L is inserted into the waveguide, with a short-circuit placed at its back (Fig. 8.33).

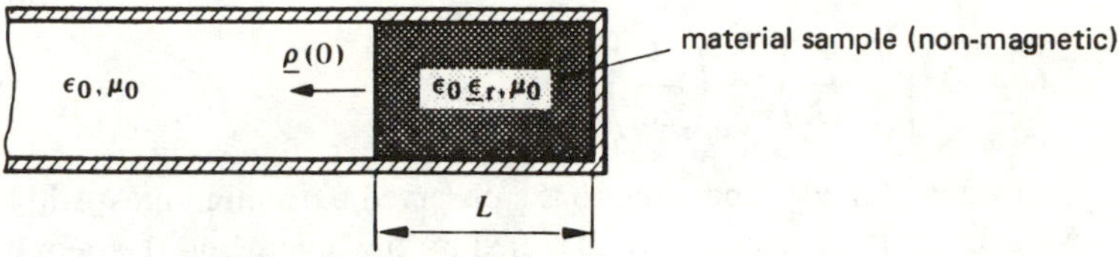

Fig. 8.33 Short-circuited sample in waveguide.

The input impedance to the section becomes, for a rectangular waveguide (9.20):

$$\frac{\underline{Z}_t(0)}{Z_{ev}} = \frac{\underline{Z}_{ec}}{Z_{ev}} \tanh(\underline{\gamma}_c L) = \sqrt{\frac{\omega^2 \epsilon_0 \mu_0 - (\pi/a)^2}{\omega^2 \epsilon_0 \underline{\epsilon}_r \mu_0 - (\pi/a)^2}} \tanh\left(\sqrt{(\pi/a)^2 - \omega^2 \epsilon_0 \underline{\epsilon}_r \mu_0} L\right) \qquad 1 \qquad (8.36)$$

where $\underline{\gamma}_c = \alpha_c + j\beta_c$ is the propagation factor in the loaded waveguide.

The value of $\underline{Z}_t(0)/Z_{ev}$ is given by a reflection measurement (Sec. 7.2 or 7.3). The determination of $\underline{\epsilon}_r$ from (8.36) is here rather involved, as it requires the resolution of a complex transcendental equation involving the function $\tanh(\underline{\gamma}_c L)/\underline{\gamma}_c$ [198]. For a lossless sample, this function becomes $\tan(\beta_c L)/\beta_c$, which can be determined from a table of functions.

8.5.5 Technique of the Minimum Reflection

A simple approach, which permits the quick determination of the permittivity of a ***low-loss material***, places a sample of known length L into the waveguide and determines at which frequencies the transmission goes through a maximum and the reflection through a minimum (Fig. 8.34).

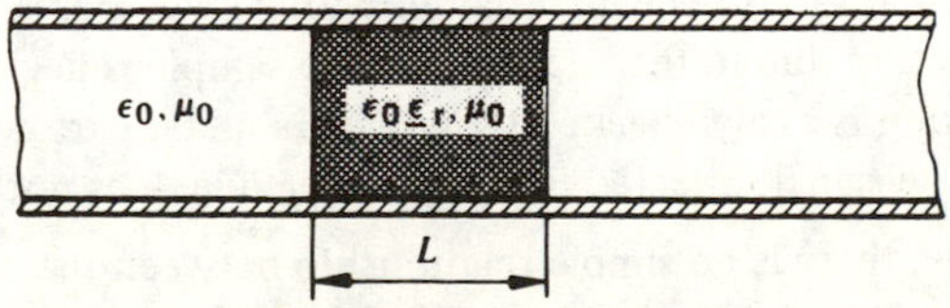

Fig. 8.34 Method of reflection minimum.

The transmission is maximal and the reflection minimal for:

$$L = \frac{n\lambda_{g\epsilon}}{2} \qquad \text{m} \qquad (8.37)$$

where $\lambda_{g\epsilon}$ is the wavelength in the loaded waveguide, defined as:

$$\lambda_{g\epsilon} = \frac{\lambda}{\sqrt{\epsilon_r - (\lambda/2a)^2}} \qquad \text{m} \qquad (8.38)$$

The value of ϵ_r is extracted from these two relations:

$$\epsilon_r = \lambda^2\left[\left(\frac{n}{2L}\right)^2 + \left(\frac{1}{2a}\right)^2\right] \qquad 1 \qquad (8.39)$$

The value of n corresponding to the observed extremum must still be determined. Measurement frequencies are specified by the length and the permittivity of the material.

8.5.6 Sources of Error

The following errors reduce the accuracy of the waveguide measurements:

1. poor joint between the sample and the waveguide (gaps);
2. inaccuracy in the determination of L (§ 8.5.4 and 8.5.5);
3. inaccuracy in the measurement of the reflection factor $\underline{\rho}(0)$, and in particular of its argument.

8.5.7 Non-Destructive Reflection Methods

The sample to measure is not placed within the waveguide itself, but at the end of it. The interface is a flat surface, then it is not necessary to machine a part having a section adjusted to that of the waveguide, but only to provide one flat face. The area of application for reflection methods can be widened, measurements can be carried out faster, and measurement errors reduced. On the other hand, the determination of permittivity is more difficult.

8.5.8 Long Sample

A rectangular or a circular waveguide is terminated by a large flat flange, placed just against the flat face of the material measured (Fig. 8.35). As in Section 8.5.3, the reflection due to the back end of the sample is neglected. The validity of this assumption is easily checked by moving a metal part near that face and checking that the signal reflected into the waveguide does not vary.

In this structure, there is no simple relationship between the reflection factor $\underline{\rho}$ and the permittivity $\underline{\epsilon}_r$. The study of radiation of the waveguide modes in the

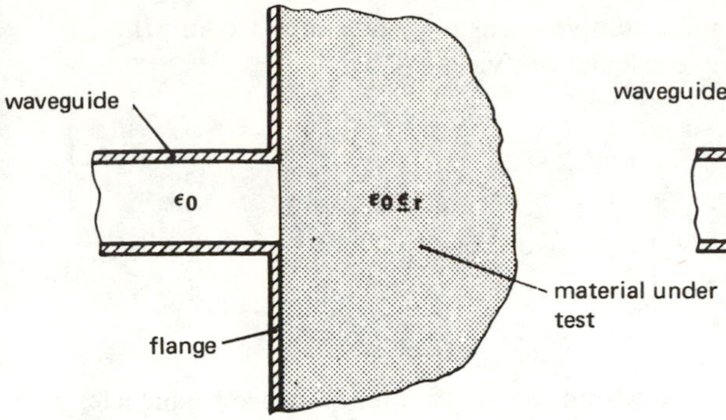

Fig. 8.35 Non-destructive reflection method.

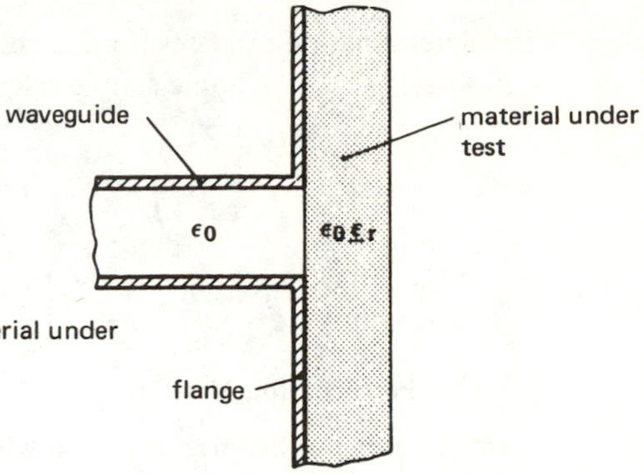

Fig. 8.36 Non-destructive reflection method for slabs.

right-hand half-space and then the application of boundary conditions for the fields across the waveguide opening yield an expression for $\underline{\rho}(\underline{\epsilon}_r)$, which involves a summation over the infinite set of higher order modes in the waveguide. A computer program solves this equation by truncating the series and leads to a graphical representation. The values of ϵ_r' and ϵ_r'' are determined by plotting the measured value of $\underline{\rho}$ in the graph [199, 200].

An open coaxial line may also be utilized, covering a broader frequency range for the measurements [201].

8.5.9 Sheet Sample

The same approach may also be utilized to measure sheets of materials (Fig. 8.36). The method of images [214] is used, giving an infinite series to be solved by computer. The convergence of the calculations is however slow when the plate is thin [202].

8.5.10 Cavity Filled with Material

The properties of a ***low loss material*** can be determined by filling a resonant cavity with it. The resonant frequency and the loaded quality factor are measured, both for the filled and the empty cavity (Sec. 7.5), for the same mode of resonance. With a loose coupling, the loaded quality factor Q_c is approximately equal to the unloaded quality factor Q_0.

One then finds, making use of (3.17) and assuming that the material is non-magnetic,

$$\epsilon_r' \cong \left(\frac{f_{0v}}{f_{0p}}\right)^2 \qquad 1 \qquad (8.40)$$

where f_{0v} and f_{0p} are, respectively, the resonant frequencies of the empty and the filled cavity.

The determination of losses is more complex: one must take into account the wall losses (§ 3.4.4). Some simple calculations yield (3.20):

$$\epsilon_r'' \cong \epsilon_r' \left(\frac{1}{Q_p} - \frac{1}{Q_v \epsilon_r'^{1/4}} \right) \qquad 1 \qquad (8.41)$$

where Q_p and Q_v are, respectively, the unloaded quality factors of the cavity, filled and empty.

8.5.11 Perturbation Method

A small sample of material is placed within the cavity, the resonance frequencies and the quality factors are measured, both without and with the sample. The permittivity is obtained from (3.81).

The perturbation method is limited on both sides:

1. when the sample is too small, the measurement of the difference of frequencies and quality factors becomes inaccurate;
2. when the sample is too large, the perturbation method is not valid.

8.5.12 Symmetrically Loaded Circular Cavity

Samples of larger sizes may be measured, outside of the limitations of the perturbation method, when an adequate computation scheme is available to analyze the loaded structure. Such a method allows one to determine $\underline{\epsilon}_r$ in terms of the measured values of ***resonant frequency*** and ***quality factor***. The cavity filled with concentric tubes, shown in Figure 8.37, was studied for the symmetrical quasi-TE_{011} mode [203]. The material measured fills the central tube; its temperature may be varied by circulating air within the second tube.

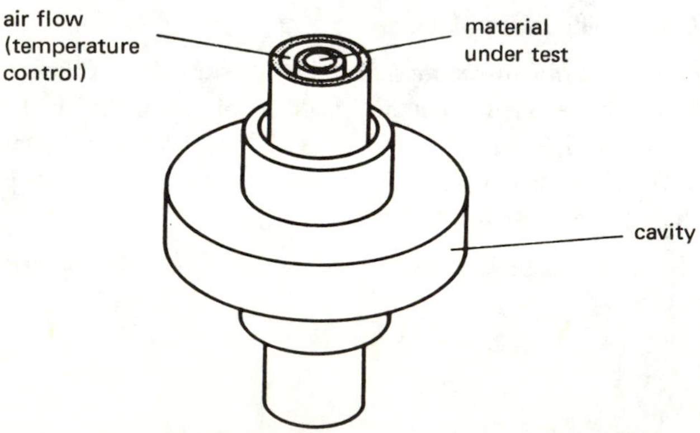

Fig. 8.37 Circular cavity loaded with tubes.

8.5.13 External Perturbation of a Cavity

The open-ended waveguides used in the non-destructive methods of Sections 8.5.8 and 8.5.9 can be terminated by a short-circuit at a distance *d* from the aperture (Fig. 8.38). In this manner, a cavity is realized, which is terminated by a short-circuit at one end, and by a reactance at the other (§ 3.3.15). The equivalent impedance of the aperture is obtained with (3.57), introducing the values measured at the resonance. The remainder of the computations is the same as for the open waveguide [202].

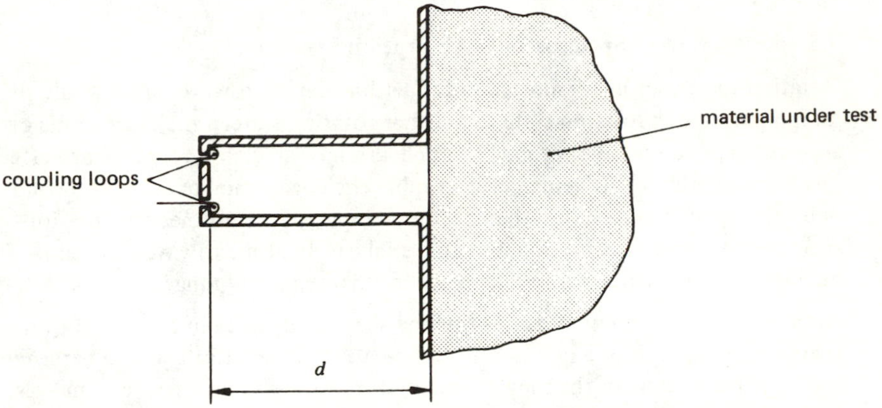

Fig. 8.38 Open-ended cavity for external perturbation measurements.

8.5.14 Transmission Measurements

Certain material properties can be determined by *transmission methods.* The measurement setup is similar to the one utilized to measure antennas (Fig. 7.29), with two fixed radiating elements. The material measured, generally in sheet form, is placed between the two antennas and the power ratio is determined (Fig. 8.39).

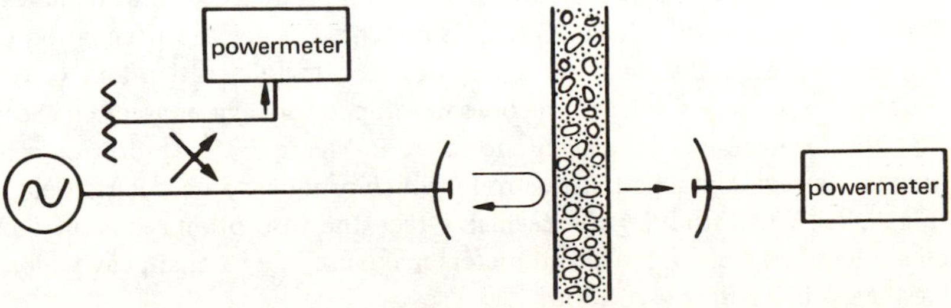

Fig. 8.39 Transmission measurement setup.

The attenuation measured in this manner is produced by two effects (§ 7.4.2): absorption of the signal within the material, and reflections at the two interfaces. The two distributions may be distinguished by running measurements with identical materials having different thicknesses. The reflected power can also be measured.

In practice, this approach is mostly experimental. The measurement setup is calibrated using known materials; the characteristics of the material measured are then determined by extrapolation of the values obtained. This approach is used in particular to determine moisture.

8.5.15 Microwave Spectroscopy, Absorption Spectrum

Rotation spectra of many molecules fall within the microwave range: while jumping from one rotation state to another rotation state, a molecule emits or absorbs one microwave photon. The application of an electric field (Stark effect) or of a magnetic field (Zeeman effect) splits the degeneracy of some spectrum lines, the hyperfine structure then becomes apparent. For molecular rotation spectra, these two effects are observable exclusively at microwaves. Certain nuclear resonance effects also take place in this frequency range.

Microwave spectroscopy measures the line shape and width in the absorption spectrum of a gas or of a liquid. This information then permits one to determine the basic parameters of the molecular structure: interatomic distances, masses, angles of chemical bonds, dipole and quadrupole moments, nuclear moments (magnetic and quadrupole), nuclear spin.

Many liquids and solutions exhibit absorption bands at microwaves. These result from the orientation of molecular dipoles in the radiation field [204].

Microwave spectroscopy has significant applications for the analysis of chemical compounds. The substance analyzed must be polar, and the analysis takes place in the gaseous phase. The resolution available at microwaves is far superior to the one of infrared spectroscopy, allowing the identification of thousands of molecules [205].

Standard spectrometers measure the attenuation produced by a waveguide section or *absorption cell* filled with the gas measured. Since the attenuation per unit length is generally small, the cell must accordingly be rather long (several meters). Techniques similar to the ones developed for cavities are used (Sec. 7.5), the absorption cell replacing the cavity in Figure 7.35. Absorption cells for the Stark effect contain a flat electrode, which produces a dc electric field (Fig. 8.40). When studying the Zeeman effect, the absorption cell is surrounded by a solenoidal coil. Other spectrometers make use of a resonant cavity containing the gas being measured [206].

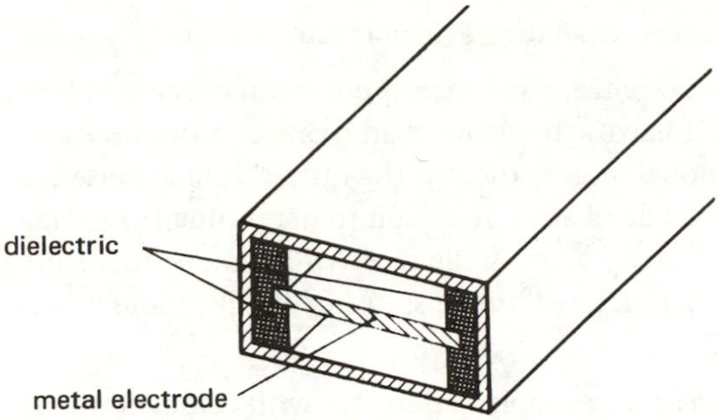

Fig. 8.40 Absorption cell for the study of the Stark effect.

8.5.16 Paramagnetic Resonance, Zeeman Effect

Within the microwave region, the *paramagnetic resonance* is also observed. When a paramagnetic substance is placed within a static induction field, the electron spins rotate in a Larmor precession around this field (§ 6.7.2). Due to the quantum character of the phenomenon, only discrete energy levels are allowed, transitions between these levels being the cause of the *Zeeman effect.* With the measurement of transition frequencies, the crystalline structures, the chemical bonds, and the electronic state of ions in crystals can be studied. This effect is utilized for analysis purposes. An important practical application is the MASER (§ 4.10.5).

The paramagnetic resonance is measured by placing a small sample of material in a cavity and determining the resulting perturbation (§ 3.4.10). A sample may also be placed within a circularly polarized magnetic field in a rectangular waveguide (Fig. 6.56). A large dc magnetic field is applied, perpendicular to the magnetic field of the microwave signal [207].

8.6 RADIOMETRY

8.6.1 Principle

Radiometry measures the noise picked up by an antenna (Sec. 7.6). The equivalent noise temperature, T_a, of the antenna is a function of the media towards which it is aimed. This way, it is possible to measure, sometimes in conjunction with other techniques, a physical temperature, the moisture of the soil, atmospheric pollution, the attenuation due to precipitations (§ 8.2.11), etc. For a large scale study, radiometers are placed aboard aircraft or satellites. They may be utilized, for instance, to study earth resources. Microwave radiometry presents certain advantages, with respect to radiometry in the visible or infrared range: measurements can be made also by night or through atmospheric perturbations (cloud cover).

8.6.2 Radiometers: Operating Principle and Principal Types

A radiometer is, in essence, a ***specialized microwave receiver***. While usual receivers are required with a narrow frequency band, one chooses here a broad bandwidth, to yield a large enough noise power at the output. Since noise is a random process, an integrator is added after detection to determine the average noise level. The receiver can directly amplify the received signal, or first shift it in frequency. The choice is mainly dictated by the status of the technology within the frequency range selected.

Figure 8.41 schematizes a simple radiometer with frequency shift. T_a is the antenna temperature (function of the target temperature), while T_r is the noise temperature of the receiver itself. In the case of Figure 8.41, called a ***full power radiometer***, it is difficult to separate T_a from T_r. The system can be calibrated; it is, however, subjected to drifts and instabilities of the different elements within the system.

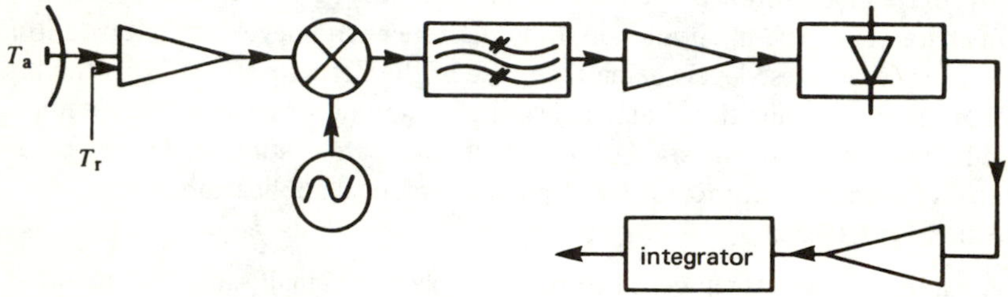

Fig. 8.41 Full power radiometer.

The minimum detectable temperature step (which is a measure of the system's sensitivity) is given by [161]:

$$\Delta T_{\min} = (T_a + T_r)\sqrt{\frac{1}{B\tau} + \left(\frac{\Delta G}{G}\right)^2} \quad \text{K} \qquad (8.42)$$

where: B = receiver bandwidth

τ = integration time

$\Delta G/G$ = normalized variation of the receiver's gain.

The instabilities of the system can be palliated by a periodical calibration. ***Dicke's radiometer*** utilizes this principle (Fig. 8.42): the radiometer is alternately connected by an electronic switch to the antenna and to a calibrated noise source. A synchronous detection is carried out at the output, which is alternately a function of T_a and of T_c.

The minimum detectable temperature step is then [161]:

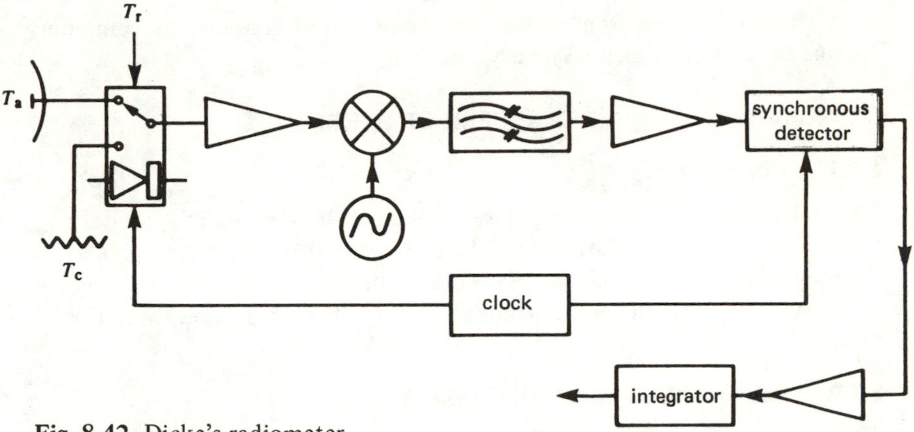

Fig. 8.42 Dicke's radiometer.

$$\Delta T_{min} = \frac{C(T_a + T_r)}{\sqrt{B\tau}} + |T_a - T_c| \frac{\Delta G}{G} \qquad \text{K} \qquad (8.43)$$

where C is the radiometer constant. In general, $C = 2$.

As may be seen in (8.43), gain instabilities have no effect when $T_c = T_a$. This fact is utilized in certain radiometers, in which the reference source temperature T_c is automatically adjusted to the value T_a. The driving signal of the adjustable reference noise source is then measured. This signal is directly related to the antenna noise temperature T_a.

8.6.3 Thermography

In the millimeter wave range (for instance at 68 GHz), it is possible to determine the temperature inside of the human body by measuring the emitted noise radiation. In this non-invasive manner, internal inflammations may be detected, or the effect of drugs may be determined [208]. The same approach might be useful to detect cancer.

8.6.4 Radioastronomy

Radiometers also detect the radiation from celestial bodies, such as the Sun and other noise sources located either within our galaxy, or further out. In 1962 an intense radioastronomical emitter was observed, which tallied with low optical emission, located at an extremely large distance: $0.16 \cdot 10^{10}$ light-years. Since then, many similar sources, called quasi-stellar objects, or ***quasars***, were observed. Other new heavenly bodies, called ***pulsars***, were discovered in 1967.

Incidentally, the possibility that other intelligent beings may exist somewhere else within the universe is investigated within several projects (OZMA, Cyclops, SETI, etc.). The noise coming from radioastronomical sources is studied by computers, looking for patterns which may carry information.

8.7 TRANSFER OF ENERGY

8.7.1 Circular Waveguide

The utilization of circular waveguides operating in the low-loss TE_{01} mode (§ 2.7.6) was investigated by researchers at Stanford University [186]. In theory, it should be possible to replace high voltage transmission lines by metallic pipes of about 1.5 meters in diameter. The resulting advantages are not obvious.

8.7.2 Solar Satellite Power Station (SSPS)

The gathering of solar energy in space, and its transformation and projection down to earth by a microwave beam were proposed in 1968 by P.E. Glaser [209]. In space, the solar energy is available on a continuous basis at a flow rate of 1.4 kW/m^2 in the vicinity of the earth. Solid state solar cells would transform this radiant energy into electricity with an efficiency of about 15%, a microwave beam would then carry this energy to a receiving antenna on the ground, with a global transfer efficiency evaluated between 58 and 72% [210]. The project has seen several different versions over the years. It is being technically studied in the United States, within the scope of new renewable energy resources.

Under its present form [211], the project foresees the launching in a stationary orbit (§ 8.2.13) of stations formed mainly of solar panels, 5200 x 10,400 meters, feeding a microwave phased array antenna having a diameter of about 1 kilometer. Generators might be klystrons, 100,000 of them each providing 70 kW CW. The beam would be collected on the ground by an array of rectifying antennas (rectennas), having an elliptical shape with axes of 10 x 13 km. The power collected on the ground would be around 5 GW.

The power density over the center of the receiving antenna would be 23 mW/cm^2, for operation at 2.45 GHz (ISM band). This power level is superior to the one presently tolerated for long term exposure (§ 8.4.1).

This project being quite gigantic, its realizability was seriously in doubt. Several experiments carried out by NASA have now shown that, technically speaking, it is feasible. ***Nevertheless, the cost and the power expenditure required to launch it with present techniques are unacceptable.*** These two parameters should be considerably reduced for the project to become cost effective.

Many questions are still to be answered concerning costs, maintenance, hazards produced by an error in the beam direction, operational safety, pollution produced by the multiple launchings, and interferences with communications and radioastronomy. Besides, the stationary orbit is already rather filled up with communication satellites (§ 8.2.13).

The World Administrative Radiocommunications Conference (WARC) held at the end of 1979 in Geneva ***refused to authorize the use of the 2450 MHz band, as requested by the U.S.***, for an experimental study of the solar power station. This proposal may only be reconsidered when the operational safety of the concept has been duly demonstrated.

8.8 PARTICLE ACCELERATORS

8.8.1 Definition

A *particle accelerator* is an apparatus in which charged elementary particles, electrons or protons, are accelerated by electromagnetic fields. Methods and structures utilized present a similarity to those utilized in microwave generators and amplifiers (Ch. 4). In both situations, energy transfer takes place between a beam of charged particles and an electromagnetic signal. In an amplifying tube, the microwave signal is amplified and the beam electrons slowed down. In an accelerator, on the contrary, the energy from a microwave source accelerates the particles. These may reach quite important speeds, very close to the velocity of light.

8.8.2 Linear Accelerator

The structure of a *linear accelerator* is basically the same as the one of a distributed coupling tube amplifier for microwaves (Sec. 4.4). The beam of particles circulates along the axis of a slow-wave structure, generally a periodically loaded transmission line, formed by a succession of resonant cavities. The transfer of power is the largest when the particle speed and the wave velocity are identical, that is at ***synchronism*** (§ 4.4.8). As the speed of the particles varies along the structure, the dimensions must be determined accordingly. Among other things, the relativistic mass increase of the particles as they approach light velocity must be taken into account. Particular precautions are also required to focus the beam, as linear accelerators may become quite long.

8.8.3 Circular Accelerators: Cyclotron, Synchrotron

The size of an accelerator can be considerably reduced by letting the charged particles rotate under the influence of a magnetic field (§ 4.2.1). In a *cyclotron,* a microwave signal at the cyclotron frequency accelerates the particles, which follow spiraling trajectories when certain phase conditions are satisfied. When the particle speed is close to the velocity of light, its mass increases due to relativistic effects, so that its cyclotron frequency decreases. This property is taken into account in the *synchro-cyclotron,* which operates with pulses of particles and variable frequency. In the *synchrotron,* particles follow a ring trajectory, the magnetic field varying with time to maintain stability [60].

8.8.4 Application: High Energy Physics

The best known uses of accelerators are certainly those related to research in physics. The very high energy (1 to 10 GeV) atom smashers have permitted the advancement of elementary particle physics (mesons, leptons, hyperions, etc.). At lower energy levels, accelerators opened the study of the internal structure of the atomic nucleus. Research in chemistry and biology also makes use of accelerators.

8.8.5 Medical Applications

Electron accelerators around 6 MeV are operated on a routine basis to produce energetic X-rays in the treatment of cancer. The X-ray radiation produced by the bombardment of a target with a particle beam is easier to control and focus than gamma rays emanating from nuclear sources such as Cobalt 60. In a similar manner, neutron beams or π-meson beams are obtained to treat cancer cells. In addition, the electron beam itself can be directly utilized to take care of surface lesions.

8.8.6 Industrial Applications

High energy X-rays, produced by electron accelerators, are also used in various industrial applications [186]:

1. sterilization of drugs and medical equipment;
2. sterilization and pasteurization of foodstuffs;
3. polymerization of plastics;
4. inspection of welded joints, in the construction of nuclear reactors and rockets.

8.9 PROBLEMS

8.9.1 At which distance does a radar of 1 kW pulse power reliably detect a person, when its receiver has a sensitivity threshold of 1 pW, its antenna has a 16 dB gain, and its operating frequency is 10 GHz?

8.9.2 A radar emits a 400 kW signal, with a 72 dB gain at 8.5 GHz, towards Ganymede, one of the Jovian satellites. The reflected signal comes back 1 hour and 7 minutes after emission. The diameter of Ganymede is 2635 km and its power reflection (albedo) is only 12% of the one for a metal sphere. Determine the received power.

8.9.3 A chirp radar has a 1 GHz/s frequency variation. What is the difference between transmitted and received signal frequencies when the target is at a distance of 100 meters?

8.9.4 A Doppler radar emits a 9 GHz signal. What is the Doppler frequency produced by an aircraft flying at 820 km/h towards the radar?

8.9.5 Determine the maximum range of the following microwave link:

1. generated power: 10 MW
2. minimum detectable received power: - 140 dBm
3. gains of both antennas: 3.49×10^8
4. wavelength: 3 mm.

What does the calculated distance correspond to?

8.9.6 At what height should the two antennas of a microwave link be located, when transmitting over 150 km of flat terrain, the signal frequency being 4 GHz? Consider the general situation, then the particular case when both antennas have the same height.

8.9.7 Determine the propagation velocities within a simple plasma (phase and group velocities) in the following situation:

1. signal frequency: 2 GHz,
2. plasma density: 10^{16} electrons per cubic meter.

8.9.8 What is the upper latitude at which a stationary satellite signal may be received:

1. under a grazing incidence,
2. under a 10° elevation angle?

The curvature of the ray produced by the atmosphere may be neglected.

8.9.9 A 0.86 μm laser signal has a 1 μW power level when reaching the receiver. Determine the photon flow per second. Assuming that 20 photons are required to transfer 1 bit of information, determine the data rate Q in bits per second.

8.9.10 The permittivity of a material is measured by placing a long sample of it in an X-band rectangular waveguide (2.286 x 1.016 cm cross section). At 9 GHz, the reflection factor in the front plane of the sample is $\underline{\rho} = -0.627 + j\,0.036$. Calculate the permittivity of the material.

8.9.11 A very low-loss material is to be measured. A 1 cm long sample is placed in an X-band waveguide (2.286 x 1.016 cm), its section being adjusted to the guide cross section. A matched load terminates the guide. The reflection goes through a minimum at 8 GHz. Determine ϵ_r.

8.9.12 Determine the minimum integration time required for a ± 2K accuracy with a radiometer having a 0.4 GHz bandwidth. The noise temperatures are, respectively, $T_a = 250$ K and $T_r = 1046$ K. The relative gain variation of the receiver is evaluated at $\Delta G/G = 10^{-3}$. Consider the two following situations:

1. total power radiometer,
2. Dicke radiometer with a source at $T_c = 77$ K, and a radiometer constant $C = 2$.

CHAPTER 9

TRANSMISSION LINE FUNDAMENTALS

9.1 INTRODUCTION

9.1.1 Definitions

When the length of an electrical conductor is of the same order of magnitude as the wavelength, the transit time of a signal is comparable to its period, and cannot be neglected. The length and size of the conductors and their geometrical setting become important parameters at high frequencies.

9.2 LOSSLESS TWO-CONDUCTOR LINE

9.2.1 Model Considered

A line made of two straight pec wires is considered (Fig. 9.1). It is uniform along the direction of propagation z (§ 2.1.1). The currents $I(z,t)$ and $-I(z,t)$ flow in the two conductors. The *transverse voltage* $U(z,t)$ is defined by integrating the electric field along a transverse path joining the two conductors. The path must lie ***entirely within a plane perpendicular to the conductors.***

9.2.2 First Line Equation

Applying Maxwell curl equation to the contour indicated in Fig. 9.1 yields, after some calculations, the first line equation:

$$\frac{\partial U(z,t)}{\partial z} = -L' \frac{\partial I(z,t)}{\partial t} \qquad \text{V/m} \qquad (9.1)$$

where L' is the complete inductance per unit length of the transmission line (which contains the mutual inductance and the self inductances of the two conductors). The space variation of the line voltage is proportional to the time variation of the current.

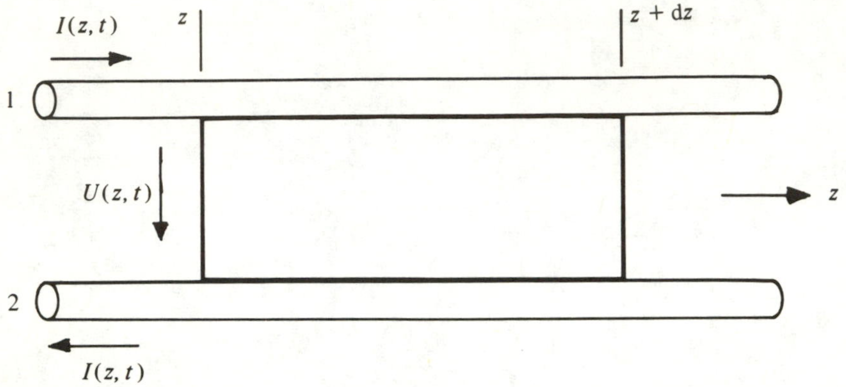

Fig. 9.1 Section of a uniform two-conductor line.

9.2.3 Second Line Equation

Applying the charge conservation equation to a box of length dz surrounding one of the conductors and integrating yields, in a similar way, the second line equation:

$$\frac{\partial I(z,t)}{\partial z} = -C' \frac{\partial U(z,t)}{\partial t} \qquad \text{A/m} \qquad (9.2)$$

where C' is the capacitance per unit length of the transmission line. Here, the space variation of the current is proportional to the time variation of the line voltage.

9.2.4 Equivalent Circuit

The expressions (9.1) and (9.2) link current and voltage in the equivalent circuit of Figure 9.2. It must be noted that this equivalent circuit applies only to a section of transmission line having an infinitesimal length dz.

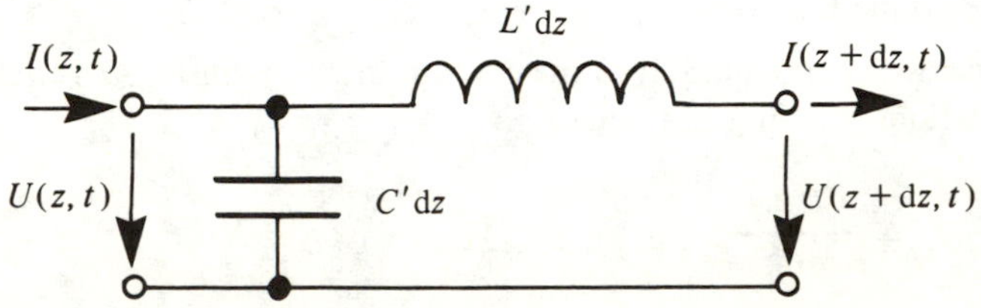

Fig. 9.2 Equivalent circuit of a section of lossless L′C′ transmission line of infinitesimal length dz.

9.2.5 Propagation along a Lossless Two-Conductor Line

Equations (9.1) and (9.2) are combined, yielding the d'Alembert equation:

$$\frac{\partial^2 U(z,t)}{\partial z^2} = L'C' \frac{\partial^2 U(z,t)}{\partial t^2} \qquad \text{V/m}^2 \qquad (9.3)$$

The solution to this equation has the general form

$$U(z,t) = f_+(z - vt) + f_-(z + vt) \qquad \text{V} \qquad (9.4)$$

with

$$v = (L'C')^{-1/2} \qquad \text{m/s} \qquad (9.5)$$

The two functions f_+ and f_- are arbitrary univocal (single-valued) functions of their argument. The term f_+ corresponds to a *forward wave,* propagating towards the $+z$ direction. The term f_- is the *backward wave,* which moves towards decreasing values of z. Both waves propagate without change of amplitude or of shape (without distortion). They do not interact with each other.

9.3 GENERALIZED TRANSMISSION LINE

9.3.1 Extension to Other Propagating Structures

When studying real lines, to the simple case of Section 9.2 must be added a series resistance, accounting for the conductor losses, and a shunt conductance, representing possible leakage through the insulator material. The model may then be further generalized, covering many other propagating structures, among them waveguides (Ch. 2). The study must then be restricted to sine waves, keeping in mind however that any signal may be represented by a combination of sine waves [215] (discrete or continuous expansion). The voltage and the current along the line are then given by phasors, in the same manner as the fields in Section 1.4:

$$U(z,t) = \sqrt{2}\, U(z) \cos[\omega t + \varphi_u(z)] = \mathrm{Re}[\sqrt{2}\, \underline{U}(z) \exp(\mathrm{j}\omega t)] \qquad \text{V} \qquad (9.6)$$

$$I(z,t) = \sqrt{2}\, I(z) \cos[\omega t + \varphi_i(z)] = \mathrm{Re}[\sqrt{2}\, \underline{I}(z) \exp(\mathrm{j}\omega t)] \qquad \text{A} \qquad (9.7)$$

where the phasors are defined as:

$$\underline{U}(z) = U(z) \exp[\mathrm{j}\varphi_u(z)] \qquad \text{V} \qquad (9.8)$$

$$\underline{I}(z) = I(z) \exp[\mathrm{j}\varphi_i(z)] \qquad \text{A} \qquad (9.9)$$

These quantities are time independent, but vary with position along the line. The derivative with respect to time is replaced, in phasor notation, by the multiplication with a factor $\mathrm{j}\omega$.

9.3.2 Equivalent Circuit

The equivalent circuit of a section of transmission line of length dz is simply obtained by extending the one of Fig. 9.2, defining a series impedance per unit length $\underline{Z}'$ and a shunt admittance per unit length $\underline{Y}'$ (Fig. 9.3).

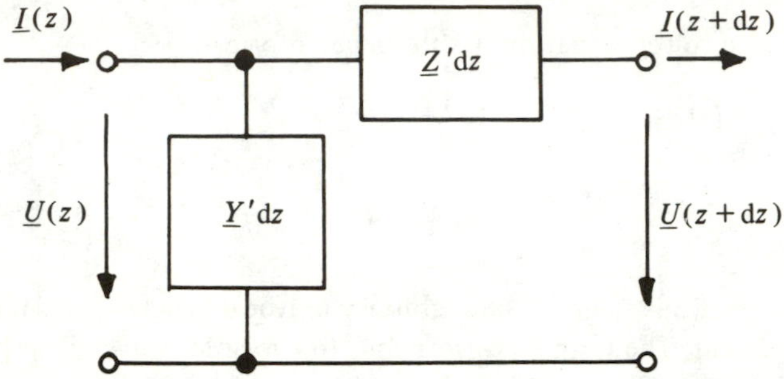

Fig. 9.3 Equivalent circuit of a section of generalized transmission line of infinitesimal length dz.

9.3.3 Line Equations

The two equations for the generalized line become:

$$\frac{\mathrm{d}\underline{U}(z)}{\mathrm{d}z} = -\underline{Z}'\underline{I}(z) \qquad \text{V/m} \qquad (9.10)$$

$$\frac{\mathrm{d}\underline{I}(z)}{\mathrm{d}z} = -\underline{Y}'\underline{U}(z) \qquad \text{A/m} \qquad (9.11)$$

9.3.4 Propagation

The two line equations (9.10) and (9.11) are combined and solved in the same manner as for the lossless line (§ 9.2.5), yielding the voltage and current along the line:

$$\underline{U}(z) = \underline{U}_+ \exp(-\underline{\gamma}z) + \underline{U}_- \exp(+\underline{\gamma}z) \qquad \text{V} \qquad (9.12)$$

$$\underline{I}(z) = \underline{Y}_c\underline{U}_+ \exp(-\underline{\gamma}z) - \underline{Y}_c\underline{U}_- \exp(+\underline{\gamma}z) \qquad \text{A} \qquad (9.13)$$

forward wave   backward wave

with

$$\underline{\gamma} = \sqrt{\underline{Z}'\underline{Y}'} = \alpha + \mathrm{j}\beta \qquad \text{m}^{-1} \qquad (9.14)$$

The square root is taken so that β is positive. When it vanishes, then a positive value is taken for α.

The *characteristic admittance* $\underline{Y}_c$ and the *characteristic impedance* $\underline{Z}_c$ of the line are defined by:

$$\underline{Y}_c = 1/\underline{Z}_c = \underline{\gamma}/\underline{Z}' = \sqrt{\underline{Y}'/\underline{Z}'} \qquad \text{S} \qquad (9.15)$$

They specify the current-to-voltage ratio of the forward wave on the line. The two complex terms $\underline{U}_+$ and $\underline{U}_-$ are integration constants, to be determined from the boundary conditions at the two ends of the line (generator and load).

9.4 BOUNDARY CONDITIONS

9.4.1 Load Impedance

A transmission line always has a finite length: at some point $z = L$ it is terminated into a load of complex impedance $\underline{Z}_L$ (Fig. 9.4). Both the current and the voltage at that point must be continuous, so that:

$$\underline{Z}_c \frac{\underline{U}_+ \exp(-\underline{\gamma}L) + \underline{U}_- \exp(+\underline{\gamma}L)}{\underline{U}_+ \exp(-\underline{\gamma}L) - \underline{U}_- \exp(+\underline{\gamma}L)} = \underline{Z}_L \qquad \Omega \qquad (9.16)$$

This expression provides a relationship for the two constants $\underline{U}_+$ and $\underline{U}_-$. Generally, one determines $\underline{U}_-$ in terms of $\underline{U}_+$.

9.4.2 Reflection Factor

To simplify the notation, the ratio of backward to forward wave voltages is called the *reflection factor* $\underline{\rho}(z)$, which is determined from (9.16):

$$\underline{\rho}(z) = \frac{\underline{U}_- \exp(+\underline{\gamma}z)}{\underline{U}_+ \exp(-\underline{\gamma}z)} = \frac{\underline{Z}_L - \underline{Z}_c}{\underline{Z}_L + \underline{Z}_c} \exp[2\underline{\gamma}(z - L)] = \underline{\rho}_L \exp[2\underline{\gamma}(z - L)] \quad (9.17)$$

where $\underline{\rho}_L = \underline{\rho}(L)$ is the load reflection factor.

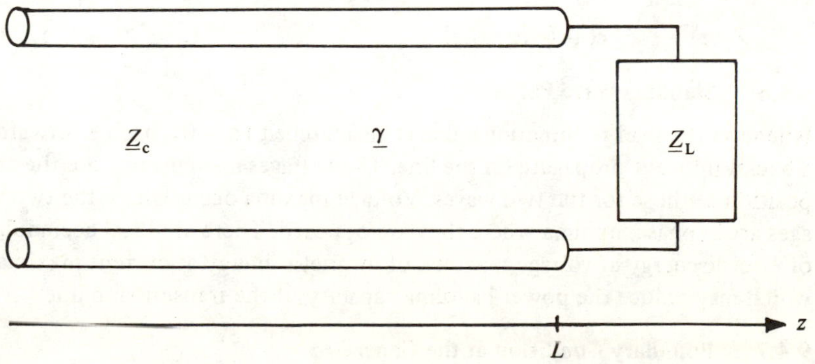

Fig. 9.4 Transmission line terminated into a load impedance.

9.4.3 Local Impedance

The *local impedance* is defined, at every point z along the line, by the ratio of the voltage to the current at that point:

$$\underline{Z}_t(z) = \frac{\underline{U}(z)}{\underline{I}(z)} = \underline{Z}_c \frac{1+\underline{\rho}(z)}{1-\underline{\rho}(z)} = \underline{Z}_c \frac{\underline{Z}_L - \underline{Z}_c \tanh[\underline{\gamma}(z-L)]}{\underline{Z}_c - \underline{Z}_L \tanh[\underline{\gamma}(z-L)]} \qquad \Omega \qquad (9.18)$$

The various forms of $\underline{Z}_t(z)$ given in (9.18) are obtained using (9.12), (9.13) and (9.17).

9.4.4 Matched Reflectionless Load

A particularly interesting load is the one for which $\underline{Z}_L = \underline{Z}_c$. In this case, as is apparent from (9.17), $\underline{\rho}(z) = 0$, so that $\underline{U}_- = 0$: there is no backward (reflected) wave. All the power carried by the forward wave is absorbed by the load, which is called a matched load. The local impedance (9.18) is then everywhere equal to the characteristic impedance of the line.

Several ways to realize a matched load are shown in Figure 6.8.

9.4.5 Reactive Load

Towards the other extreme, a purely reactive load ($\underline{Z}_L = jX_L$) cannot absorb any power. Two particular cases of reactive loads are the short-circuit and the open-circuit:

1. short-circuit. In this case, $\underline{Z}_L = 0$ and

$$\underline{\rho}(z) = -\exp[2\underline{\gamma}(z-L)] \qquad \underline{\rho}_L = -1 \qquad (9.19)$$

$$\underline{Z}_t(z) = -\underline{Z}_c \tanh[\underline{\gamma}(z-L)] \qquad \Omega \qquad (9.20)$$

2. open-circuit. The load impedance $\underline{Z}_L$ is infinite, so that

$$\underline{\rho}(z) = \exp[2\underline{\gamma}(z-L)] \qquad \underline{\rho}_L = 1 \qquad (9.21)$$

$$\underline{Z}_t(z) = -\underline{Z}_c \coth[\underline{\gamma}(z-L)] \qquad \Omega \qquad (9.22)$$

9.4.6 Standing Wave Pattern

Whenever the load terminating a line is mismatched ($\underline{\rho} \neq 0$), both a forward and a backward wave propagate on the line. The voltages and currents are the superposition of those for the two waves. Voltage maxima occur where the two voltages are in phase, minima where they are opposed. There are local accumulations of electric energy at voltage maxima, of magnetic energy at current maxima, which may reduce the power handling capacity of the transmission line.

9.4.7 Boundary Condition at the Generator

Another boundary condition is required to determine the constant $\underline{U}_+$. It is provided by the generator, connected to the line at $z = 0$ (Fig. 9.5). The generator is represented by an ideal voltage source $\underline{U}_0$ and an internal impedance $\underline{Z}_G$. As the voltage and the current must be continuous at $z = 0$, one then has, after some calculations:

$$\underline{U}_+ = \frac{\underline{U}_0}{(\underline{Z}_G/\underline{Z}_c)[1 - \underline{\rho}(0)] + 1 + \underline{\rho}(0)} \quad \text{V} \tag{9.23}$$

Another formulation is obtained by defining a generator reflection factor $\underline{\rho}_G$ by

$$\underline{\rho}_G = \frac{\underline{Z}_G - \underline{Z}_c}{\underline{Z}_G + \underline{Z}_c} \tag{9.24}$$

one then obtains:

$$\underline{U}_+ = \frac{\underline{U}_0}{2} \frac{1 - \underline{\rho}_G}{1 - \underline{\rho}_G \underline{\rho}_L \exp(-2\underline{\gamma}L)} \quad \text{V} \tag{9.25}$$

9.4.8 Power Transfer from Generator to Load

The power absorbed by the load is given by

$$\underline{S}(L) = \underline{U}(L)\underline{I}^*(L) = \frac{U_0^2}{4\underline{Z}_c^*} \frac{\exp(-2\alpha L)|1 - \underline{\rho}_G|^2(1 - \underline{\rho}_L^*)(1 + \underline{\rho}_L)}{|1 - \underline{\rho}_G \underline{\rho}_L \exp(-2\underline{\gamma}L)|} \quad \text{VA} \tag{9.26}$$

The active power is the real part of $\underline{S}$, the reactive power its imaginary part.

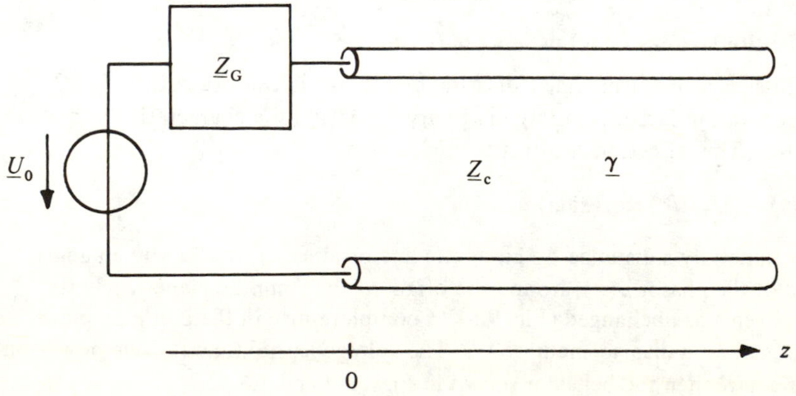

Fig. 9.5 Transmission line connected to a generator.

9.4.9 Conjugate Match

The study of (9.26) shows that the largest transfer of power does not take place for a matched reflectionless load ($\underline{\rho}_L = 0$), but when the term $\underline{\rho}_G \underline{\rho}_L \exp(-2\underline{\gamma}L)$ in the denominator of (9.26) is a real positive quantity. This is called the *conjugate match* condition. The load impedance $\underline{Z}_L$, seen through the transmission line of length L, is then the complex conjugate of the source impedance $\underline{Z}_G$.

9.4.10 Drawbacks of the Conjugate Match at High Frequencies

Even though it provides maximum power transfer, the conjugate match is seldom used at microwaves because it presents several disadvantages:

1. It may be realized only at discrete frequencies, but not over a broad frequency range, mostly because of the term $\underline{\gamma}L$ which is quite frequency sensitive. When the line is long ($L \gg \lambda$), the power at the load varies quite rapidly with frequency, so that a modulated signal would be distorted.
2. The presence of a backward wave returning towards the generator may have effects much stronger than those determined here by considering the simple generator model of Fig. 9.5. Mismatches may produce frequency shifts, and for some particularly severe reflections, simultaneous oscillations at different frequencies (Fig. 4.43).
3. The simultaneous existence of both a forward and a backward wave on the line produces a standing wave pattern (§ 9.4.6) with local accumulation of energy, which reduces the power capacity of the line.

For these reasons, conjugate match is not recommended at microwaves. One most generally tries to match the load to the transmission line, as close as possible to the reflectionless load condition ($\underline{\rho}_L \cong 0$, § 9.4.4). Techniques used for matching are outlined in Sections 7.2.12, 9.5.7, and 9.6.

9.5 PARTICULAR CASE: LOSSLESS PROPAGATING LINE

9.5.1 Reflections on a Lossless Line

On a lossless propagating line (for instance a waveguide above cutoff, § 2.2.31) the propagation factor is purely imaginary $\underline{\gamma} = \mathrm{j}\beta$ and the characteristic impedance Z_c is real. The reflection factor is then given by

$$\underline{\rho}(z) = (\underline{U}_+/\underline{U}_-) \exp(2\mathrm{j}\beta z) \tag{9.27}$$

A translation by a distance Δz along the line produces a rotation by an angle $2\beta\,\Delta z$ of the phasor $\underline{\rho}(z)$, around the origin of the complex plane, while its module remains unchanged (Fig. 9.6). A complete turn in the complex plane is obtained for a displacement of $\lambda/2$. The voltage, current, and active power on the line have then the behavior shown in Figs. 9.7 and 7.6.

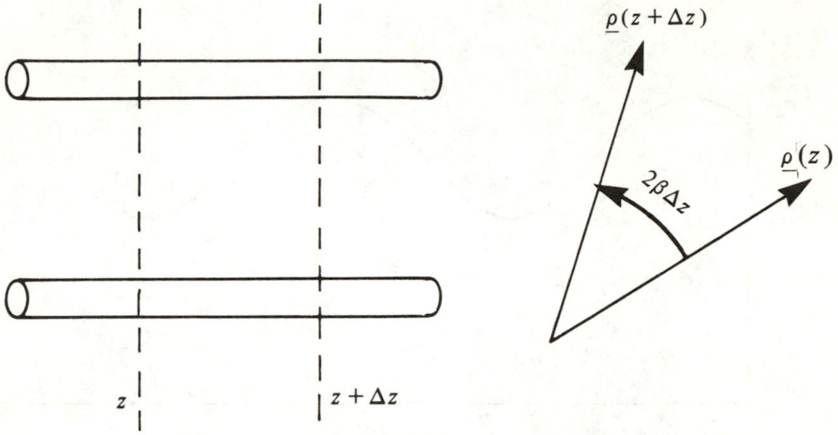

Fig. 9.6 Effect of a lateral shift along the line on the reflection factor.

$$\underline{U}(z) = \underline{U}_+ \exp(-j\beta z)[1 + \underline{\rho}(0)\exp(2j\beta z)] \qquad \text{V} \tag{9.28}$$

$$\underline{I}(z) = Y_c \underline{U}_+ \exp(-j\beta z)[1 - \underline{\rho}(0)\exp(2j\beta z)] \qquad \text{A} \tag{9.29}$$

$$P(z) = \mathrm{Re}[\underline{U}(z)\underline{I}^*(z)] = Y_c(|\underline{U}_+|^2 - |\underline{U}_-|^2) \qquad \text{W} \tag{9.30}$$

$$= Y_c|\underline{U}_+|^2(1 - \rho_0^2)$$

with $\underline{\rho}(0) = \rho_0 \exp(j\varphi_\rho)$.

The term within brackets in (9.28) and (9.29) is the sum of two phasors, the second one rotating around the tip of the first one (Fig. 9.8). The magnitude of the resulting phasor is:

$$|1 + \underline{\rho}(z)| = \sqrt{1 + \rho_0^2 + 2\rho_0 \cos(2\beta z + \varphi_\rho)} \tag{9.31}$$

The maxima and minima of this expression occur for:

$$\underline{\rho}[(n/2 - \varphi_\rho/4\pi)\lambda] = 1 + \rho_0 \tag{9.32}$$

$$\underline{\rho}[(n/2 + 1/4 - \varphi_\rho/4\pi)\lambda] = 1 - \rho_0 \tag{9.33}$$

where n is an arbitrary integer.

9.5.2 Voltage Standing Wave Ratio

The ratio of maximum to minimum voltages along the transmission line defines the voltage standing wave ratio s (VSWR) as

$$s = \frac{U_{max}}{U_{min}} = \frac{|\underline{U}_+| + |\underline{U}_-|}{|\underline{U}_+| - |\underline{U}_-|} = \frac{1 + \rho_0}{1 - \rho_0} \tag{9.34}$$

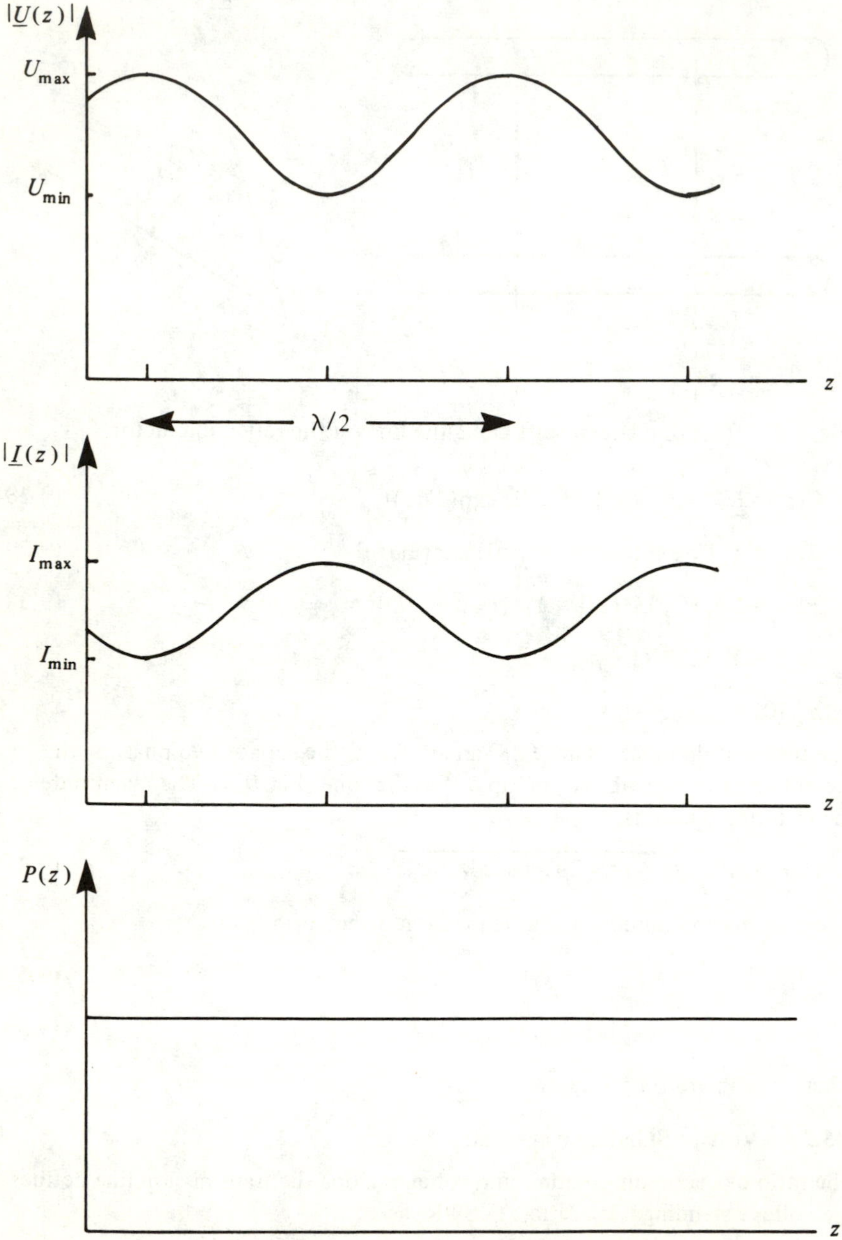

Fig. 9.7 Voltage, current, and power as a function of position along a lossless transmission line.

it was here assumed that $|\underline{U}_+| > |\underline{U}_-|$ (passive load).

The VSWR varies from 1 (matched load) to ∞ (reactive load). The magnitude of the reflection factor ρ_0 is then given by

$$\rho_0 = \frac{s - 1}{s + 1} \tag{9.35}$$

Note: in Great Britain, the VSWR is defined by the inverse ratio U_{min}/U_{max}, and therefore varies between 1 (matched load) and 0 (reactive load) [216].

9.5.3 Smith Chart for Impedances

The local impedance given by (9.18) becomes, for the lossless line:

$$\underline{Z}_t(z) = Z_c \frac{\underline{Z}_L - jZ_c \tan \beta (z - L)}{Z_c - j\underline{Z}_L \tan \beta (z - L)} \quad \Omega \tag{9.36}$$

While this expression is markedly simpler than (9.18), it still is by no means easy to use in practice. To overcome the difficulties involved in complex calculations, Phillip Smith utilized the properties of rotation of $\underline{\rho}(z)$ (Fig. 9.6) and mapped the coordinate system for the complex relative impedance $\underline{Z}_t(z)/Z_c$ onto the complex plane of $\underline{\rho}$ [217].

The impedance coordinates in the $\underline{\rho}$ plane may be utilized to determine the impedance resulting from a series connection, while the effect of a section of trans mission line is always a rotation around the center (point $\underline{\rho} = 0$). The mapping function is obtained from (9.18):

$$\underline{\rho}(z) = \frac{\underline{Z}_t(z) - Z_c}{\underline{Z}_t(z) + Z_c} = \frac{\underline{Z}_t(z)/Z_c - 1}{\underline{Z}_t(z)/Z_c + 1} = 1 - \frac{2}{\underline{Z}_t(z)/Z_c + 1} \tag{9.37}$$

This transformation is the sequence of a linear translation, followed by an inversion, and then by a second linear translation. It is a conformal mapping, which retains the angles and the circles (a straight line being a particular circle having one point at infinity).

Let us consider first the three particular loads of Sections 9.4.4 and 9.4.5, and then the purely resistive and purely reactive loads (real and imaginary axes in the plane $\underline{Z}_t/Z_c$).

1. For the matched load, $\underline{Z}_L = Z_c$. The point $\underline{Z}_t/Z_c = 1$ becomes the center of the $\underline{\rho}$ plane, at $\underline{\rho} = 0$.
2. The short-circuit corresponds to the point $\underline{Z}_t/Z_c = 0$ in the impedance plane, which is mapped onto the point $\underline{\rho} = -1$.
3. The open-circuit is the point at infinity in the impedance plane

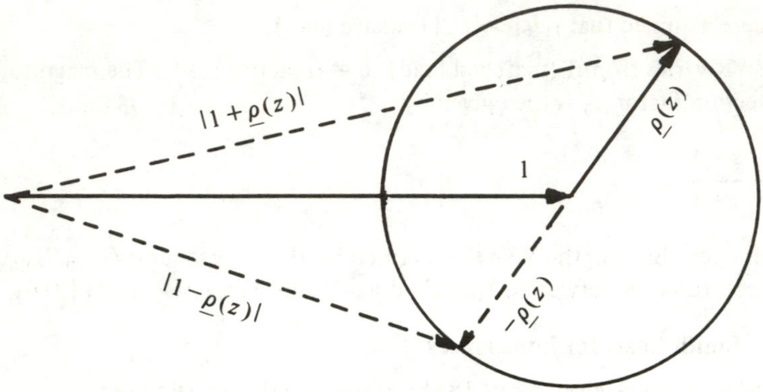

Fig. 9.8 Graphical representation of $1 + \underline{\rho}$ and $1 - \underline{\rho}$.

($\underline{Z}_t/Z_c = \infty$). It is mapped onto the point $\underline{\rho} = +1$.

4. The half-real axis $\underline{Z}_L/Z_c = R$ with R positive extends from 0 to infinity, going through point $R = 1$. Its transform in the $\underline{\rho}$ plane is located between the corresponding transformed points: it is the section of the real axis extending between −1 and +1 (Fig. 9.9).
5. The imaginary axis $\underline{Z}_L/Z_c = jX$ goes through the points $-\infty$, 0 and $+\infty$. It is furthermore perpendicular to the real axis at $\underline{Z}_t/Z_c = 0$. Its transform must go through the corresponding transformed points and retain the right angle at the short-circuit position: it is the unit circle in the $\underline{\rho}$ plane (Fig. 9.9).

The whole right-hand half plane $\underline{Z}_t(z)/Z_c$, which corresponds to passive loads, is mapped onto the inside of the unit circle in the $\underline{\rho}$ plane. All the straight lines in the $\underline{Z}_t/Z_c$ plane become circles going through the point +1 in the $\underline{\rho}$ plane.

To determine the equations of the circles corresponding respectively to the horizontal and vertical lines in the $\underline{Z}_t/Z_c$ plane, we let

$$\underline{Z}_t/Z_c = R + jX \qquad \underline{\rho} = a + jb \tag{9.38}$$

Introducing these expressions into (9.37), identifying real and imaginary parts and eliminating respectively X and R yields the loci R = constant and X = constant:

$$\left(a - \frac{R}{R+1}\right)^2 + b^2 = \left(\frac{1}{R+1}\right)^2 \tag{9.39}$$

$$(a-1)^2 + (b - 1/X)^2 = (1/X)^2 \tag{9.40}$$

The two sets of circles in the $\underline{\rho} = a + jb$ plane form the Smith chart (Fig. 7.11).

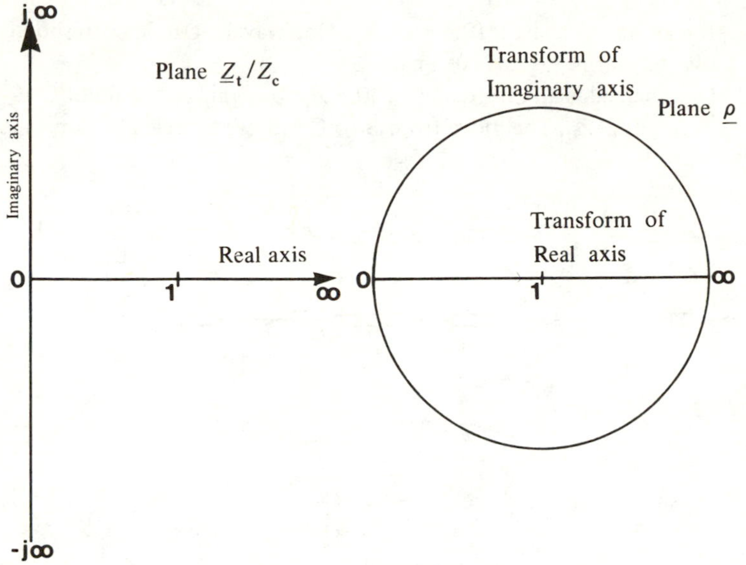

Fig. 9.9 Transformation of the real and imaginary axes of the Z_t/Z_c plane into the plane of the reflection factor $\underline{\rho}$.

9.5.4 How to Use the Smith Chart for Impedances

A sample of how to use the Smith chart is presented in Fig. 9.10. One wishes to determine the input impedance of a transmission line into which have been inserted series components, and which is terminated into a mismatched load with complex impedance. A computation would require several successive applications of (9.36): one for every section of line. In the Smith chart, the sequence of operations is carried out graphically: every section of line produces a rotation around the center of the chart, the rotation angle being related to the length of the section, divided by the wavelength. To insert a series impedance, one simply adds its value (normalized to the line impedance) to the one at the line's input, obtained on the Smith chart. When adding a reactance, one moves on a circle R = constant; when adding a resistance, on a circle X = constant.

Considering the Smith chart and the circuit at the top of Figure 9.10, the procedure followed is:

1. The load terminating the line (at right) has an impedance $\underline{Z}_L = (2.15 - j3)Z_c$. The corresponding point A in the chart is located at the coordinates 2.15 - j3.
2. The line section, long of 0.087 λ, produces a rotation from A to B by an angle 0.087 x 720° = 62.5° around the chart's center, moving towards the generator (decreasing values of z), i.e., clockwise. The impedance at point

B read on the chart is $(0.3 - j1)Z_c$. One wavelength, λ, corresponds to two full turns, i.e., a 720° angle.

3. One then adds an impedance of $j0.8\,Z_c$, bringing to the point C at $(0.3 - j0.2)Z_c$. The move from B to C follows a circle R = constant.

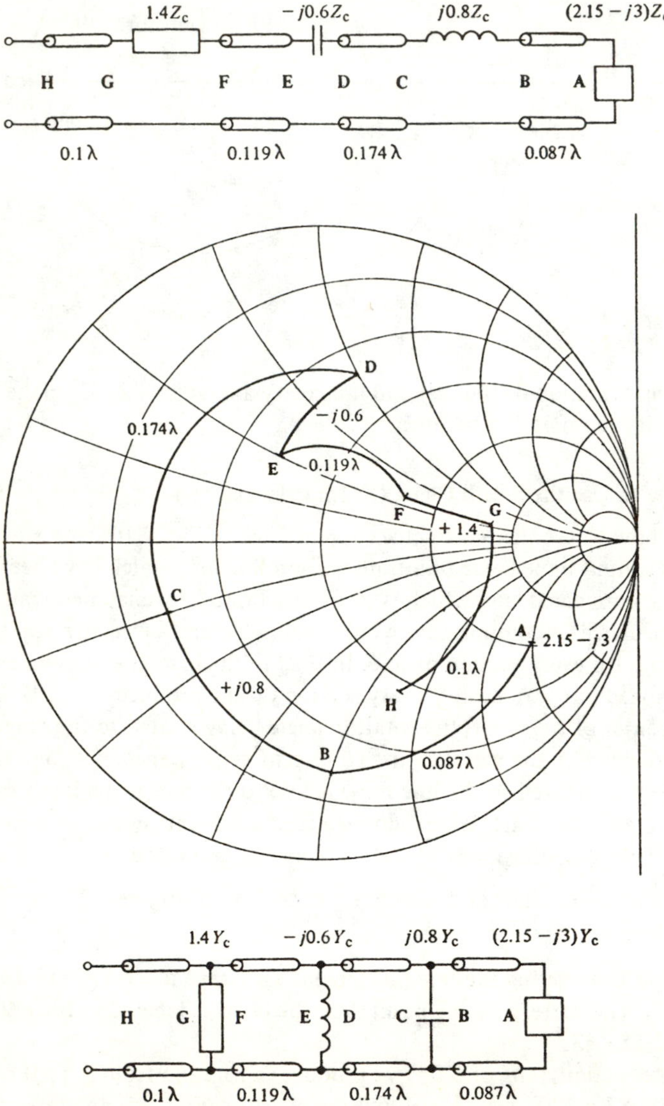

Fig. 9.10 Example of utilization of the Smith chart.

4. The section of line between C and D produces a second rotation around the chart's center, by an angle 0.174 x 720° = 125.3°. The input impedance of this section of line at point D is $(0.65 + j1)Z_c$.
5. To this last value is added the impedance of the series capacitor $-j\,0.6\,Z_c$, yielding point E at $(0.65 + j0.4)Z_c$. The displacement is done along a circle R = constant (as between B and C).
6. The addition of a series resistance between F and G leads to a move along a circle X = constant, between coordinates $R = 1.6\,Z_c$ and $R = 3\,Z_c$.
7. Finally, the line section between G and H produces a last rotation around the center of the chart, by an angle 0.1 x 720° = 72°, yielding the input impedance to the whole assembly as $(0.9 - j1.15)Z_c$.

9.5.5 Smith Chart for Admittances

When connecting components in shunt on the line (which is most often preferred at microwaves), one wishes to use local admittances, rather than impedances. The relationship of the local admittance, $\underline{Y}_t$, normalized to the characteristic admittance Y_c of the line, with the reflection factor, is given by:

$$\frac{\underline{Y}_t(z)}{Y_c} = \frac{Z_c}{\underline{Z}_t(z)} = \frac{1 - \underline{\rho}(z)}{1 + \underline{\rho}(z)} = \frac{1 + [-\underline{\rho}(z)]}{1 - [-\underline{\rho}(z)]} \tag{9.41}$$

It is the same relationship as the one for the impedance (9.18), in which $\underline{\rho}$ is simply replaced by $-\underline{\rho}$. In the Smith chart, the value of the normalized admittance $\underline{Y}_t/Y_c$ is obtained by taking the image of the point $\underline{Z}_t/Z_c$ across the center of the chart. This actually provides a simple graphical method to determine the inverse of a complex number.

9.5.6 How to Use the Smith Chart for Admittance

The input admittance of a transmission line, into which a capacitor, an inductor and a resistor have been shunt-connected between sections of line, is to be determined (circuit at the bottom of Fig. 9.10).

The procedure followed is exactly the same as the one outlined in Section 9.5.4 for series-connected elements in the Smith chart for impedances. The very same sequence A-H is followed, this time for admittances in the Smith chart, i.e., in the complex plane $-\underline{\rho}$. The input admittance obtained is $\underline{Y}_H = (0.9 - j1.15)\,Y_c$.

9.5.7 Simultaneous Use of the Smith Chart for Shunt and Series Components

One may, of course, utilize the Smith chart to connect both series impedances and shunt admittances, taking advantage of the central imaging property indicated in Section 9.5.5. An example is given in Figure 9.11.

A circuit in T, having one shunt susceptance between two series reactances, is

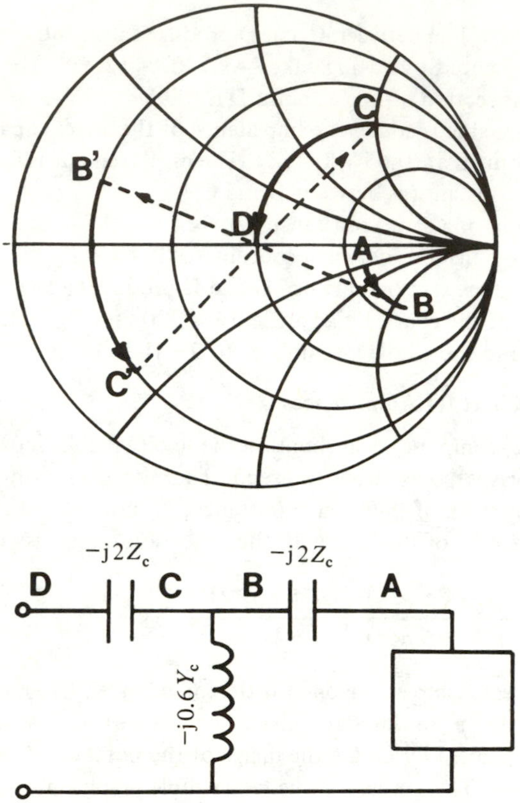

Fig. 9.11 Use of the Smith chart with both series and shunt-connected elements: application to matching.

connected to a load $\underline{Z}_L$. This is roughly the equivalent circuit of an E-H tuner (§ 6.5.28, § 7.2.12).

1. The load impedance is here $\underline{Z}_L = (2.5 - j0.5)Z_c$ so that the starting point A is located at the coordinates 2.5 − j0.5 in the Smith Chart for impedances.
2. A series capacitance is then added. Remaining on the impedance chart, one moves along an R = constant circle, by a distance corresponding to the reactance $-j2Z_c$; i.e., from −j0.5 to −j2.5, reaching point B.
3. The next component to be connected is a shunt susceptance. This means that the input admittance must be determined. This is obtained by jumping over the center of the chart, from point B to its image B′, this time in the admittance chart.
4. The addition of a shunt susceptance $-j0.6\,Y_c$ is obtained by moving along

a circle of constant conductance G = constant, reaching point C′ (still in the admittance chart).

5. The next component is connected in series, so that one has to return to the impedance chart. This is done by going to point C, symmetrically located across the center of the chart.
6. The second series reactance $-j2Z_c$ is then added, in the same way as the first one (moving along a circle R = constant). The point D reached is, for the particular set of values selected, the center of the chart.

The circuit in T, containing only lossless reactive components, has actually matched the load $\underline{Z}_L$ to the transmission line of characteristic impedance Z_c (reflectionless match).

9.5.8 Use of the Smith Chart with Lossy Lines

It is also possible, at least in principle, to use the Smith chart with lossy lines. In this case, a displacement along the line produces a shift of the corresponding point along a spiral (and no longer on a circle) in the chart. Also, the reflection factor can then become larger than unity, even for a passive load, so that the chart must be extended accordingly [218].

9.6 MATCHING

9.6.1 Definition

At microwaves, matching a load means compensating its reflection, most generally by adding other reflections which cancel out the original one. Reflections are undesirable for the reasons outlined in Section 9.4.10. Of course, matching is to be done, whenever possible, with lossless components. The power of the incident signal should be dissipated in the load, and only there. One shall thus use reactive components and sections of lossless transmission lines.

The use of reactive components for matching has been presented in the example of Section 9.5.7. One may further check that the technique presented, using one shunt susceptance between two reactances of same value, is capable to match any load – with the exception of purely reactive loads, on the outer edge of the Smith chart.

9.6.2 Matching with Reactive Elements and Sections of Transmission Line

The insertion of reactive components between sections of transmission line was considered in Sections 9.5.4 and 9.5.6. The same procedure is followed when matching, with one additional requirement: the input impedance (or admittance) must be the one of the line, so that the last point reached should be at the center of the chart.

This may only occur if the previous point was on the circle R = 1 (in the impedance chart) or $G = 1$ (in the admittance chart).

Two elements are therefore necessary:

1. a section of line providing the necessary rotation to reach the circle $R = 1$ or $G = 1$, and
2. a reactive component bringing the point to the center of the Smith chart.

At microwave frequencies, matching is most often done by means of shunt susceptances. This approach is further outlined (graphically) in Fig. 7.12.

The condition of perfect match ($\underline{\rho} = 0$) may be obtained at one particular frequency only. When operating over a frequency band, more involved matching with more than two components may be called for.

9.6.3 Matching with a Quarter-Wave Transformer

Transmission lines having different characteristic impedances are also utilized for matching, either for purely resistive loads R_L or to connect transmission lines of different kinds.

We consider here a resistive load R_L, to be matched to a line of characteristic impedance Z_c: the corresponding point A is on the real axis of the Smith chart (Fig. 9.12). A section of line of unknown characteristic impedance Z_x and length L is connected to the load. The load resistance R_L, normalized now with respect to the characteristic impedance Z_x, produces point B on the Smith chart, located at R_L/Z_x. After one half turn on the chart, another real impedance value is obtained at point C: it is Z_x/R_L. This corresponds to a length of line $L = \lambda/2$, which acts as an impedance inverter. The input impedance at point C is $(Z_x/R_L)\,Z_x$. Letting this value equal Z_c, the match is obtained and the representative point reaches the center of the chart D. The condition on Z_x is

$$Z_x = \sqrt{Z_c R_L} \qquad \Omega \qquad (9.42)$$

It is simply the geometrical mean of the two values Z_c and R_L.

The device obtained is a *quarter-wave transformer.*

This technique may be extended to loads having complex impedances: an additional section of line is first required to reach the real axis of the chart.

When matching is to be done over a broad frequency range, several intermediate sections are needed [123].

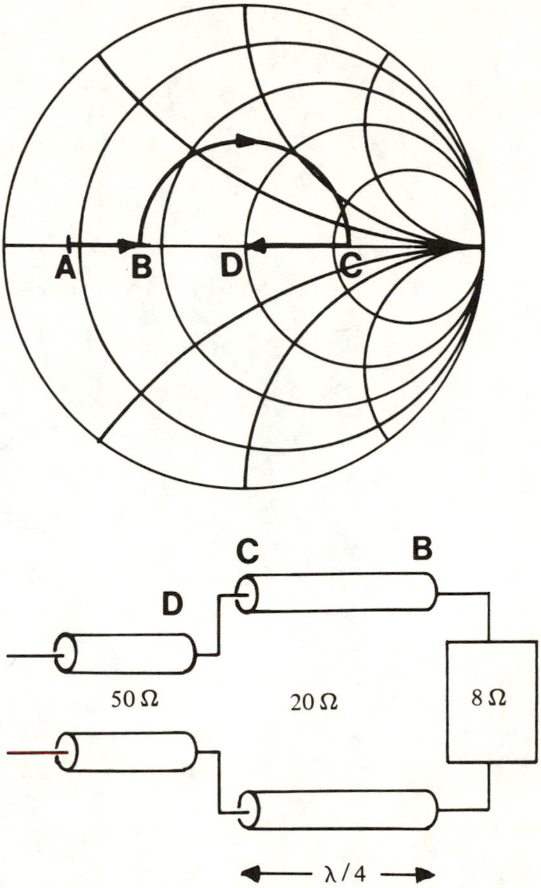

Fig. 9.12 Matching with a quarter-wave transformer: an 8 Ω load is matched to a 50 Ω line through a 20 Ω transformer.

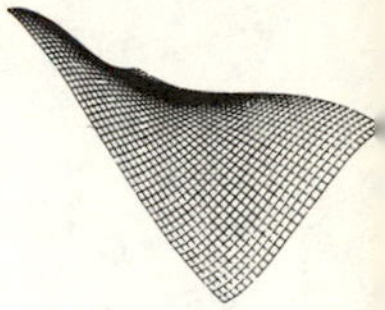

CHAPTER 10

APPENDICES

10.1 VECTOR CALCULUS

10.1.1 Basic Vector Relationships

When $\boldsymbol{A}, \boldsymbol{B}$ and $\boldsymbol{C}$ are three vectors, having components along three unit vectors $\boldsymbol{e}_1, \boldsymbol{e}_2$ and $\boldsymbol{e}_3$ in a right-hand rotating system of orthogonal coordinates (such as rectangular, circular cylindrical, or spherical coordinates), vector operations are defined as follows:

1. vector addition and subtraction

$$\boldsymbol{A} \pm \boldsymbol{B} = (A_1 \pm B_1)\boldsymbol{e}_1 + (A_2 \pm B_2)\boldsymbol{e}_2 + (A_3 \pm B_3)\boldsymbol{e}_3 \tag{10.1}$$

2. scalar (dot) product

$$\boldsymbol{A} \cdot \boldsymbol{B} = |\boldsymbol{A}|\,|\boldsymbol{B}| \cos\theta = A_1B_1 + A_2B_2 + A_3B_3 = \boldsymbol{B} \cdot \boldsymbol{A} \tag{10.2}$$

where θ is the angle formed by vectors $\boldsymbol{A}$ and $\boldsymbol{B}$.

3. vector (cross) product

$$\boldsymbol{A} \times \boldsymbol{B} = \boldsymbol{e}_1(A_2B_3 - A_3B_2) + \boldsymbol{e}_2(A_3B_1 - A_1B_3) + \boldsymbol{e}_3(A_1B_2 - A_2B_1)$$

$$= \begin{vmatrix} \boldsymbol{e}_1 & \boldsymbol{e}_2 & \boldsymbol{e}_3 \\ A_1 & A_2 & A_3 \\ B_1 & B_2 & B_3 \end{vmatrix} \tag{10.3}$$

$$|\boldsymbol{A} \times \boldsymbol{B}| = |\boldsymbol{A}|\,|\boldsymbol{B}| \sin\theta \tag{10.4}$$

4. triple scalar product

$$A \cdot B \times C = A \times B \cdot C = C \times A \cdot B = \begin{vmatrix} A_1 & A_2 & A_3 \\ B_1 & B_2 & B_3 \\ C_1 & C_2 & C_3 \end{vmatrix} \qquad (10.5)$$

5. triple vector product

$$A \times (B \times C) = (A \cdot C)B - (A \cdot B)C \qquad (10.6)$$

$$(A \times B) \times C = (A \cdot C)B - (B \cdot C)A \qquad (10.7)$$

These two operations are different: the position of the bracket is significant for the triple vector product.

10.1.2 Definition of the Differential Operator "del" or "nabla"

In rectangular coordinates, the *differential operator "del" or "nabla"* is defined by:

$$= e_x \frac{\partial}{\partial x} + e_y \frac{\partial}{\partial y} + e_z \frac{\partial}{\partial z} \qquad (10.8)$$

where e_x, e_y and e_z are, respectively, the three unit vectors along the three coordinate axes. This single operator, applied to scalar or vector functions, allows one to represent the gradient, the divergence, and the curl.

$$\mathbf{grad}\, f = \nabla f = e_x \frac{\partial f}{\partial x} + e_y \frac{\partial f}{\partial y} + e_z \frac{\partial f}{\partial z} \qquad (10.9)$$

$$\mathrm{div}\, A = \nabla \cdot A = \frac{\partial A_x}{\partial x} + \frac{\partial A_y}{\partial y} + \frac{\partial A_z}{\partial z} \qquad (10.10)$$

$$\mathbf{curl}\, A = \nabla \times A = e_x \left(\frac{\partial A_z}{\partial y} - \frac{\partial A_y}{\partial z} \right) + e_y \left(\frac{\partial A_x}{\partial z} - \frac{\partial A_z}{\partial x} \right) + e_z \left(\frac{\partial A_y}{\partial x} - \frac{\partial A_x}{\partial y} \right) \qquad (10.11)$$

Since the operator ∇ is differential, the usual rules of derivation do apply. For instance, when applied to a product of functions

$$\nabla (fg) = f \nabla g + g \nabla f \qquad (10.12)$$

10.1.3 Divergence of a Curl

This operation takes the form:

$$\nabla \cdot \nabla \times A \tag{10.13}$$

It is in fact a triple scalar product (10.5), in which the two first terms are ∇ operators. The result is a scalar quantity, in which appear second-order derivatives. When the components of the function A can be doubly differentiated, one may write:

$$\frac{\partial}{\partial x}\frac{\partial}{\partial y} A_z = \frac{\partial}{\partial y}\frac{\partial}{\partial x} A_z \tag{10.14}$$

This means that the components of the operator ∇ commute, in the same manner as usual scalar quantities commute in a product. Since two terms of the triple scalar product (10.5) are identical, the product vanishes, so that:

$$\nabla \cdot \nabla \times A = 0 \tag{10.15}$$

10.1.4 Curl of a Curl

This composite operation has the form of a triple vector product (10.6), in which the two first terms are ∇ operators. When developing the product, one must keep in mind that the operator always operates on the quantity to its right. One then obtains

$$\nabla \times \nabla \times A = \nabla(\nabla \cdot A) - (\nabla \cdot \nabla) A \tag{10.16}$$

10.1.5 Definition of the Vector Laplace Operator

The second term of the right-hand part of (10.16) may be written, in rectangular coordinates:

$$\left(e_x \frac{\partial}{\partial x} + e_y \frac{\partial}{\partial y} + e_z \frac{\partial}{\partial z}\right) \cdot \left(e_x \frac{\partial}{\partial x} + e_y \frac{\partial}{\partial y} + e_z \frac{\partial}{\partial z}\right) A$$
$$= \left(\frac{\partial^2}{\partial x^2} + \frac{\partial^2}{\partial y^2} + \frac{\partial^2}{\partial z^2}\right) A \tag{10.17}$$

This quantity is called the *Laplacian* of A. It may be expressed in different ways

$$(\nabla \cdot \nabla) A = \nabla^2 A = \Delta A \tag{10.18}$$

In rectangular coordinates, the Laplacian of a vector A is simply another vector, with its three components the scalar Laplacians of the corresponding components of the vector A:

$$\nabla^2 A = e_x \nabla^2 A_x + e_y \nabla^2 A_y + e_z \nabla^2 A_z$$

where the scalar Laplacian is

$$\nabla^2 f = \frac{\partial^2 f}{\partial x^2} + \frac{\partial^2 f}{\partial y^2} + \frac{\partial^2 f}{\partial z^2} \tag{10.19}$$

It is most important to note, at this point, that this definition ***is only valid in the rectangular coordinate system.*** In circular cylindrical coordinates, for instance:

$$\nabla^2 A \neq e_\rho \nabla^2 A_\rho + e_\varphi \nabla^2 A_\varphi + e_z \nabla^2 A_z \tag{10.20}$$

In this last situation, the Laplacian of a vector function must be determined from the general relationship obtained with (10.16):

$$\nabla^2 A = \nabla(\nabla \cdot A) - \nabla \times \nabla \times A \tag{10.21}$$

10.1.6 Differential Operations in Circular Cylindrical Coordinates

The coordinate system (ρ, φ, z) is defined in Figure 10.1. The differential operations in this system are given by:

$$\textbf{grad}\, f = \nabla f = e_\rho \frac{\partial f}{\partial \rho} + e_\varphi \frac{1}{\rho} \frac{\partial f}{\partial \varphi} + e_z \frac{\partial f}{\partial z} \tag{10.22}$$

$$\text{div}\, A = \nabla \cdot A = \frac{1}{\rho} \frac{\partial(\rho A_\rho)}{\partial \rho} + \frac{1}{\rho} \frac{\partial A_\varphi}{\partial \varphi} + \frac{\partial A_z}{\partial z} \tag{10.23}$$

$$\textbf{H curl}\, A = e_\rho \left(\frac{1}{\rho} \frac{\partial A_z}{\partial \varphi} - \frac{\partial A_\varphi}{\partial z} \right) + e_\varphi \left(\frac{\partial A_\rho}{\partial z} - \frac{\partial A_z}{\partial \rho} \right) + e_z \left(\frac{1}{\rho} \frac{\partial(\rho A_\varphi)}{\partial \rho} - \frac{1}{\rho} \frac{\partial A_\rho}{\partial \varphi} \right) \tag{10.24}$$

$$\Delta f = \nabla^2 f = \frac{1}{\rho} \frac{\partial}{\partial \rho} \left(\rho \frac{\partial f}{\partial \rho} \right) + \frac{1}{\rho^2} \frac{\partial^2 f}{\partial \varphi^2} + \frac{\partial^2 f}{\partial z^2} \tag{10.25}$$

$$\begin{aligned} \Delta A = \nabla^2 A = {} & e_\rho \left[\frac{1}{\rho} \frac{\partial}{\partial \rho} \left(\rho \frac{\partial A_\rho}{\partial \rho} \right) - \frac{A_\rho}{\rho^2} - \frac{2}{\rho^2} \frac{\partial A_\varphi}{\partial \varphi} + \frac{1}{\rho^2} \frac{\partial^2 A_\rho}{\partial \varphi^2} + \frac{\partial^2 A_\rho}{\partial z^2} \right] \\ & + e_\varphi \left[\frac{1}{\rho} \frac{\partial}{\partial \rho} \left(\rho \frac{\partial A_\varphi}{\partial \rho} \right) - \frac{A_\varphi}{\rho^2} + \frac{2}{\rho^2} \frac{\partial A_\rho}{\partial \varphi} + \frac{1}{\rho^2} \frac{\partial^2 A_\varphi}{\partial \varphi^2} + \frac{\partial^2 A_\varphi}{\partial z^2} \right] \\ & + e_z \left[\frac{1}{\rho} \frac{\partial}{\partial \rho} \left(\rho \frac{\partial A_z}{\partial \rho} \right) + \frac{1}{\rho^2} \frac{\partial^2 A_z}{\partial \varphi^2} + \frac{\partial^2 A_z}{\partial z^2} \right] \end{aligned} \tag{10.26}$$

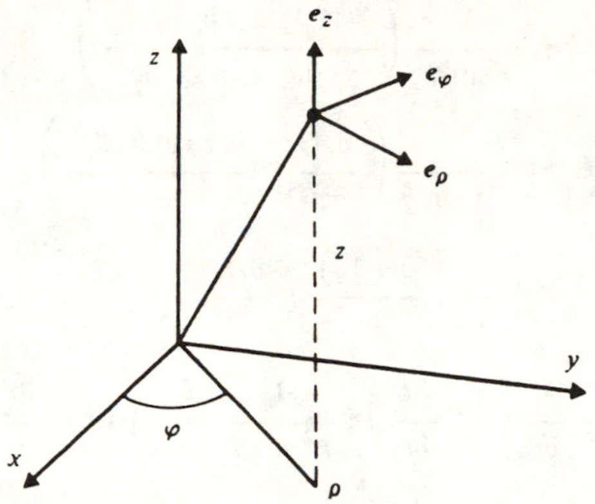

Fig. 10.1 Circular cylindrical coordinate system.

10.1.7 Differential Operations in Spherical Coordinates

The coordinate system (r, θ, φ) is defined in Figure 10.2. The differential operations in spherical coordinates are given by

$$\text{grad}\, f = \nabla f = e_r \frac{\partial f}{\partial r} + e_\theta \frac{1}{r} \frac{\partial f}{\partial \theta} + e_\varphi \frac{1}{r \sin\theta} \frac{\partial f}{\partial \varphi} \tag{10.27}$$

$$\text{div}\, A = \nabla \cdot A = \frac{1}{r^2} \frac{\partial (r^2 A_r)}{\partial r} + \frac{1}{r \sin\theta} \frac{\partial (\sin\theta\, A_\theta)}{\partial \theta} + \frac{1}{r \sin\theta} \frac{\partial A_\varphi}{\partial \varphi} \tag{10.28}$$

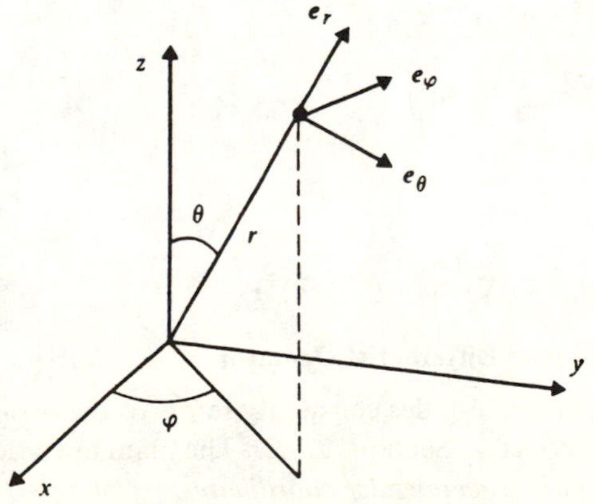

Fig. 10.2 Spherical coordinate system.

$$\text{curl}\, A = \nabla \times A = e_r \frac{1}{r \sin\theta}\left(\frac{\partial(A_\varphi \sin\theta)}{\partial\theta} - \frac{\partial A_\theta}{\partial\varphi}\right)$$

$$+ e_\theta \frac{1}{r \sin\theta}\left(\frac{\partial A_r}{\partial\varphi} - \frac{\partial(r \sin\theta\, A_\varphi)}{\partial r}\right)$$

$$+ e_\varphi \frac{1}{r}\left(\frac{\partial(rA_\theta)}{\partial r} - \frac{\partial A_r}{\partial\theta}\right) \tag{10.29}$$

$$\Delta f = \nabla^2 f = \frac{1}{r^2}\frac{\partial}{\partial r}\left(r^2 \frac{\partial f}{\partial r}\right) + \frac{1}{r^2 \sin\theta}\frac{\partial}{\partial\theta}\left(\sin\theta \frac{\partial f}{\partial\theta}\right)$$

$$+ \frac{1}{r^2 \sin^2\theta}\frac{\partial^2 f}{\partial\varphi^2} \tag{10.30}$$

10.1.8 Some Vector Identities

$$\nabla(fg) = f\nabla g + g\nabla f \tag{9.12}$$

$$\nabla \exp(f) = \exp(f)\nabla f \tag{10.31}$$

$$\nabla \cdot (fA) = f\nabla\cdot A + A\cdot\nabla f \tag{10.32}$$

$$\nabla \cdot (A \times B) = (\nabla\times A)\cdot B - (\nabla\times B)\cdot A \tag{10.33}$$

$$\nabla \times (fA) = f\nabla\times A + \nabla f \times A \tag{10.34}$$

$$\nabla \times (A \times B) = A\nabla\cdot B - B\nabla\cdot A + (B\cdot\nabla)A - (A\cdot\nabla)B \tag{10.35}$$

$$\nabla(A\cdot B) = (A\cdot\nabla)B + (B\cdot\nabla)A + A\times(\nabla\times B) +$$

$$+ B\times(\nabla\times A) \tag{10.36}$$

$$\nabla\cdot\nabla f = \nabla^2 f \tag{10.37}$$

$$\nabla\cdot\nabla\times A = 0 \tag{9.15}$$

$$\nabla\times\nabla f = 0 \tag{10.38}$$

$$\nabla\times\nabla\times A = \nabla(\nabla\cdot A) - (\nabla\cdot\nabla)A \tag{9.16}$$

10.1.9 Transverse Differential Operator

The study of the waveguides constantly refers to the two-dimensional transverse operator ∇_t, defined in Section 2.1.2 . The main operations making use of this operator become, ***in rectangular coordinates:***

$$\nabla_t f = e_x \frac{\partial f}{\partial x} + e_y \frac{\partial f}{\partial y} \tag{10.39}$$

$$e_z \times \nabla_t f = -e_x \frac{\partial f}{\partial y} + e_y \frac{\partial f}{\partial x} \tag{10.40}$$

$$\nabla_t \cdot A_t = \frac{\partial A_x}{\partial x} + \frac{\partial A_y}{\partial y} \tag{10.41}$$

$$\nabla_t \times A_t = e_z \left(\frac{\partial A_y}{\partial x} - \frac{\partial A_x}{\partial y} \right) \tag{10.42}$$

$$\nabla_t^2 f = \frac{\partial^2 f}{\partial x^2} + \frac{\partial^2 f}{\partial y^2} \tag{10.43}$$

$$\nabla_t^2 A_t = e_x \nabla_t^2 A_x + e_y \nabla_t^2 A_y \tag{10.44}$$

where f is any scalar function and A_t an arbitrary transverse vector. The same operations become ***in circular cylindrical coordinates:***

$$\nabla_t f = e_\rho \frac{\partial f}{\partial \rho} + e_\varphi \frac{1}{\rho} \frac{\partial f}{\partial \varphi} \tag{10.45}$$

$$e_z \times \nabla_t f = -e_\rho \frac{1}{\rho} \frac{\partial f}{\partial \varphi} + e_\varphi \frac{\partial f}{\partial \rho} \tag{10.46}$$

$$\nabla_t \cdot A_t = \frac{1}{\rho} \frac{\partial (\rho A_\rho)}{\partial \rho} + \frac{1}{\rho} \frac{\partial A_\varphi}{\partial \varphi} \tag{10.47}$$

$$\nabla_t \times A_t = e_z \quad \frac{1}{\rho} \left(\frac{\partial (\rho A_\varphi)}{\partial \rho} - \frac{1}{\rho} \frac{\partial A_\rho}{\partial \varphi} \right) \tag{10.48}$$

$$\nabla_t^2 f = \frac{1}{\rho} \frac{\partial}{\partial \rho} \left(\rho \frac{\partial f}{\partial \rho} \right) + \frac{1}{\rho^2} \frac{\partial^2 f}{\partial \varphi^2} \tag{10.49}$$

$$\begin{aligned} \nabla_t^2 A_t = {} & e_\rho \left[\frac{1}{\rho} \frac{\partial}{\partial \rho} \left(\rho \frac{\partial A_\rho}{\partial \rho} \right) - \frac{A_\rho}{\rho^2} - \frac{2}{\rho^2} \frac{\partial A_\varphi}{\partial \varphi} + \frac{1}{\rho^2} \frac{\partial^2 A_\rho}{\partial \varphi^2} \right] \\ & + e_\varphi \left[\frac{1}{\rho} \frac{\partial}{\partial \rho} \left(\rho \frac{\partial A_\varphi}{\partial \rho} \right) - \frac{A_\varphi}{\rho^2} + \frac{2}{\rho^2} \frac{\partial A_\rho}{\partial \varphi} + \frac{1}{\rho^2} \frac{\partial^2 A_\varphi}{\partial \varphi^2} \right] \end{aligned} \tag{10.50}$$

10.2 INTEGRAL RELATIONSHIPS

10.2.1 Theorem of the Gradient

Several expressions link the volume integral of a quantity with a surface integral over the boundary of the volume: one may, in this manner, reduce the complexity of some problems. Other expressions similarly replace a surface integral by a line integral, taken over the contour bounding the surface. This may be done when the derivatives of the functions are defined within the regions considered and when the integrals actually exist.

A volume V is considered (formed by the volume elements d V), surrounded by a surface S (having surface elements dA). The normal vector $\boldsymbol{n}$ points towards the outside of the surface (Fig. 10.4). The volume integral of the gradient of a scalar function may then be replaced by the surface integral of the function itself:

$$\int_V \nabla f \, \mathrm{d}V = \oint_S f\boldsymbol{n} \, \mathrm{d}A \tag{10.51}$$

10.2.2 Theorem of the Divergence

In the same manner, the volume integral of a divergence may be replaced

$$\int_V \nabla \cdot \boldsymbol{A} \, \mathrm{d}V = \oint_S \boldsymbol{A} \cdot \boldsymbol{n} \, \mathrm{d}A \tag{10.52}$$

10.2.3 Theorem of the Curl

A similar expression is obtained for the curl of a vector function

$$\int_V \nabla \times \boldsymbol{A} \, \mathrm{d}V = \oint_S \boldsymbol{n} \times \boldsymbol{A} \, \mathrm{d}A \tag{10.53}$$

10.2.4 Green's First Identity

Two integral expressions involving products of scalar and vector functions are known as *Green's identities*

1. scalar functions

$$\int_V (\nabla f \cdot \nabla g + g\nabla^2 f) \, \mathrm{d}V = \oint_S g\nabla f \cdot \boldsymbol{n} \, \mathrm{d}A \tag{10.54}$$

2. vector functions

$$\int_V \nabla \cdot (\boldsymbol{A} \times \nabla \times \boldsymbol{B}) \, \mathrm{d}V = \oint_S \boldsymbol{A} \times (\nabla \times \boldsymbol{B}) \cdot \boldsymbol{n} \, \mathrm{d}A \tag{10.55}$$

10.2.5 Green's Second Identity

1. scalar functions

$$\int_V (g\nabla^2 f - f\nabla^2 g) \, \mathrm{d}V = \oint_S (g\nabla f - f\nabla g) \cdot \boldsymbol{n} \, \mathrm{d}A \tag{10.56}$$

2. vector functions

$$\int_V (B \cdot \nabla \times \nabla \times A - A \cdot \nabla \times \nabla \times B)\, dV$$
$$= \oint [A \times (\nabla \times B) - B \times (\nabla \times A)] \cdot n\, dA \qquad (10.57)$$

10.2.6 Stokes Theorem

A surface S surrounded by a line contour C is considered, the contour element being dl (Fig. 10.3). The Stokes theorem is then

$$\int_S (\nabla \times A) \cdot n\, dA = \oint_C A \cdot dl \qquad (10.58)$$

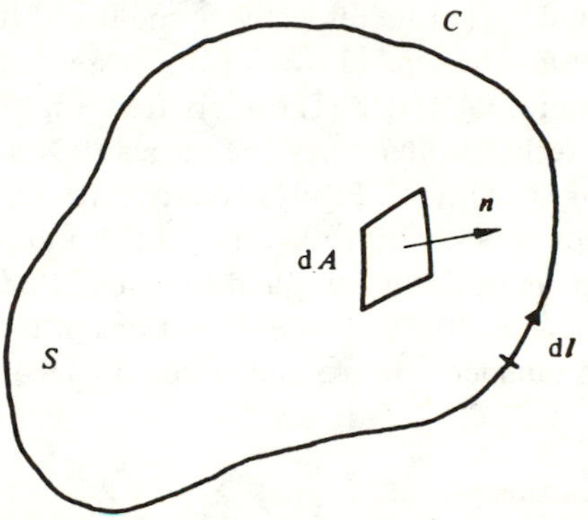

Fig. 10.3 Surface S enclosed by contour C. Definition of the integration surface element and of the curvilinear integration element.

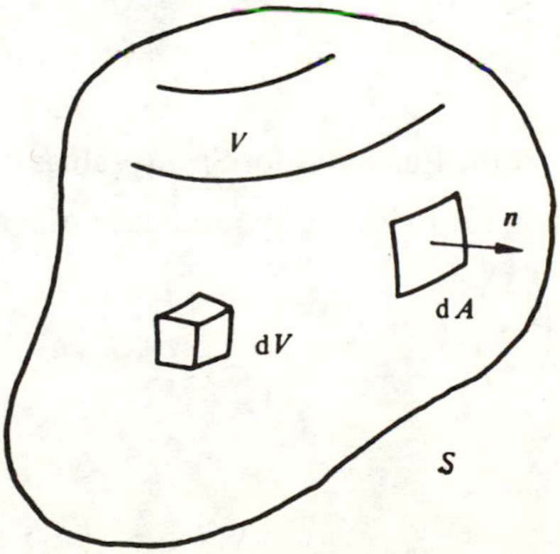

Fig. 10.4 Volume V enclosed in surface S. Definition of the integration volume and surface, and normal direction.

10.3 BESSEL FUNCTIONS

10.3.1 Differential Equation

The *Bessel functions* of first and second kind of order m are particular solutions of the differential equation

$$x^2 \frac{d^2y}{dx^2} + x\frac{dy}{dx} + (x^2 - m^2)y = 0 \tag{10.59}$$

The general solution to this equation takes the form

$$y = C_m(x) = A\,J_m(x) + B\,N_m(x) \tag{10.60}$$

in which $J_m(x)$ and $N_m(x)$ are respectively the Bessel functions of first and second kind of order m. The symbol N_m for the Bessel function of second kind is specified by the norm ISO/TC12. The alternate form Y_m is sometimes encountered in the U.S. technical literature. The constants A and B must be determined from the boundary conditions. Figures 10.5 and 10.6 show the behavior of Bessel functions of both kinds for the first integer values of m The values of the argument corresponding to the zeros and extrema of Bessel functions of the first kind are given in Tables 10.7 and 10.8. Functions of the second kind are all singular for $x = 0$. A number of books and tables are devoted to Bessel functions [10].

10.3.2 Series Expansion

Only positive integer values of m are considered

$$J_m(x) = \left(\frac{x}{2}\right)^m \left[\frac{1}{m!} - \frac{(x/2)^2}{1!(m+1)!} + \frac{(x/2)^4}{2!(m+2)!} - \frac{(x/2)^6}{3!(m+3)!} + \ldots\right] \tag{10.61}$$

10.3.3 Behavior of the Functions for Small Values of x

When x is very small, the following approximations are valid

$$J_m(x) \cong \frac{1}{m!}\left(\frac{x}{2}\right)^m \tag{10.62}$$

$$N_0(x) \cong \frac{2}{\pi}\ln x \tag{10.63}$$

$$N_m(x) \cong \frac{1}{\pi}(m-1)!\left(\frac{x}{2}\right)^{-m} \tag{10.64}$$

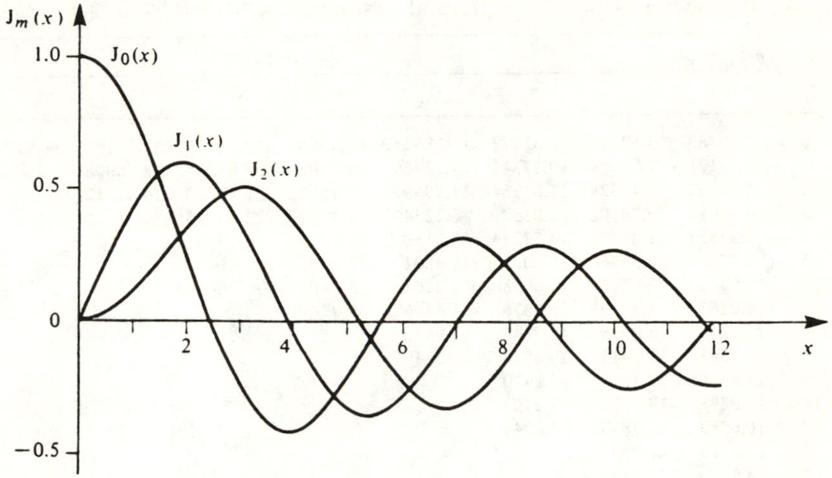

Fig. 10.5 Bessel functions of first kind for $m = 0$, 1, and 2.

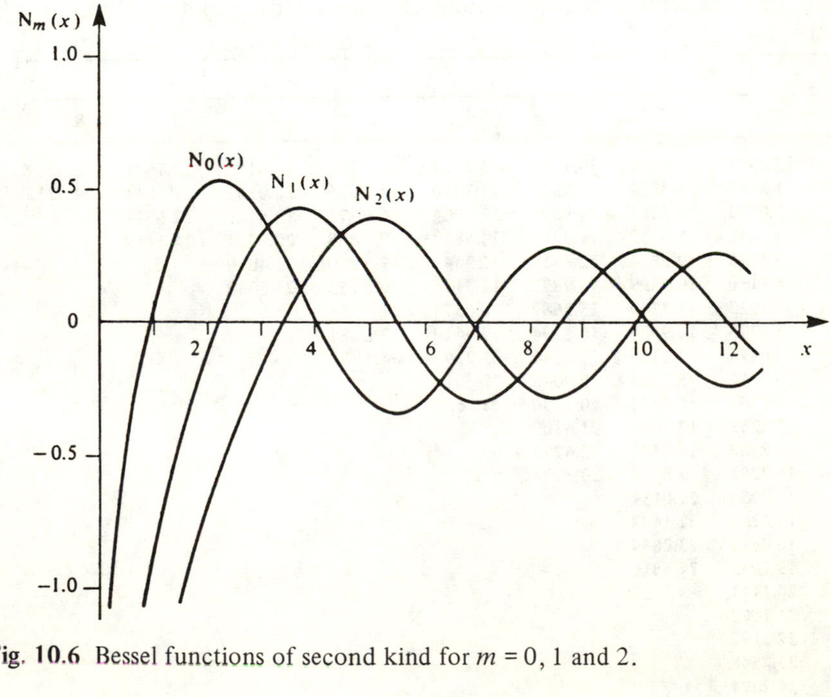

Fig. 10.6 Bessel functions of second kind for $m = 0$, 1 and 2.

Table 10.7 Zeros of Bessel functions of first kind: $J_m(x) = 0$ for $0 < x < 25$.

m	n = 1	2	3	4	5	6	7	8
0	2.40483	5.52008	8.65373	11.795153	14.93092	18.07106	21.21164	24.35247
1	3.83171	7.01559	10.17347	13.32369	16.47063	19.61586	22.76008	
2	5.13562	8.41724	11.61984	14.79595	17.95982	21.11700	24.27112	
3	6.38016	9.76102	13.01520	16.22347	19.40942	22.58273		
4	7.58834	11.06471	14.37254	17.6160	20.8269	24.1990		
5	8.77142	12.33860	15.70017	18.9801	22.2178			
6	9.93611	13.58929	17.0038	20.3208	23.5861			
7	11.08637	14.82127	18.2876	21.6416	24.9349			
8	12.22509	16.0378	19.5545	22.9452				
9	13.35430	17.2412	20.8070	24.2339				
10	14.47550	18.4335	22.0470					
11	15.58985	19.6160	23.2759					
12	16.6983	20.7899	24.4949					
13	17.8014	21.9562						
14	18.9000	23.1158						
15	19.9944	24.2692						
16	21.0851							
17	22.1725							
18	23.2568							
19	24.3383							

Table 10.8 Extrema of Bessel functions of first kind: $dJ_m(x)/dx = 0$ for $0 < x < 25$.

m	n = 1	2	3	4	5	6	7	8
0	3.8317	7.0156	10.1735	13.3237	16.4706	19.6159	22.7601	25.9037
1	1.8412	5.3314	8.5363	11.7060	14.8636	18.0155	21.1644	24.3113
2	3.0542	6.7061	9.9695	13.1704	16.3475	19.5129	22.6721	
3	4.2012	8.0152	11.3459	14.5859	17.7888	20.9724	24.1469	
4	5.3175	9.2824	12.6819	15.9641	19.1960	22.4010		
5	6.4156	10.5199	13.9872	17.3128	20.5755	23.8033		
6	7.5013	11.7349	15.2682	18.6374	21.9318			
7	8.5778	12.9324	16.5294	19.9419	23.2681			
8	9.6474	14.1156	17.7740	21.2291	24.5872			
9	10.7114	15.2868	19.0045	22.5014				
10	11.7709	16.4479	20.2230	23.7608				
11	12.8265	17.6003	21.4309					
12	13.8788	18.7451	22.6293					
13	14.9284	19.8832	23.8194					
14	15.9754	21.0154						
15	17.0203	22.1423						
16	18.0683	23.2644						
17	19.1045	24.3819						
18	20.1441							
19	21.1823							
20	22.2192							
21	23.2548							
22	24.2894							

10.3.4 Asymptotical Behavior for x Tending towards Infinity

$$J_m(x) \cong \sqrt{\frac{2}{\pi x}} \cos(x + \alpha) \tag{10.65}$$

$$N_m(x) \cong \sqrt{\frac{2}{\pi x}} \sin(x + \alpha) \tag{10.66}$$

with

$$\alpha = -\frac{\pi}{4}(2m + 1) \tag{10.67}$$

10.3.5 Recurrence Formulas

$$C_{m+1}(x) = \frac{2m}{x} C_m(x) - C_{m-1}(x) \tag{10.68}$$

$$\frac{dC_m(x)}{dx} = -\frac{m}{x} C_m(x) + C_{m-1}(x) = \frac{m}{x} C_m(x) - C_{m+1}(x) \tag{10.68}$$

$$= \frac{1}{2}[C_{m-1}(x) - C_{m+1}(x)] \tag{10.69}$$

10.3.6 Bessel Integral

The Bessel function of first kind is related to circular functions by the *Bessel integral:*

$$J_m(x) = \frac{1}{\pi} \int_0^\pi \cos(x \sin\varphi - m\varphi)\, d\varphi \tag{10.70}$$

10.3.7 Hankel Functions

The two *Hankel functions* are complex functions of a variable x, whose real part is the Bessel function of first kind, the imaginary part the Bessel function of second kind, either positive or negative.

$$H_m^{(1)}(x) = J_m(x) + jN_m(x) \tag{10.71}$$

$$H_m^{(2)}(x) = J_m(x) - jN_m(x) \tag{10.72}$$

These two functions satisfy the recurrence formulas (10.68) and (10.69).

10.3.8 Other Expressions

The following relations link together functions of opposite arguments or orders

$$C_{-m}(x) = (-1)^m C_m(x) \tag{10.73}$$

$$C_m(-x) = (-1)^m C_m(x) \tag{10.74}$$

$$C_{-m}(x) = C_m(-x) \tag{10.75}$$

with integer values for m.

Integrals involving Bessel functions

$$\int_0^x C_m^2(kx)x\,dx = \frac{1}{2}x^2\,[C_m^2(kx) - C_{m-1}(kx)\,C_{m+1}(kx)] \tag{10.76}$$

$$\int_0^x C_m(kx)\,\bar{C}_m(kx)\,x\,dx = \frac{1}{4}x^2[2C_m(kx)\,\bar{C}_m(kx) - C_{m-1}(kx)\,\bar{C}_{m+1}(kx) - \bar{C}_{m-1}(kx)\,C_{m+1}(kx)] \tag{10.77}$$

where C_m and $\bar{C}_m$ are two arbitrary linear combinations of Bessel functions of order m. When C_m is equal to $\bar{C}_m$, Equation (10.76) is obtained.

10.4 MODIFIED BESSEL FUNCTIONS

10.4.1 Differential Equation

The *modified Bessel functions* of first and second kind of order m are particular solutions of the differential equation:

$$x^2\frac{d^2y}{dx^2} + x\frac{dy}{dx} - (x^2 + m^2)\,y = 0 \tag{10.78}$$

The general solution to this equation takes the form

$$y = Z_m(x) = A\,I_m(x) + B\,K_m(x) \tag{10.79}$$

in which $I_m(x)$ and $K_m(x)$ are, respectively, the modified Bessel functions of first and second kind of order m. The constants A and B are to be determined from the boundary conditions to the problem.

The modified Bessel functions are proportional to the Bessel functions of imaginary argument. For integer values of m, one obtains (with real x)

$$I_m(x) = \exp(-jm\pi/2)J_m(jx) \tag{10.80}$$

$$K_m(x) = j\pi/2\,\exp(jm\pi/2)\,H_m^{(1)}(jx) \tag{10.81}$$

Figures 10.9 and 10.10 show the behavior of the modified Bessel functions of both kinds for the first integer values of m.

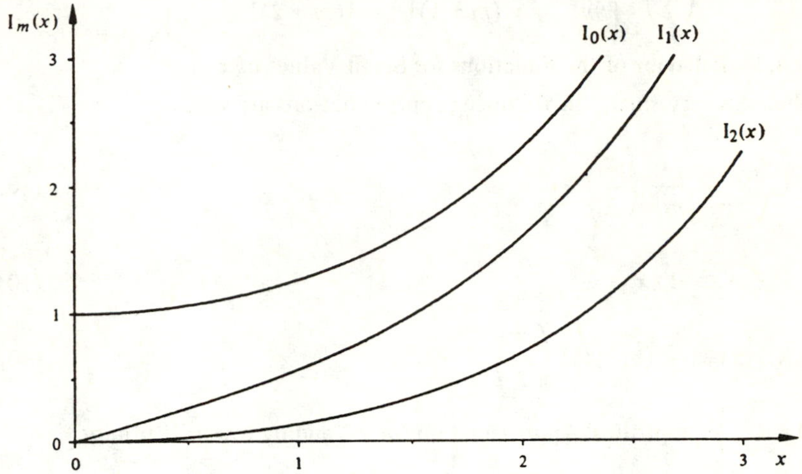

Fig. 10.9 Modified Bessel functions of first kind for $m = 0, 1$ and 2.

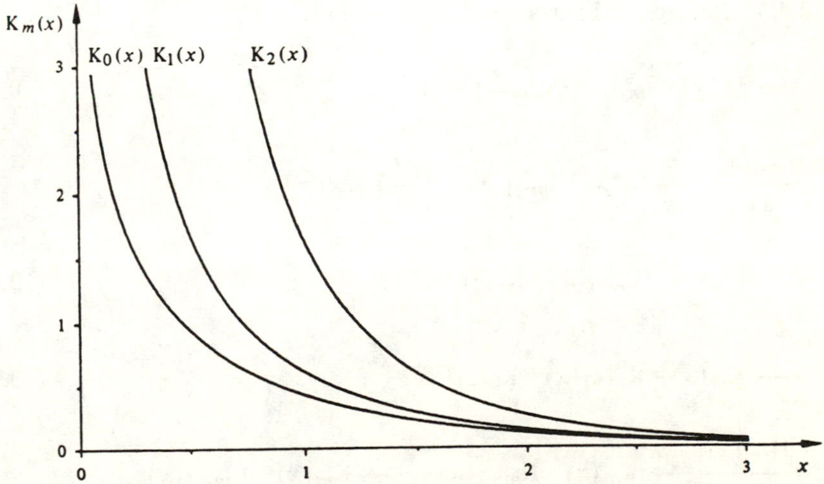

Fig. 10.10 Modified Bessel functions of second kind for $m = 0, 1$ and 2.

10.4.2 Series Expansion

Only integer values of m are considered

$$\mathrm{I}_m(x) = \left(\frac{x}{2}\right)^m \left[\frac{1}{m!} + \frac{(x/2)}{1!(m+1)!} + \frac{(x/2)^2}{2!(m+2)!} + \ldots\right] \tag{10.82}$$

10.4.3 Behavior of the Functions for Small Values of *x*

When x is very small, the following approximations are valid.

$$\mathrm{I}_m(x) \cong \frac{1}{m!}\left(\frac{x}{2}\right)^m \tag{10.83}$$

$$\mathrm{K}_0(x) \cong -\ln x \tag{10.84}$$

$$\mathrm{K}_m(x) \cong \frac{1}{2}(m-1)!\left(\frac{x}{2}\right)^{-m} \tag{10.85}$$

10.4.4 Asymptotical Approximation for *x* Tending towards Infinity

$$\mathrm{I}_m(x) = \frac{\exp(x)}{\sqrt{2\pi x}} \tag{10.86}$$

$$\mathrm{K}_m(x) = \sqrt{\frac{\pi}{2x}}\exp(-x) \tag{10.87}$$

10.4.5 Recurrence Expressions

$$\frac{2m}{x}\mathrm{I}_m(x) = \mathrm{I}_{m-1}(x) - \mathrm{I}_{m+1}(x) \tag{10.88}$$

$$\frac{\mathrm{dI}_m(x)}{\mathrm{d}x} = \frac{m}{x}\mathrm{I}_m(x) + \mathrm{I}_{m+1}(x) = -\frac{m}{x}\mathrm{I}_m(x) + \mathrm{I}_{m-1}(x)$$

$$= \frac{1}{2}\left[\mathrm{I}_{m-1}(x) + \mathrm{I}_{m+1}(x)\right] \tag{10.89}$$

$$\frac{2m}{x}\mathrm{K}_m(x) = \mathrm{K}_{m+1}(x) - \mathrm{K}_{m-1}(x) \tag{10.90}$$

$$\frac{\mathrm{dK}_m(x)}{\mathrm{d}x} = \frac{m}{x}\mathrm{K}_m(x) - \mathrm{K}_{m+1}(x) = -\frac{m}{x}\mathrm{K}_m(x) - \mathrm{K}_{m-1}(x)$$

$$= -\frac{1}{2}\left[\mathrm{K}_{m+1}(x) + \mathrm{K}_{m-1}(x)\right] \tag{10.91}$$

10.4.6 Bessel Integral

The modified Bessel function of first kind is related to circular functions by the *Bessel integral*

$$I_m(x) = \frac{1}{\pi} \int_0^{\pi} \exp(x \cos \varphi) \cos m\varphi \, d\varphi \tag{10.92}$$

for integer values of m.

10.4.7 Other Formulas

$$I_{-m}(x) = I_m(x) \tag{10.93}$$

$$I_m(-x) = (-1)^m I_m(x) \tag{10.94}$$

$$K_{-m}(x) = K_m(x) \tag{10.95}$$

for integer values of m.

10.5 SIMILARITIES BETWEEN MICROWAVES AND ACOUSTICS

Similarities between microwaves and acoustics were briefly mentioned in Section 1.1.10. A list of the main equivalent notions within the two domains is presented here in table form (Table 10.11).

Table 10.11 List of some corresponding devices and phenomena.

Microwaves	Acoustics
waveguide, optical fiber	acoustic tube, vibrating string
wave on dielectric slab	surface acoustic wave
resonant cavity	reverberating chamber, tube (musical instrument), string (musical instrument).
tunable cavity	trombone
horn antenna	musical horn
antenna (transmission)	loud-speaker
antenna (reception)	microphone
radar	active sonar
microwave link	transmitting sonar
microwave hyperthermia	ultrasonic hyperthermia
electromagnetic noise	acoustic noise
power limitation due to breakdown	power limitation due to cavitation

10.6 TABLES FOR RECTANGULAR AND CIRCULAR WAVEGUIDES

Table 10.12 Rectangular waveguide.

	TE_{mn}		TM_{mn}
E_z	0		$\frac{-j}{\omega\epsilon} p_{mn}^2 \underline{I}_e C_{mn} \sin\left(\frac{m\pi x}{a}\right) \sin\left(\frac{n\pi y}{b}\right)$
E_x	$\underline{U}_e C_{mn} n \frac{\pi}{b} \cos\left(\frac{m\pi x}{a}\right) \sin\left(\frac{n\pi y}{b}\right)$		$-\underline{U}_e C_{mn} m \frac{\pi}{a} \cos\left(\frac{m\pi x}{a}\right) \sin\left(\frac{n\pi y}{b}\right)$
E_y	$-\underline{U}_e C_{mn} m \frac{\pi}{a} \sin\left(\frac{m\pi x}{a}\right) \cos\left(\frac{n\pi y}{b}\right)$		$-\underline{U}_e C_{mn} n \frac{\pi}{b} \sin\left(\frac{m\pi x}{a}\right) \cos\left(\frac{n\pi y}{b}\right)$
H_z	$\frac{-j}{\omega\mu} p_{mn}^2 \underline{U}_e C_{mn} \cos\left(\frac{m\pi x}{a}\right) \cos\left(\frac{n\pi y}{b}\right)$		0
H_x	$\underline{I}_e C_{mn} m \frac{\pi}{a} \sin\left(\frac{m\pi x}{a}\right) \cos\left(\frac{n\pi y}{b}\right)$		$\underline{I}_e C_{mn} n \frac{\pi}{b} \sin\left(\frac{m\pi x}{a}\right) \cos\left(\frac{n\pi y}{b}\right)$
H_y	$\underline{I}_e C_{mn} n \frac{\pi}{b} \cos\left(\frac{m\pi x}{a}\right) \sin\left(\frac{n\pi y}{b}\right)$		$-\underline{I}_e C_{mn} m \frac{\pi}{a} \cos\left(\frac{m\pi x}{a}\right) \sin\left(\frac{n\pi y}{b}\right)$
p_{mn}		$\sqrt{\left(\frac{m\pi}{a}\right)^2 + \left(\frac{n\pi}{b}\right)^2}$	
C_{mn}	$\frac{1}{p_{mn}} \frac{\sqrt{2}}{\sqrt{ab}} \sqrt{1 + \delta(mn)}$	$\delta(i) = \begin{cases} 0 \text{ if } i \neq 0 \\ 1 \text{ if } i = 0 \end{cases}$	$\frac{1}{p_{mn}} \frac{2}{\sqrt{ab}}$
λ_{mn}		$\frac{2\pi}{p_{mn}}$	
$\underline{U}_e$		$\underline{U}_{e+} \exp(-j\beta z) + \underline{U}_{e-} \exp(j\beta z)$	
$\underline{I}_e$		$\frac{1}{\underline{Z}_e} \{\underline{U}_{e+} \exp(-j\beta z) - \underline{U}_{e-} \exp(j\beta z)\}$	
$\underline{Z}_e$	$\frac{\omega\mu}{\beta}$		$\frac{\beta}{\omega\epsilon}$

Table 10.13 Circular waveguide.

	TE_{mn}		TM_{mn}
E_z	0		$-j\frac{p_{mn}^2}{\omega\epsilon}\underline{I}_e C_{mn} J_m(p_{mn}\rho)\begin{Bmatrix}\cos(m\phi)\\ \sin(m\phi)\end{Bmatrix}$
E_ρ	$-\underline{U}_e C_{mn}\frac{m}{\rho}J_m(p_{mn}\rho)\begin{Bmatrix}-\sin(m\phi)\\ \cos(m\phi)\end{Bmatrix}$		$-\underline{U}_e C_{mn} p_{mn} J'_m(p_{mn}\rho)\begin{Bmatrix}\cos(m\phi)\\ \sin(m\phi)\end{Bmatrix}$
E_ϕ	$\underline{U}_e C_{mn} p_{mn} J'_m(p_{mn}\rho)\begin{Bmatrix}\cos(m\phi)\\ \sin(m\phi)\end{Bmatrix}$		$-\underline{U}_e C_{mn}\frac{m}{\rho}J_m(p_{mn}\rho)\begin{Bmatrix}-\sin(m\phi)\\ \cos(m\phi)\end{Bmatrix}$
H_z	$\frac{-j}{\omega\mu}\underline{U}_e C_{mn} p_{mn} J_m(p_{mn}\rho)\begin{Bmatrix}\cos(m\phi)\\ \sin(m\phi)\end{Bmatrix}$		0
H_ρ	$-\underline{I}_e C_{mn} p_{mn} J'_m(p_{mn}\rho)\begin{Bmatrix}\cos(m\phi)\\ \sin(m\phi)\end{Bmatrix}$		$\underline{I}_e C_{mn}\frac{m}{\rho}J_m(p_{mn}\rho)\begin{Bmatrix}-\sin(m\phi)\\ \cos(m\phi)\end{Bmatrix}$
H_ϕ	$-\underline{I}_e C_{mn}\frac{m}{\rho}J_m(p_{mn}\rho)\begin{Bmatrix}-\sin(m\phi)\\ \cos(m\phi)\end{Bmatrix}$		$-\underline{I}_e C_{mn} p_{mn} J'_m(p_{mn}\rho)\begin{Bmatrix}\cos(m\phi)\\ \sin(m\phi)\end{Bmatrix}$
p_{mn}	$J'_m(p_{mn}a) = 0$		$J_m(p_{mn}a) = 0$
C_{mn}	$\frac{\sqrt{2-\delta(m)}}{ap_{mn}\sqrt{\pi}\left[1-\left(\frac{m}{p_{mn}a}\right)^2\right]^{1/2} J_m(p_{mn}a)}$	$\delta(i) = \begin{cases}0 \text{ if } i \neq 0\\ 1 \text{ if } i = 0\end{cases}$	$\frac{\sqrt{2-\delta(m)}}{p_{mn}a\sqrt{\pi}\,J'_m(p_{mn}a)}$
λ_{mn}		$\frac{2\pi}{p_{mn}}$	
$\underline{U}_e$		$\underline{U}_{e+}\exp(-j\beta z) + \underline{U}_{e-}\exp(j\beta z)$	
$\underline{I}_e$		$\frac{1}{\underline{Z}_e}\{\underline{U}_{e+}\exp(-j\beta z) - \underline{U}_{e-}\exp(j\beta z)\}$	
$\underline{Z}_e$	$\frac{\omega\mu}{\beta}$		$\frac{\beta}{\omega\epsilon}$

10.7 PROPERTIES OF MATERIALS CURRENTLY USED AT MICROWAVES

Table 10.14 Resistivity and conductivity of various materials [212].

Materials	**Resistivity** (20°C) [Ωm x 10^{-8}]	**Conductivity** (20°C) [S/m x 10^{6}]
Aluminum	2.62	38.16
Bismuth	115.00	0.87
Brass	3.90	25.64
Graphite	1400.00	0.07
Chromium	2.60	38.46
Copper	1.72	58.13
Germanium	45.00	2.22
Gold	2.44	40.98
Lead	21.90	4.56
Nickel	6.90	14.49
Platinum	10.50	9.52
Silver	1.62	61.73
Distilled Water	100.00	1.00
Silica	$1.00 \cdot 10^{10}$	$1.0 \ 10^{-8}$

Table 10.15 Properties of some microwave dielectrics [213].

Materials	ϵ_r' 10 GHz	$\tan\sigma = \epsilon_r''/\epsilon_r'$ 25°C	**Thermal Conductivity k** cal/(cm s°C) a 25°C	**Expansion Factor** α $\alpha \cdot 10^6$/°C 25-300°C
Alumina Al_2O_3 99.5%	9.5-10	$3 \ 10^{-4}$	0.088	6
Alumina Al_2O_3 96%	8.9	$6 \ 10^{-4}$	0.084	6.4
Alumina Al_2O_3 85%	8.0	$1.5 \ 10^{-3}$	0.055	6.5
Beryllia BeO	6.4	$3 \ 10^{-4}$	0.055	6.0
Rutile TiO_2	85	$4 \ 10^{-3}$	0.06	—
Silicon (10^3 Ωm)	11.9	$4 \ 10^{-3}$	0.25	4.2
GaAs (> 10^3 Ωm)	13.0	$6 \ 10^{-3}$	0.095	5.7
Sapphire (anisotropic)	9.4-11.5	$1 \ 10^{-4}$	0.09	5-6.66
Quartz	3.75	$1 \ 10^{-4}$	0.0033	0.55
Corning Glass 7059	5.75	$3.6 \ 10^{-4}$	0.002	4.6
Glazed ceramic	7.2	$8 \ 10^{-3}$	0.002	4.6

10.8 GRAPHICAL SYMBOLS UTILIZED AT MICROWAVES

These symbols are those generally recommended by the IEC (Publication No. 117-11 and 117-11a) and the IEEE (Publication No. 76-ANSI/IEEE Y 32E, New York, 1976).

They are presented in Table 10.17 according to number of ports and, within each category, by increasing complexity.

Table 10.16 Properties of microstrip substrates [48].

Materials	ϵ_r	tan δ at 10 GHz	Temperature °C	Frequency Dispersion	Mechanical Stability	Chemical Resistance	Physical Properties	Adherence	Thermal Match	Cost
Polystyrene	2.54	$5 \cdot 10^{-4}$	-100 + 70	☆☆☆	☆☆	☆☆	☆	☆☆	★	☆
Glass-reinforced polystyrene	2.62	$1 \cdot 10^{-3}$	-100 + 70	☆☆	☆☆	☆☆	☆	☆☆	☆	☆☆
Quartz-reinforced polystyrene	2.60	$5 \cdot 10^{-4}$	-100 + 70	☆☆	☆☆	☆☆	☆	☆☆	☆	★
Polystyrene + Ceramic	3-15	$4 \cdot 10^{-3}$	-100 + 70	☆	☆	☆	★	☆☆	☆	★
Teflon (DuPont) PTFE	2.10	$4 \cdot 10^{-4}$	100 + 220	☆☆	★	☆☆☆	★★	★★★	★	–
Glass-reinforced teflon	2.55	$1.5 \cdot 10^{-3}$	-100 + 220	☆☆	☆☆	☆☆☆	☆	☆☆☆	☆	☆
Quartz-reinforced teflon	2.47	$6 \cdot 10^{-4}$	-100 + 220	☆☆	☆☆	☆☆☆	☆	☆☆☆	☆	★★
Ceramic-reinforced teflon	2.3	$1 \cdot 10^{-3}$	-95 + 220	☆☆	☆	☆☆☆	☆	☆☆☆	★	★
Silicon resin + ceramic (powder)	3-25	$4 \cdot 10^{-3}$	-100 + 230	☆	☆	☆☆	★	☆☆	☆	★★
Polyvinyl oxyde	2.55	$1.6 \cdot 10^{-3}$	-100 + 150	☆	☆	★	☆	☆	☆	☆
Polyolefin	2.32	$5 \cdot 10^{-4}$	-100 + 60	☆☆	★	☆☆☆	★	☆	★	☆☆
Glass-reinforced polyolefin	2.42	$1 \cdot 10^{-3}$	-100 + 60	☆☆	☆	☆☆☆	★	☆	☆	☆
Glass	7.5	$2 \cdot 10^{-3}$	-100 + 550	☆	☆☆☆	☆☆☆	☆☆	★	☆☆	★
Ceramic	6.5	$6 \cdot 10^{-4}$	→1600	☆☆	☆☆☆	☆☆☆	☆☆	–	☆☆	★
Polyolefin + ceramic (powder)	3-10	$1 \cdot 10^{-3}$	-100 + 60	☆	★	☆☆☆	☆	☆☆	★	★★
Polyester + ceramic (powder)	6	$1.7 \cdot 10^{-2}$	-100 + 160	☆	☆☆☆	☆☆☆	☆☆☆	☆☆	☆☆	☆

☆☆☆ = excellent or very inexpensive
★★★ = very poor or very expensive

Table 10.17 Graphical symbols utilized at microwaves.

Transmission lines	
	transmission line (general)
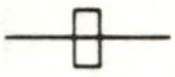	rectangular waveguide
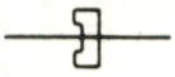	ridge waveguide
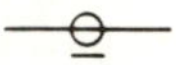	coaxial line
	stripline
	microstrip
One-port devices	
	short-circuit
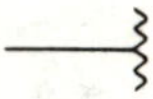	matched load
	reflecting cavity
	antenna
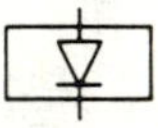	detector
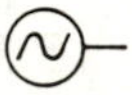	oscillator
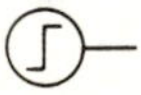	generator of unit step
Two-port devices	
	fixed attenuator
	variable attenuator
	fixed phaseshifter
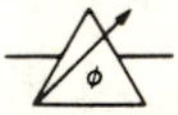	variable phaseshifter
	transmission cavity

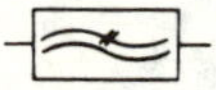	low-pass filter
	bandpass filter
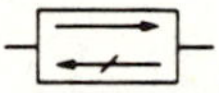	isolator
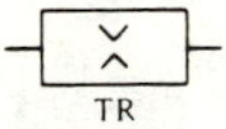	TR cell
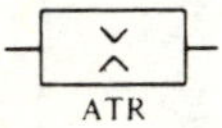	ATR cell
	varactor
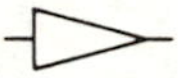	amplifier

Three-port devices

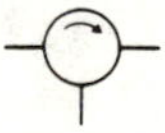	circulator
	mixer
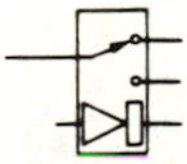	PIN-diode-switch
	power divider

Four-port devices

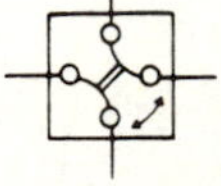	switch
	directional coupler
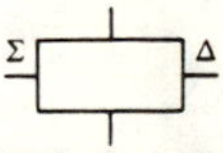	hybrid T

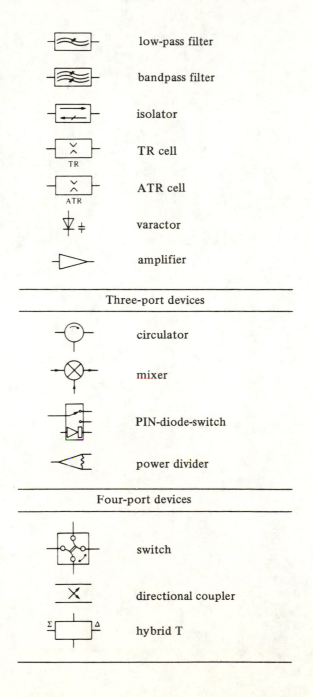

SOLUTIONS TO THE PROBLEMS

CHAPTER 2

2.12.1 $f = 238.3$ MHz, $v_\varphi = 3.75 \cdot 10^8$ m/s, $v_g = 2.4 \cdot 10^8$ m/s, $Z_e^{TE} = 471\ \Omega$, $Z_e^{TM} = 301\ \Omega$. A lossless air-filled waveguide was considered.

2.12.2 $\epsilon_r = 25$.

2.12.3 $Z_e = Z_{PU} = 414\ \Omega$, $Z_{UI} = 325\ \Omega$, $Z_{PI} = 255\ \Omega$, $\Delta\varphi = 171$ rad, $t = 36.7$ ns.

2.12.4

TABLE 2.57

Modes	f_c GHz	v_φ @ 15 GHz m/s	v_g @ 15 GHz m/s	λ_g @ 15 GHz cm
TE_{10}	5	$3.18 \cdot 10^8$	$2.83 \cdot 10^8$	2.12
TE_{20}, TE_{01}	9.99	$4.02 \cdot 10^8$	$2.24 \cdot 10^8$	2.68
TE_{11}, TM_{11}	11.17	$4.49 \cdot 10^8$	$2.00 \cdot 10^8$	3.00
TE_{21}, TM_{21}	14.13	$8.95 \cdot 10^8$	$1.00 \cdot 10^8$	5.96
TE_{30}	14.99	$817 \cdot 10^8$	$0.01 \cdot 10^8$	547.4

Note: if calculations are made taking $c_0 = 3 \cdot 10^8$ m/s, the TE_{30} mode is exactly at cutoff. On the other hand, if the actual value of the velocity of light is taken, it propagates.

2.12.5 A standard R 140 waveguide is taken, for which $a = 15.8$ mm, $b = 7.9$ mm, $\alpha_{100m} = 3.04$ Np $= 26.38$ dB.

2.12.6 $\alpha = 57.4$ Np/m $= 510$ dB/m.

2.12.7 This triangle is exactly one-half of a square: its propagating modes are obtained by superposition of the degenerate modes of the square waveguide and applying the boundary condition on the diagonal. The first two cutoff frequencies obtained are $f_{10} = c_0/(\sqrt{2}\,a)$ and $f_{11} = c_0/a$.

2.12.8 $a = 10.6$ mm.

2.12.9 $2a = 22.6$ mm.

2.12.10 The modes are those of a circular waveguide, with $m = 0, 3, 6, 9$ for the TE modes and $m = 3, 6, 9$ for the TM modes. The first two cutoff frequencies are given by f_{01}^{TE} (GHz) = $182/a$ (mm) and f_{31}^{TE} (GHz) = $200/a$ (mm). The ratio of the two cutoff frequencies is 1.096.

2.12.11 The cutoff frequencies of the perturbed modes, obtained through perturbation theory, are, respectively, $f'_{10} = 0.968\, f_{10}$, $f'_{01} = 2.028\, f_{10}$, $f'_{20} = 2.133\, f_{10}$. The calculated decrease in the TE_{10} cutoff frequency is too small. In fact, a much larger decrease is actually evidenced. The perturbation method does not take into account the drastic constraint imposed by the ridge on the electric field. For the TE_{10} mode $\boldsymbol{E}$ is quite different from $\boldsymbol{E}_0$, so that the perturbation method is not applicable.

2.12.12 The wave on the dielectric plate is no longer guided when $p_a = 0$, which is the cutoff condition. Then $p_d = n\pi(2d) = (\omega/c_0)\sqrt{\epsilon_r - 1}$. The cutoff frequency is then $f = m \cdot 12.7$ GHz.

2.12.13 $f_c = 0.721\, f_{10}$.

2.12.15 $a \leqslant 1.504\ \mu m$ for $\lambda_0 = 850$ nm, $a \leqslant 2.654\ \mu m$ for $\lambda_0 = 1.5\ \mu m$.

2.12.16 All modes for which $v_c \leqslant 8$, *i.e.*: HE_{11} (0); HE_{21}, TE_{01}, TM_{01} (2.405); HE_{12}, EH_{11}, HE_{31} (3.83); EH_{21}, HE_{41} (5.14); TE_{02}, TM_{02} (5.52); HE_{51}, EH_{31} (6.38); EH_{12}, HE_{13}, HE_{32} (7.016); EH_{41}, HE_{61} (7.59), altogether 18 modes.

2.12.17 Width $w = 1.92$ mm, length $d = 7.35$ mm, attenuation $\alpha_d = 0.0214$ dB.

2.12.18 $f_{max} = 3.9$ GHz, $f_d = 27.6$ GHz. A correction for the dispersion is not required, because radiation limits the utilization of this microstrip long before dispersion takes place.

2.12.19 $Z'_c = 72.5\ \Omega$, $\epsilon_e = 3.39$, $\epsilon'_e = 3.22$.

CHAPTER 3

3.7.1 $f_{pr} = 2.86$ GHz, $Q_0 = 90$, $\tau = 10^{-8}$ s = 10 ns.

3.7.2 $f_c = 7.48$ GHz, $d = 6.1$ cm.

3.7.3 TE_{011}: 3.16 GHz. TE_{103}, TM_{110}, TE_{012}: 3.61 GHz. TM_{111}, TE_{111}: 3.74 GHz.

3.7.4 Six degenerate modes at 3.35 GHz: TE_{012}, TE_{021}, TE_{102}, TE_{201}, TM_{120} and TM_{210}.

3.7.5 $a = 3.54$ cm, $b = 2.89$ cm, $d = 5$ cm.

3.7.7 Lines of magnetic field: $\sin(\pi x/a) \cdot \sin(\pi z/d) = \text{constant}$. Lines of current on the upper and lower cavity walls:

$$[\cos(\pi x/a)]^{a^2} / [\cos(\pi z/d)]^{d^2} = \text{constant}.$$

On the sides of the cavity, the lines of current are vertical.

3.7.8 $d = 3.08$ cm for the TE_{111} mode.

3.7.9 $f_{112}^{TM} = 10.9$ GHz.

3.7.10 $\Lambda_{mnl} = A_{mn} e_z J_m (p_{mn}^{TE} \rho) \begin{Bmatrix} \cos m\varphi \\ \sin m\varphi \end{Bmatrix} \sin(l\pi z/d)$.

3.7.11 $\epsilon_e = 7.23$; $\Delta d = 0.231$ mm; $f_n = n \cdot 3.86$ GHz.

3.7.12 $\epsilon_e = 2.98$; $f_n = n \cdot 2.21$ GHz.

3.7.13 $Q_{om} = 868 \cdot 10^6/\sqrt{f}$

3.7.14

$$Q_{om} = 24128 \frac{(x^2 + 0.3435)^{3/2}}{x^3 + 0.157x^2 + 0.397} \qquad \text{when } x = 2a/d$$

3.7.15 $f_{pr} \cong 3/[1 + 0.02 \sin^2(\pi y/a)]$ GHz

3.7.16 $f_{pr} = 7.935$ GHz; $Q_0 = 469$; $Q_e = 1563$; $Q_c = 361$; $\beta_c = 0.3$; $\tau = 19$ ns.

3.7.17 $f_{pr} = 7.935$ GHz; $Q_0 = 1563$; $Q_e = 469$; $Q_c = 361$; $\beta_c = 3.333$; $\tau = 62.7$ ns.

CHAPTER 4

4.12.1 $t = 10^{-10}$ s $= 0.1$ ns.

4.12.2 $U_0 = m\omega^2 b^2/(-2q)$; $B_0 = 2m\omega b^2/[-q(b^2 - a^2)]$; $U_0 \sim f^2$; $B_0 \sim f$; constant electronic efficiency on line $U/B^2 =$ constant.

4.12.3 A reflex klystron produces the largest power at the peaks of the negative conductance, which occur for $t_r/T = 0.782$; 1.764; 2.759; 3.757; 4.755; 5.754; 6.753; in the vicinity of $n + 3/4$.

4.12.4 $\beta_+ = 12.86 \cdot 10^6 \text{ m}^{-1}$; $\beta_- = 238.5 \cdot 10^6 \text{ m}^{-1}$; $v_{\varphi+} = 977$ m/s; $v_{\varphi-} = 52.7$ m/s.

4.12.5 When the term under the square root is positive, one has:

$$\beta^2 - (a_1 + a_2)\beta\omega + a_1 a_2 \omega^2 - (b_1 + b_2)\beta + (a_1 b_2 + a_2 b_1)\omega + b_1 b_2 - \frac{p_2}{p_1} K^2 = 0$$

This is the equation of a hyperbola having the two straight lines β_1 and β_2 as asymptotes. When the term under the square root is negative, which happens for $p_2/p_1 = -1$, one obtains:

$$\beta = [(a_1 + a_2)\omega + (b_1 + b_2)]/2 \qquad \text{straight line}$$

$$\alpha^2 + [(a_1 - a_2)\omega + (b_1 - b_2)]^2/4 = K^2 \qquad \text{ellipse}$$

4.12.6 Dissipated power: 1.04 W. Efficiency: 0.95%. Density of power dissipated in device: $13.24 \cdot 10^{12}$ W/m^3.

4.12.7 $C_{max} = (0.5)^{-\gamma}$; $C_{min} = (167.7)^{-\gamma}$; $C_{max}/C_{min} = (335.3)^{\gamma}$. For an abrupt junction, $\gamma = 0.5$, $C_{max} = 1.41\ C_0$; $C_{min} = 0.077\ C_0$; $C_{max}/C_{min} = 18.31$. For a graded junction, $\gamma = 0.333$; $C_{max} = 1.26\ C_0$; $C_{min} = 0.181\ C_0$; $C_{max}/C_{min} = 6.95$.

4.12.8 $a = 5.93$ cm; $b = a/2 = 2.96$ cm; $d = 9.27$ cm; $f_{pump} = 8.469$ GHz.

4.12.9 $\Delta W = 6.624 \cdot 10^{-24}$ J; $N_3/N_2 = 0.998$ (@ 293 K); 0.993 (@ 77 K); 0.881 (@ 4.2 K).

4.12.10 $P_e = 2$W; $P_s = 10$W; $G = 7$ dB.

CHAPTER 5

5.10.1 When the TE_{111} mode is utilized: $a = 2.07$ cm, 1.48 cm $\leqslant d \leqslant$ 1.89 cm. The TM_{011} and the TE_{211} modes may also resonate over the frequency range of the wavemeter and must be suppressed. When utilizing the TE_{011} mode (largest quality factor, Fig. 3.18), $a = 2.41$ cm, 2.3 cm $\leqslant d \leqslant$ 3.1 cm. In this case, the modes TM_{111} (degenerate with the TE_{011}) and the TE_{311} are those that must not be excited.

5.10.2 $\Delta f = f_0/(2Q_c)$.

5.10.3 $f = 11.79$ GHz.

5.10.4 $f = 10.253$ GHz.

5.10.5 $f = 7.2784$ GHz.

5.10.6 Values of f_0: 2; 2.5; 3.333; 4 GHz.

5.10.7 There are two possible solutions: either $f = 7.22$ GHz and $f_2 = 3.74$ GHz or $f = 3.74$ GHz and $f_2 = 7.22$ GHz. Other combinations of frequencies produce additional spectral lines.

5.10.8 $F = 4.238 \cdot 10^{-8}$ N = 42.38 nN.

5.10.9 $\Delta T = 0.1435 \cdot 10^{-3}$ K.

5.10.10 $P = 0.067$ W = 67 mW.

5.10.11 4.56 mW $\leqslant P \leqslant$ 6.81 mW. The matched load must be capable of dissipating the larger value.

5.10.12 When the measurements yield $P \geqslant 10.55$ mW, the oscillator meets the requirement for any load condition. When $P \leqslant 7.98$ mW, it definitely does not, and should be rejected.

5.10.13 The power fed to a non-reflecting load is $P = 21.8$ W. The generator VSWR is 1.933.

5.10.14 $P_i = 10^6$ W = 1 MW (90 dBm).

5.10.15 $P_i = 80$ W (49 dBm).

CHAPTER 6

6.10.1

$$\begin{pmatrix} 0.367 & 0.5 \\ 0.5 & 0.437 \end{pmatrix}$$

The two-port is non-symmetrical, reciprocal, lossy, and mismatched.

6.10.3 $(\underline{Z})^{\mathrm{d}} = (F)^{-1}[(1) + (\mathrm{diag}\exp(\mathrm{j}\varphi))(F)\{(\underline{Z}) - (G)\}\cdot$
$\{(\underline{Z}) + (G)\}^{-1}(F)^{-1}(\mathrm{diag}\exp(\mathrm{j}\varphi))]\cdot[(1) - (\mathrm{diag}\exp(\mathrm{j}\varphi))(F)\{(\underline{Z}) - (G)\}\cdot$
$\{(\underline{Z}) + (G)\}^{-1}(F)^{-1}(\mathrm{diag}\exp(\mathrm{j}\varphi))]^{-1}(F)(G)$.

6.10.4 For the resulting two-port, one obtains: $|\underline{s}_{11}| = |\underline{s}_{22}| = 0.422$; $3.88 \leqslant |\underline{s}_{21}| \leqslant 4.08$; $0.9 \leqslant |\underline{s}_{12}| \leqslant 0.989$.

6.10.5 The attenuator is assumed to be matched and the reference planes are selected to have real transfer functions. The reflection of the sliding short-circuit is $-\exp(-2\mathrm{j}\beta L)$. At the input to the assembly, one then has

$$\underline{\rho} = -10^{-\mathrm{L}A/10}\exp(-2\mathrm{j}\beta L).$$

6.10.6 $R_1 = 54.87\ \Omega$; $R_2 = 5.37\ \Omega$.

6.10.7 The device which should be used is an isolator. Its attenuation LA must be as small as feasible (technical requirement). The sum of the attenuation (insertion loss) and the isolation must be:

$$\mathrm{L}A + \mathrm{L}I \geqslant 10\log_{10}\frac{100}{0.2} = 27\ \mathrm{dB}.$$

6.10.8 $X(f) = 24.3\ \mathrm{cotg}(0.295 \cdot 10^{-9} f)\ \Omega$.

6.10.9 There is no solution outside of the trivial case of two unconnected two-ports.

6.10.10 $\mathrm{L}C_{13} \geqslant 8.85$ dB.

6.10.11 The device is a five-branched star with a resistor $R = 0.6\ Z_c = (1 - 2/n)Z_c$ series-connected within each branch. The scattering matrix then has $\underline{s}_{ii} = 0$ diagonal terms (match) and $\underline{s}_{ij} = 0.25 = 1/(n-1)$ off-diagonal terms (for $i \neq j$).

6.10.12 $\mathrm{L}D_{13} = 32.256$ dB + LA.

6.10.13 The reflections of the two sliding short-circuits are, respectively, $-\exp(-2j\beta L_1)$ and $-\exp(-2j\beta L_2)$. The scattering matrix of the resulting two-port has the form

$$\frac{1}{2}\begin{pmatrix} \exp(-2j\beta L_2)-\exp(-2j\beta L_1) & -j[\exp(-2j\beta L_1)+\exp(-2j\beta L_2)] \\ -j[\exp(-2j\beta L_1)+\exp(-2j\beta L_2)] & \exp(-2j\beta L_1)-\exp(-2j\beta L_2) \end{pmatrix}$$

if $L_1 = L_2$, then $\underline{s}_{11} = \underline{s}_{22} = 0$ and $\underline{s}_{12} = \underline{s}_{21} = -j\exp(-2j\beta L_1)$: one obtains in this manner a matched variable phaseshifter (§ 6.3.11).

6.10.14 The matrix terms of the resulting two-port are:

$$\underline{s}'_{11} = \underline{s}'_{33} = j\exp(j\varphi/2)\sin(\varphi/2);\quad \underline{s}'_{13} = \underline{s}'_{31} = \exp(j\varphi/2)\cos(\varphi/2).$$

The device obtained is a reactive attenuator.

6.10.15 The phaseshifter is connected between the two couplers. This yields a directional coupler having variable power ratios $\alpha^2 = \cos^2(\phi/2)$ and $\beta^2 = \sin^2(\phi/2)$ where ϕ is the phaseshift produced by the variable phaseshifter.

6.10.16 One input is connected to two outputs. The ratios of output to input power for the three cases considered are:

- 100; 0;
- 225; 25;
- 99.24; 0.76.

6.10.17 The frequency response is proportional to $(\Delta f/f_0)/[1 + (Q_c\Delta f/f_0)^2]$.

6.10.19 A tuner (either E-H, § 6.5.28, or slide-screw, § 6.3.21) is connected to the port 2 of the circulator. The signal reflected from the tuner emerges at port 3 and combines with the parasitic signal transfer from port 1 to port 3. By adjusting the tuner, it is possible to compensate the latter. An infinite isolation is obtained when the two signals are in phase opposition and when, additionally:

$$20\log|\underline{s}_{12}\underline{\rho}\,\underline{s}_{32}| = 17\text{ dB}$$

where $\underline{\rho}$ is the reflection factor from the tuner. The same approach is used to determine the directivity of a coupler in Section 7.4.17.

CHAPTER 7

7.7.1 $\lambda_g = 4$ cm; $\underline{s}_{ii} = j\,0.286$.

7.7.2 Relative error of 1%: $\text{VSWR}_{min} = 8.15$; $d/\lambda_g = 0.039$;
Absolute error of 1%: $\text{VSWR}_{min} = 58.58$; $d/\lambda_g = 0.0054$.

7.7.3 A second iris, identical to the first one, is placed at a distance $3\lambda_g/8$ from the first one. In an X-band waveguide, $L = 1.49$ cm.

7.7.4 $|\underline{\rho}| = 0.2 \pm 0.005$.

7.7.5 The measurement must yield VSWR < 1.2 to be quite certain that the requirement is actually met.

7.7.6 $\underline{b}_r/\underline{b}_s = 5.385 \exp(-j\ 158.2°)$.

7.7.7 A Purcell junction cannot be used: in a six-port for the measurement of reflection (§ 7.3.15), two input ports are each connected to four other ports (outputs, where the detectors are connected). In the Purcell junction, any one port is connected to only three other ports (even ⇌ odd).

7.7.8 $|\underline{s}_{11}| = |\underline{s}_{22}| = 0.6$; $|\underline{s}_{12}| = |\underline{s}_{21}| = 0.663$; $LA = 3.565$ dB.

7.7.9 VSWR = 37.97; B = ±6.

7.7.10 0.041 dB $\leqslant LA \leqslant$ 3.011 dB.

7.7.11 $G_1 = 20$; $G_2 = 35$; $G_3 = 45$. For condition (7.50) to be satisfied, the largest antenna dimension must be smaller than 0.354 cm.

7.7.12 $LA = 1.76$ dB.

7.7.13 1.225 dB $< LA <$ 7.38 dB.

7.7.14 $LD_{13} = 21.828$ dB.

7.7.15 $F_{21} = 1.957$.

7.7.16 $F = 3.044$; $T_r = 593$ K.

CHAPTER 8

8.9.1 $R = 774$ m.

8.9.2 $P = 3.12 \cdot 10^{-22}$ W.

8.9.3 $\Delta f = 666.7$ Hz.

8.9.4 $f_D = 13.67$ kHz.

8.9.5 $L = 83.3 \cdot 10^{15}$ m = 8.81 light-years = $557 \cdot 10^3$ astronomical units = 2.5 parsecs. This is the distance from the Earth to the star Sirius (Big Dog).

8.9.6 For two antennas of the same height $h = 383$ m.

8.9.7 $v = 3.36 \cdot 10^8$ m/s; $v_g = 2.68 \cdot 10^8$ m/s.

8.9.8 Grazing incidence: 81.28° latitude. For a 10° incidence: 71.41° latitude.

8.9.9 Photon flux: $4.33 \cdot 10^{12}\ s^{-1}$; $Q = 216 \cdot 10^9$ bits/s.

8.9.10 $\underline{\epsilon}_r = 9.12 - j\ 2.07$.

8.9.11 $\epsilon_r = 3.515 \cdot n^2 + 0.672$. For $n = 1$, $\epsilon_r = 4.188$; for $n = 2$, $\epsilon_r = 14.733$; for $n = 3$, $\epsilon_r = 32.308$ and so on.

8.9.12 For a full power radiometer: $\tau = 1.81$ ms. For a Dicke radiometer: $\tau = 5.03$ ms.

BIBLIOGRAPHY

[1] Massachusetts Institute of Technology, *Radiation Laboratory Series,* New York: McGraw-Hill Book Co., Inc., 1948-1950.

[2] S.A. Schelkunoff, *Electromagnetic Waves,* New York: Van Nostrand, 1943.

[3] International Electrotechnical Commission, *Hollow Metallic Waveguides, Part 2: Relevant Specifications for Ordinary Rectangular Waveguides,* IEC Publication 153-2, Geneva, Switzerland.

[4] International Electrotechnical Commission, *Hollow Metallic Waveguides, Part 4: Relevant Specifications for Circular Waveguides,* IEC Publication 153-4, Geneva, Switzerland.

[5] F.L. Ng, "Tabulation of Methods for the Numerical Solution of the Hollow Waveguide Problem," *IEEE Trans. Microwave Theory Tech.,* vol. MTT-22, No. 3, March 1974, pp. 322-329.

[6] J.G. Kretzschmar, "Wave Propagation in Hollow Conducting Elliptical Waveguides," *IEEE Trans. Microwave Theory Tech.,* vol. MTT-18, No. 9, September 1970, pp. 547-554.

[7] J.R. Pyle, "The Cutoff Wavelength of TE_{10} Mode in Ridged Rectangular Waveguide of any Aspect Ratio," *IEEE Trans. Microwave Theory Tech.,* vol. MTT-14, No. 4, April 1966, pp. 175-183.

[8] J.P. Montgomery, "On the Complete Eigenvalue Solution of Ridged Waveguide," *IEEE Trans. Microwave Theory Tech.,* vol. MTT-19, No. 6, June 1971, pp. 547-555.

[9] M.A.R. Gunston, *Microwave Transmission Line Impedance Data.* New York: Van Nostrand Reinhold, 1972.

[10] M. Abramowitz, I.A. Stegun, *Handbook of Mathematical Functions.* New York: Dover, 1972.

[11] A.E. Karbowiak, *Trunk Waveguide Communication.* London: Chapman and Hall, 1965.

[12] D.A. Dunn, W. Loewenstern, *Economic Feasibility of Microwave Power Transfer in Circular Waveguide,* Report SU-SEL-66-109, Stanford University, October 1966.
[13] S. Lefeuvre, *Hyperfréquences.* Paris: Dunod, 1969.
[14] N.S. Kapany, J.J. Burke, *Optical Waveguides.* New York: Academic Press, 1972.
[15] F.J. Tischer, "The H-Guide, a Waveguide for Microwaves," *IRE Convention Record,* 1956, Pt. 5, pp. 44-51.
[16] L.G. Chambers, "Propagation in waveguides filled longitudinally with two or more dielectrics," *Brit. J. Appl. Phys.,* vol. 4, February 1953, pp. 39-45.
[17] P.H. Vartanian, W.P. Ayres, A.L. Helgesson, "Propagation in dielectric slab loaded rectangular waveguide," *IRE Trans. Microwave Theory Tech.,* vol. MTT-6, No. 2, April 1958, pp. 215-222.
[18] F.E. Gardiol, "Higher-Order Modes in Dielectrically Loaded Rectangular Waveguides," *IEEE Trans. Microwave Theory Tech.,* vol. MTT-16, No. 11, November 1968, pp. 919-924.
[19] F.E. Gardiol, "Comment on the Design of Dielectric Loaded Waveguides," *IEEE Trans. Microwave Theory Tech.,* vol. MTT-25, No. 7, July 1977, pp. 624-625.
[20] R. Olshansky, D. Keck, "Pulse Broadening in Graded Index Optical Fibers," *Applied Optics,* vol. 15, No. 2, February 1977, pp. 483-491.
[21] R. Yamada, Y. Inabe, "Guided Waves along Graded Index Dielectric Rod," *IEEE Trans. Microwave Theory Tech.,* vol. MTT-22, No. 8, August 1974, pp. 813-814.
[22] J.D. Decotignie, F.E. Gardiol, "Méthodes d'analyse de la Propagation dans les Fibres Optiques," *Bulletin ASE/UCS,* vol. 70, No. 15, August 1979, pp. 830-837.
[23] D. Gloge, E. Marcatili, "Multimode Theory of Graded-Core Fibers," *Bell System Technical Journal,* vol. 52, No. 7, November 1973, pp. 1563-1578.
[24] T. Okoshi, *Optical Fibers.* New York: Academic Press, 1982.
[25] D. Marcuse, *Light Transmission Optics.* New York: Van Nostrand Reinhold, 1972.
[26] D. Marcuse, *Theory of Dielectric Optical Waveguides.* New York: Academic Press, 1974.
[27] P.J.B. Clarricoats, *Optical Fibre Waveguides,* IEE Reprint Series. Stevenage, England: Peter Peregrinus, 1975.
[28] J.A. Arnaud, *Beam and Fiber Optics.* New York: Academic Press, 1976.
[29] H.G. Unger, *Planar Optical Waveguides and Fibres,* Engineering Science Series, Oxford: Clarendon Press, 1977.

[30] E.O. Hammerstad, "Equations for Microstrip Circuit Design," *Proceedings of the 5th European Microwave Conference,* Hamburg, September 1975.

[31] M.V. Schneider, "Microstrip Lines for Microwave Integrated Circuits," *Bell System Technical Journal,* vol. 48, No. 5, May-June 1969, pp. 1421-1444.

[32] H.A. Wheeler, "Transmission Line Properties of Parallel Strips Separated by a Dielectric Sheet," *IEEE Trans. Microwave Theory Tech.,* vol. MTT-13, No. 3, March 1965, pp. 172-185.

[33] R.A. Pucel, D. Masse, C.P. Hartwig, "Losses in Microstrip," *IEEE Trans. Microwave Theory Tech.,* vol. MTT-16, June 1968, pp. 342-350.

[34] E.O. Hammerstad, F. Bekkadal, *Microstrip Handbook,* ELAB Report STF44 A74169, University of Trondheim, Norwegian Institute of Technology, 1975.

[35] W. Janssen, *Hohlleiter und Streifenleiter*, Heidelberg: Ed. Hüthig, 1977.

[36] W.J. Getsinger, "Microstrip Dispersion Model," *IEEE Trans. Microwave Theory Tech.,* vol. MTT-21, January 1973, pp. 34-39.

[37] K.C. Gupta, R. Garg, I.J. Bahl, *Microstrip Lines and Slot Lines.* Dedham, MA: Artech House, 1979.

[38] K.C. Gupta, A. Singh, *Microwave Integrated Circuits.* New Delhi: Wiley Eastern, 1974.

[39] G.K. Grünberger, H.H. Meinke, "Experimenteller und theoretischer Nachweis der Längsfeldstärken in der Grundwelle der Mikrowellen-Streifenleitung," *Nachrichtentech. Zeitung,* vol. 24, 1971, pp. 364-368.

[40] S.B. Cohn, "Slotline on a Dielectric Substrate," *IEEE Trans. Microwave Theory Tech.,* vol. MTT-17, October 1969, pp. 768-778.

[41] E.A. Mariani, C.P. Heinzman, J.P. Agrios, S.B. Cohn, "Slot Line Characteristics," *IEEE Trans. Microwave Theory Tech.,* vol. MTT-17, December 1969, pp. 1091-1096.

[42] R. Garg, K.C. Gupta, "Expressions for the Wavelength and Impedance of a Slotline," *IEEE Trans. Microwave Theory Tech.,* vol. MTT-24, August 1976 p. 532.

[43] C.P. Wen, "Coplanar Waveguide: A Surface Strip Transmission Line Suitable for Non-Reciprocal Gyromagnetic Device Applications," *IEEE Trans. Microwave Theory Tech.,* vol. MTT-17, December 1969, pp. 1087-1090.

[44] H.E. Brenner, "Numerical Solution of TEM-Line Problems Involving Inhomogeneous Media," *IEEE Trans. Microwave Theory Tech.,* vol. MTT-15, August 1967, pp. 485-487.

[45] F. Gardiol, "Careful MIC Design Prevents Waveguide Modes," *Microwaves,* May 1977, pp. 188-191.

[46] A.M.K. Saad, K. Schunemann, "A Simple Method for Analyzing Fin-Line Structures," *IEEE Trans. Microwave Theory Tech.*, vol. MTT-26, December 1978, pp. 1002-1007.

[47] W.J.R. Hoefer, A. Ros, "Fin Line Parameters Calculated with the TLM Method," *1979 IEEE International Microwave Symposium*, Orlando, FL, 20 April to 2 May 1979.

[48] "Characteristics of Strip-Line Laminates," *Microwaves*, January 1968 pp. 105-112.

[49] N. Marcuvitz, *Waveguide Handbook*. New York: McGraw-Hill Book Co., 1950.

[50] P. Silvester, P. Benedek, "Microstrip Discontinuity Capacitances for Right-Angle Bends, T-Junctions and Crossings," *IEEE Trans. Microwave Theory Tech.*, vol. MTT-21, May 1973.

[51] J. Van Bladel, "On the Resonances of a Dielectric Resonator of Very High Permittivity," *IEEE Trans. Microwave Theory Tech.*, vol. MTT-23, February 1975, pp. 199-208.

[52] T.S. Saad (Editor), *Microwave Engineer's Handbook*. Dedham, MA: Artech House, 1971.

[53] E. Ramis, C. Deschamps, J. Odoux, *Cours de Mathématiques Spéciales: Algèbre et Application à la Géométrie,* Tome 2, Masson, 1979, pp. 96-97.

[54] W. Meyer, "Dielectric Measurements on Polymeric Materials by Using Superconducting Microwave Resonators," *IEEE Trans. Microwave Theory Tech.*, vol. MTT-25, December 1977, pp. 1092-1097.

[55] K. Agyeman *et al.*, "New Materials and Surface Treatments for Practical Superconducting Microwave Cavities," *Research in Materials, Annual Report,* Massachusetts Institute of Technology, January 1978, p. 36.

[56] International Electrotechnical Commission, *Recommended Graphical Symbols, Part 11: Microwave Techniques,* IEC Publication 117-11, Geneva, Switzerland.

[57] R. Beringer, "Resonant Cavities as Microwave Circuit Elements," in *Principles of Microwave Circuits,* edited by C.G. Montgomery, R H. Dicke, E.M. Purcell, New York: McGraw-Hill Book Co., 1948.

[58] C.C. Johnson, *Field and Wave Electrodynamics*. New York: McGraw-Hill Book Co., Inc., 1965.

[59] K. Hinkel, *Magnetrons*. New York: J.F. Rider, Philips Technical Library, 1961.

[60] J.C. Slater, *Microwave Electronics*. New York: Van Nostrand, 1950 (Dover, New York, 1969).

[61] G. Collins, *Microwave Magnetrons*. New York: McGraw-Hill Book Co., Inc., 1949 (M.I.T. Rad. Lab. Series).

[62] M. Weinstein, "Voltage Tunable Magnetron," *Microwave Journal,* vol. 21, November 1978, pp. 64-65.

[63] J.R.G. Twistleton, "Twenty Kilowatt 980 Mc/s Continuous Wave Magnetron," *Proc. IEE,* vol. 111, January 1974, pp. 51-56.

[64] W.C. Brown, "High Power Microwave Generators of the Crossed-Field Type," *Journal Microwave Power,* vol. 5, 1970, pp. 245-259.

[65] D. Hamilton, J. Knipp, J. Kuper, *Klystrons and Microwave Triodes.* New York: McGraw-Hill Book Co., Inc., 1949 (M.I.T. Rad. Lab. Series).

[66] R.E. Collin, *Foundations for Microwave Engineering.* New York: McGraw-Hill Book Co., Inc., 1966.

[67] M. Chodorow, C. Susskind, *Fundamentals of Microwave Electronics.* New York: McGraw-Hill Book Co., Inc., 1964.

[68] W.H. Louisell, *Coupled Mode and Parametric Electronics.* New York: J. Wiley and Sons, 1960.

[69] J. Voge, *Les Tubes aux Hyperfréquences.* Paris: Eyrolles, 1973.

[70] R. Kompfner, "Backward-Wave Oscillator," *Bell Labs. Record,* vol. 31, August 1953, pp. 281-285.

[71] J.R. Pierce, *Traveling Wave Tubes.* New York: Van Nostrand, 1950.

[72] J.M. Osepchuk, "Life Begins at Forty: Microwave Tubes," *Microwave Journal,* vol. 21, November 1978, pp. 51-60.

[73] R.S. Elliott, "Some Limitations on the Maximum Frequency of Coherent Oscillations," *Journal Applied Physics,* vol. 23, August 1952, pp. 812-818.

[74] J.L. Hirschfield, V.L. Granatstein, "The Electron Cyclotron Maser, An Historical Survey," *IEEE Trans. Microwave Theory Tech.,* vol. MTT-25, June 1977, pp. 522-527.

[75] V.A. Flyagin, A.V. Gaponov, M.I. Petelin, V.K. Yulpatov, "The Gyrotron," *IEEE Trans. Microwave Theory Tech.,* vol. MTT-25 June 1977, pp. 514-521.

[76] N.I. Zaytsev, T.B. Pankratowa, M.I. Petelin, V.A. Flyagin, "Millimeter and Submillimeter Gyrotrons," *Radio Engineering and Electronic Physics,* vol. 19, May 1974, pp. 103-106.

[77] J.E. Carroll, *Hot Electron Microwave Generators.* London: Arnold, 1970.

[78] J.B. Gunn, "Microwave Oscillations of Current in III-V Semiconductors," *IBM Journal of Research and Development,* vol. 8, 1964, pp. 141-159.

[79] F.K. Manasse, J.A. Ekiss, G.R. Gray, *Modern Transistor Electronic Analysis and Design.* Englewood Cliffs, New Jersey: Prentice-Hall, Inc., 1967.

[80] J.A. Copeland, "LSA Oscillator Diode Theory," *Journal Applied Physics,* vol. 38, 1967, pp. 3096-3101.

[81] P. Jeppesen, *Gallium Arsenide Transferred Electron Devices.* Lyngby, Denmark: Technical University, 1978.

[82] G. Gibbons, *Avalanche Diode Microwave Oscillators.* Oxford: Clarendon Press, 1973.

[83] W.T. Read, "A Proposed High Frequency Negative Resistance Diode," *Bell System Technical Journal,* vol. 37, 1958, pp. 401-446.

[84] A.S. Clorfeine, R.J. Ikola, L.S. Napoli, "A Theory for the High-Efficiency Modes of Operation in Avalanche Diodes," *RCA Review,* vol. 30, September 1969, pp. 397-421.

[85] D.J. Coleman, S.M. Sze, "A Low Noise Metal-Semiconductor-Metal (MSM) Microwave Oscillator," *Bell System Technical Journal,* vol. 50, May-June 1971, pp. 1695-1699.

[86] E J. Colussi, "Internally Matched RF Power Transistors," *Microwave Journal,* vol. 21, April 1978, pp. 81-84.

[87] R.S. Carson, *High Frequency Amplifiers.* New York: Wiley Interscience, 1975.

[88] H.F. Cooke, "Microwave Transistors, Theory and Design," *Proceedings IEEE,* vol. 59, August 1971, pp. 1163-1181.

[89] W. Baechtold, "X and Ku-Band Amplifiers with GaAs Schottky Barrier Field-Effect Transistors," *IEEE Journal Solid-State Circuits,* vol. SC-8, February 1973, pp. 54-58.

[90] J. Millman, C. Halkias, *Electronic Devices and Circuits.* New York: McGraw-Hill Book Co., Inc., 1967, pp. 384-417.

[91] K. Sekido, J.A. Arden, "Recent Advances in FET Devices Performance and Reliability," *Microwave Systems News,* vol. 6, April 1976, pp. 71-81.

[92] G. Bechtel, W. Hooper, P. Hower, "Design and Performance of the GaAs FET," *IEEE Journal of Solid State Circuits,* vol. SC-5, December 1970, pp. 319-323.

[93] L.O. Chua, P.M. Lin, *Computer Aided Analysis of Electronic Circuits.* Englewood Cliffs, New Jersey: Prentice Hall, 1975.

[94] M. Uenohara, J.W. Gewartowski, "Varactor Applications," in *Microwave Semiconductors and their Circuit Applications,* H.A. Watson, editor. New York: McGraw-Hill Book Co., Inc., 1969, pp. 228-258.

[95] C.H. Page, "Frequency Conversion with Positive Nonlinear Resistors," *Journal Research National Bureau of Standards,* vol. 56, 1956.

[96] J.M. Manley, H.E. Rowe, "Some General Properties of Nonlinear Elements," *Proceedings IRE,* vol. 44, July 1956, pp. 904-913.

[97] H.T. Friis, "Analysis of Harmonic Generator Circuits for Step-Recovery Diodes," *Proceedings IEEE,* vol. 55, July 1967, pp. 1192-1194.

[98] J.C. Decroly, L. Laurent, J.C. Lienart, G. Marechal, J. Vorobeitchik, *Parametric Amplifiers.* London: Macmillan, 1973 (Philips Technical library).

[99] A.E. Siegman, *Microwave Solid-State Masers.* New York: McGraw-Hill Book Co., Inc., 1964.

[100] Pacific Measurements Inc., *Ultra-Fast RF Power Meter Model 1045,* Palo Alto, CA, 1979.

[101] R. Beringer, "The Measurement of Wavelength," in *Technique of Waveguide Measurement,* edited by C.G. Montgomery. New York: McGraw-Hill Book Co., Inc., 1947 (MIT Rad. Lab. Series).

[102] R.V. Pound, *Microwave Mixers.* New York: McGraw-Hill Book Co., Inc., 1948 (MIT Rad. Lab. Series).

[103] P. Kartaschoff, *Frequency and Time.* London: Academic Press, 1978.

[104] A.B. Carlson, *Communication Systems,* 2nd Edition. New York: McGraw-Hill Book Co., Inc., 1975.

[105] L. Frenkel, T. Sullivan, M.A. Pollack, T.J. Bridges, "Absolute Frequency Measurement of the 118.6-μm Water-Vapor Laser Transition," *Applied Physics Letters,* vol. 11, December 1967, pp. 344-345.

[106] CGPM, *Comptes rendus des séances de la treizième Conférence Générale des Poids et Mesures* (Paris, October 1967), Paris: Gautheir Villars, 1968.

[107] M. Engelson, F. Telewski, *Spectrum Analyzer, Theory and Applications.* Dedham, MA: Artech House, 1974.

[108] A.L. Cullen, "A General Method for the Absolute Measurement of Microwave Power," *Proceedings IEE,* vol. 99, part IV, 1952, pp. 112-120.

[109] C.G. Montgomery, *Technique of Microwave Measurement.* New York: McGraw-Hill Book Co., Inc., 1947 (MIT Rad. Lab. Series).

[110] J.A. Lane, *Microwave Power Measurement.* Stevenage, England: Peter Peregrinus, 1972.

[111] M. Sucher, J. Fox, *Handbook of Microwave Measurements.* Brooklyn, NY: Polytechnic Press, third edition, 1963.

[112] W.H. Jackson, "A Thin-Film/Semiconductor Thermocouple for Microwave Power Measurements," *Hewlett Packard Journal,* September 1974, pp. 16-18.

[113] E.L. Ginzton, *Microwave Measurements.* New York: McGraw-Hill Book Co., Inc., 1957.

[114] A.L. Lance, *Introduction to Microwave Theory and Measurements.* New York: McGraw-Hill Book Co., Inc., 1964.

[115] R.E. Collin, *Field Theory of Guided Waves.* New York: McGraw-Hill Book Co., Inc., 1960.

[116] C.G. Montgomery, R.H. Dicke, E.M. Purcell, *Principles of Microwave Circuits.* New York: McGraw-Hill Book Co., Inc., 1948 (MIT Rad. Lab. Series).

[117] Y. Chow, E. Cassignol, *Théorie et Application des Graphes de Transfert.* Paris: Dunod, 1965.

[118] G. Boudouris, P. Chenevier, *Circuits pour Ondes Guidées.* Paris: Dunod, 1975.

[119] W.P. Allis, S.J. Buchsbaum, A. Bers, *Waves in Anisotropic Plasmas.* Cambridge, MA: The MIT Press, 1963.

[120] F.L. Warner, *Microwave Attenuation Measurement.* Stevenage, England: Peter Peregrinus, 1977.

[121] A.G. Fox, "An Adjustable Waveguide Phase Changer," *Proceedings IRE,* vol. 35, December 1947, pp. 1489-1498.

[122] C.L. Hogan, "The Ferromagnetic Faraday Effect at Microwave Frequencies and its Applications," *Bell System Technical Journal,* 1952, pp. 1-31.

[123] G.L. Matthaei, L. Young, E.M.T. Jones, *Microwave Filters, Impedance-Matching Networks, and Coupling Structures.* New York: McGraw-Hill Book Co., Inc., 1964.

[124] R.A. Johnson, "Understanding Microwave Power Splitters," *Microwave Journal,* December 1975, pp. 49-56.

[125] C. Staeger, P. Kartaschoff, "Measurement Accuracy Hinges on Coupler Design," *Microwaves,* vol. 16, April 1977, pp. 41-46.

[126] F. Arndt, "High Pass Transmission Line Directional Coupler," *IEEE Trans. Microwave Theory Tech.,* vol. MTT-16, May 1968, pp. 310-311.

[127] G. Schaller, "Optimization of Microstrip Directional Couplers with Lumped Capacitors," *Archiv Elektronik Uebertragungstechnik,* vol. 31, July 1977, pp. 301-307.

[128] R.F. Soohoo, *Theory and Applications of Ferrites.* New York: Prentice Hall, 1960.

[129] E. Schlömann, "Microwave Behaviour of Partially Magnetized Ferrites," *Journal Applied Physics,* vol. 41, January 1970, pp. 204-214.

[130] F. Gardiol, A.S. Vander Vorst, "Computer Analysis of E-Plane Resonance Isolators," *IEEE Trans. Microwave Theory Tech.,* vol. MTT-19, March 1971, pp. 315-322.

[131] F. Gardiol, "Computer Analysis of Latching Phaseshifters in Rectangular Waveguide," *IEEE Trans. Microwave Theory Tech.,* vol. MTT-21, January 1973, pp. 57-61.

[132] H. Bosma, "On the Principle of Waveguide Circulation," *Proc. IEE,* vol. 109, part B, supplement No. 21, 1962.

[133] C.E. Fay, R.L. Comstock, "Operation of the Ferrite Junction Circulator," *IEEE Trans. Microwave Theory Tech.,* vol. MTT-13, January 1965, pp. 15-27.

[134] M.E. Hines, "Reciprocal and Nonreciprocal Modes of Propagation in Ferrite Strip-Line and Microstrip Devices," *IEEE Trans. Microwave Theory Tech.,* vol. MTT-19, May 1971, pp. 442-451.

[135] B. Chiron, G. Forterre, "Emploi des Ondes de Surface Electromagnétique pour la Réalisation de Dispositifs Gyromagnétiques à très Grande Largeur de Bande," *Premier Séminaire International sur les Dispositifs Hyperfréquences à Ferrite,* Toulouse, 27-30 March, 1972.

[136] P.S. Carter, "Magnetically Tunable Microwave Filters Using Single Crystal Yttrium-Iron Garnet Resonators," *IRE Trans. Microwave Theory Tech.,* vol. MTT-9, May 1961, pp. 252-260.

[137] K.E. Mortenson, J.M. Borrego, *Design, Performance and Applications of Microwave Semiconductor Control Components.* Dedham, MA: Artech House, 1972.

[138] Hewlett-Packard, *1977 Diode and Transistor Designer's Catalog,* Palo Alto, CA, 1977.

[139] L. Young, H. Sobol, *Advances in Microwaves,* vol. 8. New York: Academic Press, 1974.

[140] H.A. Watson, *Microwave Semiconductor Diodes and their Circuit Applications.* New York: McGraw-Hill Book Co., Inc., 1969.

[141] I. Weitman, "Increase the Range of Crystal Detectors," *Microwaves,* vol. 18, January 1979, pp. 86-88.

[142] B.M. Oliver, J.M. Cage, *Electronic Measurements and Instrumentation,* Inter-University Electronics Series, vol. 12, New York: McGraw-Hill Book Co., Inc., 1971, pp. 654-661.

[143] H. Groll, *Mikrowellen Messtechnik,* Braunschweig: Vieweg, 1969.

[144] R.L. Thomas, *A Practical Introduction to Impedance Matching.* Dedham, MA: Artech House, 1976.

[145] P.I. Somlo, "The Locating Reflectometer," *IEEE Trans. Microwave Theory Tech.,* vol. MTT-20, February 1972, pp. 105-112.

[146] F.C. de Ronde, "A Precise and Sensitive Reflecto-'meter' Providing Full Band Matching of Reflection Coefficient," *IEEE Trans. Microwave Theory Tech.,* vol. MTT-11, July 1965, pp. 435-440.

[147] S.F. Adam, "Swept SWR Measurement in Coax," *HP Application Note 84,* February 1967.

[148] D.E. Dunwoodie, Verbesserte Reflexionsmessungen in der Mikrowellentechnik, *Mikrowellen Magazin,* 2/78, 1978, pp. 102-106.

[149] G.F. Engen, R.W. Beatty, "Microwave Reflectometer Techniques," *IRE Trans. Microwave Theory Tech.,* vol. MTT-7, July 1959, pp. 351-355.

[150] R.W. Anderson, O.T. Dennison, "An Advanced Network Analyzer for Sweep Measuring Amplitude and Phase from 0.1 to 12.4 GHz," *Hewlett Packard Journal,* vol. 18, February 1967.

[151] G.F. Engen, "The Six-Port Reflectometer: An Alternative Network Analyzer," *IEEE Trans. Microwave Theory Tech.,* vol. MTT-25, December, 1977, pp. 1075-1080.

[152] A.M. Nicholson, C.L. Bennett, D. Lamensdorf, L. Susman, "Applications of Time Domain Metrology to the Automation of Broad-Band Microwave Measurements," *IEEE Trans. Microwave Theory Tech.,* vol. MTT-20, January 1972, pp. 3-9.

[153] F.L. Warner, "New Expression for Mismatch Uncertainty When Measuring Microwave Attenuation," *IEE Proceedings,* vol. 127, part H, April 1980, pp. 66-69.

[154] Nhu Bui Hai, *Antennes Microondes.* Paris: Masson, 1978.

[155] G.A. Deschamps, "Determination of Reflection Coefficients and Insertion Loss of a Waveguide Junction," *Journal Applied Physics,* vol. 24, August 1953.

[156] P. Delogne, "Compensation à l'aide d'un Ordinateur des Mesures des Paramètres d'un Quadripole aux Hyperfréquences," *Revue HF* (Brussels), vol. 8, July 1968.

[157] R.C. Ajmera, "Microwave Measurements with Active Systems," *Proceedings IEEE,* vol. 62, January 1974, pp. 118-127.

[158] M. Ney, F. Gardiol, "Automatic Monitor for Microwave Resonators," *IEEE Trans. Instrumentation Measurement,* vol. IM-26, March, 1977, pp. 10-13.

[159] J. Rossel, *Physique Générale,* Neuchatel, Ed. du Griffon, 1970, pp. 275-279.

[160] F.J. Tischer, *Mikrowellen Messtechnik.* Berlin: Springer, 1958.

[161] M. Skolnik, *Radar Handbook.* New York: McGraw-Hill Book Co., Inc., 1970.

[162] D.K. Barton, *Radars.* Dedham, MA: Artech House, 1975.

[163] M. Carpentier, *Radars, Concepts Nouveaux.* Paris: Dunod, 1966.

[164] H.L. Van Trees, *Detection, Estimation, and Modulation Theory.* New York: John Wiley and Sons, Inc., 1968.

[165] F.V. Schultz, R.C. Burgener, S. King, "Measurement of the Radar Cross Section of a Man," *Proceedings IRE,* vol. 46, February 1958, pp. 476-481.

[166] B.L. Dalton, "Microwave Non-Contact Measurement and Instrumentation in the Steel Industry," *Journal Microwave Power,* vol. 8, November 1973, pp. 235-244.

[167] D.W. Griffin, "Microwave Interferometers for Biological Studies," *Microwave Journal,* vol. 21, May 1978, pp. 69-72.

[168] R.M. Goldstein, G.A. Morris, "Ganymede: Observation by Radar," *Science,* vol. 188, June 1975, pp. 1211-1212.

[169] *Electronic Warfare Magazine,* "The International Countermeasures Handbook," edited by H.F. Eustace, Second Edition, 1976-77.

[170] S. Yonezawa, *Microwave Communication.* Tokyo: Maruzen, 1970.

[171] J.A. Ratcliffe, *An Introduction to the Ionosphere and Magnetosphere.* Cambridge: University Press, 1962.

[172] A. Benoit, "Signal Attenuation due to Neutral Oxygen and Water Vapour, Rain and Clouds," *Microwave Journal,* vol. 11, November 1968, pp. 73-80.

[173] URSI Commission F, *Proceedings of Open Symposium.* La Baule, France, 28 April to 6 May 1977.

[174] F. Ananasso, "Coping with Rain Above 11 GHz," *Microwave Systems News,* vol. 10, March 1980, pp. 58-72.

[175] H. Brodhage, W. Hormuth, *Planning and Engineering of Radio Relay Links.* Berlin, Siemens AG, 7th edition, 1968.

[176] A.C. Clarke, "Extraterrestrial Relays," *Wireless World,* October 1945.

[177] N. Mokhoff, "Technology '80: Communications and Microwaves," *IEEE Spectrum,* vol. 17, January 1980, pp. 38-43.

[178] A.J. Brejcha, "Microwave Communication from Outer Planets: The Voyager Project," *Microwave Journal,* vol. 23, January 1980, pp. 25-44.

[179] G. Elion, H.A. Elion, *Fiber Optics in Communication Systems.* New York and Basel: Marcel Dekker, 1978.

[180] M. Kawahata, "The HI OVIS Communication System (Highly Interactive Optical Visual Information System)," *Proceedings 9th European Microwave Conference,* Brighton, GB 17-20 September 1979, pp. 83-98.

[181] N.S. Kapany, *Fiber Optics, Principles and Applications.* New York: Academic Press, 1967.

[182] D.A. Hill, *Fiber Optics.* London: Business Book, 1977.

[183] R.V. Decareau, "The Amana Story," *Microwave Energy Applications Newsletter,* vol. 10, 1977.

[184] H.J. Van Zante, *The Microwave Oven.* Boston: Houghton Mifflin, 1973.

[185] P.O. Risman, N.E. Bengtsson, "Dielectric Properties of Foods at 3 GHz as Determined by a Cavity Perturbation Technique," *Journal of Microwave Power,* vol. 6, July 1971, pp. 101-123.

[186] E.C. Okress, *Microwave Power Engineering.* New York: Academic Press, 1968.

[187] A.L. Van Koughnett, "Fundamentals of Microwave Heating," *Trans. IMPI,* vol. 1, 1973, pp. 17-40.

[188] R.G. Bosisio, M. Giroux, "Automatic Field Measurements in Microwave Applicators," *Journal Microwave Power,* vol. 10, September 1969, pp. 152-156.

[189] D.R. McConnell, "Energy Consumption: A Comparison Between the Microwave Oven and the Conventional Electric Range," *Journal Microwave Power,* vol. 9, 1974, pp. 341-347.

[190] S.O. Nelson, "Potential Insect Control Applications for Microwaves," *Proceedings 3rd European Microwave Conference,* Brussels, September 1973, p.c. 14.5.

[191] S. Licht, *Therapeutic Heat and Cold.* Baltimore: Waverly Press, 1965.

[192] J. Daels, "Microwave Heating of the Uterine Wall During Parturition," *Journal Microwave Power,* vol. 11, June 1976, pp. 166-168.

[193] F. Sterzer, *et al.,* "Microwave Apparatus for the Treatment of Cancer," *Microwave Journal,* vol. 23, January 1980, pp. 39-44.

[194] S.M. Michaelson, "Human Exposure to Nonionizing Radiant Energy, Potential Hazards and Safety Standards," *Proceedings IEEE*, vol. 60, April 1972, pp. 389-421.

[195] A.W. Guy, J.C. Lin, P.O. Kramar, A.F. Emery, "Effect of 2450 MHz Radiation on the Rabbit Eye," *IEEE Trans. Microwave Theory Tech.*, vol. MTT-23, June 1975, pp. 492-498.

[196] C.H. Bonney, P.L. Rustan, G.E. Ford, "Evaluation of Effects of the Microwave Oven and Radar Electromagnetic Radiation on Noncompetitive Cardiac Pacemakers," *IEEE Trans. Biomedical Eng.*, vol. BME-20, 1975, pp. 357-364.

[197] F. Gardiol, "Nomograms Save Time in Determining Permittivity," *Microwaves*, vol. 12, November 1973, pp. 68-70.

[198] A.R. von Hippel, *Dielectric Materials and Applications.* New York: J. Wiley and Sons, 1954.

[199] M.C. Decreton, F. Gardiol, "Simple Nondestructive Method for the Measurement of Complex Permittivity," *IEEE Trans. Instrumentation Measurement*, vol. IM-23, December 1974, pp. 434-438.

[200] M. Gex-Fabry, J.R. Mosig, F.E. Gardiol, "Reflection and Radiation of an Open-Ended Circular Waveguide: Application to Nondestructive Measurement of Materials," *Archiv für Elektronik und Uebertragungstechnik*, vol. 33, December 1979, pp. 473-478.

[201] J.C.E. Besson, J.R. Mosig, F.E. Gardiol, "Reflection of an Open-Ended Coaxial Line: Application to Nondestructive Measurement of Materials," *International URSI Symposium on Electromagnetic Waves*, Munich, August 26-29, 1980.

[202] M.S. Ramachandraiah, M.C. Decreton, "A Resonant Cavity Approach for the Nondestructive Determination of Complex Permittivity at Microwave Frequencies," *IEEE Trans. Instrumentation Measurement*, vol. IM-24, December 1975, pp. 287-291.

[203] G. Mur, "Field Analysis and Complex Resonance Frequency of the Quasi-TE_{011} Mode in an Inhomogeneously Filled Resonator with Losses," *Applied Science Research*, vol. 29, April 1974, pp. 137-144.

[204] W. Gordy, W. Smith, R. Trambarullo, *Microwave Spectroscopy.* New York: J. Wiley and Sons, 1953.

[205] D.J.E. Ingram, *Free Radicals, as Studied by Electron Spin Resonance.* London: Butterworths, 1958.

[206] C. Townes, A. Schawlow, *Microwave Spectroscopy.* New York: McGraw-Hill Book Co., Inc., 1955.

[207] G.A. Pake, *Paramagnetic Resonance.* New York: W.A. Benjamin, 1962.

[208] J. Edrich, C.J. Smyth, "Millimeter Wave Thermography as Subcutaneous Indicator of Joint Inflammation," *Proceedings 7th European Microwave Conference*, Copenhagen, September 5-8, 1977, pp. 713-717.

[209] P.E. Glaser, "Satellite Solar Power Station," *Solar Energy,* vol. 12, Pergamon Press, 1969, pp. 353-361.

[210] W.C. Brown, "Solar Power Satellites: Microwaves Deliver the Power," *IEEE Spectrum,* vol. 16, June 1979, pp. 36-42.

[211] K.W. Billman, "Radiation Energy Conversion in Space," *Progress in Astronautics and Aeronautics,* vol. 61, New York: American Institute of Aeronautics and Astronautics, 1978.

[212] I.T.T., *Reference Data for Radio Engineers.* Indianapolis, IN: Howard W. Sams Inc., 1970.

[213] I. Wolff, *Einführung in die Mikrostrip-Leitungstechnik,* I. Wolff, Aachen (no date given).

[214] S. Ramo, J.R. Whinnery, T. Van Duzer, *Fields and Waves in Communication Electronics.* New York: J. Wiley and Sons, 1965.

[215] A. Papoulis, *The Fourier Integral and its Applications.* New York: McGraw-Hill Book Co., Inc., 1962.

[216] A.E. Booth, *Microwave Data Tables.* London: Iliffe, 1959.

[217] P.H. Smith, *Electronic Applications of the Smith Chart in Waveguide, Circuit and Component Analysis.* New York: McGraw-Hill Book Co., Inc., 1969.

[218] J.G. Kretzschmar, D. Schoonaert, "Smith Chart for Lossy Transmission Lines," *Proceedings IEEE,* vol. 57, September 1969, pp. 1658-1660.

GLOSSARY

Symbol	Units	Description	Page	Paragraph
a_i	$W^{1/2}$	Complex normalized wave	237	6.1.7
$\boldsymbol{A}$	A/m	Surface current density	11	1.4.4
A_e	m^2	Effective reception area	367	8.1.2
b_i	$W^{1/2}$	Complex normalized wave	237	6.1.7
$\underline{\boldsymbol{B}}$	T	Induction phasor vector	10	1.4.2
c_0	m/s	Velocity of light in vacuum	12	1.4.5
c_p	J/kg · K	Specific heat	220	5.6.3
$\underline{\boldsymbol{D}}$	As/m^2	Displacement phasor vector	10	1.4.2
$\underline{\boldsymbol{E}}$	V/m	Electrical phasor vector	10	1.4.2
f_B	Hz	Cyclotron frequency	150	4.2.1
f_c	Hz	Cutoff frequency	34	2.2.28
f_D	Hz	Doppler frequency	377	8.1.13
f_t	Hz	Transition frequency	186	4.8.2
F	1	Noise figure	359	7.6.6
F_m	N	Lorentz force	149	4.2.1
g	1	Landé factor	288	6.7.2
G	1	Power gain	348	7.4.11
$\underline{\boldsymbol{H}}$	A/m	Magnetic phasor vector	10	1.4.2
H_m	1	Hankel function	451	10.3.7
$\underline{I}_e$	A	Equivalent line current	26	2.2.7
$\underline{I}_g$	A	Waveguide current	33	2.2.25
$\underline{\boldsymbol{J}}$	A/m^2	Current density phasor vector	10	1.4.2
$\underline{\boldsymbol{J}}_e$	A/m^2	Perturbating electrical current density	63	2.7.3
$\underline{\boldsymbol{J}}_m$	V/m^2	Perturbating magnetic current density	63	2.7.3
J_m	1	Bessel function of first kind of order m	448	10.3.1
k	m^{-1}	Wave number	24	2.2.4
k_B	J/K	Boltzmann constant	5	1.2.4

k_p	m^{-1}	Resonance wave number	112	3.2.4
K	1	Distributed coupling factor	169	4.4.7
$K(u)$	1	Elliptical integral of first order	60	2.6.3
K_m	1	Modified Bessel function of order m	452	10.4.1
$\mathrm{L}A$	dB	Attenuation level	277	6.5.10
$\mathrm{L}C$	dB	Coupling level	277	6.5.11
$\mathrm{L}D$	dB	Directivity	278	6.5.15
m	Am^2	Spin magnetic moment	288	6.7.2
M	A/m	Magnetization	289	6.7.3
N	W	Average noise power	359	7.6.3
NA	1	Numerical aperture	82	2.10.4
N_m	1	Bessel function of second kind of order m	448	10.3.1
p	m^{-1}	Transverse wave number	24	2.2.4
P_i	W	Pulse power	229	5.8.3
Q_0	1	Unloaded quality factor	114	3.2.9
		Overvoltage factor	135	3.4.12
Q_0m	1	Metallic quality factor	130	3.4.4
Q_0e	1	Sample quality factor	133	3.4.9
Q_c	1	Loaded quality factor	138	3.5.9
Q_e	1	External quality factor	138	3.5.8
s	1	VSWR: Voltage Standing Wave Ratio	324	7.2.2
$(\underset{\sim}{s})$	1	Scattering matrix	238	6.1.12
T_a	K	Antenna noise temperature	359	7.6.4
T_r	K	Receiver noise temperature	359	7.6.5
$\underline{U}_\mathrm{e}$	V	Equivalent line voltage	26	2.2.7
v_g	m/s	Group velocity	17	2.1.5
v_φ	m/s	Phase velocity	17	2.1.5
$\underline{Y}_\mathrm{f}$	S	Beam equivalent admittance	162	4.3.7
Z_0	Ω	Characteristic impedance of vacuum	12	1.4.5
Z_c	Ω	Characteristic impedance of microstrip	93	2.11.7
$\underline{Z}_\mathrm{e}$	Ω	Wave impedance	33	2.2.25
$\underline{Z}_\mathrm{m}$	Ω	Metal characteristic impedance	63	2.7.3
Z_{UI}	Ω	Waveguide impedance (voltage – current)	33	2.2.25
Z_{PI}	Ω	Waveguide impedance (power – current)	33	2.2.25
Z_{PU}	Ω	Waveguide impedance (power – voltage)	33	2.2.25
α	Np/m	Attenuation per unit length	17	2.1.5
β	rad/m	Phaseshift per unit length	17	2.1.5
β_c	1	Coupling factor	138	3.5.6
γ	m^{-1}	Propagation factor	17	2.1.5
γ_g	$(\mathrm{sT})^{-1}$	Gyromagnetic factor	288	6.7.2

δ	m	Skin depth	17	2.1.5
δ_{jk}	1	Kronecker delta symbol	242	6.1.23
$\underline{\epsilon}$	As/Vm	Complex permittivity	10	1.4.2
ϵ_0	As/Vm	Electrical constant	12	1.4.5
$\underline{\epsilon}_r$	1	Relative permittivity	10	1.4.2
η_d	1	Mismatch efficiency	224	5.7.2
η_e	1	Electronic efficiency	155	4.2.8
η_p	1	Partial efficiency	198	4.11.4
η_{pa}	1	Power added efficiency	198	4.11.4
η_r	1	Coupling factor	141	3.5.13
η_s	1	Substitution efficiency	227	5.7.8
η_t	1	Total efficiency	197	4.11.4
λ_c	m	Cutoff wavelength	34	2.2.28
λ_g	m	Waveguide wavelength	35	2.2.32
$\lambda_{g\epsilon}$	m	Loaded waveguide wavelength	406	8.5.5
Λ_{mnl}	Am	Hertz potential	121	3.3.8
$\underline{\mu}$	Vs/Am	Complex permeability	10	1.4.2
μ_0	Vs/Am	Magnetic constant	12	1.4.5
μ_p	1	Carrier mobility	188	4.8.4
$\underline{\mu}_r$	1	Relative permeability	12	1.4.2
$\underline{\underline{\mu}}$	Vs/Am	Permeability tensor	291	6.7.5
Π_{mnl}	Vm	Hertz potential	121	3.3.8
$\underline{\rho}$	1	Reflection factor	244	6.2.1
$\underline{\rho}$	C/m^3	Charge density phasor	10	1.4.2
σ	S/m	Conductivity	10	1.4.2
σ	m^2	Effective scattering cross section	369	8.1.4
ϕ	1	Transverse potential of TM mode	31	2.2.21
ψ	1	Transverse potential of TE mode	28	2.2.13
ω_L	rad/s	Larmor angular frequency	290	6.7.4
ω_M	rad/s	Magnetization angular frequency	291	6.7.4
ω_p	rad/s	Complex eigen-angular frequency	113	3.2.6
ω_{pr}	rad/s	Eigen-angular frequency	114	3.2.8

INDICES

a	Air
c	Characteristic
	Loaded
	Cutoff
d	Dielectric
	Dissipated
e	Sample
	Effective
	Electrical
	Equivalent
	External
g	Guide
H	Hartree
in	Input
	Incident
L	Load
m	Magnetic
	Material
r	Relative
t	Transverse component
T	Transversely-dependent part of the transverse component
v	Vacuum
	Volume

SUBJECT INDEX